INTRODUCTORY ASTRONOMY AND ASTROPHYSICS

Second Edition

MICHAEL ZEILIK
University of New Mexico

ELSKE v. P. SMITH
Virginia Commonwealth University

SAUNDERS COLLEGE PUBLISHING
Philadelphia   New York   Chicago
San Francisco   Montreal   Toronto
London   Sydney   Tokyo   Mexico City
Rio de Janeiro   Madrid

Address orders to:
383 Madison Avenue
New York, NY 10017

Address editorial correspondence to:
210 West Washington Square
Philadelphia, PA 19105

Text Typeface: Palatino
Compositor: Progressive Typographers
Acquisitions Editor: John Vondeling
Project Editor: Margaret Mary Kerrigan
Copy Editor: Irene Nunes
Art Director: Carol C. Bleistine
Text Designer: Detta Penna
Cover Designer: Lawrence R. Didona
Text Artwork: J&R Technical Services, Inc.
Production Manager: Tim Frelick
Assistant Production Manager: JoAnn Melody

Cover Credit: Space Grid by Hans Wendler/©1987 THE IMAGE BANK

Endpaper Credit: *Front:* Horsehead Nebula (Orion). *Back:* Orion Nebula (M 42). Both photographs were taken with the Astromak™ telescope. Courtesy Jim Riffle, Astro Works Corporation, White Rock, New Mexico.

Photographs provided by the National Optical Astronomy Observatories (NOAO) are not used to state or imply the endorsement by NOAO, or by any NOAO employee, of any individual, philosophy, commercial product, process, or service.

Library of Congress Cataloging-in-Publication Data

Zeilik, Michael.
   Introductory astronomy and astrophysics.

   Rev. ed. of: Introductory astronomy and
astrophysics / Elske v.P. Smith, Kenneth C. Jacobs.   1973.
   Bibliography: p.
   Includes index.
      1. Astronomy.   2. Astrophysics.   I. Smith, Elske v. P.
(Elske van Panhuys), 1929–   .   II. Smith Elske v. P. (Elske
van Panhuys), 1929–   .   Introductory astronomy and astro-
physics. III. Title.
QB45.Z43   1986   520   86-10047
ISBN 0-03-004499-5

INTRODUCTORY ASTRONOMY AND
ASTROPHYSICS, SECOND EDITION   ISBN 0-03-004499-5

789   032   9876543

CBS COLLEGE PUBLISHING
Saunders College Publishing
Holt, Rinehart and Winston
The Dryden Press

PREFACE

Astronomy can have it both ways: as the oldest observational science, born in the wonder about the starry skies; and as the youngest science, when combined with the theoretical discipline of astrophysics. Astronomy and astrophysics span the Cosmos — from the nearby planets to the distant reaches of quasars. Indeed, they cover all of time, from the beginning of the Universe to its possible end. So astronomy encompasses both the cosmic and the human in time as well as in space.

We have written this book to convey the content and outlook of modern astronomy and astrophysics for the serious science student. By this, we mean an undergraduate willing to combine basic mathematical skills (trigonometry, algebra, and calculus) with an elementary knowledge of physics. (We assume that this physics background includes electromagnetism as well as mechanics.) Our ideal student has the goal of using these intellectual tools to broaden his or her understanding of the contents and structure of the Universe. We have aimed to provide for this person a solid coverage in modern astronomy with a solid foundation in astrophysics. This slant requires the frequent use of physical concepts that pertain to astronomical entities and situations.

This book follows the traditional Earth-out approach, which has the advantage of starting out with the familiar and moving to the fantastic. These realms fall into the four parts of the book: Part One, The Solar System; Part Two, The Stars; Part Three, The Milky Way Galaxy; and Part Four, The Universe. Within the context of Part One, we introduce many of the basic physical concepts, especially those of mechanics, used throughout the text. We also examine the planets as places that have evolved since the ori-

gin of the solar system. In Part Two, we introduce essential concepts about light and its remote sensing by astronomers. We then examine the light of a special star, the Sun, which serves as a model for other stars. In Part Three, we explore the local conglomerate of stars — the Milky Way Galaxy — whose structure and evolution rests so heavily on the stars that make it up. The Galaxy also serves as a model for other galaxies. Part Four then turns to the distant Universe to examine its contents and evolution on a grand scale.

Along with bringing the material up-to-date, we have made major changes in this edition to improve its usefulness as a learning tool. Some of these changes are consequences of new approaches resulting from recent developments in the science. In Part One, we emphasize comparative planetology as a technique to probe the evolution of the planets. We have also more cleanly divided the planets physically into the terrestrial and Jovian groups. Material on the origin of the Solar System has been integrated with that on the many small bodies, which give crucial clues as to how the Solar System formed. Part Two now includes a brief chapter on telescopes and detectors but does not dwell on the technology as such. Rather, we emphasize the purposes of observations. We have also compared other stars more directly to the Sun. Part Three includes a new chapter that pulls together starbirth with the interstellar medium, in the context of stellar evolution and the structure of the Galaxy. Part Four is largely new material reflecting the explosive growth of extragalactic astronomy in the past decade.

We have tried to develop the material consistently along physical lines of argument rather than purely astronomical ones. To this end, we have expanded the number and increased the range of the chapter-end problems. We expect that they will challenge the students and deepen their understanding of the material in the chapter. Some of these problems were provided by Alan P. Marscher of Boston University. The Instructor's Manual for this book provides detailed solutions of the end-of-chapter problems.

A word about units. After wrestling between ourselves and with our colleagues, we have decided to break with astronomical tradition and use SI rather than cgs units. Why? First, SI units, despite some problems, *are* the international standard; we should follow it, as teachers. Second, many students will have had or will be taking a physics course, where such units are standard. We hope that following the same units will cause less confusion. Third, some equations—notably those in electromagnetism—are simpler in the SI formulation. We will confess, however, that in some areas we have used units of convenience rather than SI ones—such as atmospheres for atmospheric pressure. And we could not avoid parsecs! Appendix 6 supplies information to help with conversions between SI and cgs units.

We expect that instructors will use this book for one-year (two-semester) course sequences, perhaps with some additional materials. For a one-semester course, the selection is best left up to the instructor. We do urge, though, that Chapters 1 and 8 be included, because they contain so much of the basic physical concepts.

Before the credits, let us briefly define astronomy and astrophysics and explain how they differ. They are scientific disciplines in which we collect, correlate, and interpret data from the observable Universe. Astronomy has its basis in observations of the sky, while astrophysics is more theoretical, applying and extrapolating the knowledge gained from laboratory physics to astronomical phenomena. Most astronomers are also astrophysicists and *vice versa*. The two

fields differ but together provide an in-depth understanding of the physical Universe. Astronomy works from observations toward an interpretation, while astrophysics provides a physical framework for understanding the observations by devising theoretical models. Where these two approaches meet marks the ground of our knowledge of a Cosmos vast and wonderful to our eyes and our minds.

We received special help from the following colleagues: Jack Burns of the University of New Mexico; Tom Balonek at Colgate University; Richard Teske of the University of Michigan; Alma Zook at Pomona College; Alan Marscher, Boston University; John Cowan, the University of Oklahoma; Alan Bentley, Eastern Montana College; Roger Chevalier, University of Virginia; and Bill Tifft, University of Arizona. We particularly want to acknowledge the part played by Dr. Ken C. Jacobs, Hollins College, who coauthored the first edition and whose contributions may still be detected in this second version. For the task of typing, Ona Bailey justly deserves the credit.

We are responsible for any errors that you may find. Please send any to Zeilik at the address below —with our thanks! Small changes can be made in future printings of this edition.

Michael Zeilik
Department of Physics and Astronomy
The University of New Mexico
Albuquerque, New Mexico 87131

Elske v. P. Smith
Virginia Commonwealth University
900 Park Avenue
Richmond, Virginia 23284

Note that due to the lead time in the production process, we have decided to add the Halley's Comet material as an update at the end of the chapters before the Appendices.

CONTENTS

PART ONE: THE SOLAR SYSTEM   1

Chapter 1   Celestial Mechanics and the Solar
            System   3

1–1 The Historical Basis   4
1–2 Planetary Orbits   7
1–3 Newton's Mechanics   10
1–4 Newton's Law of Universal
     Gravitation   13
1–5 Physical Interpretations of Kepler's
     Laws   15
1–6 Applications to the Solar System   18
     Problems   20

Chapter 2   The Solar System in Perspective   22

2–1 Contents of the Solar System   23
2–2 Unanswered Questions   33
     Problems   33

Chapter 3   The Dynamics of the Earth   35

3–1 Time and the Seasons   36
3–2 Evidence of the Earth's Rotation   39
3–3 Evidence of the Earth's Revolution About
     the Sun   42
3–4 Differential Gravitational Forces   45
     Problems   50

Chapter 4   The Earth–Moon System   52

4–1 Dimensions   53
4–2 Dynamics   54
4–3 Interiors   56

4–4 Surface Features   58
4–5 Atmospheres   66
4–6 Magnetic Fields   70
4–7 The Evolution of the Earth–Moon
     System   74
     Problems   76

Chapter 5   The Terrestrial Planets: Mercury,
            Venus, and Mars   77

5–1 Modern Planetology   78
5–2 Mercury   78
5–3 Venus   83
5–4 Mars   88
5–5 Comparative Evolution of the Terrestrial
     Planets   94
     Problems   96

Chapter 6   The Jovian Planets   97

6–1 Jupiter   98
6–2 Saturn   104
6–3 Uranus   106
6–4 Neptune   109
6–5 Pluto and Charon   110
     Problems   112

Chapter 7   Small Bodies and the Origin of the
            Solar System   113

7–1 Moons and Rings   114
7–2 Asteroids   127
7–3 Comets   128

7–4 Meteoroids and Meteorites   132
7–5 Interplanetary Gas and Dust   134
7–6 The Formation of the Solar System   135
     Problems   140

PART TWO: THE STARS   143

Chapter 8   Electromagnetic Radiation and Matter   145

8–1 Electromagnetic Radiation   146
8–2 Atomic Structure   152
8–3 The Spectra of Atoms, Ions, and Molecules   157
8–4 Spectral-Line Intensities   158
8–5 Spectral-Line Broadening   161
8–6 Blackbody Radiation   162
     Problems   165

Chapter 9   Telescopes and Detectors   167

9–1 Optical Telescopes   168
9–2 Invisible Astronomy   170
9–3 Detectors and Image Processing   174
     Problems   177

Chapter 10   The Sun: A Model Star   179

10–1 The Structure of the Sun   180
10–2 The Photosphere   181
10–3 The Chromosphere   185
10–4 The Corona   189
10–5 The Solar Wind   192
10–6 Solar Activity   193
      Problems   202

Chapter 11   Stars: Distances and Magnitudes   204

11–1 The Distances to Stars   205
11–2 The Stellar Magnitude Scale   206
11–3 Absolute Magnitude and Distance Modulus   207
11–4 Magnitudes at Different Wavelengths   208
      Problems   211

Chapter 12   Stars: Binary Systems   212

12–1 Classification of Binary Systems   213
12–2 Visual Binaries   213
12–3 Spectroscopic Binaries   216
12–4 Eclipsing Binaries   218
12–5 Interferometric Stellar Diameters   223
      Problems   224

Chapter 13   Stars: The Hertzsprung–Russell Diagram   226

13–1 Stellar Atmospheres   227
13–2 Classifying Stellar Spectra   229
13–3 Hertzsprung–Russell Diagrams   233
      Problems   241

PART THREE: THE MILKY WAY GALAXY   243

Chapter 14   Our Galaxy: A Preview   245

14–1 The Shape of the Galaxy   246
14–2 The Distribution of Stars   249
14–3 Stellar Populations   252
14–4 Galactic Dynamics: Spiral Features   252
14–5 A Model of the Galaxy   253
      Problems   255

Chapter 15   Galactic Rotation: Stellar Motions   257

15–1 Components of Stellar Motions   258
15–2 The Local Standard of Rest   260
15–3 Moving Clusters   261
15–4 Galactic Rotation   262
      Problems   268

Chapter 16   The Evolution of Stars   270

16–1 The Physical Laws of Stellar Structure   271
16–2 Theoretical Stellar Models   277
16–3 Stellar Evolution   278
      Problems   291

Chapter 17   Star Deaths   292

17–1 White Dwarfs   293
17–2 Neutron Stars   297

17–3 Black Holes   303
      Problems   306

Chapter 18   Variable and Violent Stars   308

18–1 Naming Variable Stars   309
18–2 Pulsating Stars   310
18–3 Nonpulsating Variables   312
18–4 Extended Stellar Atmospheres: Mass
      Loss   316
18–5 Cataclysmic Variables   320
18–6 X-Ray Sources: Binary and Variable   328
      Problems   333

Chapter 19   The Interstellar Medium and Star
              Birth   335

19–1 Interstellar Dust   336
19–2 Interstellar Gas   342
19–3 Star Formation   352
      Problems   356

Chapter 20   The Evolution of Our Galaxy   358

20–1 The Structure of Our Galaxy from Radio
      Studies   359
20–2 The Distribution of Stars and Gas in Our
      Galaxy   363
20–3 Evolution of the Galaxy's Structure   369
20–4 Cosmic Rays and Galactic Magnetic
      Fields   371
      Problems   373

PART FOUR: THE UNIVERSE   375

Chapter 21   Galaxies Beyond the Milky
              Way   377

21–1 The Classification of Galaxies   378
21–2 Characteristics of Galaxies   383
21–3 The Distance Scale and the Hubble
      Constant   385
      Problems   392

Chapter 22   Large-Scale Structure in the
              Universe   394

22–1 Clusters of Galaxies   395
22–2 Superclusters   401
22–3 Intergalactic Matter   405
22–4 The Missing(?) Mass   406
      Problems   407

Chapter 23   Active Galaxies and Quasars   408

23–1 Active Galaxies   409
23–2 Quasars: Discovery and
      Description   419
23–3 Problems With Quasars   421
23–4 Astrophysical Jets   426
      Problems   428

Chapter 24   Cosmology: The Big Bang and
              Beyond   430

24–1 Fundamental Observations   431
24–2 Models of the Universe   433
24–3 The Primeval Fireball   436
24–4 The Standard Big Bang Model   439
24–5 The Inflationary Universe   443
      Problems   446

Halley's Comet Update   448

Appendices
      Appendix 1   Bibliography   450
      Appendix 2   Constellations   452
      Appendix 3   Solar System Data   455
      Appendix 4   Stellar Data   459
      Appendix 5   Atomic Elements   462
      Appendix 6   Conversion of Units   464
      Appendix 7   Constants and Units   466
      Appendix 8   The Greek Alphabet   468
      Appendix 9   Mathematical Operations   469
      Appendix 10 The Celestial Sphere   480

Index   485

The
Solar
System

Celestial
Mechanics
and
the
Solar
System

Viewed from the Earth, the Sun slides eastward on its annual journey relative to the stars of the Zodiac while the wandering planets perform periodic gyrations within the Zodiac. The planets fall into two groups, based on their observed motions. The first, Mercury and Venus, can move eastward of the Sun, appearing higher and higher in the western sky at sunset as an *evening star*. At greatest eastern elongation, the planet attains maximum angular distance east of the Sun; it then moves closer to the western horizon at sunset as it sweeps toward the Sun. Moving westward of the Sun, it rises as a predawn *morning star* in the eastern sky.

The second group visible to the naked eye are Mars, Jupiter, and Saturn. They progress steadily eastward across the sky relative to the stars until they approach *opposition*—180° away from the Sun in the sky. Then they slow to a halt, trace a *retrograde* loop toward the west, and finally resume their eastward march (Figure 1–1). These phenomena fascinated and frustrated the ancient astronomers. This chapter briefly discusses the historical efforts to explain these planetary motions—efforts that gave birth to physics and astrophysics and culminated in the creation of *celestial mechanics* by Isaac Newton in the seventeenth century.

1–1 The Historical Basis

(a) The Heliocentric Model of Copernicus

In the sixteenth century, the Polish astronomer Nicolaus Copernicus grew dissatisfied with the traditional *geocentric* (Earth-centered) model of the Solar System and introduced a new *heliocentric* (Sun-centered) one. This model forms the foundation of the one we use today. Copernicus placed the Sun at the center of the Solar System and had the planets (including the Earth) orbit it along circles. This heliocentric model was slow to be accepted because its predictions were not significantly better than those of the geocentric model, but it eventually caught on, primarily for aesthetic reasons of simplicity and harmony.

To understand the explanatory purposes of the Copernican model, you need a little background on naked-eye observations of the Sun, Moon, and planets. These are naturally geocentric but still useful. Daily, the sky turns *westward* with respect to the horizon. The Sun appears to move *eastward* with respect to the stars and circles the sky in a year. The imaginary path traced out by the Sun is called the *ecliptic*; behind it lies the special set of 12 constellations of the *Zodiac*. The planets (and Sun and Moon) move with respect to the stars within the band of the Zodiac. In distance from the Sun in the heliocentric model, the planets range from Mercury (the closest) through Venus, Earth, Mars, Jupiter, Saturn, Uranus, and Neptune to Pluto (the most distant). The planets closer to the Sun than the Earth are termed *inferior planets*; these are Mercury and Venus. The planets orbiting farther from the Sun than the Earth are called *superior planets*; these are Mars through Pluto. The motions of the inferior planets as seen in our night sky differ markedly from the motions of the superior planets.

We define *elongation* (Figure 1–2) as the angle seen at the Earth between the direction to the Sun's center and the direction to a planet. We speak of

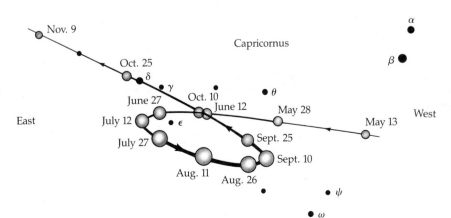

Figure 1–1

Retrograde motion of Mars. The planet's path is indicated relative to the stars of the constellation Capricornus. The stars are labeled in order of their brightness by Greek letters. The sizes of the open circles indicate the relative brightness of the planet; note that it shone the brightest on August 11.

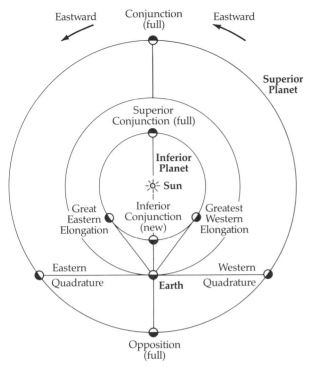

Figure 1–2

Heliocentric planetary configurations. Arrows indicate the direction of orbital motion as well as the rotational direction of the Earth.

eastern or western elongation according to whether the planet lies east or west of the Sun, as seen from the Earth. Elongations of particular geocentric significance are given special names: an elongation of 0° is termed *conjunction* (*inferior conjunction* when the planet lies between the Earth and the Sun and *superior conjunction* when the planet lies on the opposite side of the Sun from the Earth). An elongation of 180° is called *opposition*, and one of 90° is *quadrature*. When an inferior planet attains its maximum elongation, we refer to *greatest elongation*. Only inferior planets may be at inferior conjunction or at greatest elongation (28° for Mercury and 48° for Venus), but they may never be at either quadrature of opposition. Inferior conjunction can never occur for superior planets, and their greatest elongation is 180° (when they are in opposition). Note that our Moon may pass through inferior conjunction, quadrature, and opposition (its greatest elongation) because it is a satellite of the Earth.

Copernicus correctly stated that the farther a planet lies from the Sun, the slower it moves around the Sun. When the Earth and another planet pass each other on the same side of the Sun, the apparent retrograde loop occurs (Figure 1–3) from the relative motions of the other planet and the Earth. As we view the planet from the moving Earth, our line of sight reverses its angular motion twice, and the loop comes about because the orbits of the two planets are not coplanar.

Copernicus derived a relationship between the synodic and sidereal periods of a planet in the heliocentric model. The *synodic period S* is the time it takes the planet to return to the same position in the sky relative to the Sun, as seen from the Earth. For the inferior planets, Mercury and Venus, this time is the interval between successive inferior conjunctions; for the superior planets, it is the interval between successive oppositions. The *sidereal period P* is the time it takes the planet to complete one orbit of the Sun with respect to the stars (Figure 1–4). The Earth's sidereal period E is 365.26 days. The Earth moves at the rate of $360/E$ degrees per day in its orbit, while a planet's rate of angular motion is $360/P$ as viewed from the Earth. For a superior planet, the Earth completes one orbit and must then traverse the angle $S(360/P)$ in the time $S - E$ to return the superior planet to opposition. Hence,

$$(S - E)(360/E) = S(360/P)$$

or

$$1/S = 1/E - 1/P$$

For an inferior planet, the Earth is a superior planet, and so we interchange E and P to arrive at Copernicus' result.

$$1/S = \begin{array}{l} 1/P - 1/E \text{ (inferior)} \\ 1/E - 1/P \text{ (superior)} \end{array}$$

As an example, consider Venus, an inferior planet with an observed synodic period of $S = 583.92$ days. The appropriate relationship is

$$1/583.92 = 1/P - 1/365.26$$

$$1/P = 0.00171 + 0.00274$$
$$= 0.00445$$

$$P = 224.7 \text{ days}$$

The telescopic observations reported by Galileo Galilei in his *Siderius Nuncius* in 1610 strongly sup-

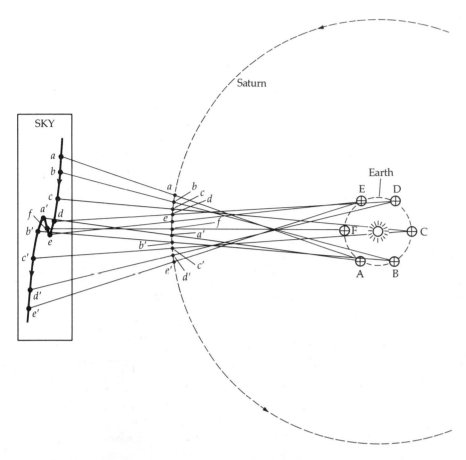

Figure 1–3

Retrograde motion in a heliocentric model. As the Earth passes a superior planet, that planet appears to move opposite its normal eastward direction with respect to the stars. Here the Earth passes Saturn at point *f*, which marks the middle of the retrograde motion.

ported the heliocentric model of the Solar System. His drawings of the wrinkled lunar surface and the moving sunspots on the Sun weakened the ancient belief in perfect and immutable heavens. Galileo also discovered the four largest moons of Jupiter and showed that they orbited Jupiter, not the Earth. This crack in the wall of geocentricism became an irreparable breach with his discovery of the phases of Venus.

The planets and the Moon shine only because of the sunlight they reflect. Half of a planet is always sunlit while the other half is dark. The fraction of the sunlit hemisphere seen from the Earth, however, varies with the configuration. So the phase termed *new* occurs when we see only the dark hemisphere (at inferior conjunction for the Moon, Mercury, and Venus), and *full* phase takes place at opposition when the entire sunlit hemisphere faces us. The superior planets can never be in *crescent* phase (when less than half the observable hemisphere is sunlit) and are almost always in *gibbous* phase (when more than half

the planet appears sunlit). Galileo observed that Venus shows *all* phases — hence it must orbit the Sun. This observation directly confirmed Copernicus' model.

(b) The Methods of Kepler

Using the heliocentric model of Copernicus and the positional observations of Tycho Brahe painstakingly accumulated over 20 years, Johannes Kepler discovered the need for elliptical planetary orbits (Section 1–2). In 1609 and 1619, he published his three empirical laws of planetary motion. These set the stage for Newton's great scheme of gravitation.

Let's examine Kepler's methods for determining distance to the planets. We will use the Sun–Earth distance — called the *astronomical unit* (AU) — as the unit of distance.

Th Sun–planet distance *r* (in AUs) may be found when an inferior planet reaches greatest elongation (Figure 1–5A). The angle *SEP* is observed (call it α)

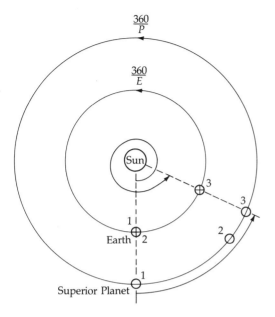

Figure 1-4

Synodic and sidereal periods in a heliocentric model. As the Earth orbits the Sun at an angular speed of $360/E$ degrees per day, a superior planet moves at $360/P$ degrees per day (as seen from the Sun). The Earth moves from position 1 to position 2 after one orbit and has $S - E$ days to reach the next opposition (at position 3). During this time, the superior planet has moved from position 1 to position 3.

and the angle EPS is $90°$; hence, trigonometry yields $r = \sin \alpha$. (This method was first worked out by Copernicus.)

Kepler's method for finding the distance to a superior planet is more complicated than the preceding procedure (Figure 1-5B). The planet is at P at the

beginning and end of one sidereal period, and the Earth is at E and E' at these two times. Note that point P is on the planet's orbit but is otherwise arbitrary. Because we know the planet's sidereal period, we also know the angle ESE'; we must observe the angles PES and $PE'S$. We can solve the triangle ESE' using the law of cosines and trigonometry (Mathematical Appendix) to obtain EE' and the angles SEE' and $SE'E$. Hence, by subtraction, the angles PEE' and $PE'E$ are known, and we may solve the triangle EPE'. Enough information is now available to solve for r, using either triangle SEP or triangle $SE'P$. This process was the method by which Kepler first traced out the orbit of Mars to find that it is elliptical — a significant break with the astronomical tradition of circular orbits.

1-2 Planetary Orbits

(a) Kepler's Three Empirical Laws

Kepler labored for more than 20 years to understand the orbits of the planets. He tested many models of orbital shapes — even ovals — but discarded them all. He finally showed that the orbital planes of the planets pass through the Sun and discovered that the orbital shape was an *ellipse* [Section 1-2(b)]. These findings were announced in 1609 as Kepler's first law — the law of ellipses: *the orbit of each planet is an ellipse with the Sun at one foci* (Figure 1-6A).

Kepler also investigated the speeds of the planets and found that the closer in its orbit a planet is to the Sun, the faster it moves. Drawing a straight line connecting the Sun and the planet (the *radius vector*), he discovered that he could express this fact in Kepler's

Figure 1-5

Distance determinations in a heliocentric model. (A) When an inferior planet reaches greatest elongation (*P*), we know angle SEP and can find r because angle SPE is a right angle. (B) A superior planet is at P at the start and end of one sidereal period; at these times, the Earth is at E and E'. Angles PES and $PE'S$ are observed; they are the elongations of the planet from the Sun.

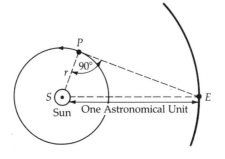

A

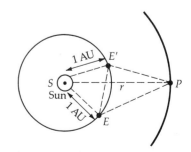

B

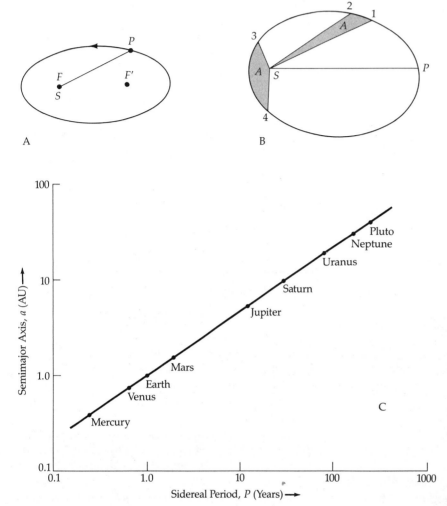

Figure 1–6

Kepler's laws of planetary motion. (A) Each planet *(P)* traces an elliptical orbit *(E)* around the Sun *(S)*, which is at one focus *(F)* of the ellipse. (B) Consider two equal time intervals, that from 1 to 2 and that from 3 to 4. The radius vector to the planet *(SP)* sweeps out the same area *(A)* during these times. (C) For all major planets, this log–log plot of semimajor axes *(a)* versus sidereal periods *(P)* falls very close to a straight line of slope 3/2, confirming Kepler's third law.

second law—the law of areas: *the radius vector to a planet sweeps out equal areas in equal intervals of time* (Figure 1–6B).

Ever seeking a greater harmony in the motions of the planets, Kepler toiled for another decade and in 1619 put forth his third law—the harmonic law: *the squares of the sidereal periods of the planets are proportional to the cubes of the semimajor axes (mean radii) of their orbits* (Figure 1–6C).

The third law may be written algebraically as

$$P^2 = ka^3$$

where P is a planet's sidereal period and a is its average distance from the Sun (the semimajor axis of an elliptical orbit; see below); the constant k has the same value for every body orbiting the Sun. By 1621,

Kepler had shown that the four moons of Jupiter discovered by Galileo obeyed the third law (with a different value of k), confirming its wide applicability.

(c) Geometric Properties of Elliptical Orbits

An *ellipse* is defined mathematically as the locus of all points such that the sum of the distances from two foci to any point on the ellipse is a constant (Figure 1–7); hence

$$r + r' = 2a = \text{constant} \qquad (1-1)$$

The line joining the two foci F and F' intersects the ellipse at the two vertices A and A'. When r and r' lie along this line, we find that a is half the distance between the vertices; we call a the *semimajor axis* of the ellipse. The shape of the ellipse is determined by its *eccentricity e,* such that the distance from each focus to the center of the ellipse is ae. When $e = 0$, we have a circle. One-half of the perpendicular bisector of the major axis is the *semiminor axis b*. Using the dashed lines ($r = r' = a$) in Figure 1–7 and the Pythagorean theorem, we find

$$b^2 = a^2 - a^2e^2 = a^2(1 - e^2) \qquad (1-2)$$

Kepler's first law places the Sun at one focus F. Then vertex A is termed the *perihelion* of the orbit (point nearest the Sun), and vertex A' is called the *aphelion* (point farthest from the Sun). The perihelion distance AF is $a - ae = a(1 - e)$, and the aphelion distance $A'F$ is $a(1 + e)$. The *mean (average) distance* from the Sun to a planet in elliptical orbit is just the semimajor axis a. We prove this fact by noting that for each point P on the ellipse at a distance r from focus F, there is a symmetrical point P' a distance r' from F; the average of these distances is $(r + r')/2 = a$. This result holds for any arbitrary but symmetrical pair of points.

It is extremely useful to know the distance from one focus to a point on the ellipse (such as the Sun–planet or planet–satellite distance) as a function of the position of that point. Center a polar coordinate system (r, θ) at F and let the line FA correspond to $\theta = 0$. Now r measures the distance FP, and θ—the *true anomaly*—measures the counterclockwise angle AFP. Using

$$\cos(\pi - \theta) = -\cos\theta$$

and the law of cosines, we have

$$r'^2 = r^2 + (2ae)^2 + 2r(2ae)\cos\theta$$

From Equation 1–1, however, $r' = 2a - r$, and so we find that

$$r = a(1 - e^2)/(1 + e\cos\theta) \qquad (1-3)$$

Equation 1–3 is the equation for an ellipse in polar coordinates for $0 \le e < 1$.

To derive the *area* of an ellipse, we find the analog of Equation 1–3 in Cartesian coordinates (x, y) positioned at the center of the ellipse. Figure 1–6 and the Pythagorean theorem give

$$r'^2 = (x + ae)^2 + y^2$$
$$r^2 = (x - ae)^2 + y^2$$

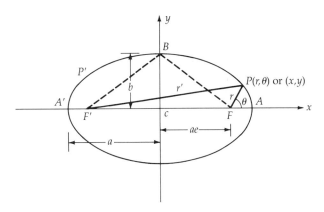

Figure 1–7

An ellipse. Important properties labeled here are AF, perihelion distance; $A'F$, aphelion distance; a, semimajor axis; b, semiminor axis; and c, center.

Subtracting these two equations and using Equation 1–1, we find $r' = a + ex$. Substituting back into the first of the above two equations and employing Equation 1–2, we obtain

$$(x/a)^2 + (y/b)^2 = 1 \qquad (1-4)$$

which is the equation for an ellipse in Cartesian coordinates. The area of the ellipse is given by the double integral

$$A = 4\int_0^b dy \int_0^x dx$$

where, from Equation 1–4,

$$x = a[1 - (y/b)^2]^{1/2}$$

The integration is easy if we use the substitution $y = b\sin z$ (so that $dy = b\cos z\, dz$) and remember the relationship $\sin^2 z + \cos^2 z = 1$; the final answer is

$$A = \pi ab \qquad (1-5)$$

The ellipse is one example of a class of curves called *conic sections*. This family of curves, all of which result from slicing a cone at different angles with a plane, includes the circle, ellipse, parabola, and hyperbola (Figure 1–8). From Equation 1–3, note that the ellipse degenerates to a *circle* of radius $r = a$ when $e = 0$. If we increase e, the foci move apart. When $e = 1$, one of the foci is at infinity and we have the *parabola* specified by

$$r = 2p/(1 + \cos\theta) \qquad (1-6)$$

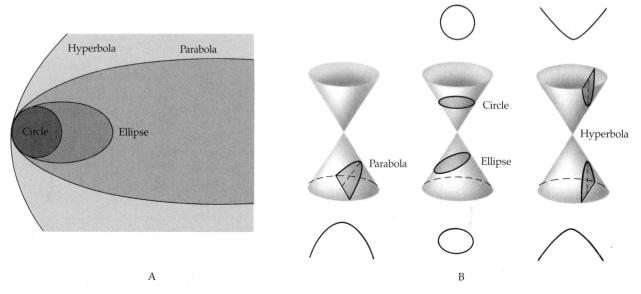

Figure 1–8

Conic sections. (A) The family of conic-section curves includes the circle ($e = 0$), the ellipse ($0 < e < 1$), the parabola ($e = 1$), and the hyperbola ($e > 1$). (B) Conic sections are formed when a cone is cut with a plane. When the plane is perpendicular to the cone's axis, the result is a circle; when it is parallel to one side, the result is a parabola; intermediate angles result in ellipses. A hyperbola results when the angle the plane makes with the cone's side is greater than the opening angle of the cone.

where p is the distance of closest approach (at $\theta = 0$) to the remaining focus. When the eccentricity is greater than unity, the open *hyperbola* results:

$$r = a(e^2 - 1)/(1 + e \cos \theta) \qquad (1-7)$$

Its distance of nearest approach to the sole focus is $a(e - 1)$.

When one body moves under the gravitational influence of another, the relative orbit of the moving body must be a conic section. (The relative orbit is that seen from an observer on the more massive body.) Planets, satellites, and asteroids have elliptical orbits; many comets have eccentricities so close to unity that they follow essentially parabolic orbits. A few comets have nonperiodic hyperbolic orbits; after one perihelion passage, such comets leave the Solar System forever. Space probes have been launched into hyperbolic orbits with respect to the Earth, but they are nearly always captured into elliptical orbits about the Sun. Pioneer 10 was the first spacecraft with an orbit that, when perturbed by Jupiter, led to an escape from the Solar System.

1–3 Newton's Mechanics

Using Kepler's empirical deductions about planetary orbits, Sir Isaac Newton created his unified scheme of dynamics and gravitation, which he published in the *Principia* in 1687. Newton's brilliant insight and elegant formulation laid the foundations for the *Newtonian physics* that we know today. This section presents the first half of his unified structure — *mechanics*.

Newton assumed that the arena within which motions take place is three-dimensional, Euclidean space — *absolute space*. These motions occur in *absolute time*, which passes steadily and which is unaffected by any phenomenon in the Universe. The basic entity in the scheme is the *point particle*, which has mass but no extent (Figure 1–9A). The *position* of the particle at time t, relative to some origin, is indicated by a vector $\mathbf{x}(t)$; the length of this vector is measured in units such as meters. At a slightly later time $t + \Delta t$, the particle has moved to $\mathbf{x} + \Delta \mathbf{x}$ at ap-

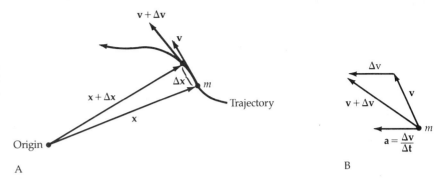

Figure 1–9

Motion of a particle. (A) A mass m at time t is at position x in its trajectory. At $t + \Delta t$, it is at position $x + \Delta x$. The instantaneous velocity at $x + \Delta x$ is $v + \Delta v$. (B) The change in velocity between t and Δt is Δv, from which instantaneous acceleration a is defined.

proximately the *velocity*

$$\mathbf{v} \approx [(\mathbf{x} + \Delta \mathbf{x}) - \mathbf{x}]/[(t + \Delta t) - t] = \Delta \mathbf{x}/\Delta t$$

As we let $\Delta t \to 0$, the velocity vector becomes parallel to the trajectory at point $\mathbf{x}$, where it is defined by the derivative

$$\mathbf{v} \equiv d\mathbf{x}/dt \qquad (1-8)$$

The magnitude of the velocity vector is called the *speed,* and its units are distance/time, such as meters per second (m/s) or kilometers per second (km/s). Noting that the velocity of the particle at t is $\mathbf{v}$, while at $t + \Delta t$ it is $\mathbf{v} + \Delta \mathbf{v}$, we may express the change in velocity by the *acceleration* vector (Figure 1–9B):

$$\mathbf{a} \approx [(\mathbf{v} + \Delta \mathbf{v}) - \mathbf{v}]/[(t + \Delta t) - t] = \Delta \mathbf{v}/\Delta t$$

$$\mathbf{a} = d\mathbf{v}/dt \qquad (\text{as } \Delta t \to 0) \qquad (1-9)$$

The units of acceleration are speed/time, which is distance/time², such as meters/second² or kilometers/ second². Newton also considered the *linear momentum* vector (units of mass times speed, for example, kg · m/s) of the particle, defined by the product

$$\mathbf{p} = m\mathbf{v} \qquad (1-10)$$

where m is the particle's mass and $\mathbf{v}$ is its instantaneous velocity. With these kinematic fundamentals in mind, let's turn to Newton's laws of motion.

(a) The Law of Inertia

In his *Physics,* Aristotle attempted to show that the natural state of a body is one of rest. Our daily experience seems to verify his observation that all moving objects eventually slow to a halt. Indeed, to explain the flight of an arrow, Aristotle said that the diverted air rushing in upon the tail of the arrow "pushed" it along.

Galileo came to a quite different conclusion. He released balls so that they rolled down smooth inclined planes and observed that they rolled up adjacent inclined planes to approximately the same height as that from which he had released them. As he made the planes smoother and inclined the second plane less to the horizontal, he found that the balls rolled farther. Galileo attributed any slowing down of the balls to friction and conjectured that a smooth ball on a horizontal plane would roll forever at a constant speed.

René Descartes later formulated this principle in the form Newton adopted as his first law of motion (the law of inertia): *the velocity of a body remains constant (in both magnitude and direction) unless a force acts upon the body.* For a freely moving body, the first law may be written $\mathbf{v} = $ constant; when the constant is zero, a body initially at rest will remain at rest unless acted upon by a force.

The modern form of Newton's first law is known as the law of conservation of linear momentum. For a body of mass m, we may write $\mathbf{p} = m\mathbf{v} = $ constant, which is equivalent to

$$d\mathbf{p}/dt = 0 \qquad (\text{force-free}) \qquad (1-11)$$

Note that this is true even when a body's mass changes, as in the case of a rocket ship. In the case of a constant mass, then,

$$m \, d\mathbf{v}/dt = 0$$

and

$$m\mathbf{a} = 0 \qquad (\text{force-free, constant mass})$$

(b) The Definition of Force

Newton's second law is already implied by the proviso "unless a force acts upon the body" in the first law, for a force causes a change in velocity. Such a change in velocity (in either speed or direction or both) is indicated by the acceleration vector (Equation 1–9). An important special case of accelerated motion is circular motion, in which the speed remains constant while the direction of motion changes.

The concept of *force* was defined in Newton's second law of motion (the force law): *the acceleration imparted to a body is proportional to and in the direction of the force applied and inversely proportional to the mass of the body.* So we may write $\mathbf{a} = \mathbf{F}/m$ or, more commonly,

$$\mathbf{F} = m\mathbf{a} \qquad (1-12)$$

Note that force is a vector, with the units mass times acceleration, such as $kg \cdot m/s^2$. If several forces act upon a single body, the resultant acceleration is determined by Equation 1–12 using the force that is the *vector sum* of the individual forces—this is the *principle of superposition.* Two forces $\mathbf{F}_1$ and $\mathbf{F}_2$ add to give the resultant force F (Figure 1–10); we may reverse this procedure to decompose the force $\mathbf{F}$ into any two or more component forces; two are indicated here by F_x and F_y.

In Equation 1–12, the body's mass m must remain constant. This restriction vanishes in the modern statement of the second law, which is formulated using the body's linear momentum:

$$\mathbf{F} = d\mathbf{p}/dt \qquad (1-13)$$

We may recover Newton's form of the second law by using Equations 1–9, 1–10, and 1–13 when m is

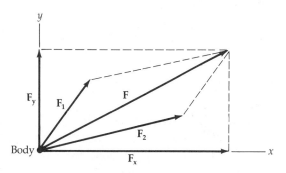

Figure 1–10

Superposition of forces. Two forces $\mathbf{F}_1$ and $\mathbf{F}_2$ act on a body. The resulting motion depends on $\mathbf{F}$, the vector sum of $\mathbf{F}_1$ and $\mathbf{F}_2$. The orthogonal components $\mathbf{F}_x$ and $\mathbf{F}_y$ yield the same motion because their sum results in $\mathbf{F}$.

constant. Note that the first law of motion (Equation 1–11) is now a consequence of the second.

What is the meaning here of *mass*? In dynamics, mass represents the *inertia* of a body, that is, that body's resistance to any change in its state of motion. If we apply the same force to two bodies, the more massive body will change velocity at a lower rate than the less massive body. In everyday terms, think of mass as the amount of material comprising a body; thus two identical lead balls constitute twice the mass of one such ball. Mass is a scalar quantity characterizing a body, and it does not depend on the body's location or state of motion [Section 1–4(b)].

(c) Action and Reaction

To complete his dynamic theory and make it applicable to a collection of point particles, Newton developed his third law of motion (the law of action–reaction): *for every force acting on a body (in a closed system), there is an equal and opposite force exerted by that body.* A simple example will at least illustrate why the third law is necessary. The weight (a force) of a book lying on a table must be exactly balanced by the force that the table exerts on the book; otherwise, according to the second law, the book would accelerate off or through the table. The third law describes the static situation of balanced forces.

The modern version of the third law is the law of conservation of *total linear momentum.* Though we may treat any number of bodies, consider only two bodies here. The total linear momentum of the system is given by $\mathbf{P} = \mathbf{p}_1 + \mathbf{p}_2 =$ constant when no external force acts upon the system (the first two laws). Considering two instants of time (the latter instant indicated by primes), we have

$$\mathbf{p}_1 + \mathbf{p}_2 = \mathbf{p}_1' + \mathbf{p}_2'$$

If the time interval is Δt and if we call $\Delta\mathbf{p}_1 = \mathbf{p}_1' - \mathbf{p}_1$ and $\Delta\mathbf{p}_2 = \mathbf{p}_2' - \mathbf{p}_2$, we can rearrange the above equation and divide by Δt to get

$$\Delta\mathbf{p}_1/\Delta t = -\Delta\mathbf{p}_2/\Delta t$$

For arbitrarily small Δt, the deltas become differentials and Equation 1–13 yields the third law:

$$\mathbf{F}_1 = -\mathbf{F}_2$$

(d) A Summary: Newton's Laws of Motion

We collect here the modern version of Newton's three laws of mechanics. Recall that the Newtonian context is absolute space and time, wherein a point particle of mass m describes a trajectory $\mathbf{x}(t)$ with instantaneous velocity $\mathbf{v}(t)$, linear momentum $\mathbf{p} = m\mathbf{v}$, and acceleration $\mathbf{a}(t)$.

First law (inertia):
  $\mathbf{v}$ and $\mathbf{p} =$ constant

The velocity and linear momentum of a body remain constant (in both magnitude and direction) unless a force acts upon the body.

Second law (force):
  $\mathbf{F} = d\mathbf{p}/dt$

The time rate of change of the linear momentum of a body (or system of bodies) equals the force acting upon the body (or system).

Third law (action–reaction)
  $\mathbf{F}_{\text{exerted}} = -\mathbf{F}_{\text{acting}}$
  $\mathbf{P} =$ constant

In a closed system, the force exerted by a body is equal to and opposite the force acting upon the body, or, the total linear momentum of a closed system of bodies is constant in time.

1–4 Newton's Law of Universal Gravitation

Before Newton, Kepler suspected that some force acted to keep the planets in orbit about the Sun; he attributed the elliptical orbits to the force of magnetic attraction. Following a train of thought similar to the simplified derivation that follows, Newton discovered his law of universal gravitation, tested it on the motion of the Moon, and then explained the motions of the planets in detail.

(a) Centripetal Force and Gravitation

Consider a body moving in a circular orbit of radius r about a center of force (Figure 1–11A). From symmetry, the speed v of the body must be constant, but the direction of the velocity vector is constantly changing. Such a changing velocity represents an acceleration — the *centripetal acceleration* that maintains the circular orbit; from the geometry of the figure, we may deduce the acceleration. The time is t at point A, where the body's velocity is $\mathbf{v}$. At an infinitesimal time interval Δt later, the body has traversed the angle $\Delta\theta$ to B, where the velocity is $\mathbf{v}'$. In Figure 1–11B, we show the change in velocity $\Delta\mathbf{v} = \mathbf{v}' - \mathbf{v}$ by joining the tails of the two velocity vectors; the angle between $\mathbf{v}$ and $\mathbf{v}'$ is clearly $\Delta\theta$. Recalling that the magnitude of both $\mathbf{v}$ and $\mathbf{v}'$ is the speed v, we use trigonometry to deduce, for small values of $\Delta\theta$ (Figure 1–11A),

$$\Delta\theta = s/r = v\,\Delta t/r$$

and (Figure 1–11B)

$$\Delta\theta = \Delta v/v$$

where the arc length s approximates the chord joining points A and B. So the centripetal acceleration has the magnitude

$$a = \Delta v/\Delta t = v^2/r$$

and it points directly toward the center of the circle as Δt vanishes. If m is the mass of the orbiting body, Newton's second law gives the magnitude of the *centripetal force* as

$$F_{\text{cent}} = ma = mv^2/r \qquad \textbf{(1–14)}$$

If P is the orbital period of the body, then its speed is

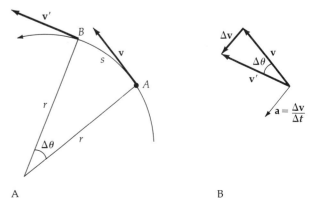

Figure 1–11

Centripetal acceleration. (A) An object moves in a circular orbit of radius r at speed v along a short arc S during the time interval Δt. (B) The centripetal acceleration $\mathbf{a}$ results from the change in the direction of the velocity during time Δt.

A   B

$$v = 2\pi r/P$$

but Kepler's third law relates the period and the orbital radius by

$$P^2 = kr^3$$

where k is the proportionality constant. Substituting these two results into Equation 1–14, we find

$$F = 4\pi^2 m/kr^2$$

that is, the force maintaining the orbit is *inversely proportional to the square of the radius.* According to Newton's third law, the body (of mass M) at the center of the orbit feels an equal but opposite force. Because the centripetal force acting on the central body must be proportional to M, the *mutual* gravitational force is proportional to the product of the two masses; redefining the constant of proportionality, we have

$$\mathbf{F}_{\text{grav}} = GMm/r^2 \qquad (1\text{–}15)$$

which is Newton's law of universal gravitation; the direction of the gravitational force is along the line joining the two bodies (from the third law of motion). The gravitation constant G has the measured value 6.67×10^{-11} m^3/kg · s^2 in SI units.

Equation 1–15 expresses the attractive gravitational force between two *point* masses. To find the gravitational attraction from an extended body, we must sum the vectorial contributions from each small piece of the body. In general, this process is rather difficult, but for a spherically symmetric body acting on a point mass, symmetry arguments alone tell us that the gravitational force must act along the line joining the centers of the bodies. By integrating the effects of all parts of the spherical body, we can show that *it behaves gravitationally as though its entire mass were concentrated at its center.* This important result will be used many times in this book.

(b) Weight and Gravitational Acceleration

Near the Earth's surface, bodies have a constant downward gravitational acceleration of magnitude (from Equation 1–15)

$$g = GM_\oplus/R_\oplus^2 \approx 9.8 \text{ m/s}^2$$

where $M_\oplus$ is the mass of the Earth and $R_\oplus$ is its radius. The Earth's oblate surface and rapid rotation cause variations in the measured value of g from 9.781 m/s^2 at the equator to 9.832 m/s^2 at the poles.

The *weight* of a body is the force necessary to hold it motionless in a gravitational field. At the Earth's surface, we comply with this definition by weighing the body on a spring scale, and if the mass of the body is m, we find

$$\text{Weight} = mg$$

In contrast to its constant mass, the weight of a body depends upon its *location.* An astronaut on the Moon's surface will weigh approximately one-sixth her normal Earth weight; in a satellite orbit, her weight will be zero since she is falling freely in the gravitational field. Weight and force have the same units, and the conventional SI unit is the *newton* (Appendix 6), where

$$1 \text{ newton} = 1 \text{ kg} \cdot \text{m/s}^2$$

Keep in mind that *weight is a force.*

(c) Determination of G and $M_\oplus$

Let's describe how Henry Cavendish measured the gravitational constant G in 1798 and how Phillip von Jolly established the mass of the Earth $M_\oplus$ in 1881. These two methods represent early procedures to determine these important constants.

To find G, Cavendish used an apparatus consisting of two small balls of equal mass m hung from a torsion beam and two larger balls of equal mass M attached to an independently suspended but coaxially aligned beam (Figure 1–12A). Each adjacent M–m pair is initially placed a distance D apart, but the gravitational force between the balls twists the torsion bars to a static equilibrium distance d between each M–m pair. From symmetry, the gravitational force causing the deflection is $F_{\text{tot}} = 2GMm/d^2$; by directly measuring F_{tot}, M, m, and d, Cavendish found G.

Von Jolly's apparatus (Figure 1–12B) consisted of a balance bearing two small masses m, with the rotation axis of the balance beam aligned horizontally. When he placed a large mass M below one of these small masses, the balance tipped; to restore the original balance, he placed a small mass n on the pan alongside the other mass m. If the equilibrium M–m distance is d, the forces acting on each side of the torsion beam are

$$F_1 = GMm/d^2 + GM_\oplus m/R_\oplus^2$$
$$F_2 = GM_\oplus n/R_\oplus^2 + GM_\oplus m/R_\oplus^2$$

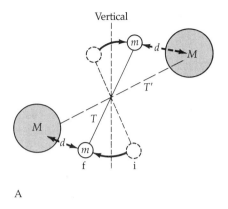

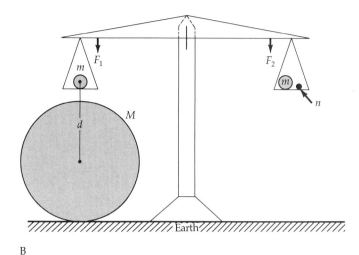

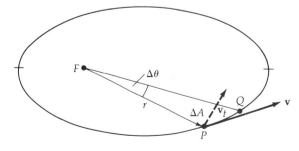

Figure 1–12

Measuring Newton's constant of gravitation G. (A) The Cavendish experiment uses two torsion bars T and T', which are free to rotate about a vertical axis. After the gravitational force between adjacent M–m spheres swings the bars to their equilibrium positions (to position f from position i), the separations between M and m are d. (B) Von Jolly's experiment has a horizontal balance that is in an equilibrium horizontal position when masses M and n are absent. When mass M is placed below the left-hand mass m, mass n must be added to the right-hand mass m to restore equilibrium.

The beam is horizontal again when $F_1 = F_2$, and so we find

$$M_\oplus = (Mm/n)(R_\oplus/d)^2 = 5.976 \times 10^{24} \text{ kg}$$

1–5 Physical Interpretations of Kepler's Laws

Newton combined his laws of motion and gravitation to *derive* all three of Kepler's empirical laws. We too could derive the elliptical orbits of Equation 1–3, using Equations 1–13 and 1–15, but since this requires a knowledge of vector differential equations, we won't do so. We will accept Kepler's first law and follow Newton's footsteps in deducing Kepler's second and third laws.

(a) The Law of Areas and Angular Momentum

Let's illustrate Kepler's law of areas for an elliptical orbit (Figure 1–13). A body orbits the focus F at the position r and velocity $\mathbf{v}$. During an infinitesimal time interval Δt, the body moves from P to Q and the radius vector sweeps through the angle $\Delta\theta$. This small angle is $\Delta\theta \approx v_t \Delta t/r$, where v_t is the component of $\mathbf{v}$ perpendicular to r. During this time, the radius vector has swept out the triangle FPQ, the area of which is $\Delta A \approx r v_t \Delta t/2$. Therefore, for $\Delta t \to 0$,

$$dA/dt = r v_t/2 = r^2(d\theta/dt)/2 = H/2 \qquad (1\text{--}16)$$

where the constant H (the angular momentum per unit mass) appears because Kepler's second law

Figure 1–13

Law of areas for an elliptical orbit. A body at r moves at velocity $\mathbf{v}$ in an elliptical orbit about focus F and moves from P to Q in time Δt. The component of $\mathbf{v}$ perpendicular to r is $\mathbf{v}_t$.

states that the rate of change of area with time is a constant.

Note that $A/P = H/2$, where $A = \pi ab$ is the total area of the ellipse and P is the orbital period; we can prove this by integrating Equation 1–16. By combining this result with Equation 1–16 and noting that v_t is the total speed at perihelion or aphelion, we can deduce the perihelion and aphelion speeds of a planet orbiting the Sun. For example, at perihelion we have

$$v = H/r = 2A/Pr = 2\pi ab/Pa(1 - e)$$

where Equation 1–3 with $\theta = 0$ is used in the last equality. Carrying through the derivation at aphelion and using Equation 1–2, we get

$$v = \begin{bmatrix} (2\pi a/P)[(1 + e)/(1 - e)]^{1/2} & \text{(perihelion)} \\ (2\pi a/P)[(1 - e)/(1 + e)]^{1/2} & \text{(aphelion)} \end{bmatrix}$$
$$(1-17)$$

For the Earth, a is 1 AU (1.496×10^8 km), P is one year (3.156×10^7 s), and the orbital eccentricity is $e = 0.0167$; hence, the orbital speed varies from 30.3 km/s at perihelion to 29.3 km/s at aphelion.

A modern Newtonian derivation of Kepler's second law requires the concept of the orbiting body's *angular momentum*:

$$\mathbf{L} = \mathbf{r} \times \mathbf{p} = m(\mathbf{r} \times \mathbf{v}) \qquad (1-18)$$

where m is the body's mass, $\mathbf{r}$ its position vector, and $\mathbf{p}$ its linear momentum (Equation 1–10). The *vector cross product* in Equation 1–18 is an operation that yields the product of the *perpendicular* components of the two vectors under consideration (Mathematical Appendix); hence, if $\mathbf{r}$ and $\mathbf{p}$ are parallel, then $\mathbf{r} \times \mathbf{p} = 0$. Angular momentum is a vector quantity $\mathbf{L}$ with the units kg $\cdot$ m²/s. Differentiating Equation 1–18, we have

$$d\mathbf{L}/dt = \mathbf{v} \times \mathbf{p} + \mathbf{r} \times (d\mathbf{p}/dt) = \mathbf{r} \times \mathbf{F} \quad (1-19)$$

since $\mathbf{v}$ is parallel to $\mathbf{p}$ and $d\mathbf{p}/dt$ defines force. We call $d\mathbf{L}/dt$ the *torque* (with units kg $\cdot$ m²/s²) and see that when $\mathbf{F}$ is collinear with $\mathbf{r}$—a *central force*, such as gravitation—the torque vanishes. Hence, $\mathbf{L}$ is constant in time so that *angular momentum is conserved for all central forces*. Applying Equation 1–18 to the situation in Figure 1–12, we find

$$\mathbf{L}/m = \mathbf{r}v_t = H = \text{constant}$$

which is Kepler's second law.

(b) Newton's Form of Kepler's Third Law

The external forces that act upon the Solar System are essentially negligible; hence, the total linear momentum of the Solar System is constant. If the Sun did not move in response to the orbiting planets, this would clearly not be the case, and so the Sun must move about the center of mass of the Solar System. We apply this idea to an isolated system of two bodies moving in circular orbits from their mutual gravitational attraction; our final result, Newton's form of Kepler's third law, is also applicable to elliptical orbits.

Consider two bodies of masses m_1 and m_2, orbiting their stationary center of mass at distances r_1 and r_2 (Figure 1–14). Because the gravitational force acts only along the line joining the centers of the bodies, both bodies must complete one orbit in the same period P (though they move at different speeds v_1 and v_2). The centripetal forces of the orbits are therefore

$$F_1 = m_1 v_1^2/r_1 = 4\pi^2 m_1 r_1/P^2 \quad \textbf{(1-20a)}$$

$$F_2 = m_2 v_2^2/r_2 = 4\pi^2 m_2 r_2/P^2 \quad \textbf{(1-20b)}$$

Newton's third law requires $F_1 = F_2$, and so we obtain

$$r_1/r_2 = m_2/m_1 \qquad \textbf{(1-21)}$$

The more massive body orbits closer to the center of mass than does the less massive one; Equation 1–21 defines the position of the center of mass.

The total separation of the two bodies $a = r_1 + r_2$ is also the radius of their relative orbits. Now Equation 1–21 may be expressed in the form

$$r_1 = m_2 a/(m_1 + m_2) \qquad \textbf{(1-22)}$$

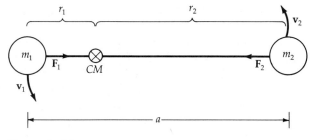

Figure 1–14

Center of mass. Two masses m_1 and m_2 orbit a common center of mass CM with a total separation a.

However, the mutual gravitational force $F_{grav} = F_1 = F_2$ is

$$F_{grav} = Gm_1m_2/a^2 \qquad (1-23)$$

so that combining Equations 1–20a, 1–22, and 1–23 gives *Newton's form of Kepler's third law:*

$$P^2 = 4\pi^2a^3/G(m_1 + m_2) \qquad (1-24)$$

If body 1 is the Sun and body 2 any planet, then $m_1 \gg m_2$; therefore $k = 4\pi^2/GM_\odot)$ is the proportionality "constant" (to a good approximation) in Kepler's third law. Note that the third law presents a way to describe gravitational effects without discussing forces.

(c) Orbital Velocity

To better understand elliptical orbits, consider the *orbital velocity* **v**. We may decompose this velocity into two perpendicular components (Figure 1–15): v_r, the radial speed, and v_θ, the "angular" speed. From Equation 1–16 and the discussion following it, we have

$$d\theta/dt = (2\pi/P)(a/r)^2(1 - e^2)^{1/2} \qquad (1-25)$$

Using Equation 1–3, which is the polar equation of an ellipse, and Equation 1–25, we compute the time derivatives indicated below to find:

$$v_r \equiv dr/dt = (2\pi a/P)(e\sin\theta)(1 - e^2)^{-1/2} \qquad (1-26a)$$

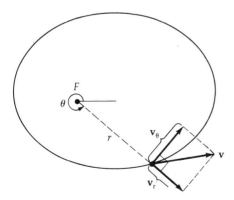

Figure 1–15

Components of orbital velocity. At any point in an elliptical orbit, the orbital velocity **v** may be decomposed into two perpendicular components, the radial speed parallel to the radius vector and the angular speed perpendicular to it.

$$v_\theta \equiv r(d\theta/dt) = (2\pi a/P)(1 + e\cos\theta)(1 - e^2)^{-1/2} \qquad (1-26b)$$

Note that Equation 1–26b reduces to Equation 1–17 at perihelion and aphelion. The total orbital speed now follows from Equations 1–26 as

$$\begin{aligned} v^2 &= v_r^2 + v_\theta^2 \\ &= (2\pi a/P)^2(1 + 2e\cos\theta + e^2)/(1 - e^2) \end{aligned} \qquad (1-27)$$

Rearranging the polar equation for an ellipse, we get

$$e\cos\theta = [a(1 - e^2) - r]/r$$

and substituting this result into Equation 1–27, we finally obtain (with the help of Equation 1–24)

$$v^2 = G(m_1 + m_2)[(2/r) - (1/a)] \qquad (1-28)$$

Therefore, for given masses, *the total orbital speed depends only upon the separation and the orbit's semimajor axis.* This useful result will appear frequently in this book.

(d) The Conservation of Total Energy

The concept of energy provides an alternative approach to that of forces and Newtonian mechanics for problems relating to orbits. Energy often proves a more powerful way to gain insight into orbital problems.

Energy is a quantity assigned to one body that indicates that body's ability to change the state of another body. *Heat* is a form of energy, for a hot body will warm a cold body if the two are brought into contact. Electrical energy causes the filament of a lightbulb to glow (become hot), showing that one type of energy may be converted to another. *Kinetic energy (KE)* is a body's energy of *motion,* but if we decide to move along with the body, it has no motion; it has no kinetic energy. *Potential energy (PE)* is due to the *position* of the body; if the body is free to move, this energy may be converted to kinetic energy. In celestial mechanics, we sum the kinetic and potential energies to obtain the *total energy: TE = KE + PE.*

Assume that a force **F** acts upon a body of mass m that is moving in the trajectory $x(t)$ about a central force. In the infinitesimal time dt, the body moves through the vector distance $d\mathbf{x}$. As the body moves from position A to position B, we define the work W (SI units $= kg \cdot m^2/s^2 = joules$; see Appendix 6) done *on* the body by the force as

$$W = \int_A^B \mathbf{F} \cdot d\mathbf{x} \qquad (1-29)$$

The *vector dot product* operation in Equation 1–29 yields the product of the *parallel* components of $\mathbf{F}$ and $d\mathbf{x}$ (Mathematical Appendix); when $\mathbf{F}$ and $d\mathbf{x}$ are mutually perpendicular, their dot product vanishes. To evaluate Equation 1–29, we note the following:

$$\mathbf{F} \cdot d\mathbf{x} = m(d\mathbf{v}/dt) \cdot \mathbf{v}\, dt = m(\mathbf{v} \cdot d\mathbf{v}) = d(mv^2/2)$$

where Newton's second law and the definitions of velocity and speed have been employed. Thus Equation 1–29 may be integrated directly to give

$$W = (mv^2/2)_B - (mv^2/2)_A = KE_B - KE_A \quad (1-30)$$

where the kinetic energy is specified by $KE = mv^2/2$. Therefore, *the work done by the force on the body changes the body's kinetic energy.* In a system of two bodies of masses m_1 and m_2, the kinetic energy is $KE = (m_1 v_1^2/2) + (m_2 v_2^2/2)$. If the force is the gravitational force between the two bodies, we have

$$\mathbf{F} \cdot d\mathbf{x} = -Gm_1 m_2/r^2\, dr = d(Gm_1 m_2/r)$$

where r is the radial separation of the bodies. Equation 1–29 is again integrated:

$$W = (Gm_1 m_2/r)_B - (Gm_1 m_2/r)_A = PE_A - PE_B \qquad (1-31)$$

where we have inserted the definition of the mutual gravitational potential energy $PE = -Gm_1 m_2/r$. Thus the potential energy is the *negative* of the work done by the gravitational force as m_1 moves from $r = \infty$ to $r = r$ with m_2 held fixed. Note that the mutual gravitational potential energy vanishes when the two bodies are infinitely separated.

Because both kinetic and potential energy have the same units (joules) as work, we combine Equations 1–30 and 1–31 to obtain

$$(KE + PE)_B = (KE + PE)_A$$

or, in terms of the total energy of the two-body system,

$$TE_B = TE_A = \text{constant}$$

Therefore, the total energy of our gravitating system is conserved:

$$TE = (m_1 v_1^2/2) + (m_2 v_2^2/2) - (Gm_1 m_2/r)$$
$$= \text{constant} \qquad (1-32)$$

As the bodies move, kinetic and potential energy may

be interchanged but the total energy of the system remains constant.

Now to evaluate the constant TE in Equation 1–32 for elliptical orbits. Refer back to Section 1–3(c) and recall that the total linear momentum of our isolated system is constant; we choose this constant to be zero so that $m_1 \mathbf{v}_1 = -m_2 \mathbf{v}_2$ and, in terms of the speeds of the bodies,

$$m_1 v_1 = m_2 v_2$$

Since $v = v_1 + v_2$ is the relative speed of either body with respect to the other, we find

$$v_1 = m_2 v/(m_1 + m_2)$$
$$v_2 = m_1 v/(m_1 + m_2)$$

Substituting this result into Equation 1–32 yields

$$TE = m_1 m_2 [v^2/2(m_1 + m_2) - (G/r)] \quad (1-33)$$

Using Equation 1–24, let's evaluate Equation 1–33 at the perihelion of the orbit, where $r = a(1 = e)$ and v is the perihelion speed given by Equation 1–17. Our result, $TE = -Gm_1 m_2/2a$, shows that the total energy is negative—the orbit is *bound*. Now Equation 1–33 takes its final form:

$$v^2 = G(m_1 + m_2)[(2/r) - (1/a)] \quad (1-34)$$

exactly the expression found in Equation 1–28. This useful and classic result, which is a statement of total energy conservation, is sometimes referred to as the *vis viva equation.*

1–6 Applications to the Solar System

To fully understand and appreciate the formal manipulations just carried out, you should use the results to solve a few physical problems. We illustrate this point by considering several simple examples of the application of Newton's form of Kepler's third law.

(a) Using Kepler's Third Law

The modern form of the harmonic law is Equation 1–24:

$$P^2 = 4\pi^2 a^3/G(m_1 + m_2) \qquad (1-35)$$

In applying this relation to the Sun ($M_\odot$) and planets (m_p), considerable simplification occurs when we measure the sidereal periods P in years and the semimajor axes a in AU, for then

$$a^3/P^2 = 1 + (m_p/M_\odot)$$

Note that we have simplified by using Earth units; this clue tells you to always use the units most appropriate to the system under consideration. Implement this idea by forming *ratios* of Equation 1–24 as

$$[(m_1 + m_2)/(m_1' + m_2')](P/P')^2 = (a/a')^3$$

where the system m_1 and m_2 (with period P and semimajor axis a) is compared with the *standard system* m_1' and m_2' (with P' and a'). For objects orbiting the Sun or for binary stars (Chapter 12), the standard system is the Sun–Earth system: P is expressed in years, a in AU, and all masses m in solar masses ($M_\odot$). In these units, $P^2 = a^3$; that is, $k = 1$, so $G = 4\pi^2$. For planetary satellites (either natural moons or artificial satellites), we use the Earth–Moon system: set $P' = 27.3$ days, $a' = 3.84 \times 10^5$ km, and $(m_1' + m_2') = M_\oplus$ (or 5.976×10^{24} kg); we then obtain P in days, a in kilometers, and the masses m in Earth masses (or kilograms).

As an example, consider a comet with an elliptical orbital period of seven years; we want to find the semimajor axis of its orbit. Since $M_\odot + m_{comet} \approx M_\odot$, Equation 1–35 gives $a = 7^{2/3} = 3.7$ AU. Another example is to find the mass of Uranus (M_U) in terms of the Earth's mass. We observe that Miranda (whose mass is M_m), a moon of Uranus, orbits the planet in 1.4 days at a mean distance of 128,000 km. Using the Earth–Moon standard system, we have

$$[(M_U + M_m)/(M_\oplus + M_m)](P_m/P)^2$$
$$\approx (M_U/M_\oplus)(P_m/P)^2 = (a_m/a)^3$$

where we have neglected the masses of Miranda and the Moon with respect to the much larger masses of their primary planets, Uranus and Earth. Substituting the relevant data, we obtain

$$M_U \approx (27.3/1.4)^2(128,000/384,000)^3 M_\oplus \approx 14 M_\oplus$$

(b) Launching Rockets

Consider projectiles launched vertically upward from the Earth's surface. Neglecting atmospheric friction, which continuously decreases the projectile's speed, we can use the conservation of total energy to find the height to which the projectile finally rises.

Near the Earth's surface, the strength of the downward gravitational force is $F = mg$, where m is the mass of the projectile. Hence, the potential energy at altitude h is $PE = mgh$. If the speed of the body is v at ground level, its kinetic energy there is $KE = mv^2/2$. Conservation of total energy says that

$$TE = (KE + PE)_{ground} = \text{constant} = (KE + PE)_h$$

so that by evaluating TE at $h = 0$ and at maximum height h, we find $mv^2/2 = mgh$, or,

$$h = v^2/2g \qquad (1\text{–}36)$$

with $g = 9.8$ m/s^2, a rock thrown upward with speed $v = 14$ m/s will rise to a height of 10 m before falling back to the ground.

When we consider altitudes greater than about one Earth radius ($h \gtrsim R_\oplus$), our previous approximation breaks down and we must use the correct formula, Equation 1–32. Evaluating this expression at the ground ($R_\oplus$) and at maximum height ($R_\oplus + h$), we find

$$(mv^2/2) - (GM_\oplus m/R_\oplus) = -GM_\oplus m/(R_\oplus + h)$$

or

$$h = R_\oplus\{(v^2 R_\oplus/2GM_\oplus)/[1 - (v^2 R_\oplus/2GM_\oplus)]\}$$
$$= (v^2/2g)\{R_\oplus/[R_\oplus - (v^2/2g)]\} \qquad (1\text{–}37)$$

Equation 1–36 follows from Equation 1–37 in the limit $v^2/2g \ll R_\oplus$, that is, $h \ll R_\oplus$. Note that when $v^2/2g = R_\oplus$, at the speed $v = (2gR_\oplus)^{1/2} = 11.2$ km · s^{-1}, the projectile escapes to $h = \infty$. This critical speed is called the *escape speed*.

(c) Orbits of Artificial Satellites and Space Probes

Most major space vehicles are placed into Earth orbit — a *parking orbit* from which deep-space probes are accelerated toward the Moon and the planets. Multistage rockets lift the vehicle beyond the Earth's atmosphere, where the final stages accelerate the payload horizontally to the desired orbital speed. The Earth's rotational speed (0.46 km/s at the equator) aids launchings toward the east. For a circular orbit at distance r from the center of the Earth ($r = R_\oplus + h$, if h is the altitude of the orbit), the circular speed v_c may be found by equating the centripetal and gravitational forces:

$$v_c = (GM_\oplus/r)^{1/2} = v_0(1 + h/R_\oplus)^{-1/2} \qquad (1\text{–}38)$$

where $v_0 = (GM_\oplus/R_\oplus)^{1/2} = 7.86$ km/s is the circular speed at the Earth's surface (neglecting atmospheric friction). Rearranging Equation 1–37, we find

$$v = \sqrt{2}v_0[(R_\oplus/h + 1)]^{-1/2}$$

so that $h \to \infty$ at the *escape speed* $v_{escape} = \sqrt{2}v_0$; note that this is the escape speed from the Earth's surface, not from a satellite orbit.

The orbit of a space probe depends critically upon the velocity at burnout. Consider the simple case of a projectile moving *parallel* to the Earth's surface at this point (point A in Figure 1–16). The energy equation (Equation 1–28) gives the semimajor axis a of the orbit if we know the burnout speed v:

$$a = r/[2 - (v/v_c)^2] \qquad \textbf{(1–39a)}$$

where r is the distance from the Earth's center to point A. We can now invert Equation 1–39a to find the injection speed needed to attain an orbit of semimajor axis a:

$$v = v_c(2 - r/a)^{1/2} \qquad \textbf{(1–39b)}$$

A circular orbit of radius $r = a$ results when $v = v_c$, as expected. When we have $v = \sqrt{2}v_c$, the semimajor axis becomes infinite and the projectile escapes along a parabolic orbit; hence, this is the escape speed appropriate to radius r. When $v > \sqrt{2}v_c$, the space probe escapes on a hyperbolic trajectory. If $v < \sqrt{2}v_c$, the projectile enters an elliptical orbit and point A must clearly be either the *perigee* (point closest to Earth) or *apogee* (point farthest from Earth) of the orbit, since the velocity is perpendicular to the radius vector at point A. For $v_c < v < \sqrt{2}v_c$, Equation 1–39a tells us

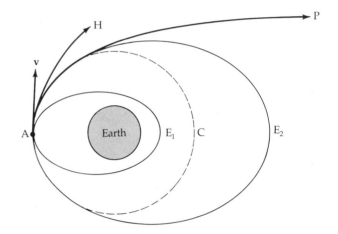

Figure 1–16

Orbits around the Earth. At *A*, a spacecraft has burnout velocity **v**. If the velocity equals that for a circular orbit at that distance from the Earth's center, a circular orbit C results. When **v** = 2 **v**$_C$, a parabolic orbit P ensues; for a greater velocity, the hyperbolic escape orbit H is hyperbolic. When **v** falls between these two limits, *A* is the perigee of elliptical orbit E$_2$; when 0 < **v** < **v**$_C$, *A* is at the apogee of elliptical orbit E$_1$.

that $a > r$ at A; hence, we insert at perigee, and the orbit never approaches the Earth closer than A. For $0 < v < v_c$, we have $a < r$, so that A is the apogee. The satellite will collide with the Earth if the perigee distance is less than $R_\oplus$.

Problems

1. Assume that the orbital plane of a superior planet is inclined 10° to the ecliptic and that the planet crosses the ecliptic moving northward at opposition. Make a diagram similar to Figure 1–3, showing the retrograde path of this superior planet.

2. Imagine you are observing the Earth from Jupiter. What would you observe the Earth's synodic orbital period to be? What would it be from Venus?

3. (a) Explicitly carry out the derivation of Equation 1–4, showing all the appropriate steps.
   (b) On graph paper, plot the following polar equations: Equation 1–3 for an ellipse, Equation 1–6 for a parabola, and Equation 1–7 for a hyperbola.

4. In terms of the gravitational acceleration g at the surface of Earth, find the surface gravitational accelera-

tion of
   (a) the Moon ($M_m = 0.0123 M_\odot$, $R = 1738$ km),
   (b) the Sun ($M_\odot = 2 \times 10^{30}$ kg, $R_\odot = 7 \times 10^7$ m), and
   (c) Jupiter ($M_J = 318 M_\oplus$, $R = 11.2 R_\oplus$).

5. What are the perihelion and aphelion speeds of Mercury? What are the perihelion and aphelion distances of this planet? Compute the product vr (speed times distance) at each of these two points and interpret your result physically.

6. Find the relative position of the center of mass for (a) the Sun–Jupiter system and (b) the Earth–Moon system.

7. A television satellite is in circular orbit about the Earth, with a sidereal period of exactly 24 h. What is the distance from the Earth's surface for such a satellite?

(*Hint:* Use Kepler's laws.) If the satellite appears stationary to an earthbound observer, what is the orientation of its orbital plane?

8. Using orbital data for Titan (Appendix 3), find the mass of Saturn.

9. A stone is released from rest at the Moon's orbit and falls toward Earth. What is the stone's speed when it is 192,000 km from the center of Earth?

10. (a) What is the semimajor axis of the *least-energy* elliptical orbit of a space probe from Earth to Venus?
    (b) Relative to the Earth, what is the velocity of such a probe at the Earth's orbit?
    (c) When the probe reaches Venus, what is its velocity relative to that planet?

11. Suppose that a projectile has a burnout speed $\sqrt{2}v$ (where $v_c/\sqrt{2} < v < v_c$) at a distance r from the Earth's center ($r > R_\oplus$). If the velocity vector points $45°$ *above* the local parallel to the Earth's surface, find the semimajor axis a, the sidereal period P, and the eccentricity e of the resulting elliptical orbit in terms of r, v, and constants. Can you also find θ, the angle from burnout to the orbit's perigee? (*Hint:* Section 1–5(c) is very helpful.)

12. Compare the escape velocity of a rocket launched from the Earth with the escape velocity of one at a distance of 1 AU from the Sun (that is, the escape velocity from the Solar System at the Earth's distance from the Sun).

The
Solar
System
in
Perspective

Our Earth is one member of the family of the Sun—the *Solar System.* Nine major planets, many natural satellites (moons), multitudes of asteroids, an unknown number of comets, and an abundance of meteoroids orbit the Sun. Each class of objects has distinguishing characteristics (this chapter) and individual idiosyncrasies (Chapters 3–6) as well as systematic behavior that requires explanation in terms of the origin of the Solar System (Chapter 7). Here we will preview the physical aspects of the Solar System; the specifics will follow later. Celestial mechanics applied to Solar System bodies forms the physical basis of this perspective.

2–1 Contents of the Solar System

All objects bound in orbits by the Sun's gravity comprise the Solar System. These bodies make up a hierarchy based on mass: the Sun crowns the top, then the planets, their moons and rings, interplanetary debris (comets, asteroids, and meteoroids), down to interplanetary gas and dust. (See Appendix 3 for a summary of Solar System data.)

(a) Planets

Today we place the Sun at the center of the heliocentric system of nine *planets* (in order of average distance from the Sun): Mercury, Venus, Earth, Mars, Jupiter, Saturn, Uranus, Neptune, and Pluto. Together, the planets have a total mass of but $0.0014M_\odot$, and they shine only by reflected sunlight. By mass, the Solar System is mostly the Sun (Figure 2–1).

Motions

The planets obey the laws of Kepler and Newton as they move in elliptical orbits about the Sun. Their

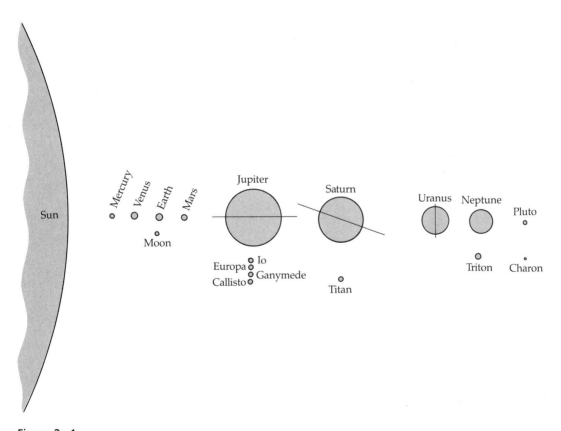

Figure 2–1

Relative sizes of the planets. On this scale, only a part of the limb of the Sun can be shown. Included with the planets are the eight largest moons.

orbital semimajor axes are spaced in a fairly regular pattern. In 1766 (before Uranus, Neptune, and Pluto were discovered), Titius of Wittenberg found an approximate empirical rule relating mean Sun–planet distances; Johann Bode publicized this relationship in 1772, and it is now called *Bode's Law* or the *Titius-Bode Rule*. The prescription is to write down the series of numbers

$$4, 4 + (3 \times 2^0), 4 + (3 \times 2^1), 4 + (3 \times 2^2),$$
$$4 + (3 \times 2^3), \ldots$$

divide each result by 10, and arrive at the sequence

0.4, 0.7, 1.0, 1.6, 2.8, 5.2, 10.0, 19.6, 38.8, 77.2, . . .

With the Sun–Earth distance (1 AU) as the unit of length, the mean Sun–planet distances (Table A3–2) are

Mercury Venus Earth Mars—
   0.39   0.72   1.00 1.52

Jupiter Saturn Uranus Neptune Pluto
  5.20   9.54   19.2   30.1   39.5

Except for the gap at 2.8 AU (where the asteroids lie), Bode's law works surprisingly well for the inner seven planets — it is a regularity that hints at a physical process. Note the close spacing from the Sun to

Mars (Figure 2–2) and the more open and regular distribution from Jupiter outward.

The orbits exhibit three striking features of planetary motion (Table A3–1). First, all planets orbit the Sun counterclockwise as seen from directly above the Earth's orbital plane—the orbits are *direct*. Second, the orbital planes lie very close to the ecliptic plane (the plane of the Earth's orbit), so that all the planets except Pluto are always found in the 16°-wide band of the Zodiac when viewed from the Earth (the inclination of Pluto is 17°). Third, the orbital eccentricities are less than 0.1 except for the closest planet to the Sun, Mercury, and the most distant, Pluto. Because of its large eccentricity ($e = 0.249$), the orbit of Pluto ranges from 29.68 AU at *perihelion* (closest point to the Sun) to 49.36 AU at *aphelion* (farthest point from the Sun). Pluto passes closer to the Sun than Neptune on occasion (Figure 2–3). The conjecture that Pluto is an escaped satellite of Neptune rests on this observation.

A planet's *sidereal rotation period* refers to its rotation with respect to the stars; these periods are determined in a variety of ways (Table A3–2). One main technique involves the Doppler effect, described in detail in Section 8–1(a). The essence of the Doppler shift is that when light is emitted by (or radio

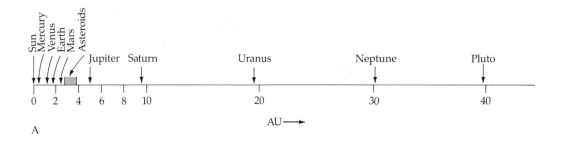

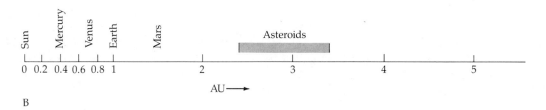

Figure 2–2

Relative orbital distances of the planets. (A) The distance scale in AUs from the Sun to Pluto. (B) An expanded scale shows the inner Solar System.

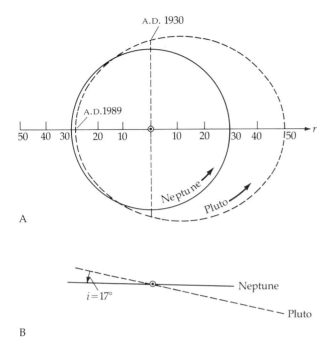

Figure 2–3

The orbits of Neptune and Pluto. (A) Face-on view of the orbits, with the scale in AUs. Note that at perihelion (1989), Pluto lies inside the orbit of Neptune. (B) Side view of the orbits, showing their relative inclinations.

waves are bounced off of) a body, a radial component of velocity will be shifted in wavelength (or frequency) relative to its rest wavelength: a redshift for recession and a blueshift for approach. The amount of shift depends on the magnitude of the radial velocity (for nonrelativistic speeds):

$$\Delta\lambda/\lambda = v_r/c$$

where $\Delta\lambda$ is the wavelength change, λ the rest wavelength, v_r the radial part of the velocity, and c the speed of light.

Mercury and Venus were assumed to be in synchronous rotation about the Sun (that is, their sidereal rotation periods were believed to be *equal* to their sidereal orbital periods) until the mid-1960s, when Doppler radar measurements finally fixed their periods. For the Moon and Mars, clear and abundant surface markings exist that may be followed to deduce the rotation rates. The gigantic planets Jupiter, Saturn, Uranus, and Neptune show only their upper atmospheres, so that their rotation is found by following atmospheric features or by observing the

Doppler-shifted radiation from their limbs. The atmospheres of these big planets rotate slightly faster at their equators than at their poles; hence, no unambiguous rotation rate exists. Sometimes that of the magnetic field is given. Pluto's surface is not uniformly reflective; the amount of sunlight it reflects varies as it rotates. By observing these brightness fluctuations, we infer Pluto's rotation period.

We define a planet's *oblateness* ϵ by

$$\epsilon = (r_e - r_p)/r_e$$

where r_e is the equatorial radius and r_p the polar radius. A perfect sphere has $\epsilon = 0$. If all planets adjust to an equilibrium fluid shape, we expect that ϵ will increase as the rotation rate increases; the trend (Table A3–2) corroborates this idea. Accurate measurements of the Earth yield an oblateness of 1/298.3. For the other planets, oblateness is determined by measuring the visible disk of the planet or by analyzing perturbations on the orbits of the planet's moons. The large liquid planets are very oblate, so much so that Jupiter and Saturn appear distinctly oval in telescopes.

The equators of the planets are inclined to their orbital planes by varying amounts. The rotation axes of Mercury, our Moon, and Jupiter are nearly aligned with their revolution axes, whereas those of the Earth, Mars, Saturn, and Neptune are tilted about 25°. The *equatorial inclination* is less than 90° for these bodies, so that they all rotate from west to east, or *direct* (in the same sense that all the planets orbit the Sun). Venus and Uranus rotate from east to west, or *retrograde* (Fig. 2–4A). The rotation axis of Uranus lies essentially in its orbital plane, so that if Uranus presents its pole toward us now, in 21 years we will lie in its equatorial plane (Figure 2–4B).

Mercury and Venus have low rotation rates, which may be explained by *spin–orbit coupling and resonance*. The synodic rotation period of Venus (its rotation period with respect to the Earth) is 146 days, which is one-quarter of its synodic orbital period. The sidereal rotation period of Mercury is exactly two-thirds its sidereal orbital period (Table A3–2). Hence, Venus appears to be in a roughly 4:1 synodic resonance with the Earth, and Mercury seems to exhibit a 3:2 sidereal lock on the Sun. Both planets are slightly prolate (cigar-shaped). Solar tidal forces interact with these deformations, slowing each planet's direct rotation until a resonance is achieved. Mercury's highly eccentric orbit resulted in the Sun–Mercury reso-

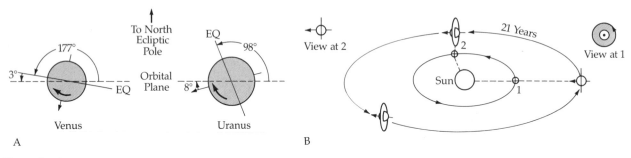

Figure 2–4

Equatorial inclinations. (A) Venus and Uranus rotate retrograde because the inclinations of their equators (EQ) to their orbital planes are greater than 90°. (B) The rotation axis of Uranus and its orbital plane are nearly aligned. At 1, we see the planet pole-on; 21 years later at 2, we see its equatorial plane edge-on. The satellites' orbits are face-on at 1 and side-on at 2.

nance (Figure 2–5). Mercury points alternate ends of its long axis toward the Sun at each perihelion, rotating three times in two sidereal orbital periods.

Interiors

The planets naturally divide into two groups (Table A3–3): (1) the small, solid *terrestrial* planets (Mercury, Venus, Earth, Moon, and Mars), with masses no greater than the Earth's, and (2) the large, liquid *Jovian* planets (Jupiter, Saturn, Uranus, and

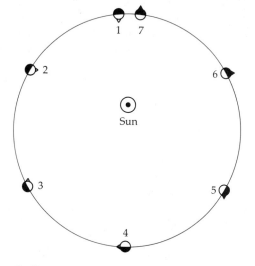

Figure 2–5

Mercury's orbit. Because this planet's rotation rate is two-thirds its orbital period, it completes three sidereal rotations in two revolutions.

Neptune), which range in mass from $15M_\oplus$ to $318M_\oplus$. Pluto, a binary planet, does not fall cleanly into either class.

Planetary masses are determined (1) by applying Kepler's third law to satellite orbits and, when there are no natural satellites, (2) by observing a planet's gravitational perturbations on the orbits of other planets, asteroids, comets, and artificial space probes. Planetary radii are deduced (1) by measuring the apparent optical size of the planetary disk, (2) by accurately timing the *occultations* (when the planet passes in front of an object) of stars, the planet's moons, and space probes, and, for the closer planets, (3) by precisely timing radar pulses reflected from various points on the planet's surface. By dividing a planet's mass M by its total volume $4\pi R^3/3$, we may define its *average density* ρ (SI units = kg/m³):

$$\bar{\rho} = M/(4\pi R^3/3) \qquad (2-1)$$

The density of pure water is 1000 kg/m³. The high densities of the terrestrial planets, in the range of 3400 to 5500 kg/m³, reflect the fact that they are composed of heavy, nonvolatile elements such as iron, silicon, and magnesium. The very low densities of the Jovian planets (Saturn could *float* on water!) imply a composition similar to that of our Sun, with hydrogen and helium dominating.

The internal structure of a planet depends upon the distribution of its chemical composition, density, temperature, and pressure. In general, the pressure increases closer to the center of a planet, and the temperature also rises as a result of the greater pressure and the heat released by decaying radioactive elements. Different materials are stable under different conditions, so that the chemical composition is radially layered. The Earth has a molten metallic core above which float lighter silicates; we expect Venus to exhibit a similar interior because its mass and composition are similar to the Earth's. The Jovian planets have thick interiors of mostly hydrogen and helium, probably in the form of a slush.

Using the observed average density, composition, and oblateness of a planet, we may construct physically consistent models of its interior (just as we do for stars in Chapter 16). Usually a range of models accommodate our data, so that only improved theory and observation can lead us to a unique picture. Because we cannot probe planetary interiors directly, we proceed by inference and induction.

Surfaces

Whereas the Jovian planets and Venus show only their upper cloudy atmospheres, the terrestrial planets and larger moons reveal surface markings. The most important general data on planetary surfaces include color, albedo, and temperature.

A planet's *color* relates to the composition of its surface and atmosphere. The oceans and continents give the Earth a blue color mottled with green, brown, and orange; large areas of cloud or snow cover appear white. The basaltic surface of the Moon looks dark gray with some tan, whereas the deserts of Mars give its characteristic brown-orange color. The surface of Io (a large moon of Jupiter) has a yellowish cast because of outflows from sulfur volcanoes.

The *albedo* of an object is the fraction of the incident sunlight reflected by it. For planets with little or no atmosphere (Mercury, our Moon, and Mars), the albedo is very low because rocks are poor reflectors. Icy surfaces, as on most of Saturn's moons, have relatively high albedos. The high reflectivity of clouds leads to the high albedos of the Jovian planets and Venus. The Earth's albedo is variable since it depends upon the season and upon snow and cloud cover; it averages about 0.35.

An important characteristic of planetary surfaces is *surface temperature.* Let's see how surface temperatures are observed and computed for blackbody radiators. Chapter 8 contains a complete discussion about blackbody radiators. Here we will introduce you to the basic properties of this hypothetical object. You can think of blackbodies in two ways: by their absorption properties and by their emissive ones. As the name implies, a blackbody absorbs completely all forms of electromagnetic radiation striking it; none is reflected, and so its albedo is *zero.* When a blackbody heats up to some temperature, its spectrum of emitted light (Figure 2–6) has a characteristic shape, called a *Planck curve,* with one maximum. A blackbody radiates most of its energy at wavelengths near

$$\lambda_{\text{max}} = (0.002898\text{m})/T = (2898\ \mu\text{m})/T \quad \textbf{(2–2)}$$

where T is the absolute temperature of the blackbody; one micrometer (μm) equals 10^{-6} m. This relation is *Wien's law,* which tells us that solar radiation peaks at 0.5 μm (green light); at normal room temperature ($T \approx 290$ K), the peak is at the longer infrared wavelength $\lambda_{\text{max}} = 10\ \mu$m (Figure 2–6). Hence, we can determine a planet's surface temperature by observing its radiation in the infrared.

Radiation of different wavelengths comes from different parts of a planet's surface and atmosphere. For example, centimeter radio waves originate from the solid surface of Venus, millimeter waves can escape only from its lower atmosphere, and infrared waves come from its upper atmosphere. Therefore, we can probe the temperature and composition of varying levels of a planet's surface and atmosphere by observing the amount of radiation that escapes at different wavelengths. Because the Moon has no atmosphere, infrared rays come directly from its visible surface, and longer-wavelength radiation originates several centimeters beneath the surface. We can use these to probe subsurface conditions.

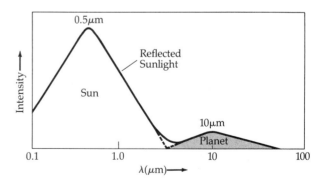

Figure 2–6

Continuous emission of the Sun and a planet. Radiating like a blackbody, the Sun's emission peaks near 0.5 μm. For a planet at 290 K, the peak occurs at 10 μm.

A planet that radiates more energy per second than it receives from the Sun must have an internal heat source. Delicate measurements at the Earth's surface reveal heat flowing from the hot interior; the same occurs with the Moon. The other planets known (from infrared observations) to produce excess heat are Jupiter, Saturn, and Neptune.

Now we show how theoretical blackbody temperatures are found (Table A3–3). The basic idea is to find the temperature at which a small blackbody (a planet) must radiate to balance the energy input from the Sun. We use *Stefan's law* of blackbody radiation (Chapter 8). This law relates the *energy flux E* (energy radiated per unit area per unit time, or W/m^2) to the temperature T of a blackbody:

$$E = \sigma T^4 \; W/m^2 \qquad (2-3)$$

where the proportionality constant is $\sigma = 5.67 \times 10^{-8} \; W \cdot m^{-2} \cdot K^{-4}$. Because the area of a spherical surface of radius R is $4\pi R^2$, our Sun radiates energy like a blackbody at the rate ($T_\odot = 5800$ K)

$$4\pi R_\odot^2 E_\odot = 4\pi R_\odot^2 \sigma T_\odot^4 = 3.8 \times 10^{26} \; W$$

which is the Sun's luminosity ($L_\odot$)—its total radiative power. This same energy flows through a sphere of area $4\pi r_p^2$ at the Sun–planet distance r_p, so that the energy flux there is

$$E_p = 4\pi R_\odot^2 E_\odot / 4\pi r_p^2 = (R_\odot/r_p)^2 E_\odot$$

If 1 m^2 of blackbody material intercepts this energy head-on, it will be raised to the *subsolar temperature*:

$$T_{ss} = (R_\odot/r_p)^{1/2} T_\odot \approx 395 (r_p)^{-1/2} K \qquad (2-4)$$

since $E_p = \sigma T_{ss}^4$. In the last equality, we have inserted the appropriate solar values and expressed r_p in AU. The subsolar temperatures are essentially the equilibrium *noontime* temperatures.

Although subsolar temperatures apply to the local noon on the surfaces of very slowly rotating planets (Mercury, the Moon, and Pluto), they are not appropriate for planets with atmospheres or for planets in rapid rotation. For these, assume that the effective absorbing area is the cross section πR_p^2 and that the effective radiating area is the total surface area $4\pi R_p^2$, where R_p is the radius of the planet. The albedo A is the fraction of incident solar radiation which is reflected, so that only the fraction $1 - A$ is absorbed. Therefore, energy absorbed per second is

$$(1 - A)\pi R_p^2 E_p = (1 - A)\pi R_p^2 (R_\odot/r_p)^2 \sigma T_\odot^4$$

while the planet radiates the power $4\pi R_p^2 \sigma T_p^4$. We again use the concept of equilibrium: that the rate of absorption of energy equals the rate at which it is radiated away fixes an equilibrium temperature. (Otherwise the temperature would either rise or fall.) Equating these and solving for the temperature, we have

$$T_p = (1 - A)^{1/4} (R_\odot/2r_p)^{1/2} T_\odot$$
$$\approx 277(1 - A)^{1/4} (r_p)^{-1/2} K \qquad (2-5)$$

with r_p expressed in AU. The equilibrium blackbody temperatures (Table A3–3) follow when $A = 0$. By including the albedo effect, we may more closely approximate the observed planetary temperatures, but remember that these crude estimates neglect important complications, such as the circulation of planetary atmospheres and their heat retention.

Atmospheres

Of the terrestrial planets, Mercury and the Moon have essentially no atmosphere, Venus and Mars possess a carbon dioxide (CO_2) atmosphere, and the Earth's atmosphere is primarily molecular nitrogen (N_2) and oxygen (O_2). The principal constituents of the atmospheres of the Jovian planets are molecular hydrogen (H_2) and helium (He). In general, planetary atmospheres are densest near the planet's surface and thin rapidly with increasing altitude. The composition of an atmosphere may be stratified, with the heaviest gases residing closest to the surface of the planet, but turbulent mixing and winds can lead to regions of homogeneous composition. Far from the planet's surface, incoming solar ultraviolet rays and X-rays usually ionize atmospheric atoms or dissociate molecules to form the layered *ionosphere* (Chapter 4).

To gain some understanding of planetary atmospheres, consider the simplest model for their retention. To a first approximation, an atmosphere behaves like a *perfect gas,* that is, as particles that interact only through elastic collisions. Such a gas obeys a special relationship between pressure, temperature, and density:

$$P = nkT$$

where P is the pressure (the rate of change of the particles' momenta from collisions) in units of force per unit area (N/m^2), n is the number density of particles ($\#/m^3$), T is the absolute temperature (K), and k is Boltzmann's constant, equal to 1.38×10^{-23}

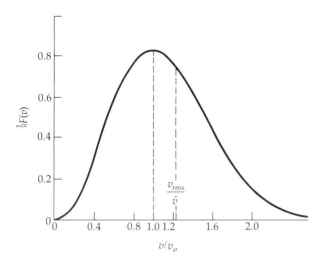

Figure 2–7

Maxwellian distribution of particles in a gas. The velocities of the particles spread around the most probable velocity v_p, which marks the peak of the distribution. The numbers on the vertical axis have been normalized to the total n in the gas.

J/K. Now, from the continuous collisions, the particles of the gas achieve, at a given temperature, an equilibrium distribution of velocities so that

$$F(v)dv \propto \exp(-\tfrac{1}{2}mv^2/kT)v^2dv$$

known as the *Maxwellian distribution* of velocities in a gas (Figure 2–7). Note that, because of the exponential decrease, the distribution has a long tail at large velocities—a few of the particles have been boosted to high speeds by the collisions. The peak of this distribution defines a *most probable speed*:

$$v_p = (2kT/m)^{1/2}$$

where m is the mass of a gas particle. While the speeds of these particles (atoms or molecules) are distributed over a large range and change violently in each collision, the *average* kinetic energy per particle is

$$\overline{KE} = (m/2)\overline{v^2} = 3kT/2 \qquad \textbf{(2–6)}$$

Here m is the particle's mass and T is the kinetic absolute temperature of the gas. From Equation 2–6 we obtain the *root mean square speed* v_{rms}:

$$v_{rms} = (\overline{v^2})^{1/2} = (3kT/m)^{1/2} \qquad \textbf{(2–7)}$$

which tells us that the mean speed of the particles increases with temperature and decreases with mass.

In the very thin upper regions of atmospheres, a particle that moves outward with the *escape speed* v_e has an excellent chance of leaving the atmosphere, with v_e being

$$v_e = (2GM/R)^{1/2} \qquad \textbf{(2–8)}$$

where M is the planet's mass and R is its radius. If $v_{rms} = v_e$ for a given particle species, that gas will leave the atmosphere in only a few days. To retain an atmosphere for several billion years (approximately the age of the Solar System), a planet must have $v_e \geq 10v_{rms}$. (The factor of 10 takes into account the high-speed tail of the Maxwellian distribution of speeds.) Therefore a given type of molecule is retained indefinitely when (Equations 2–7 and 2–8)

$$T \leq GMm/100kR \qquad \textbf{(2–9)}$$

Figure 2–8 shows points corresponding to the equilibrium blackbody temperature (Table A3–3) and v_e

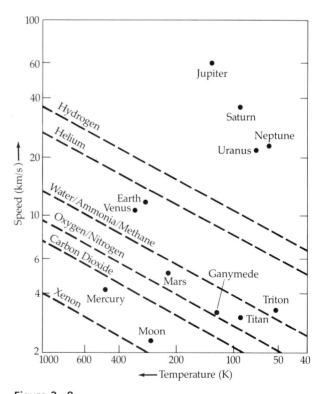

Figure 2–8

Retention of atmospheric gases. Mean molecular speeds are given as a function of temperature, along with the escape speeds for the indicated bodies. The dashed lines show ten times the mean molecular speeds, which defines an essentially infinite lifetime.

for the planets and some moons; the dashed lines represent $10v_{rms}$ for various molecular species. In terms of this crude model, a planet retains all gases with lines passing below its point, and the other gases escape. This model is reasonably consistent with the observations that the Jovian planets have retained all gases and the Earth, Venus, and Mars have lost their hydrogen and helium but retained nitrogen and carbon dioxide, Mercury and the Moon have no atmosphere, and the largest moons have some atmosphere. (Titan, in fact, has a dense atmosphere composed largely of nitrogen but made smoggy because of photochemical reactions on methane and other compounds.)

(b) Moons and Rings

Natural planetary satellites, or *moons,* are many in the Solar System, but together they comprise a total mass of only 0.1 $M_\oplus$. Eight satellites are about the size of our Moon, while the others are much smaller and resemble large asteroids. Our Moon, the two satellites of Mars, the five inner satellites of Jupiter, the eight innermost of Saturn, and five of Uranus' moons have nearly circular orbits lying essentially in their planet's equatorial plane. Observations show that these 21 moons exhibit synchronous rotation from tidal friction (as does our Moon).

Only three moons circle the terrestrial planets, whereas the Jovian planets possess at least 51. The larger the mass of a planet, the greater the range of its gravitational influence. This coupled with the Jovian planets' proximity to the asteroid belt means that they can capture asteroids gravitationally. The Jovian satellites with small masses, highly eccentric and inclined orbits, and retrograde motion are probably captured asteroids, as are the Martian moons.

Three of the four Jovian planets are known to have rings, and the fourth, Neptune, may have them also. Of these ring systems, Saturn's is by far the most spectacular. In each case, the ring system is within the Roche limit [Section 3–4(d)]. Some rings appear to be kept in line by small satellites just outside and inside their boundaries.

(c) Asteroids

Bode's law "predicted" the existence of a planet at 2.8 AU, between Mars and Jupiter, but it was not until 1801 that Giuseppe Piazzi discovered the *minor planet* Ceres in this region. By 1890, more than 300 *asteroids* had been found with semimajor orbital axes

between 2.3 and 3.3 AU—they made up the *asteroid belt.* Today, the orbital elements of more than 3000 asteroids are known, and each such body is numbered in order of orbit determination and given a name, such as asteroid 1000 Piazzia.

Asteroids are too small to retain atmospheres, for their observed diameters range from relatively large (Ceres, 1020 km; Pallas, 538 km; Vesta, 549 km) to the more abundant smaller ones (about 1 km); certainly a multitude of small rocks also swarm in the asteroid belt. The total mass in asteroids is probably about 4 percent of our Moon's mass. The largest asteroids tend to reside farthest from the Sun, and the smallest closest; they are all in direct orbit about the Sun. Asteroids exhibit orbital eccentricities up to 0.83, with most in the 0.1 to 0.3 range, and orbital inclinations as large as 68° but more typically less than 30°. The asteroid belt shows distinctly depleted regions, called *Kirkwood gaps,* at semimajor axes where the orbital period would be a simple fraction (such as $\frac{1}{2}, \frac{1}{3}, \frac{1}{4}, \frac{2}{5}, \frac{3}{7}$) of Jupiter's orbital period. Periodic gravitational perturbations from Jupiter have removed all asteroids from these gaps. Where the ratio of periods is 2/3 and 1/1, asteroids accumulate in *groups,* or *families;* the 15-member Trojan groups are situated at Jupiter's orbit on the vertices of equilateral triangles with the Sun and Jupiter at the other two vertices. The Trojans are bound here where they are stable against perturbations about their equilibrium positions. Some asteroids, which are known as the Apollo group and include Daedalus, Icarus, and Geographos, have perihelia within the Earth's orbit and may pass near the Earth at times. On June 15, 1968, Icarus careened past the Earth at a distance of a mere 6.4×10^6 km. Some people feared a collision and predicted the world would end. It did not.

Just as with Pluto, the rotation periods of asteroids are determined from fluctuations in their reflected sunlight. Some observed light curves exhibit two maxima and two minima per cycle, corresponding to oblong bodies that tumble every 3 to 20 h (the average rotation period for asteroids is 7 h). So, we picture irregular fragments of rock, several kilometers in diameter, spinning in a few hours and colliding with one another no more frequently than once every million years.

(d) Comets

Comets are small and infrequent interlopers in the inner Solar System (Figure 2–9). Bright visual comets have tails that can stretch 90° across the sky.

Figure 2–9

Comet Arend-Roland. Note the thin countertail pointing toward the Sun. *(Lick Observatory)*

A comet is named after its discoverer (Comet Oterma) or co-discoverers (Comet Ikeya-Seki) and also by year in the order of its discovery (1971a was the first discovered in 1971, 1971b the next, and so on). After an orbit has been computed, the numbering is in the order of perihelion passage (Comet 1971 I was the first to pass perihelion in 1971). All comets move in elliptical orbits (some retrograde) about the Sun and are members of the Solar System. They fall into two distinct classes: (1) the *long-period* comets, which are the great majority and which have orbital eccentricities very close to unity (almost parabolic), and (2) the much smaller group of *short-period* comets, which are periodic in their returns. Most observed comets reach perihelion around 1 or 2 AU from the Sun, and the inclinations of their orbits range through all values; hence, comets move in a spherical volume centered on the Sun while the planets move near the ecliptic plane. The aphelia of long-period comets may extend as far as 50,000 AU, with orbital periods as long as a million years or even greater; these comets are seen only during one perihelion passage. In some cases, planetary gravitational perturbations make comet orbits hyperbolic ($e > 1$) so that they escape the Solar System, while in other cases the orbits are changed into small ellipses. In this way, Jupiter has captured a family of about 45 comets that now orbit the Sun with aphelia near Jupiter's orbit. Short-period comets move around within the planetary system: Comet Encke has the shortest period, 3.3 years ($a \approx 2.2$ AU). The most famous short-period comet is *Halley's Comet*, which returns to perihelion about every 76 years.

With each perihelion passage, a comet loses material as a result of the intense solar heating and tidal forces; eventually the comet vanishes. Some short-period comets have been observed to split into several pieces or even disintegrate.

Because comets waste away, the supply must constantly be replenished if we are to see them today. We presume that cometary nuclei were formed with the Solar System about five billion years ago. To account for their continued existence, Jan Oort hypothesized a spherical *comet cloud* out to 50,000 AU from the Sun with about 10^{11} cometary bodies. At its outer edge, this reservoir loses nuclei as a result of perturbations from passing stars; at its inner edge, planetary perturbations deflect nuclei toward the Sun. Further perturbations may cause a nucleus to either escape from the Sun altogether or enter a periodic orbit, where it slowly disintegrates. Recent estimates suggest that there may be as many as 10^{14} comets, including large numbers in more or less permanent orbits at 10,000 to 20,000 AU.

(e) Meteoroids

Roaming throughout the Solar System with orbits of all inclinations are the *meteoroids*. These range in size from small asteroids (10 km) down to micrometeoroids (<1 mm) and interplanetary dust (≈1 μm). Pieces from asteroid collisions form the larger rocks, while the smaller pieces come from disintegrated comets. When a meteoroid enters the Earth's atmosphere, friction heats it to incandescence and visibility at an altitude of about 120 km—a *meteor*. From the brightness of meteors, we deduce that their average density is 200 to 1000 kg/m³ (similar to a comet's nuclear material!). A meteoroid loses mass by vaporization, melting, and fragmentation, and it ionizes the air through which it passes. Some meteors are bright enough to cast shadows—these are called *fireballs*. A typical meteor particle is about the size of a grain of sand; a fireball, the size of a pebble.

Meteoroids that are not totally consumed in the atmosphere strike the ground as *meteorites*. Large meteorites produce craters, such as those preserved on the Moon and the kilometer-wide Barringer Meteor Crater in Arizona (Figure 2–10). Smaller meteorites do not annihilate themselves upon reaching the ground, and the *micrometeorites* (with diameters of 0.5 to 200 μm) simply drift through the atmosphere to deposit a total mass of about 10⁶ kg per day on the Earth's surface.

The Earth occasionally passes through a group of meteoroids—pieces of solid material from a comet. When this intercept occurs, we see a *meteor shower* in the sky. During such a shower, the meteors appear to come from a particular point in the sky, called the *radiant*. Showers are usually named after the constellation in which their radiant lies. For instance, the Perseid meteor shower, appearing in August, seems to come from the constellation Perseus. The debris from a meteor shower comes from a comet. For instance, that for the Perseids is associated with a comet named *1862 III*.

(f) Interplanetary Dust and Gas

Finally we turn to *interplanetary dust* (size ≈ 1 to 100 μm), which probably comes from the dust tails of comets. Mariner space probes revealed dust clouds around the Earth and Mars. All planets have probably captured dust clouds by gravitational attraction. The faint band of *zodiacal light*, best seen before dawn, results from the sunlight reflected by a marked concentration of dust in the ecliptic plane; the intensity of this scattered light diminishes with distance from the Sun (Figure 2–11).

Figure 2–10

The Barringer Meteor Crater. Located in Arizona near Flagstaff, this crater was formed by an impact some 40,000 years ago. *(U.S. Air Force)*

2–2 Unanswered Questions

Our brief survey has revealed the structure and content of the Solar System, but the story is not closed. We can successfully interpret many of the observed features, such as the sources of meteorites. However, a plethora of unanswered questions remain concerning the origin of these features. These unresolved questions prompt tentative answers (Chapter 7).

The fundamental question is how and why the Solar System originated. Why are there five tiny terrestrial planets and four huge Jovian planets? What produced the distribution of planetary orbits; is Bode's law fundamental? Why do the planets move in direct orbits near the ecliptic plane?

The *angular* momentum distribution of the Solar System is also a puzzle. The rotational angular momentum of the Sun is

$$2M_\odot R_\odot^2 \omega_\odot /5 \approx 10^{42} \ \mathrm{kg} \cdot m^2 \cdot s^{-1}$$

while each planet has an orbital angular momentum $2\pi ma^2/P$, where m is the planet's mass, a is its orbital semimajor axis, and P is its sidereal orbital period; the planetary total is about $33 \times 10^{42} \ \mathrm{kg} \cdot m^2 \cdot s^{-1}$. Hence, the planets contribute 97 percent of the angular momentum of the system. Why doesn't the much more massive Sun provide the dominant contribution?

Was there a primordial rotation rate, and if so, why do the terrestrial planets rotate much more slowly than the Jovian planets? What caused the rotation axis of Uranus to lie in its orbital plane? Are any planetary atmospheres primordial? Why do Venus and Titan and the Jovian planets have such dense atmospheres? What sculpted the surfaces of planets and satellites? How did planetary satellites originate, with their observed number, masses, and orbital distributions? Why do the rings of Saturn, Jupiter, and Uranus differ so? What caused the asteroid belt? How were comets formed, and how many are out there? What produced the meteoritic compositions we observe?

We think the most basic questions must be answered by any model pretending to account for our Solar System. Until such a model is forthcoming, we can only accept what we see and attempt to decipher the riddle before us.

Figure 2–11

The zodiacal light. This short time exposure (note the stars are trailed) shows the light above Mt. Haleakala, Hawaii. *(A. Peterson and L. Kieffaber)*

Problems

1. **(a)** Describe the apparent path of the Sun across the sky of Mercury during one solar "day," as seen by an observer at that planet's equator.
   (b) Describe the seasons of Uranus, giving their durations where appropriate.

2. Consider the planets Uranus, Neptune, and Pluto.
   (a) On a graph, compare their orbital semimajor axes with the distances predicted by Bode's law.
   (b) Show that the ratios of their orbital periods are approximately *commensurable*, that is, nearly fractions such as $\frac{3}{2}$.

3. Show that the two satellites of Mars, Phobos and Deimos, obey Kepler's third law; deduce the mass of Mars from the orbits of these moons.

4. **(a)** How much does a spaceship having a mass of 10^3 kg weigh on the Earth's equator? At its poles? *Hint:* centripetal force.
   (b) Remembering Jupiter's rapid rotation, how much will this craft weigh at the equator of Jupiter's surface? At the poles?

5. Derive the formula that gives the synodic rotation pe-

riod (solar day) of Venus with respect to the Earth, and verify the number quoted in the text.

6. In the seventeenth century, Ole Romer concluded that the speed of light is finite by observing that the satellite Jupiter I (Io) was occulted by the planet approximately 16.7 min *earlier* when Jupiter was in opposition than when it was near superior conjunction. Use this information to draw an appropriate diagram and to calculate an approximate value for the speed of light (which Romer did *not* do!).

7. How far from the star would we have to be (in AU) to find a substellar temperature comparable to that of the Earth for
   (a) Rigel (surface temperature $T = 12,000$ K, radius $R = 35R_{\odot}$)?
   (b) Barnard's star ($T = 3000$ K, $R = 0.5R_{\odot}$)?

8. Formaldehyde (H_2CO) has been discovered in interstellar space.
   (a) Calculate its "mean" molecular speed for $T = 280$ K. Would our Moon retain this gas for billions of years?
   (b) Would Saturn's satellite Titan retain formaldehyde? (For Titan, radius ≈ 2600 km, mass $\approx 10^{23}$ kg.)

9. Consider a comet with an aphelion distance of 5×10^4 AU and an orbital eccentricity of 0.995.
   (a) What are the perihelion distance and orbital period?
   (b) What is the comet's speed at perihelion and at aphelion?
   (c) What is the escape speed from the Solar System at the comet's aphelion, and what do you conclude from this result?

10. Draw a diagram to explain why some meteor showers are consistent from year to year whereas others are spectacular on occasion and feeble at other times (such as the Perseids in 1985).

11. The albedo of Venus is about 0.77 because of the cloudy atmosphere. What would the noontime temperature be? (The measured temperature is 750 K.)

12. (a) At noon, Mercury's surface temperature is roughly 700 K; at midnight, 125 K. Calculate the wavelength at which the surface emits most of its energy at noon and at midnight.
    (b) Calculate the energy output per square meter of the surface at midnight and noon.

13. The Earth, Venus, and Mars all have carbon dioxide in their atmospheres. Find the ratio of the root mean square speed to escape velocity for each and make a statement about the retention of carbon dioxide for these planets.

14. What is the approximate orbital period for a cometary nucleus in the Oort cloud?

15. Calculate the blackbody equilibrium temperature of a fast-rotating asteroid with a radius of 100 km and an albedo of 0.5.

16. Roughly speaking, the spin angular momentum of a sphere is MVR, where V is the equatorial velocity. Make an approximate calculation of the spin and orbital angular momenta of Jupiter and compare them with the spin angular momentum of the Sun.

The Dynamics of the Earth

Before investigating the Solar System in detail, we present the dynamics of our Earth. Chapter 4 deals with the physical properties of Earth and the Moon. Both the dynamics and physics of the Earth lay the base for an investigation of the other terrestrial planets because we know the Earth better than any other planet.

3–1 Time and the Seasons

You are familiar with the words *second, minute, hour, day, week, month,* and *year,* but what exactly do they mean? Measuring time is arbitrary and conventional. Astronomers define the second, minute, hour, and day in terms of the Earth's rotation, the week and month in terms of the Moon's orbital motion, and the year in terms of the Earth's revolution about the Sun. (We suggest you review Appendix 10 before proceeding.)

(a) Terrestrial Time Systems

A *day* is that time interval between two successive upper transits of a given celestial reference point. An *upper transit* occurs when a celestial reference point or body crosses the celestial meridian moving

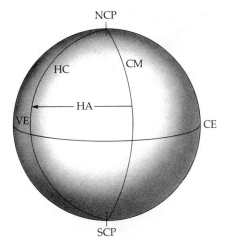

Figure 3–1

Reference circles on the sky. Shown on the celestial sphere are the observer's celestial meridian (CM), the celestial equator (CE), the north and south celestial poles (NCP and SCP), and the hour angle (HA) of the vernal equinox (VE) along its hour circle (HC).

westward; the *celestial meridian* is the imaginary line drawn through the north point of the horizon, the zenith, and the south point of the horizon. The vernal equinox is the zero point for *sidereal time,* and the sidereal day is arbitrarily divided into 24 sidereal hours of equal length, each of which consists of 60 sidereal minutes with 60 sidereal seconds per minute. Now 1^h corresponds to 15° of the Earth's rotation with respect to the stars, so that the *local sidereal time* is 0^h 0^m 0^s when the vernal equinox lies on our celestial meridian and 2^h when the vernal equinox is 30° —2 h—west of the celestial meridian. We define local sidereal time as the *hour angle of the vernal equinox* (Figure 3–1). The *hour angle* is how far west (positive) or east (negative) of the meridian a celestial object is. The vernal equinox precesses westward along the celestial equator at the rate of about 50″ per year [Section 3–4(c)], and so the actual period of the Earth's rotation (measured with respect to the stars) is 0.008^s longer than one sidereal day.

Timekeeping on the Earth is based in a complicated way upon the position of the Sun in the sky. *Apparent solar time* is the hour angle of the real Sun plus 12^h, so that the zero point is midnight at 0^h and local apparent noon always occurs at 12^h. The length of the apparent solar day is *not* constant during the year, however, even to a given observer. These variations are caused by the eccentricity of the Earth's orbit and the inclination of the Earth's equatorial plane to the ecliptic, that is, its orbital plane (Figure 3–2). The Earth's orbital speed hits a maximum at perihelion (about January 2) and a minimum at aphelion (about July 3), in accordance with Kepler's second law. The Sun reflects this variable motion by moving eastward along the ecliptic faster at perihelion than at aphelion. To return the Sun to noon, the Earth must turn through a greater angle, and so the apparent solar day is longer at perihelion than at aphelion.

Another factor is that the Sun moves on the ecliptic but apparent solar time is measured along the celestial equator. So only that component of the Sun's eastward motion parallel to the celestial equator affects apparent solar time. To avoid the inconveniences of this variable solar time, we define *mean solar time* as the hour angle of a fictitious point (the *mean Sun*) that moves eastward along the *celestial equator* at the average angular rate of the true Sun. The *mean solar day* begins at midnight (with respect to this point), and its length is 1/365.2564 year. The difference between apparent solar time and mean

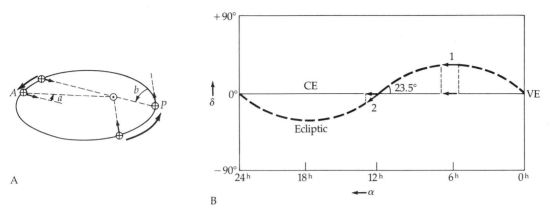

A

B

Figure 3–2

Variations in apparent solar time. (A) A schematic view of the Earth's orbital motion in one day. Near aphelion *(A)*, the Earth rotates through 360° plus angle *a* to complete one apparent solar day. Near perihelion *(P)*, the Earth completes an apparent solar day by turning through 360° plus angle *b*. Hence, the solar day is longer at perihelion than at aphelion because *b > a*. (B) A rectangular map of the sky shows the ecliptic inclined 23.5° to the celestial equator (CE). The Sun's eastward motion is shown at two points (1 and 2). Only the components of this motion parallel to the CE cause the apparent solar time to vary.

solar time is called the *equation of time;* because the effects are cumulative, the mean Sun may lead or lag the true Sun by as much as 16 min.

Now to compare sidereal time with mean solar time (Figure 3–3). The Earth's rotation returns the vernal equinox to upper transit as the Earth moves the distance *A* in its orbit—one sidereal day has passed. Since the Earth has moved about $360/365 \approx 1°$ around its orbit, it must rotate through this angle before the Sun returns to the local meridian and a mean solar day has passed (*B*). However, 1° corresponds to 4^m of sidereal time, and so the mean solar day is about 4 min *longer* than the sidereal day. The length of the sidereal day is $23^h\ 56^m\ 4.09^s$ in mean solar time units, and so stars appear to rise about 4 min *earlier* each night in terms of the mean solar time. A star that is in upper transit at midnight tonight will reach the meridian at 10 p.m. one month from now $(30 \times 4^m = 2^h)$.

Mean solar time differs at every longitude on the Earth's surface because the hour angle of the fictitious Sun depends on the observer's location. The practical difficulties of such a timekeeping system have been avoided by the establishment of 24 *time zones* around the world. Within each longitudinal zone, which are approximately 15° (or 1^h) wide, all locations have the same *standard time;* the boundaries

of each zone are adjusted for maximum convenience (for example, a city is usually placed wholly within one time zone). The reference zone is centered on Greenwich, England, at 0° longitude. Standard time at Greenwich is referred to as *Greenwich mean time* or, equivalently, *universal time* (abbreviated U.T.); astronomical events such as total eclipses are usually given in terms of universal time. New York City lies 5 h west of Greenwich in the eastern standard time

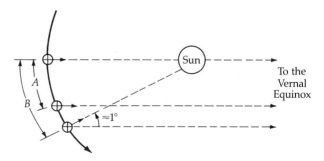

Figure 3–3

Sidereal and solar days. As the Earth rotates once with respect to the vernal equinox, it moves through distance *A* in its orbit. To complete a mean solar day, it rotates 1° more to bring the Sun back to the meridian. The Earth has then moved through distance *B*.

(E.S.T.) zone, and so we subtract 5^h from universal time to find the local New York time for such events. To take advantage of the extra hours of daylight during the summer [Section 3 – 1(b)], an hour is added to local standard time from mid-spring to mid-fall in many parts of the world. Hence, 11 p.m. *Pacific daylight savings time* (P.D.T.) in San Francisco corresponds to 10 p.m. Pacific standard time (P.S.T.), and we add 8^h to Pacific standard time to find universal time since San Francisco is eight time zones west of Greenwich (universal time equals Pacific daylight savings time plus 7^h).

The Earth's rotation rate is subject to extremely small, but unpredictable, variations (Section 3 – 4). To predict precisely the positions of bodies in the Solar System, we require a steady time standard, and so universal time is replaced by *ephemeris time* (E.T.) in celestial mechanics. At the beginning of A.D. 1900, an *ephemeris second* was defined as

$$\frac{1}{31,556,925.97474}$$

The length of the tropical year 1900 and universal and ephemeris time were in agreement. Today these times differ by about 40^s.

The *week* and *month* are of ancient origin and derive from the Moon's synodic orbital period of 29.53 days. To fit within the year, months have been given conventional lengths of 28, 30, and 31 days. The week of seven days (each named after a planet) is probably based upon the quarter phases of the Moon $(29.53/4 = 7.38 \approx 7^d)$.

The *year* is the time it takes the Earth to orbit the Sun, but different definitions give three types of years. With respect to the stars, the Earth's revolution takes one *sidereal year* of 365.2564 mean solar days $(365^d\ 6^h\ 9^m\ 10^s)$, whereas the *tropical year* of 365.2422 mean solar days $(365^d\ 5^h\ 48^m\ 46^s)$ is the period with respect to the vernal equinox that precesses about $50''$ westward along the ecliptic each year. Finally, because planetary perturbations cause the Earth's perihelion to precess in the direction of orbital motion, we call the time between successive perihelion passages the *anomalistic year* of 365.2596 mean solar days $(365^d\ 6^h\ 13^m\ 53^s)$.

The *Gregorian calendar*, which attempts to approximate the year of seasons (the tropical year), contains 365 days per common year and 366 days in years divisible by four (leap years). To achieve an accuracy of one day in 20,000 years, only those century years divisible by 400 are leap years (A.D. 2000); century years divisible by 4000 remain common (A.D. 8000).

(b) The Seasons

The Earth's seasons — spring, summer, autumn, and winter — arise because the Earth's equatorial plane is inclined 23.5° to the ecliptic plane (Figure 3 – 4). The eccentricity of the Earth's orbit is too small $(e = 0.017)$ to affect the seasons; note that perihelion occurs during the northern winter (January 2). The number of daylight hours and the noon altitude of the Sun lead to the characteristic temperatures of the

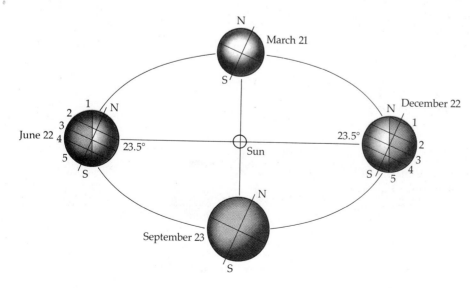

Figure 3 – 4

The Earth's equatorial inclination and the seasons. The Earth's equator inclines 23.5° to the ecliptic, the plane of its orbit. Day and night are shown for the solstices and the equinoxes. The latitudes marked are (1) Arctic Circle (66.5° N), (2) Tropic of Cancer (23.5° N), (3) Equator (0°), (4) Tropic of Capricorn (63.5° S), and (5) Antarctic Circle (63.5° S).

seasons; both depend upon the latitude of the observer and the Sun's position on the ecliptic. Let's consider these two causes independently.

When the Sun is at the *vernal equinox* (about March 21) or the *autumnal equinox* (about September 23) along the ecliptic, its declination is 0° and there are 12 h of day and 12 h of night at all points on the Earth's surface. The noon altitude of the Sun is 90° (the zenith) at the equator and diminishes to 0° at the poles. At the *summer solstice* (about June 22), the Sun attains its greatest declination, +23.5°, and passes directly overhead at noon for all observers at latitude 23.5°N—the *Tropic of Cancer.* (About 3000 years ago, the Sun was in the constellation Cancer at the summer solstice.) On this date, the days are longest in the Northern Hemisphere, and summer starts there; at the same time, winter begins in the Southern Hemisphere and the days are shortest there.

The situation is reversed at the *winter solstice* (about December 22), when the Sun's declination is −23.5°. The Sun then passes directly overhead for all observers at latitude 23.5°S—the *Tropic of Capricorn* (the winter solstice 3000 years ago was in the constellation Capricornus); it is winter in the Northern Hemisphere and summer in the Southern.

Latitude 66.5°N is called the *Arctic Circle.* As spring becomes summer there, the days lengthen until the summer solstice, at which time the Sun doesn't set for 24 h; from fall to winter, the days shorten until the 24-h night at the winter solstice. North of the Arctic Circle, the Sun does not set for many days (the *midnight Sun*) in the summer and does not rise for many days in the winter. At the north pole, a six-month "day" begins at the vernal equinox and a six-month "night" begins at the autumnal equinox. South of the *Antarctic Circle* (latitude 66.5°S), exactly the same events occur six months after they have taken place in the north.

The overall amount of solar energy received by the Earth remains constant from day to day, but the heating effectiveness of this energy—the *solar insolation*—depends upon both latitude and time. The altitude of the Sun determines over what area a given amount of radiation is spread (Figure 3–5). Suppose that a unit of energy falls upon the area A when the Sun is at the zenith. When the Sun's altitude is θ, this same amount of energy is spread over the area $A/\sin\theta$—the heating efficiency decreases as θ decreases. So the summer is warmer than the winter because the Sun is higher in the sky for more

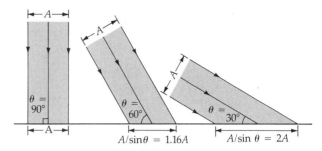

Figure 3–5

Solar insolation. A unit of solar energy strikes the Earth over area A when the Sun is at its zenith. For other altitudes θ, the sunlight is spread over area $A/\sin\theta$.

hours in the summer. The Earth's equatorial regions are always warm because the noon Sun never passes far from the zenith. The poles have *polar icecaps* because the Sun spends many months below the horizon and stays at a low altitude when it does rise.

The Earth's surface (especially the oceans because of the high heat capacity of water) and atmosphere are good thermal reservoirs and respond slowly to solar heating. As a result, temperature variations are moderated by daily and seasonal time lags between the extremes of solar insolation and the extremes of temperature. For example, the early afternoon is usually the warmest part of the day, even though solar insolation peaks at noon. February is the coldest month of northern winter, but December is the month of least insolation.

3–2 Evidence of the Earth's Rotation

How to prove that the Earth rotates? The westward circling of the celestial sphere could be a reflection of the daily eastward turning of the Earth, but this is no proof because we could equally justify the concept of a rotating celestial sphere centered upon a stationary Earth. We can prove the Earth's rotation only by basing our arguments upon the well-verified dynamical laws of Newton.

(a) The Coriolis Effect

The apparent trajectories of rockets and Earth satellites can be understood if the Earth rotates. Consider a projectile launched from the north pole to land at the equator (Figure 3–6). On a nonrotating Earth,

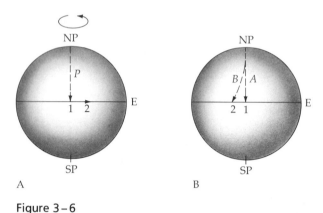

Figure 3–6

Projectile trajectories on the Earth. (A) On a rotating Earth, a rocket launched from the north pole (NP) is aimed at a target located at point 1. During the flight time, the Earth (and the target) rotate from 1 to 2. (B) The view from the Earth's surface shows the rocket's trajectory if the Earth did not rotate (A) and the actual one (B).

the projectile would clearly follow a single meridian of longitude during its entire flight. On a rotating Earth, however, the target on the equator moves eastward at 0.46 km/s and the projectile lands west of the target. Though the projectile's motion is due south, it appears to be deflected to the right with respect to the Earth's surface. The fictitious acceleration that produces this effect — the *Coriolis effect* — was deduced by Gaspard Gustave de Coriolis (1792–1843) in 1835.

Moving bodies always appear to be deflected to the right in the Northern Hemisphere and to the left in the Southern Hemisphere. If the projectile's velocity is **v** and the Earth's vector angular velocity is ω (its direction is toward the north celestial pole, and its magnitude measures the Earth's spin in units of radians/s), then these observations are summarized in terms of the *Coriolis acceleration*:

$$\mathbf{a}_{\text{Coriolis}} = 2(\mathbf{v} \times \boldsymbol{\omega}) \qquad (3–1)$$

The cross product yields the product of the perpendicular components of **v** and ω, and the direction of $\mathbf{a}_{\text{Coriolis}}$ is the direction in which your thumb points when you align the fingers of your right hand along **v** and rotate them through the smallest angle to ω (the *right-hand rule*).

Let's derive Equation 3–1. A body moves with a constant radial velocity **v** above a turntable rotating with angular speed ω (Figure 3–7). At time t, the

body leaves the origin (A) and moves a distance dr to point B in an infinitesimal time dt. Meanwhile, point B has revolved through the angle

$$d\theta = \omega \, dt$$

to B'. From the geometry, we have

$$dr = v \, dt \qquad \text{and} \qquad ds = dr \, d\theta$$

Therefore

$$ds = (v \, dt)(\omega \, dt) = v\omega(dt)^2$$

But Newton's second law implies that a body moves the distance

$$ds = a(dt)^2/2$$

in time dt when it experiences the *constant* acceleration **a**, so that

$$\mathbf{a}_{\text{Coriolis}} = 2\mathbf{v}\omega$$

In addition, the direction of the apparent deflection is to the right, as with Equation 3–1.

The Coriolis effect controls the characteristics of large-scale wind patterns in the Earth's atmosphere

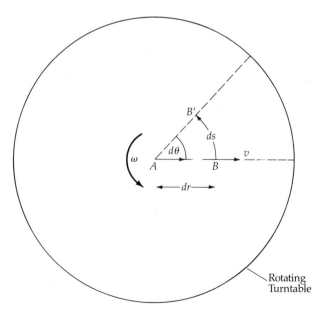

Figure 3–7

The Coriolis effect. A rotating turntable moves through angle $d\theta$ while a body on it moves from A to B (distance dr) in time dt. At the same time, B rotates to B' through distance ds. The body appears to be deflected to the right relative to the turntable.

(as well as ocean currents). A *cyclone* is a local counterclockwise circulation of air in the Northern Hemisphere (clockwise in the Southern) produced by the rightward Coriolis deflection of air flowing toward the center of a low-pressure region. An *anticyclone* arises when air flowing away from the center of a high-pressure region is deflected into a local clockwise circulation in the Northern Hemisphere (counterclockwise in the Southern). Solar heating produces large-scale vertical cells of wind motion called *Hadley cells* (Figure 3–8). At the Earth's surface, the Coriolis effect causes these winds to flow in the known directions of the easterly trade winds (5 to 30°N and S), the temperate westerlies (35 to 50°N and S), and the polar easterlies (60 to 90°N and S). At low latitudes, the bands of relatively calm air are known as the doldrums (0 to 5°N and S) and the horse latitudes (30 to 35°N and S).

Note that the Coriolis effect combined with convection creates the overall pattern of atmospheric flow. *Convection* is one way to transport thermal energy from one location to another. (The other two processes are radiation and conduction.) Convection takes place in the Earth's atmosphere when the Sun's radiation heats the ground and the air in contact with it. The air expands and its density decreases. Cooler, denser air descends to displace the hotter air, which cools as it rises. The falling air heats upon contact with the ground, and so eventually is displaced upward. This circuit of rising and falling air transfers heat from the ground into the atmosphere and also establishes the up-down flow.

(b) Foucault's Pendulum

In 1851, Bernard Léon Foucault (1819–1868) hung a pendulum (a heavy ball at the lower end of 60 m of wire) from the ceiling of the Pantheon in Paris and proved the Earth's rotation by noting that the pendulum's plane of oscillation rotated during the day. If the Earth did not rotate, this rotation of the oscillation plane would not occur because all forces acting on the ball (the Earth's gravity and the tension in the wire) would lie in the plane of oscillation.

Step off the Earth and observe the pendulum swinging as the Earth turns. At the north pole, the pendulum oscillates in a fixed plane while the Earth rotates below it every 24 sidereal hours; the pendulum appears to rotate *westward* with the period $P = 24^h$. A pendulum swinging in the equatorial plane at

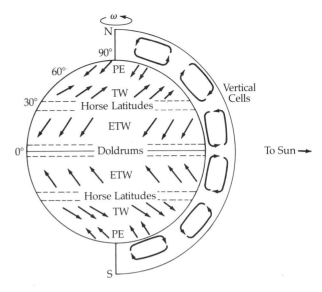

Figure 3–8

General wind patterns on the Earth. Solar heating produces the vertical cells of convecting air, and the Coriolis effect deflects this motion to the right (northern hemisphere) or left (southern hemisphere) to establish the easterly trade winds (ETW), the temperate westerlies (TW), and the polar easterlies (PE).

the Earth's equator feels no forces perpendicular to the plane of oscillation and so doesn't rotate at all ($P = \infty$). At an intermediate latitude ϕ (Figure 3–9), the vertical component ω of the Earth's angular speed is $\omega \sin \phi$. The angular speed is inversely proportional to the period of rotation ($\omega = 2\pi/P$), however, so that the pendulum appears to rotate westward with the period $P = 24^h/\sin \phi$. Precise measurements of Foucault pendula verify this relationship in detail and permit a purely dynamical determination of the Earth's period of rotation.

(c) The Oblate Earth

The shape of the Earth's surface is that of an *oblate spheroid*; the polar radius ($r_p = 6356.8$ km) is 21.4 km less than the equatorial radius ($r_e = 6378.2$ km), so that the *oblateness* $\epsilon = (r_e - r_p)/r_e = 21.4/5378.2 = 1/298.3$. This oblateness indicates that the Earth rotates. If the Earth were a fluid body, its shape would prove its rotation since a fluid must adjust its shape to all external forces—a nonrotating fluid body is spherical. The Earth is composed, however, of materials that might maintain an oblate shape even if no rotation occurs.

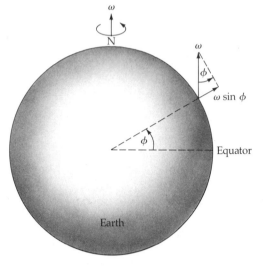

Figure 3-9

Foucault's pendulum. At latitude ϕ, the vertical component of the Earth's angular speed is $\omega \sin \phi$, and so the Earth rotates below the pendulum in a period proportional to $1/\sin \phi$.

Today we know that the materials the Earth is made of have an average strength close to that of steel, but they are plastic and so maintain the equilibrium shape of a rotating fluid body. A mass m at latitude ϕ on a spherical, rotating Earth (Figure 3–10) experiences two accelerations [Section 1–4(a)]: (1) the gravitational acceleration $GM_\oplus/r^2$ directed toward the Earth's center and (2) the centripetal acceleration $\omega^2 r \cos \phi$ in the circular orbit of radius $r \cos \phi$. If a mass is free to move (a fluid mass), the centripetal acceleration affects it. The vertical component of this centripetal acceleration, $a = \omega^2 r \cos^2 \phi$, reduces the weight of a mass while the horizontal component, $b = \omega^2 r \sin \phi \cos \phi$, causes a mass to migrate to the Earth's equator. Considering many such fluid masses (in fact, the whole Earth), we see that a bulge grows at the Earth's equator until fluid masses can no longer climb this equatorial "hill" — establishing the equilibrium oblate shape.

3-3 Evidence of the Earth's Revolution about the Sun

As with the Earth's rotation, we must be careful to prove the Earth's revolution about the Sun. The following three proofs were unavailable at the time of Copernicus and Kepler, and so the ideas of these two scientists were considered suspect and were slow to be accepted. Today these proofs are indisputable evidence for the heliocentric model of the Solar System.

(a) The Aberration of Starlight

In 1729, the English astronomer James Bradley (1693–1762) discovered the *aberration of starlight*, and using the finite speed of light $c \approx 3 \times 10^5$ km/s, he explained this phenomenon as caused by the orbital motion of the Earth.

Suppose you walk in a vertically falling rain with an umbrella over your head. The faster you walk, the farther you must lower the umbrella in front of you to prevent the rain from striking your face. When starlight enters a telescope, an analogous phenomenon occurs (Figure 3–11). If the Earth were at rest, we would point our telescope toward the zenith to see a star situated there. If the Earth is in motion at speed v, however, we must tilt the telescope in the direction of motion by an angle (where θ is in radians)

$$\theta \approx \tan \theta = v/c \qquad (3\text{-}2)$$

so that the bottom of the telescope can meet a light ray which has entered the top of the telescope. Bradley observed this very small tilt angle, $\theta = 20.49''$.

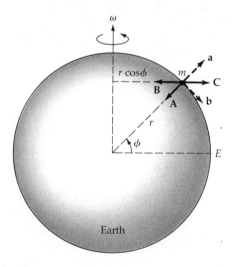

Figure 3-10

The Earth's oblateness. A mass m at latitude ϕ has a gravitational acceleration **A** and centripetal acceleration **B** (from its rotation around the Earth's axis). The vector sum of these forces results in a vertical component **a** that decreases the mass's weight and a horizontal component **b** that causes the mass to slide toward the equator *(E)*.

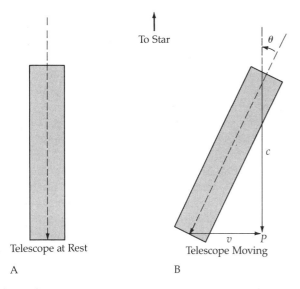

Figure 3–11

The aberration of starlight. (A) A telescope is at rest and so pointed upward to observe the light from a star. (B) When a telescope moves at speed v (as a result of the Earth's revolution), it is tilted through an angle θ, so that the starlight reaches P at the same time as the bottom of the telescope does.

Using Equation 3–2, we deduce the Earth's orbital speed as $v = \theta c = (9.934 \times 10^{-5}$ rad$)(3.0 \times 10^5$ km/s$) = 29.80$ km/s.

The direction in which we tilt our telescope constantly changes as the Earth moves around the Sun. Because the Earth's orbit is essentially circular, stars appear to trace out annual *aberration orbits* on the celestial sphere (Figure 3–12). A star at the ecliptic pole is seen to move around a circle of angular radius 20.49″ once a year. Stars on the ecliptic oscillate to and fro along lines of angular half-length 20.49″. At an intermediate celestial latitude β (angle from the ecliptic), the aberration orbit is an ellipse with semimajor axis 20.49″ and semiminor axis (20.49″) sin β. If the Earth did not revolve around the Sun, this observed behavior of the aberration orbits would be inexplicable.

(b) Stellar Parallax

As you drive along a highway in your car, notice that nearby objects seem to be moving backward with respect to more distant objects (Figure 3–13A). This perspective effect of our line of sight is termed *parallax*. According to the heliocentric model of the Solar System, nearby stars should exhibit parallax effects on the celestial sphere since the Earth is in motion about the Sun (Figure 3–13B). If the observed parallax angle of a star is π'' (in seconds of arc), then, from Figure 3–13B, that star's distance is

$$d = (206{,}265/\pi'') \text{ AU} \qquad (3–3)$$

Note that 206,265 is the number of arcseconds in 1 rad. (See Chapter 11 for a complete discussion of stellar parallaxes and distances.)

The heliocentric model of Copernicus remained on shaky ground until the first stellar parallax was observed in the nineteenth century. In 1838, Friedrich Wilhelm Bessel (1784–1846) published the first observed stellar parallax: 0.294″ for the star 61 Cygni; at about the same time, F.G.W. Struve found the parallax of Vega (Alpha Lyrae) and T. Henderson found that of Alpha Centauri. Today we know that the nearest star is Proxima Centauri with a parallax of 0.764″ and a distance of 270,000 AU (about 4×10^{13} km or 4 lightyears). All stellar parallaxes are less than 1.0″.

As the Earth moves around the Sun in its orbit, each star traces out a yearly *parallactic orbit* on the

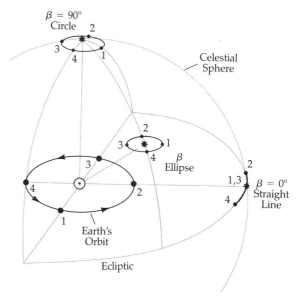

Figure 3–12

Apparent stellar orbits resulting from the aberration of starlight. The Earth is shown at four positions in orbit around the Sun (1 to 4). The apparent paths traced out by stars because of the change in the direction of aberration during a year are shown for stars directly above the orbital plane (ecliptic pole), in the plane (the ecliptic), and at an intermediate position.

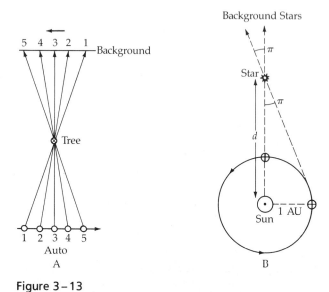

Figure 3–13

Parallax. (A) A nearby object appears to move with respect to the background as our baseline position changes from 1 to 5. (B) A stellar parallax occurs (angle π) as a result of the motion of the Earth around the Sun. The change in baseline position results in an apparent displacement of the star relative to background stars.

$$\Delta\lambda/\lambda_0 = (\lambda - \lambda_0)/\lambda_0 = v_r/c \qquad (3-4)$$

where v_r is the relative line-of-sight speed (positive for recession; negative for approach) between the observer and the observed and c is the speed of light.

For stars at the ecliptic pole, no Doppler shift occurs. At an intermediate celestial latitude β, the shift is sinusoidal with a period of one year and the amplitude $\Delta\lambda$ varies as $\cos\beta$. The maximum amplitude of this Doppler shift occurs for stars on the ecliptic, where the full magnitude of the Earth's orbital velocity comes into play; a standard method of determining the Earth's speed of revolution is to measure this maximum shift and to use Equation 3–4 to deduce $v_\oplus = 29.80$ km/s. The spatial motion of a star with respect to the Sun superimposes a constant Doppler shift on the time varying shift caused by the Earth's revolution; the two effects are easily untangled. The Earth's rotation causes a diurnal Doppler shift with an amplitude proportional to $\cos\phi \cos\delta$, where ϕ is the terrestrial latitude of the observer and δ is the declination of the observed source.

celestial sphere (Figure 3–14). Stars at the ecliptic poles move in circles with radii dependent upon their distances from the Sun, and stars on the ecliptic oscillate along lines. For the general parallactic ellipse, the ratio of semiminor to semimajor axis is $\sin\beta$, where β is the celestial latitude of the star, just as for aberration orbits. Note, however, that for a given direction β, the aberration orbit is always the same but the parallactic orbit depends upon the star's distance. Also, the aberration orbit of a star is 90° out of phase with the parallactic orbit.

(c) The Doppler Effect

Our final proof of the Earth's revolution uses the Doppler effect. Chapter 8 derives the fact that the wavelength of electromagnetic radiation (light) is shifted in proportion to the relative line-of-sight speed of the object observed. If a star emits radiation at the wavelength λ_0 and we observe this radiation at the wavelength λ, then the Doppler formula for velocities much less than that of light is

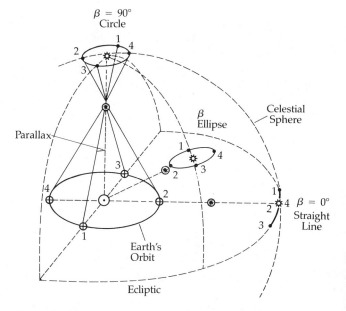

Figure 3–14

Parallactic orbits. The Earth's orbital change results in a periodic motion of stellar positions. Note that these are 90° out of phase with respect to the aberration orbits and their angular size depends on the distance to the star.

3-4 Differential Gravitational Forces

Two spherical bodies behave gravitationally as point masses. If the bodies are elastic or nonspherical, or if several more bodies are on the scene, *differential gravitational forces* may become important. This effect arises because gravitation depends upon the distance between bodies, and different parts of an extended body (or system) will therefore experience different gravitational accelerations.

(a) Tides

If you have spent time near a large body of water, you know that the level of the water rises and falls twice daily in the *tides* and that a given tide occurs about an hour later each day. The Moon is the principal cause of the tides because it returns to upper transit 53 min later each day—at about the same time relative to high tide. The solid body of the Earth, which has much greater cohesion than water, also responds quickly to the Moon's tidal forces. Albert A. Michelson (1852-1931) first measured these Earth tides in 1913 by observing water tides in long horizontal pipes. He assumed that the Earth was infinitely rigid and employed Newton's laws to calculate the expected water tides; the observed water tides were only 69 percent of what theory said they should be. This difference could be accounted for by assigning the Earth a rigidity somewhat greater than that of steel, so that it would respond to the Moon's tidal forces by raising body tides several centimeters in height.

Let's use Newton's laws to explain the tides in a general way. We assume that the Earth has its equilibrium rotation shape (an oblate spheroid), which is essentially spherical. Cover this sphere with a uniform depth of water and ask what effect the Moon's gravitational force has on this water (Figure 3-15A). By subtracting the vector acceleration at the Earth's center (C) from each of the surface vector accelerations, we obtain the *differential tidal accelerations* (Figure 3-15B). These tidal forces raise water tides about 1 m high at points A and B on the line of centers, and the Earth rotates below this configuration once each day. Actual tides arise from forced oscillations in the Earth's ocean basins, so that the height and timing of tides may differ markedly from the theoretical case. In some bays and estuaries, the tidal waters may accumulate to heights greater than 10 m.

Now to be more quantitative in our derivation of tidal accelerations. As a first approximation, we will neglect the small centripetal acceleration at the Earth's surface due to the orbital motion of the Earth and Moon about their barycenter every sidereal month (27.32^d). The centers of Earth and the Moon are separated by the distance d in Figure 3-16, and a small particle on the Earth's surface is at an angle ϕ to the line of centers. The gravitational acceleration of the Earth's center due to the Moon has the magnitude $A = GM_m/d^2$, and the particle's acceleration is $B = GM_m/r^2$. Subtract A from B vectorially. The component of **B** perpendicular to the line of centers is unaffected by this subtraction and has the magnitude

$$a = B \sin \theta = GM_m R_\oplus \sin \phi/r^3 \qquad (3-5)$$

The component of **B** parallel to the line of centers is

$$b = B \cos \theta = GM_m(d - R_\oplus \cos \phi)/r^3$$

and the remainder after subtracting **A** is the differential of the force:

$$b' = (GM_m/r^3)[d - R_\oplus \cos \phi - (r^3/d^2)] \qquad (3-6)$$

Using the law of cosines and the fact that $R_\oplus/d \ll 1$, we have

Figure 3-15

Tidal forces on the Earth. (A) The gravitational attraction of the Moon on the Earth is shown by selected force vectors. (B) The vector acceleration of the Earth's center (C) is subtracted from the surface accelerations. The remaining vectors show the tidal differential forces.

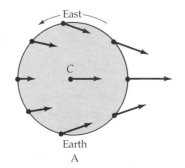

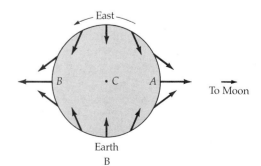

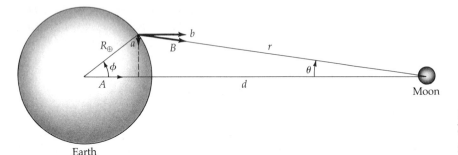

Earth

Figure 3-16

The geometry of the Moon's tidal forces on the Earth.

$$r^3 = d^3[1 - 2(R_\oplus/d) \cos \phi + R_\oplus/d^2]^{3/2} \approx d^3$$

Therefore, Equation 3-5 finally becomes (to lowest order)

$$a \approx GM_m R_\oplus \sin \phi/d^3 \qquad (3-7a)$$

Equation 3-6 may be written as

$$b' \approx (GM_m/d^2)[1 - (R_\oplus/d) \cos \phi - (r/d)^3]$$

and, expanding our result for r^3 using the *binomial theorem* (given in the Mathematical Appendix), we obtain

$$b' \approx 2GM_m R_\oplus \cos \phi/d^3 \qquad (3-7b)$$

Equations 3-7 verify the qualitative picture (Figure 3-15B) and indicate that tidal gravitational forces go as MR/d^3, where M is the mass of the source of tidal force, R is the size of the body being tidally influenced, and d is the separation of the two bodies.

We can show this relationship for tidal forces generally by differentiating Newton's law of gravitation with respect to R:

$$dF/dR = -2GM_m/R^3$$

and moving the dR to the right-hand side of the equation so that

$$dF = -(2GM_m/R^3)dR \qquad (3-8)$$

where dF is the differential gravitational force directed along R. Then dR is the diameter of a single solid body or the separation between two bodies close together that are being acted upon by the tidal forces.

The Sun also produces tidal effects on the Earth. Since the differential accelerations go as MR/d^3 and since $R = R_\oplus$ in both cases, the tide-raising force of the Sun relative to that of the Moon is

$$(M_\odot/M_m)(r_m/r_\odot)^3 =$$
$$(1.99 \times 10^{30} \text{ kg})/(7.36 \times 10^{22} \text{ kg})$$
$$\times (3.84 \times 10^5 \text{ km})^3/(1.50 \times 10^8 \text{ km})^3$$
$$\approx 5/11$$

The tidal effects of the Sun and Moon combine vectorially, so that the resultant tides depend upon the elongation of the Moon. When the Moon is at conjunction or opposition, the two forces add to produce the very high *spring tides;* when the Moon is at quadratures, the two forces partially cancel to give the unusually low *neap tides.*

(b) Consequences of Tidal Friction

When the Earth and the oceans yield to the tide-raising forces, energy is dissipated (in the form of heat) as a result of friction; most of this energy is lost in shallow seas and at shorelines, where ocean tides abut against the continents. This *tidal friction* reduces the energy of the Earth's rotation, so that the length of the day increases at the measurable rate of about 0.002^s per century. Tidal friction causes two fascinating phenomena in the Earth–Moon system: the Moon's *synchronous rotation* and *tidal evolution.*

The Earth raises body tides on the Moon that are about $M_\oplus R_m/M_m R_\oplus \approx 20$ times higher than the Earth's body tides. The enormous energy dissipation that results slowed the Moon's rotation until the Moon was forced into *synchronous rotation,* where its sidereal rotation period is exactly the same as its sidereal period of revolution about the Earth.

The average external torques exerted on the Earth–Moon system are negligible, so that the total angular momentum (Section 1.5) of the system must remain constant. The angular momentum of the Earth decreases as tidal friction slows its rotation;

hence, the Moon must increase its angular momentum by moving away from the Earth. Kepler's third law then implies that the month must be lengthening. In the distant future, the "day" and "month" will become the same and equal to about 50 present days. The tide-raising forces accelerate the water, but the oceans do not respond instantly to the Moon's tidal forces (Figure 3–17). As the Earth rotates beneath the Moon, the tidal bulges rise slightly eastward of the Earth–Moon center line. Bulge A is nearer the Moon than bulge B, and so A exerts a slightly greater force on the Moon. The resulting noncentral force accelerates the Moon, causing it to spiral away from the Earth. At the same time, the Moon's force on the bulges decelerates the Earth's rotation (Newton's third law). Going backward in time, we see that the Earth must have been spinning faster and the month must have been shorter than today. Indeed, paleontological studies of fossilized corals which lived about 10^8 years ago show that there were 400 "days" in each year and that ocean tides were more vigorous than today — the Earth was rotating faster and the Moon was closer in the past.

Let's show more explicitly that the moon must increase its orbital radius to gain orbital angular momentum. Kepler's laws require that the larger an orbit, the longer its period. So, if a body moves to a larger orbit, does its angular momentum (= mass × orbital radius × orbital speed) increase or decrease? It increases, proportional to the square root of the distance.

Here's how to see it. Assume a circular orbit of radius R. Then

$$L \propto RV$$

where

$$V = 2\pi R/P$$

so

$$V^2 = 4\pi^2 R^2/P^2$$

But Kepler's third law requires that

$$P^2 \propto R^3$$

$$1/V^2 \propto R$$

or

$$V \propto 1/R^{1/2}$$

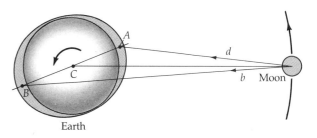

Figure 3–17

Tidal evolution. The Earth's tidal bulges (A and B) are driven by friction to be ahead of the Moon's orbital position. The friction slows the Earth's rotation, and the bulges accelerate the Moon in its orbit.

Go back to

$$L \propto RV$$

$$\propto R/R^{1/2}$$

$$\propto R^{1/2}$$

as stated.

(c) Precession and Nutation

You are probably familiar with the behavior of a spinning top or a gyroscope. When the top's spin axis is not aligned with the vertical, we expect the top to fall on its side, but instead the spin axis maintains the same angle to the vertical and simply rotates slowly about the vertical. This *precession* of the top is predicted by Newton's laws of motion.

The Earth's gravitational attraction **F** on the top produces the *horizontal* torque **N** = **r x F** [Section 1–5(a)]. Since torque is just the time rate of change of the top's vector angular momentum **L** (**N** = $d\mathbf{L}/dt$), there is no vertical component to topple the top and so it can only rotate, or *precess*, about the vertical. Differential gravitational forces acting upon the oblate Earth's (rotational) equatorial bulge produce torques that lead to a similar phenomenon — the precession of the Earth.

The Moon and Sun cause the Earth's luni–solar precession, with the Moon's effect dominating. The Moon's orbit is inclined about 5° to the ecliptic, but its average force is centered on the ecliptic. The Earth's equatorial plane is inclined 23.5° to the ecliptic (Figure 3–18), with the prominent equatorial bulges at A and B. Bulge A is more strongly attracted to the Moon than bulge B, and the differential forces are indicated.

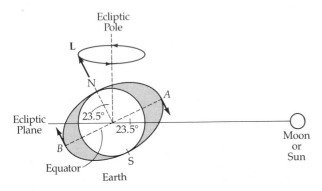

Figure 3–18

Precession of the Earth's axis. The differential tidal forces on the Earth's equatorial bulge (*A* and *B*) result in a torque (pointing into the page) that causes the Earth's angular momentum vector (**L**) to precess westward.

The torque that results points into the page, so that the Earth's angular momentum vector **L** precesses toward the west. The Sun induces a similar but slightly weaker effect, and the perturbing torques from the other planets produce a *planetary precession* that amounts to about 2 percent of the total. As a result of these torques, the celestial poles remain inclined 23.5° to the ecliptic pole but trace a circular path around the ecliptic pole once every 26,000 years. At present, the modestly bright star Polaris (Alpha Ursae Minoris) roughly marks the position of the north celestial pole, but in about A.D. 14,000, the pole star will be Vega (Alpha Lyrae).

As the celestial poles precess, the intersection of the celestial equator and the ecliptic (vernal and autumnal equinoxes) progresses westward at the rate of 360°/26,000 years ≈ 50″ per year along the ecliptic (about 50″ cos 23.5° ≈ 46″ per year along the celestial equator). This phenomenon of the *precession of the equinoxes* has several important consequences: it affects terrestrial timekeeping systems through the definition of the day and the year; it changes those stars that a given observer would consider circumpolar; and it significantly affects the celestial coordinate positions (Appendix 10) of all celestial objects. Because the precession effect is readily observable, the positions of objects in the celestial equatorial coordinate system (Appendix 10) must be constantly updated to the current epoch.

Because the Moon and Sun move above and below the Earth's equatorial plane, periodic variations occur in the torques acting on the Earth's equa-

torial bulge. These variations lead to a *nutation*, or wobbling, of the Earth's rotation axis. Small contributions to the nutation have monthly and yearly periods, but the principal contribution (discovered by James Bradley of aberration fame) of amplitude 9″ and period 18.6 years is from the regression of the nodes of the Moon's orbit (Chapter 4).

(d) The Roche and Instability Limits

We have discussed the relative orbits of two gravitating (rigid) spherical masses and obtained Kepler's laws. Here we consider the effects of differential gravitational forces upon nonrigid bodies and upon systems of more than two bodies. In general, a satellite cannot approach its primary planet too closely (the *Roche limit*) or stray too far (the *instability limit*) without dire dynamical consequences.

Consider a spherical satellite of mass m and radius r orbiting at a distance d from its very massive ($M \gg m$) primary planet of mass M and radius R (Figure 3–19A). If the satellite is large enough ($r \gtrsim 500$ km), its self-gravitation dominates all cohesive forces and determines its shape and strength. In 1850, Edouard Roche (1820–1883) demonstrated that such a satellite would be torn asunder by tidal forces if it approached the primary closer than

$$d = 2.44(\rho_M/\rho_m)^{1/3}R \qquad (3-9)$$

where ρ_M is the average density (SI units = kg/m³) of the primary and ρ_m is the average density of the satellite. This is the *Roche limit*. For example, if our Moon were to come closer than $d \approx 2.9R_\oplus \approx 18,500$ km from the center of the Earth, it would disrupt into small fragments.

Equation 3–9 was derived from a fluid satellite that assumed the shape of a *prolate* (football-shaped) spheroid in response to the primary's tidal forces. Now consider a *rigid* spherical satellite to approximately derive this result (Figure 3–19A). Since the satellite's orbital centripetal acceleration $\omega^2 d$ is produced by the gravitational attraction from the primary GM/d^2, the angular speed of the satellite about the massive primary is (Kepler's third law)

$$\omega = (GM/d^3)^{1/2}$$

The differential gravitational acceleration between the center of the satellite (point 1) and the outer edge (point 2) due to the primary is

$$A = (GM/d^2) - GM/(d + r)^2 \approx 2GMr/d^3$$

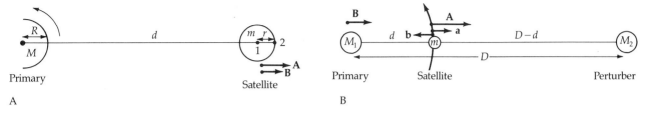

Figure 3-19

Roche and instability limits. (A) Geometry for the Roche limit, when the differential gravitational force **A** and differential centripetal acceleration **B** between points 1 and 2 exceed a body's self-gravitation. (B) A satellite m escapes from its primary planet of mass M_1 as a result of differential perturbations $\mathbf{a} = \mathbf{A} - \mathbf{B}$ arising from the perturber of mass M_2 when d exceeds the instability limit.

and the differential centripetal acceleration between these two points is

$$B = \omega^2(d + r) - \omega^2 d = \omega^2 r = GMr/d^3$$

The combination $A + B = 3GMr/d^3$ must be balanced by the satellite's self-gravitational acceleration Gm/r^2 if the satellite is not to be pulled apart, so that disruption occurs at

$$d = r(3M/m)^{1/3} \qquad (3-10)$$

Because the average density is defined as total mass divided by volume, we have $\rho_M = 3M/4\pi R^3$ and $\rho_m = 3m/4\pi r^3$. Substituting into Equation 3-10 yields

$$d = R(3\rho_M/\rho_m)^{1/3} \approx 1.44(\rho_M/\rho_m)^{1/3}R \quad (3-11)$$

The numerical difference between Equations 3-9 and 3-11 arises from our assumption of a rigid (non-fluid) satellite.

We can see this result more elegantly by considering the differential form of tidal forces (Equation 3-8) for two particles of equal size and mass that are just touching. Their attractive force is

$$F = Gmm/(dr)^2$$

and we equate this to the tidal forces to find Roche's limit d from mass M:

$$GMm/(dr)^2 = 2GMm \, dr/d^3$$

so that

$$d = (2M/m)^{1/3} \, dr$$

which becomes, for bodies with uniform densities,

$$d \approx 2.5(\rho_M/\rho_m)^{1/3}R \qquad (3-12)$$

For rocky or icy bodies greater than 40 km in diameter, the numerical coefficient is 1.38, and for a body falling right into a planet, the coefficient is 1.19.

The natural satellites in the Solar System that are of significant size orbit beyond the Roche limit of their primaries, though some of Saturn's tiny moonlets actually orbit within it. Saturn's beautiful system of rings lies about 80,000 to 136,000 km from the center of the planet and so is entirely within the Roche limit of about 150,000 km. The Jupiter and Uranus ring systems are also within the Roche limits for those planets. No rocky or icy satellite with a diameter in excess of about 40 km can exist or could have been formed within these limits without disrupting. That these rings are composed of multitudes of solid particles implies that their cohesive strength is greater than the disrupting tidal forces. Artificial satellites certainly do orbit within the Earth's Roche limit, but they are held together by the tensile strength of their materials. A solid sphere of steel 1 m in diameter can approach within 100 m of a point mass of $M_\oplus$ before the tidal forces overwhelm its internal tensile strength!

As a body orbits farther and farther from its primary, *differential perturbations* from other bodies become more important (Figure 3-19B). Beyond the *instability limit*, the body escapes from its primary. The perturber M_2 produces the differential acceleration

$$\begin{aligned} a = A - B &= [GM_2/(D - d)^2] - GM_2/D^2 \\ &\approx 2GM_2 d/D^3 \end{aligned}$$

between the orbiting body and its primary (when $d \ll D$). When this acceleration equals the gravitational acceleration from the primary $b = GM_1/d^2$, the orbiting body is at the instability limit:

$$d = (M_1/2M_2)^{1/3}D \qquad (3-13)$$

Note that Equation 3–13 is valid only when $M_1 \ll M_2$; when $M_1 \gtrsim M_2$, we must use the exact relation

$$d^3(2D - d) = (M_1/M_2)D^2(D - d)^2 \quad (3-14)$$

to solve for d in terms of D. From Equation 3–13, the instability limit for our Moon—with the Sun as perturber—is 1.7×10^6 km; this is four times the Moon's present distance from the Earth, and so our Moon is now stable against escape. Comets may escape from the Solar System as a result of perturbations from other stars $(M_1 \approx M_2)$ when the comets' aphelia lie at distances greater than about 10^5 AU (since Equation 3–14 implies $d \approx D/2$ and $D \approx 2 \times 10^5$ AU).

Problems

1. (a) How much does a sidereal clock gain (or lose) on a mean solar clock in five mean solar hours?
    (b) What is the approximate sidereal time when it is noon apparent solar time on the following days: (*i*) the first day of spring, (*ii*) the first day of summer, (*iii*) April 21, (*iv*) January 2?

2. In terms of azimuth and altitude (Appendix 10), describe the Sun's daily path across the sky during every season of the year at
    (a) the equator
    (b) latitude 35°N
    (c) the north pole
    Use such descriptive terms as *noon altitude, sunrise azimuth, sunset azimuth,* and *angle at which Sun meets horizon.*

3. Cape Canaveral is at longitude 80°23′ W and latitude 28°30′ N. A rocket is launched from there due south, and it lands on the equator 10 min later. What is the longitude of impact?

4. In discussing the Coriolis effect, we mentioned that a body which undergoes a constant acceleration a will travel a distance $s = at^2/2$ in a time t. Show that the speed of the body is proportional to time and that the body's acceleration is indeed a.

5. The Earth's rotation also produces an aberration of starlight.
    (a) What is the maximum value of this daily aberration?
    (b) Where on the Earth is this effect maximum?
    (c) For what stars (location on celestial sphere) is the effect maximum?

6. (a) If the centripetal acceleration from the Earth's rotation is $\omega^2 R_\oplus$ at the equator and if $\omega = 2\pi/P$, where P is one day, then by what percentage is a person's weight reduced as he or she walks from the north pole to the equator? Ignore the Earth's oblateness.
    (b) Now ignore the Earth's rotation and find by what percentage a person's weight increases as he or she walks from equator to pole on the oblate Earth?

    (c) Compare your results from (a) and (b) and combine them to deduce how the *effective g* varies from the Earth's poles to the equator.

7. The Earth's orbital speed is approximately 30 km/s. A star emits a spectral line at wavelength $\lambda_e = 517.3$ nm (1 nm $= 10^{-9}$ m). Over what amplitude does this wavelength oscillate as the Earth orbits the Sun when the star is located at the ecliptic (celestial latitude $\beta = 0°$)?

8. In Section 3–4(a), we computed the Moon's differential tidal forces at the Earth's surface, ignoring the motion of the Earth–Moon system about its center of mass every month. Choose a coordinate system centered upon this center of mass and rotating eastward with the angular speed ω (due to this sidereal monthly motion) and include the centripetal acceleration at the Earth's surface to deduce the correct dependence of the total tidal acceleration at the Earth's surface. Ignore the daily rotation of the Earth. (*Hint:* The Earth–Moon center of mass is located within the Earth.)

9. On a large piece of graph paper, plot *to scale* the distances of the Roche limit and orbits of the innermost planetary satellites and rings for (*i*) the Earth, (*ii*) Mars, (*iii*) Jupiter, (*iv*) Saturn, and (*v*) Uranus. Assume that $\rho_M = \rho_m$ in every case. Write a brief statement summarizing your results.

10. Compare the tidal forces the Moon exerts on the Earth (at perigee) with those the Sun exerts on the Earth (at perihelion) and those Venus exerts on the Earth (at closest approach).

11. Assume that the Earth and Moon are spherical and that the Moon orbits the Earth in a circle. Calculate the *spin* angular momenta of the Earth and Moon and compare these with the *orbital* angular momentum of the Moon. A spherical mass of uniform density has a spin angular momentum of $(2/5)MVR$, where V is the equatorial velocity and R the radius. The sum of these momenta must be a constant for the Earth–Moon system (if we ignore external torques). From the rate of

loss of the Earth's spin angular momentum from tidal friction, estimate the rate at which the Moon moves radially away from the Earth.

12. The Moon will move away from the Earth until it no longer lags the tidal bulges, and the angular momentum transfer will stop. Computer calculations indicate that this will occur at an Earth–Moon distance of 6.45×10^5 km. Calculate the Moon's orbital period then.

13. Compare the solar insolation, at noon, in Albuquerque, New Mexico (latitude about 35°N) for the day of the summer solstice and the day of the winter solstice.

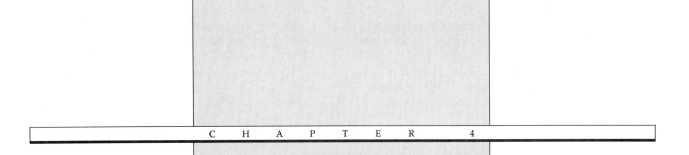

The
Earth – Moon
System

Our detailed examination of the Solar System begins with our home planet, the Earth, and its satellite, the Moon. Centuries of probing the Earth have led to a good understanding of this planet, and the study of the Moon has been revolutionized by the space age since 1957, with manned and unmanned landings. This chapter outlines our knowledge of our terrestrial neighborhood to find the Earth as the most evolved terrestrial planet and the Moon as a fossil world.

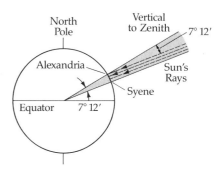

Figure 4–1

Geometry for Eratosthenes' method for determining the size of the Earth.

4–1 Dimensions

The size of the Earth was first determined by the Greek astronomer Eratosthenes (276–195 B.C.), who noted that, at the summer solstice, the noon altitude of the Sun in the city of Syene differed by $7°12'$ from the altitude in Alexandria (Figure 4–1). Assuming that the Earth was spherical and noting that Alexandria was 5000 stadia (1 stadium ≈ 0.16 km) due north of Syene, he found the Earth's circumference to be $(360°/7.2°)5000 = 250{,}000$ stadia ($\approx 40{,}000$ km). This value gives a radius within 1 percent of the Earth's equatorial radius $R_\oplus = 6378.2$ km, but we don't really know the exact value of the stadium, and so the agreement with the modern value may be an accident. Eratosthenes certainly had the right idea in using this method. Modern measurements include techniques such as satellite geodesy and radar.

The Earth is the major partner in the *Earth–Moon system*. The Moon is one of the largest and most massive satellites in the Solar System, relative to its primary planet (Figure 4–2). Pluto and Charon are now first in this regard with Charon's radius almost 0.5 Pluto's and mass 0.1 Pluto's (Chapter 6). Our Moon's radius is 1738 km ($0.272R_\oplus$), and its mass is 7.35×10^{22} kg ($0.0123M_\oplus$); the closest runners-up are Neptune's Triton ($0.090R_N$, $0.0013M_N$) and Saturn's Titan ($0.040R_S$, $0.00025M_S$). The mean distance between the centers of the Earth and the Moon is 384,405 km, or about $60.3R_\oplus$ (Figure 4–3A). The *barycenter* (center of mass) of the system is located $M_m a_m/(M_\oplus + M_m) = (0.0123)(384{,}405)/(1.0123) = 4671$ km from the Earth's center; the Earth and Moon orbit this point, which is buried 1707 km *below* the Earth's surface, once each month (Figure 4–3B).

We can estimate the mass of the Earth if we assume that the Moon's mass is small relative to that of the Earth and use Kepler's third law: $M_\oplus = 4\pi^2 a_m^3/GP_m^2$; this law may also be applied to artificial satellite orbits. The result is

$$M_\oplus = 5.98 \times 10^{24} \text{ kg}$$

(As shown below, the Moon's mass is about 1 percent of the Earth's mass, and so our approximation results in an error of this size.) The Moon's mass may be determined by observing the Earth's motion about the barycenter; by knowing the Earth's mass and the location of the barycenter, we compute

$$M_m = (d_\oplus/d_m)M_\oplus = 7.35 \times 10^{22} \text{ kg}$$
$$= (1/81.3)M_\oplus$$

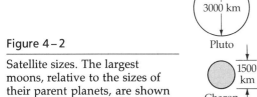

Figure 4–2

Satellite sizes. The largest moons, relative to the sizes of their parent planets, are shown to scale.

Pluto 3000 km

Charon 1500 km

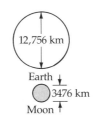

Earth 12,756 km

Moon 3476 km

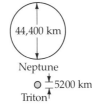

Neptune 44,400 km

Triton 5200 km

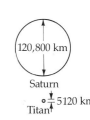

Saturn 120,800 km

Titan 5120 km

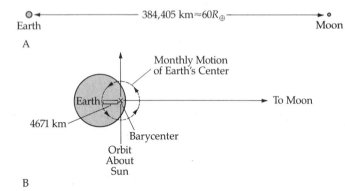

A

B

Figure 4–3

(A) The Earth–Moon system drawn to scale. (B) The location of the barycenter (center of mass) of the Earth–Moon system.

where $d_\oplus$ is the distance from the barycenter to the Earth's center and d_m is the distance from the barycenter to the Moon's center. Today the orbits of lunar spacecraft are used to accurately deduce M_m as well as the internal mass distribution of the Moon. The oblateness of the Moon is 0.006; the Moon is slightly oblong, with its long axis (tidal bulge!) pointed toward the Earth.

Today the Earth–Moon distance is determined by reflecting radar pulses from the lunar surface (accuracy of a few meters) or laser light pulses from mirrors set up by the Apollo astronauts (accuracy of about 1 cm). Using the distance to the Moon, we may find the size of the Moon by (1) timing occulations of stars and planets or of solar eclipses and (2) measuring the apparent angular size of the Moon in the sky. From its apparent diameter, which is 31' of arc (about 1/2°), and the Earth–Moon distance (384,000 km), we calculate its diameter to be 3476 km (radius of 1738 km).

4–2 Dynamics

(a) Motions

Let's summarize the principal motions of the Earth (Chapter 3). The barycenter of the Earth–Moon system orbits the Sun at about 1 AU in one sidereal year of 365.2564 mean solar days. The Earth rotates once every 24 sidereal hours, and its center orbits the barycenter in one sidereal month of 27.322 days. The Earth's rotation axis precesses about the ecliptic pole in roughly 26,000 years, and its body axis wobbles slightly about the rotation axis. Finally, tidal friction slows the Earth's rotation by roughly 0.002 s per century.

The Moon's motions are much more complicated. With respect to the stars, the Moon orbits the barycenter in one *sidereal month* (27.322^d); with respect to the Sun, this orbit takes one *synodic month* (the month of phases, 29.531^d). With respect to the Moon's line of nodes (see below), the orbital period is one *nodical* or *draconic month* (27.212^d), and with respect to its perigee, the period is one *anomalistic month* (27.555^d). The Moon's *synchronous* rotation period is one sidereal month, and so we see only one hemisphere of the Moon. The far side was not seen until photographed by the Russian spacecraft Luna 3 in October 1959.

Actually, we can see about 59 percent of the total lunar surface as a result of the Moon's *librations* (first known and interpreted by Galileo). Lunar librations are geometric effects caused by the inclination of the Moon's orbit and equator. The Moon rotates at a fairly steady rate, but it moves at different speeds in its eccentric orbit; this variation leads to an apparent east–west "rocking" of the Moon by about 6°17' — the *libration in longitude* (Figure 4–4A). Although the Moon's equator lies essentially parallel to the ecliptic plane, the lunar orbit is inclined 5°9' to the ecliptic; hence, we see an apparent north–south "nodding" motion of about 6°41' — the *libration in latitude* (Figure 4–4B).

Differential solar gravitational forces on the Earth–Moon system produce three significant effects: (1) they tend to elongate the Moon's orbit at quadrature, (2) they cause the perigee of the Moon's orbit to precess eastward (direct) with a period of 8.85 years, and (3) they produce a torque on the inclined orbit, which causes the *line of nodes* (intersection of the Moon's orbital plane and the ecliptic plane) to regress westward along the ecliptic with a period of 18.6 years. Because of its eccentricity ($e = 0.055$), the

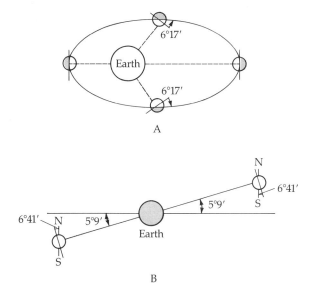

Figure 4–4

Lunar librations. (A) A libration in longitude occurs as a result of the Moon's constant rotation in an elliptical orbit, where the orbital speed varies. (B) A libration in latitude results from a combination of the inclination of the Moon's orbit to the ecliptic and the inclination of its equator to its orbital plane.

Moon's orbital distance ranges from 363, 263 km at perigee to 405,547 km at apogee. The angular diameter of the Moon varies from 32.9′ to 29.5′.

(b) Phases

The Sun always illuminates one hemisphere of the Moon, but from the Earth we see varying fractions of the sunlit hemisphere depending upon the Moon's elongation. The cycle of *geocentric phases* of the Moon lasts one *synodic* month (Figure 4–5); these phases occur in the following sequence: new (inferior conjunction), waxing crescent, first quarter (quadrature), waxing gibbous, full (opposition), waning gibbous, third quarter (quadrature), waning crescent, and back to new. The Moon's position in the sky is correlated with its phase. The first-quarter Moon rises in the east at noon, transits the local meridian at sunset, and sets in the west at midnight. As seen from the Moon, the Earth exhibits similar phases in a reverse cycle (new Moon occurs at the time of full Earth), and relative to the Moon's horizon, the Earth remains roughly at the same place in the sky.

(c) Eclipses

In general, an *eclipse* occurs when the shadow of one celestial body falls on another body; Chapter 6 mentions the eclipses of Jupiter's moons by the planet. Eclipses in the Earth–Moon system depend upon the orientation of the *line of nodes* of the Moon's inclined orbit, since the Moon must be near the ecliptic plane to pass directly between the Sun and the Earth in a solar eclipse and to pass through the Earth's shadow in a lunar eclipse (Figure 4–6).

Solar Eclipses

The Sun's apparent angular diameter is 32′, almost the same as that of the Moon! The line of nodes of the Moon's orbit points toward the Sun twice each year, so that new Moon can occur close enough to the ecliptic for the Moon to cover the Sun—a *solar eclipse*. When the Moon passes to one side of the center of the Sun, we see a *partial* solar eclipse. If the Moon is not near perigee in its orbit, its angular size is slightly smaller than that of the Sun and an *annular*

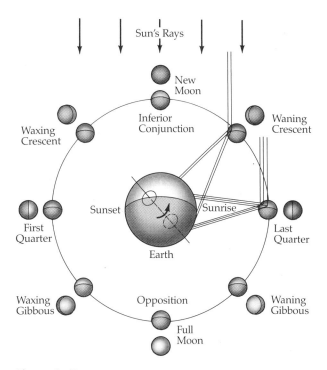

Figure 4–5

Lunar phases. The Moon is shown here at eight positions in its synodic phase cycle as it orbits the Earth. The shaded portion is the dark, unilluminated half of the Moon.

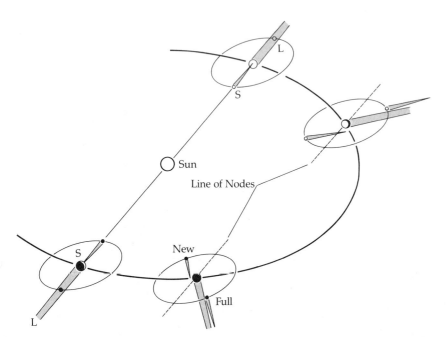

Figure 4–6

Lunar nodes and eclipses. The alignments of the Sun, Earth, and Moon are shown for one year. The shadows cast by the Earth and Moon are the shaded regions. The line of nodes (dashed line) must point to the Sun for an eclipse to occur. Lunar (L) and solar (S) eclipses are indicated.

solar eclipse can take place with a ring of the Sun's disk visible. Near its perigee, when the Moon's center crosses the Sun's center, we have a spectacular *total* solar eclipse.

Lunar Eclipses

At the Moon's orbit, the Earth's shadow (where the Sun is totally hidden) is 9212 km across, which is equivalent to an angular diameter of $1°22.4'$ as viewed from the Moon. When the line of nodes points down this shadow, a *lunar eclipse* can take place at full Moon (Figure 4–7). Both *partial* and *total* lunar eclipses are possible, with the Moon remaining dark for up to 1^h40^m in a central total lunar eclipse.

The Greek astronomer Meton (c. 400 B.C.) noted that the Moon exhibits the same phases on the same day of the month (such as full Moon on April 13) at intervals of 18.6 years—this is known as the *Metonic cycle*. A similar phenomenon, also due to the *regression of the nodes* of the Moon's orbit, is the *Saros cycle*. Similar solar and lunar eclipses take place at intervals of 223 synodic months (18 years, 10 days); since the Saros interval is 6585.32 days, we must wait three Saros cycles to see an eclipse repeat at the same place on the Earth. The Saros cycle comes about from the near equality of 223 synodic months, 242 nodical months, and 239 anomalistic months.

4–3 Interiors

(a) The Earth

The *average density* of the Earth is

$$\bar{\rho} = 3M_\oplus/4\pi R_\oplus^3 = 5520 \text{ kg/m}^3$$

(the density of pure water is 1000 kg/m³). Because the density of rocks at the Earth's surface is about 3000 kg/m³, the interior of the Earth must be *very* dense. The Earth's interior is stratified: a 35-km-thick surface layer, the crust, of density 3300 kg/m³, then the *solid mantle* of silicates such as olivine [(Mg, Fe)$_2$, SiO$_4$] with densities approximately 3000 to 6000 kg/m³ to a depth of 2900 km, next the 2200-km-thick *liquid outer core* with densities from about 9000 to 11,000 kg/m³, and finally the solid inner core of radius 1300 km with densities around 12,000 kg/m³ (Figure 4–8).

This picture of the Earth's interior has been deduced from the study of the propagation of earthquake waves (*seismology*). Earthquakes generate both *longitudinal compression* (P) waves and *transverse distortion* (S) waves, which travel on paths determined by the density and composition of the interior materials (Figure 4–9). Both types of waves are refracted by changes in the density of the medium through

Figure 4-7

A time sequence of the stages of a total lunar eclipse in July 1982. From left to right, the series of exposures on the same negative shows the moon emerging from the Earth's shadow. *(B. Walski)*

which they are traveling and reflected by density discontinuities; only the P waves, however, can propagate through a liquid region. By measuring whether the wave has compressional (P) or up–down (S) characteristics and its time of travel from the center of an earthquake, we infer the wave's path through the Earth. No S waves are seen through the core (they are absorbed by the liquid outer core); P waves are refracted at the mantle–core interface, and weak P waves reflected by the *solid* core are observed.

We do know that part of the Earth's core is liquid, but its exact composition is not really known. Most geophysicists attribute the density jump to a change from silicates in the mantle to nickel–iron or some iron alloy in the core. Whatever its exact composition, the core is metallic in contrast to the rocky mantle.

(b) The Moon

From its mass and radius, we find the Moon's average density is 3370 kg/m³, similar to the density of the Earth's crust. The Apollo missions have returned lunar surface samples with densities near 3000 kg/m³. The composition of these samples is like that of basaltic silicates, and so the density cannot increase much toward the Moon's center. Seismic stations set up by Apollo astronauts have revealed that the Moon is seismically quiet; moonquakes measured less than 2 on the Richter scale, compared with 7 or 8 for major earthquakes. Some lunar quakes are attributed to meteorite impacts. A few feeble vibrations originate at lunar perigee, and the occurrence of moonquake groups, consisting of a series of quakes

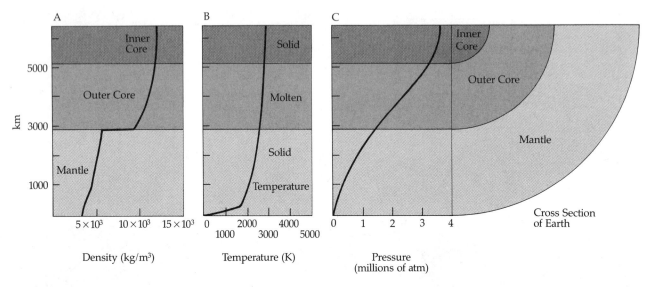

Figure 4–8

The Earth's interior. This model is inferred from seismic wave observations. (A) The run of density from the surface to the center. (B) The variation in temperature, which rises quickly then levels off. (C) The trend in pressure.

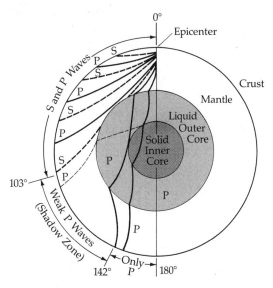

Figure 4–9

Seismic waves. Both S (dashed line) and P (solid line) waves are shown. The S waves are absorbed by the liquid outer core, and the P waves are reflected by the inner core (light dashed line).

lasting several days, has been observed. The origin of these quake groups is unknown, however, and apparently unrelated to tidal effects. Vibrations caused by the deliberate impacts of Apollo components on the lunar surface damp out very slowly (the Moon resonates for almost an hour!). One current model of the Moon's interior based on this lunar seismology includes a solid, inactive core; a mantle in which partial melting has occurred; and a crust at least 60 to 70 km deep in some places (Figure 4–10). So, the outer regions of the Moon's interior appear to be stratified and differentiated. The core is *not* metallic, as expected from the low bulk density.

4–4 Surface Features

The surfaces of the Earth and Moon are radically different from each other. The Earth's surface undergoes rapid and extensive evolution as a result of the hot interior and the thick, erosive atmosphere. Because the Moon's interior is cold and because the Moon has no atmosphere, its surface has preserved its early development. Let's briefly outline the characteristics of the Earth's surface and then examine the lunar surface in contrast.

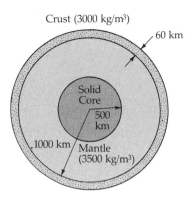

Figure 4–10

The Moon's interior. This model is based on Apollo mission results; it shows typical values for sizes and densities.

(a) The Earth's Surface and Age

The surface of the Earth is the interface between the low-density *crust* and the *atmosphere*. The crust is composed of (1) the solid *lithosphere* of average density 3300 kg/m³, with light granitic continental blocks reaching to a depth of 35 km and the denser

basaltic ocean floors extending down only 5 km, and (2) the liquid-water *hydrosphere*, which covers approximately 70 percent of the Earth's surface area to an average depth of 3.5 km (Figure 4–11A). The crust floats on the mantle, in buoyant equilibrium. Slow motions of the upper mantle drive continental drift, mountain building, and earthquakes. The Earth's surface consists of a number of separate continental and oceanic plates. Along ridges in the oceanic plates, lava from the lower crust wells up to push the plates apart and move the continents around the surface (Figure 4–11B); this evolutionary process for the crust is called *plate tectonics*.

Plate tectonics profoundly changes the face of the Earth over time scales that are astronomically short. The present continents, for instance, have moved considerably since their breakup from a single landmass 200 million years ago. Geologic evidence indicates that the rate of continental drift now amounts to only a few centimeters per year. That estimate has been confirmed by very-long-baseline interferometry (Chapter 9), which has detected a drift of 1 to 2 cm/year in real time. Plate tectonics insures that the ocean basins are the youngest part of the

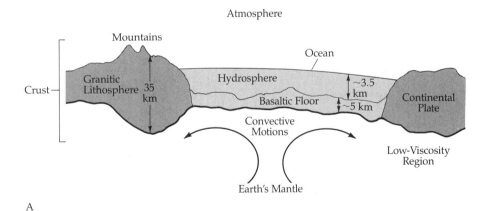

A

Figure 4–11

The Earth's crust. (A) The upper mantle, crust (lithosphere and hydrosphere), and plates, with convective motions in the mantle moving the plates. (B) Collisions between crustal and oceanic plates drive the evolution of the surface structure.

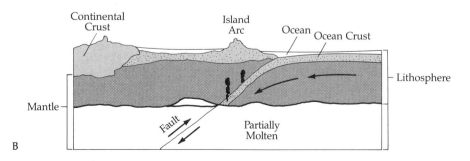

B

Earth's crust because that is where new lava flows up from the mantle.

The surface of the lithosphere, with an observed temperature range of 200 to 340 K, is the home of humankind. Seen from space, the Earth's cloud cover and polar icecaps appear bright (high albedo), and the continents show up light green and tan. The oceans are very dark blue. Our home is a beautiful but complex ship in space.

How old is the Earth? The ocean basins make up the youngest part of the crust; parts of the continental masses are the oldest. We can date the ages of these rocks by *radioactive decay* and so infer the Earth's age. The radioactive dating technique works because of the natural instability of the nuclei of radioactive elements. When these nuclei decay, they break apart into simpler nuclei. The original isotope is called the *parent*; the resulting products are *daughter* atoms. With a large number of radioactive atoms, you can determine a gross rate of disintegration. Half a piece

of uranium-238 (^{238}U) decays to lead in 4.5 billion years, half again in the next 4.5 billion years, and so on (Figure 4–12). You can calculate the amount of uranium left at any time, even though the decay time for any one uranium atom cannot be specified. Conversely, given a rock sample containing ^{238}U and knowing the halflife of uranium, you can calculate the age of the sample. Elements in addition to uranium that can serve as radioactive clocks include rubidium (^{87}Rb), which decays to strontium (^{87}Sr) with a halflife of 47 billion years, and potassium (^{40}K), which decays to the inert gas argon (^{40}Ar) with a halflife of 1.3 billion years. The derived date is the time elapsed since the rocks were last *solidified* and the parent and daughter atoms trapped. Remelting and resolidification of rocks *after* their origin would reset their dating clocks.

Let's examine the functional form of radioactive decay. Given n, a large number of radioactive atoms, the average number decaying in a time interval dt will be dn, which is proportional to the number remaining at time t. So

$$-dn = \lambda n \, dt$$

$$dn/n = -\lambda \, dt$$

where λ is the *decay constant*. We integrate this equation to get

$$\ln n = -\lambda t + \text{constant}$$

At $t = 0$, $n = n_0$, and so

$$\ln n = -\lambda t + \ln n_0$$

$$\ln n - \ln n_0 = -\lambda t$$

$$\ln (n/n_0) = -\lambda t$$

and finally,

$$n/n_0 = \exp(-\lambda t) \qquad \textbf{(4–1)}$$

The halflife occurs when $n = n_0/2$ at time τ, so that

$$\tau = (\ln 2)/\lambda = 0.693/\lambda$$

Then

$$n/n_0 = \exp(-0.693t/\tau) \qquad \textbf{(4–2)}$$

Suppose, for example, that a rock has an ^{40}Ar to ^{40}K ratio of 1/4. How old is it (assuming that the initial fraction of ^{40}Ar was negligible and no argon has escaped)?

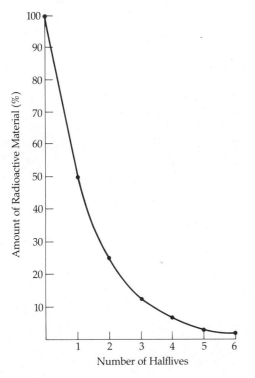

Figure 4–12

Radioactive decay. During each halflife, one half of the parent isotope decays. The form of this curve is that of an exponential decline.

$$1/4 = \exp(-0.693t/\tau)$$
$$\ln 0.25 = -0.693t/\tau$$
$$-1.39 = -0.693t/\tau$$
$$2 = t/\tau$$
$$t = 2\tau = (2)(1.3 \times 10^9 \text{ years})$$
$$= 2.6 \times 10^9 \text{ years}$$

(b) The Lunar Surface

Through remote sensing and lunar landings, the visible surface and the first few meters of the Moon's soil have been carefully studied. The surface is made up of gray-tan materials of basaltic composition with a very low average albedo (0.07). As the Moon's *terminator* (the line separating the dark and sunlit hemispheres) moves along the lunar equator at 15.4 km/h, the surface temperature drops rapidly from a maximum of 390 K to a minimum of 110 K. We can calculate the subsolar temperature by [Section 2–1(a)]

$$T = 394[(1 - A)/D^2]^{1/2}$$

where A is the Moon's albedo and D its distance from the Sun (in AU). So

$$T = 394[(1 - 0.07)/1^2]^{1/2}$$
$$= 386 \text{ K}$$

almost what is observed. (Note that for this calculation we have assumed an equilibrium situation for a blackbody.)

Figure 4–13

Moon at last quarter. The terminator between the sunlit and dark regions separates the western (day) and eastern (night) sides of the Moon. *(Lick Observatory)*

The Visible Structure

In A.D. 1609, Galileo first discerned (Figure 4–13) the dark lunar *maria*, or "seas" (which were given Latin names by Riccioli in 1651), the light lunar *highlands* with their numerous *craters*, and lunar *mountains*—all through his crude telescope. Spacecraft, manned and unmanned, have revealed a great deal about the lunar surface.

Craters The entire lunar surface is pockmarked with *craters*: concave depressions with rims elevated above the exterior level, sometimes with a peak in the center of the floor. Craters range in size from microscopic pits to the largest on the nearside, Bailly, with a diameter of 295 km. The large crater Copernicus (Figure 4–14) dominates in the Ocean of Storms. The juxtaposition of different features permits us to determine relative ages since young craters will clearly overlap older craters. The lunar maria have far fewer craters than the lunar highlands, and so the maria must be younger than the highlands. Many craters, especially Tycho and Copernicus (Figure 4–14), exhibit patterns of *rays*, which include numerous *secondary craters* from the blown-out debris.

Craters evolve and disappear slowly as material slides down their walls and as the walls themselves slump (Figure 4–15); meteorite bombardment produces new craters, which fill, obliterate, and degrade the older craters. The lifetime against such erosion has been estimated at several million years for craters 1 cm in diameter and longer than the age of the Moon for large craters (tens of kilometers in diameter). Although large craters do not disappear, their walls do slump and their rims and rays are eroded by meteoritic impact, temperature changes, and moonquakes.

The accepted model of crater origin is *meteorite impact*, with the largest craters arising from small asteroids that hit the Moon. A few lunar features, in-

Figure 4–14

Copernicus crater. Note the ray system emerging from the crater and the rugged terrain outside the rim. *(Lick Observatory)*

Figure 4-15

Slumping and ejecta blanket. This view of the north wall of Copernicus shows several ledges formed by slumping and the downward flow of the rim's material. Note the ejecta outside the rim. *(NASA)*

cluding some craters, such as Alphonsus, betray evidence of lava flows or volcanic origin, but very few are not impact craters.

An impact crater is produced when an incoming meteoroid hits the surface and converts much of its kinetic energy to explosive thermal energy; the remainder of the kinetic energy is used to fracture surface materials and create seismic waves. Lunar material scatters in analogy to chemical (TNT) or nuclear explosions. A shock wave is formed as the meteorite strikes the surface, compressing the rock so that some of it is pulverized and behaves like a fluid. This rocky material is ejected explosively, excavating a hole to form a crater, and is piled up to form an elevated rim. Some is flung out as an *ejecta blanket* of debris beyond the rim (Figure 4-15). Sprays of high-velocity particles land farther out to produce rays of secondary craters, and the rebound from the initial shock frequently raises a central peak (similar to the rebound when an object drops into a liquid). The meteoroid is destroyed in the process. An asteroid 1 km in radius

contains about 2×10^{13} kg of rock, and when it hits the Moon at a speed of 45 km/s, the kinetic energy of $mv^2/2 \cong 10^{22}$ J vaporizes the rock to create a crater about the size of Copernicus. (Note that 45 km/s is the escape velocity from the Moon and that from the Sun 1 AU away from it; this sum is the minimum speed with which an object falling in from "infinity" would hit the Moon's surface.) Large craters such as Tycho, Clavius, and Mare Orientale required energies in the range 10^{21} to 10^{26} J for their excavation.

Highlands Most of the lunar surface consists of the light-colored, heavily cratered *lunar highlands*, especially the lunar farside (Figure 4-16A). The highlands have been dated at 4.0×10^9 years; this represents the oldest surviving lunar surface. In general, the highlands have a greater elevation than the maria, with an average difference of 3 km.

Lowlands (Maria) The large dark areas of the Moon's surface are called *maria* or "seas" (the singular is *mare*), though they are really flat plains of dark

A

basaltic lava. Of the 30 maria known, only four are found on the farside of the Moon. Several are roughly circular and range in diameter from about 300 to 1000 km. These circular *mare basins* are believed to be impact features that subsequently filled with molten lava (Figure 4–16B). The presence of old, flooded craters implies that some cratering occurred between the time of formation of the mare basin and the extrusion of the lava. The impact model of mare basins is supported by the curved shape of some of the

mountain chains (such as the Apennines) that border and rise above the maria and by the observation of ejecta sculpting on the nearby highlands. The Apennine Mountains resulted from uplifting of the crust at the time of the impact that produced Mare Imbrium.

The maria have fewer craters than the lunar highlands (Figure 4–16C); this lower crater density implies a younger age—an idea confirmed by the dating of lunar samples, which are about 3.2×10^9 years old. Fresh, young craters, such as Copernicus

B

C

Figure 4-16

The lunar surface. (A) This photograph of the lunar farside shows the cratered highland regions. The dark-floored crater on the east (right) is Tsiolkovsky. *(NASA)* (B) A lunar mare basin; note the circular shape. *(Lick Observatory)* (C) A closer view of the surface of a mare as the Apollo lunar module descends for a landing.

(Figures 4-13 and 4-14), are seen superimposed on the maria. In addition, the maria exhibit wrinkled ridges, domes, hills, and valleys, many of which are evidence of some type of volcanic activity.

Gravitational perturbations of the lunar orbiter spacecraft imply that the mare basins on the nearside are filled with dense material—these gravitational anomalies are called *mascons* (*mas*s *con*centration*s*). They occur because the lava filling the basins is of higher density and different composition than the rest of the lunar surface. Their existence indicates that the Moon's crust is so rigid and deep that the mascons do not sink into it.

The Composition of Surface Materials

American and Russian landers have probed and analyzed the lunar surface; the manned Apollo missions also returned hundreds of kilograms of surface soil and rocks. The bedrock of the lunar surface is covered by a thin, well-mixed layer of loose soil and rocks—the *regolith.* The regolith was produced by meteorite impacts that fragmented and scattered the bedrock and pulverized the soil over the eons. This soil ranges in depth from 2 to 10 m in the maria and is estimated to be 10 m deep in the highlands because of extensive cratering; large chunks of bedrock are frequently seen as boulders at the bottoms of and near deep impact craters. The lunar soil has the texture and strength of cohesive sand, and its color is dark gray with a shade of tan. Churning by meteorite impacts mixes the regolith thoroughly in about 50 million years.

Materials brought back from the Moon (Figure 4-17) include (1) samples of the cohesive, fine-grained soil containing particles of average size 10 μm, with an admixture of small (50 μm) glass or obsidian spheres; (2) small *igneous* rocks from maria that show evidence of melting, recrystallization, and shock fracture; (3) small chunks of coarse-grained composite *breccias,* which resemble a compressed conglomerate of many materials; and (4) light-colored igneous rocks from the highlands, called *anorthosites.* Note that both highland and lowland rocks are igneous and so formed by the solidification of lava. They differ strikingly in composition, however. The anorthosites are low-density aluminum and calcium silicates; the mare basalts are high-density magnesium and iron silicates. The anorthosites show large crystals, which indicate that they cooled more slowly than the mare basalts. The igneous rocks are fragments of the lunar bedrock brought to the surface by cratering, and the breccias were produced when the regolith was compressed into agglomerates by meteorite impact. Most samples returned from the maria have been dated to 3.2 billion years, while some frag-

A

B

C

Figure 4–17

Lunar surface samples. (A) A mare basalt, showing the holes from gas bubbles in the lava. The pit in the center resulted from the impact of a small meteorite. (B) A highlands rock containing anorthosite (white areas). This sample from Apollo 15 is known as the Genesis Rock. (C) A breccia, in which large pieces of other rocks are imbedded in the darker material that had melted from an impact. *(NASA)*

ments in the highlands have ages as high as 4.6 billion years. Generally, highland rocks have ages of 3.8 to 4.0 billion years.

One important conclusion derived from these ages is that the major part of the evolution of the Moon's surface must have occurred within a very short time span—within the first billion years after the origin of the Moon itself. The mare basins must have been formed well after the highlands, and their flooding by lava came later still. The dating of maria material gives it an age of 3.2 billion years.

The chemical composition of the lunar surface is basically *basaltic*, with the following key features: (1) the maria are lavas of igneous (molten) origin; (2) the content of free iron is low, with the maria containing more iron, cobalt, and nickel than the highlands; (3) most of the volatile elements are underabundant, and some chemical differentiation has apparently taken place; (4) the lunar lavas have a higher content of iron and titanium bound into minerals than do terrestrial basalts; (5) the lunar materials are anhydrous; that is, they were not formed in the presence of water or water vapor. The areas sampled indicate that the maria are chemically homogeneous and differ somewhat from the highlands in that they have a considerably lower abundance of aluminum.

4–5 Atmospheres

(a) The Moon

No lunar atmosphere has ever been detected. Indeed, the Moon appears to have too little mass to undergo extensive internal heating, though rare instances of outgassing (like volcanic vapors) have been observed. Any atmosphere generated by the lava flows of the maria has long since escaped as a result of the Moon's low surface gravity and high surface temperatures. In addition, the solar wind is very efficient in sweeping away any trace gases which might emerge at the surface. Even the exhaust gases from the Apollo landers did not linger for long.

The following example demonstrates this point. Consider water dumped on the lunar surface by the Apollo astronauts. At lunar noon ($T \approx 400$ K), the root mean square speed of water molecules (Equation 2–7) is

$$V_{rms} = (3kT/m)^{1/2}$$
$$= [3(1.4 \times 10^{-23})(400)/(18)(1.7 \times 10^{-27})]^{1/2}$$
$$\approx 740 \text{ m/s}$$

so that $V_{esc}/V_{rms} \approx 3$, which means that the lifetime is only a few years as the molecules escape from the high-speed tail of the Maxwellian velocity distribution. (Recall that in Chapter 2 we used the criterion that $V_{esc}/V_{rms} \geq 10$.)

(b) The Earth

The Earth lost its original atmosphere (of H, He, and H compounds) but produced a *secondary* atmosphere as a result of vulcanism and outgassing. The large mass and gravitational attraction of the Earth permit it to retain this atmosphere, which was modified into the atmosphere we breathe today (Table 4–1).

Atmospheric Structure

The Earth's atmosphere is layered (Figure 4–18). Nearest the surface is the dense, well-mixed *troposphere*, where most weather takes place. As evidenced by snow-capped mountains, the temperature decreases steadily with height until we reach the *tropopause* at 15 km. Then the temperature rises slightly in the thin, stable *stratosphere*, which extends to 40 km into the *mesosphere*. A second temperature minimum occurs near 90 km (at about 190 K), and then the temperature increases steadily through the *thermosphere* (90 to 250 km) until it levels off near 1500 to 2000 K at the base of the *exosphere*. The exosphere marks the region where the atmosphere can escape into space.

Note that atmospheric density decreases rapidly in the troposphere and then more gradually at higher

Table 4–1 Chemical Composition of the Earth's Atmosphere Near Ground Level (by Volume)

Gas	Percentage
Nitrogen (N_2)	78.08
Oxygen (O_2)	20.95
Argon (Ar)	0.934
Carbon dioxide (CO_2)	0.033
Neon (Ne)	0.0018
Helium (He)	0.00052
Methane (CH_4)	0.00015
Krypton (Kr)	0.00011
Hydrogen (H_2), carbon monoxide (CO), xenon (Xe), ozone (O_3), radon (Rn), etc.	<0.0001
Water vapor (H_2O)	Variable (0 to 4)

altitudes, the type of decline typical of an exponential function. We can describe this characteristic as follows: note that the Earth's atmosphere does not suddenly rise up off the surface or collapse to the ground. The downward force of gravity on every layer is just balanced by the upward gas pressure—a condition called hydrostatic equilibrium. Consider a small slab of the atmosphere of thickness Δr and area A. The gravitational force pulling the slab down is

$$F_g = ma = \rho(r) \, A \, \Delta r \, (GM/r^2)$$

where $\rho(r)$ is the density of the air in the slab; note we calculate the mass of this air from its density times $A \, \Delta r$, the volume of the slab. The *net* upward pressure on the slab is $A \, \Delta P$, where ΔP is the pressure *difference* between the top and bottom of the slab. In equilibrium, these forces balance so that

$$A \, \Delta P = -\rho(r) \, A \, \Delta r \, (GM/r^2)$$

and

$$\Delta P/\Delta r = -\rho(r) \, (GM/r^2)$$

Take the limit as $\Delta r \rightarrow 0$; then

$$dP/dr = -\rho(r) \, (GM/r^2) \qquad \textbf{(4-3)}$$

is the *equation of hydrostatic equilibrium*. It applies to any situation of a pressure–gravity balance.

Assume that the atmosphere is an ideal gas, so that it obeys the equation of state

$$P = nkT$$

where $n = \rho/m$. Here n is the number density (#/m^3), ρ the mass density (kg/m^3), and m the *mean molecular weight* (in units of hydrogen masses) of the atmosphere gas. So

$$P = \rho kT/m$$

or

$$\rho = mP/kT$$

We substitute this for ρ in Equation 4–3:

$$dP/dr = -P(m/kT) \, (GM/r^2)$$

or

$$dP/P = -(m/kT) \, (GM/r^2) \, dr$$

Now recall that

$$g(r) = GM/r^2$$

where g is the acceleration from gravitation at dis-

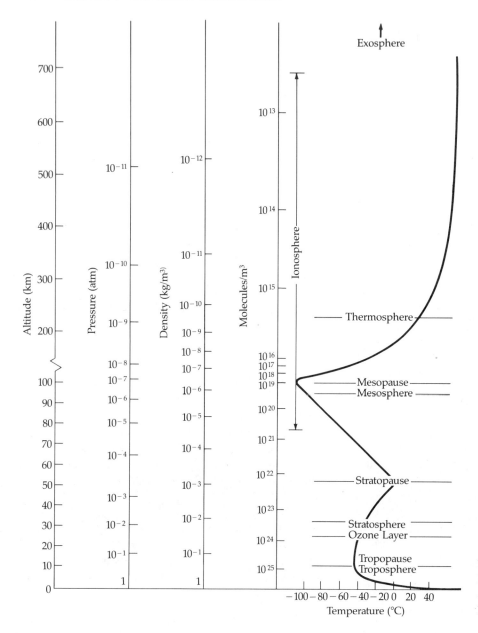

Figure 4–18

Pressure, density, and temperature change with altitude through each of the layers of the Earth's atmosphere.

tance r from the Earth's center and M is the Earth's mass. Then

$$dP/P = -g(r)(m/kT)\, dr$$

and, if we integrate this equation from some r_0 to r:

$$P(r)/P(r_0) = \exp[-g(m/kT)(r - r_0)]$$

where we have assumed that g, T, and m are roughly constant over the range $r - r_0$. Now define $r - r_0 = h$, the height above the surface (r_0), and

$$H = kT/gm \simeq \text{constant}$$

as the *scale height*. Then

$$P(h) = P(h_0)\exp(-h/H)$$

where h is any height above a reference level h_0. This relation is called the *barometric equation;* it applies to regions in planetary atmospheres where the temperature and mean molecular weight do *not* change rapidly. Note that H, the scale height, has the units of length; it is the distance to move up in the atmosphere for the pressure to decrease by $1/e$. At the Earth's surface, $H \approx 8$ km, and so the pressure (and density) at an 8-km elevation is roughly $1/e$, 2.7 times less than at the surface. The pressure at the Earth's surface is *1 atmosphere* (atm) and equals 1.01×10^5 N/m².

Whereas atmospheric density decreases steadily with height, the temperature profile exhibits two minima and three maxima. Solar ultraviolet rays and X-rays interact with the compositionally stratified atmosphere to dissociate oxygen and nitrogen molecules and to ionize some atoms to produce the stratified density of free electrons in the *ionosphere* (about 50 to 300 km). The ionization balance of the ionosphere is self-regulating since free electrons recombine faster with ions as more electrons and ions become available; hence, a steady-state equilibrium is set up. Concentrations of atomic oxygen (O), atomic nitrogen (N), ozone (O_3), and nitric oxide (NO, NO^+) are found at different levels of the ionosphere, and the entire system acts as a shield that absorbs most of the solar radiation dangerous to life.

Observed Effects of the Atmosphere

Electromagnetic radiation is absorbed, scattered, and refracted by the Earth's atmosphere, and all these phenomena depend upon the wavelength of the incoming radiation. The atmosphere is opaque to most wavelengths, transmitting (1) radio waves with wavelengths in the range 1 mm to 20 m (which are detected by radio telescopes) and (2) visible light from 290 nm (near ultraviolet) to about 1 μm (near infrared) — the "window" through which we see the Universe with our eyes and telescopes. Short-wavelength radiation (less than 290 nm) dissociates and ionizes the atmosphere and is absorbed in the process. For example, the ozone layer (10 to 40 km) absorbs radiation from 300 to 210 nm. In the infrared region (about 1 μm to 1 mm), radiation is absorbed when it excites molecules such as N_2, O_2, CO_2, and especially H_2O. At wavelengths longer than about 20 m, the ionosphere acts like a conducting shield that absorbs and reflects incoming radiation.

How the atmosphere affects visible light, from violet (410 nm) to red (650 nm), directly influences our view of the Universe. The most important effects are scattering, extinction, refraction (and seeing), and dispersion. Light is *scattered* when it interacts with a particle, and how it scatters depends upon its wavelength and on the size L of the scatterer. When light is scattered by atmospheric molecules ($L \ll \lambda$), the *in-tensity* of the scattered radiation obeys the *Rayleigh scattering law:*

$$I_{scat} \propto 1/\lambda^4$$

Hence, more blue light is scattered out of incident sunlight than red, and we see a blue sky. This preferential scattering causes the Sun (or any other star) to appear *reddened* when viewed through a considerable thickness of atmosphere, such as at sunset. The scattering depletes the blue light along the line of sight. Some radiation is scattered at every wavelength, even the red, so that a star's brightness is dimmed by the atmosphere — this we term *extinction* of the light. Astronomical observations must be corrected to account for reddening and extinction, although errors can be reduced by viewing near the zenith or from a mountaintop. When $L \approx \lambda$, such as for 1-μm dust particles scattering red light, the scattering law becomes

$$I_{scat} \propto 1/\lambda$$

Finally, when $L \gg \lambda$, as for water droplets in clouds, the scattering is essentially *independent of wavelength* — clouds appear white since they scatter sunlight this way.

When light passes from one medium to another, it is *refracted,* or bent. We characterize a medium by its *index of refraction.* The index of refraction of air increases with its density. Hence, starlight coming from a star at a true altitude θ is refracted downward toward the Earth, so that it appears to reach the surface at the altitude angle $\theta' > \theta$ (Figure 4–19). This effect, which vanishes at the zenith, becomes more important as we approach the horizon. At the horizon, a celestial object appears to be elevated about 35' above its true position, so that the setting Sun is actually *below* the horizon when we see its bottom resting on the horizon. Density inhomogenities in the atmosphere cause randomly fluctuating amounts of refraction for incoming light, so that stars appear to dance about (*twinkle*) in the sky. This effect limits the *angular resolution* of telescopes to about 1" (0.25" in exceptional cases), and we term this *astronomical seeing.* The seeing is good when the atmosphere is stable and the twinkling is small. Planets and satellites with angular diameters exceeding the seeing limit shine steadily since only their edges twinkle.

With all of these effects interfering with light from space (especially with the opaqueness of the Earth's atmosphere to most wavelengths of radia-

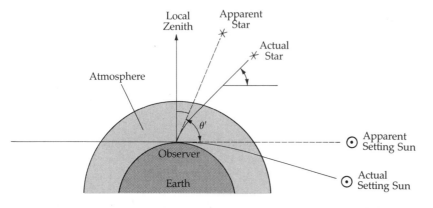

Figure 4–19

Atmospheric refraction. Light rays from celestial objects, such as the Sun or any other star, are bent by the atmosphere so that they appear higher than they actually are.

tion), you should not be surprised that observers locate their instruments on mountaintops or strive to view the Universe from orbiting satellites. The ideal observatory is clearly one located above the Earth's obscuring atmosphere.

4–6 Magnetic Fields

(a) Lunar Magnetism

Whereas the Earth's magnetic field, with a strength of about 0.4×10^{-4} T at the surface, has been studied for several hundred years, lunar magnetism could be investigated only with space probes to the Moon's surface. Magnetometers placed there by the Apollo astronauts reveal that the intrinsic lunar magnetic field is less than 10^{-9} T, but localized sources of magnetism exist on the Moon with strengths around 10^{-8} T. In any case, the Moon's *global* magnetism is negligibly small now. If the dynamo model [Section 4–6(b)] for the origin of a planet's magnetic field is correct, then the Moon's core cannot be molten or composed mainly of iron and nickel. (Also, the Moon's density is too low for it to have a metallic core.) On the other hand, some surface rock samples are magnetized much more than you would expect from such a weak magnetic field. Iron minerals in an igneous rock preserve the magnetic field present at the time of solidification. So in the past the Moon's magnetic field must have been stronger than it is now.

(b) The Earth's Magnetosphere

The Earth has a *dipolar* magnetic field whose axis is inclined 12° to the Earth's rotation axis, with field lines that emerge at the north magnetic pole and re-enter at the south magnetic pole. The close alignment of the magnetic and spin axes implies that the Earth's rotation helps produce the magnetism. Fluid motions in the metallic (electrically conducting) outer core are believed responsible for the magnetic field, through some type of *dynamo* (moving electric charges produce a magnetic field). Paleomagnetic investigations have shown that the field reverses its direction in a random way with an average period of about 10^4 to 10^5 years, and today we observe that the strength of the field is slowly decreasing.

Because the strength of a dipolar magnetic field drops roughly as $1/r^3$ with distance r from the Earth whereas the density of the atmosphere falls exponentially, the magnetic field is important far beyond the atmosphere. The domain of the Earth's magnetic field is the *magnetosphere,* which ends at the *magnetopause* boundary, where it meets and interacts with the solar wind (Figure 4–20). The magnetosphere has a cometlike structure caused by the solar wind's pushing the Earth's magnetic field away from the Sun. Within $8R_\oplus$ to $10R_\oplus$ of the Earth, the dipolar field is evident, and beyond there is an abrupt shock-and-transition region on the sunward side and the magnetic tail longer than $1000R_\oplus$ pointing away from the Sun.

The Earth's magnetosphere contains complex physical processes involving plasmas and electromagnetic fields, which we are just beginning to understand. Luckily, this magnetosphere is the closest one and has been closely studied since the beginning of space exploration. It serves as the model for other magnetospheres in the Solar System, which are found around other planets and with comets. The key point is this: the interaction of the solar wind and the magnetosphere creates a planetary electromagnetic generator that converts the kinetic energy of the solar wind into electricity at a rate of some 10^6 megawatts!

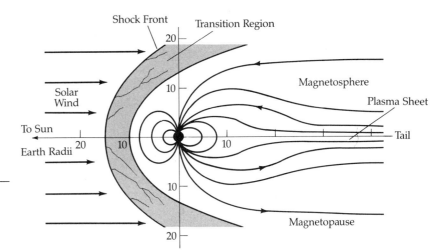

Figure 4-20

The Earth's magnetosphere. The solar wind encounters the Earth's dipolar magnetic field to create a shock wave behind which lies the magnetosphere. Units are in Earth radii.

(c) The Van Allen Radiation Belts

The magnetopause deflects the solar wind away from the Earth, but many protons and electrons leak into the magnetosphere. There they are trapped by the Earth's dipolar magnetic field in toroidal radiation belts, concentric to the magnetic axis (Figure 4-21). The discovery of these belts (*Van Allen radiation belts*) by a group headed by James A. Van Allen came from satellite observations in 1958.

Satellites carrying an assortment of particle detectors have now mapped the radiation belts in detail. Though both protons and electrons are found throughout the magnetosphere, there are two belts of particular concentration: (1) the small *inner belt* be-tween $1R_\oplus$ and $2R_\oplus$, where protons of energy 50 MeV and electrons with energies greater than 30 MeV tend to reside, then a distinct gap, and (2) the large *outer belt* from $3R_\oplus$ to $4R_\oplus$, where less-energetic protons and electrons are concentrated. (The *electron volt*, eV, is a convenient energy unit in atomic and particle physics; 1 eV $= 1.6021 \times 10^{-19}$ J, 1 keV $= 10^3$ eV, and 1 MeV $= 10^6$ eV. It is the energy gained by an electron accelerating between two plates separated by a distance of 1 m and a potential difference of 1 V.) The inner belt is relatively stable, but the outer belt varies in its number of particles by as much as a factor of 100. The particles trapped in the belts come from the solar wind and from cosmic ray interactions in the Earth's upper atmosphere.

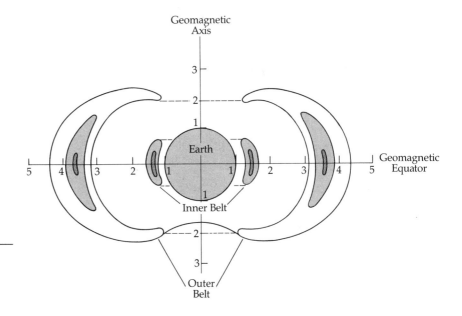

Figure 4-21

The Van Allen radiation belts. Dimensions are in Earth radii. The shaded contours represent the relative densities of particles.

The charged particles trapped in the radiation belts spiral along lines of magnetic force while bouncing (with periods from 0.1 to 3 s) between the northern and southern *mirror points* (Figure 4–22). Particles in the inner belt may interact with the thin upper atmosphere at these mirror points to produce the aurorae [Section 4–6(d)]; such particles are lost from the belts. In addition to spiraling and oscillating north–south, the particles drift in longitude owing to the decreasing strengths of both magnetic and gravitational fields with increasing distance from the Earth. High-energy protons drift westward around the Earth in about 0.1 s, and low-energy electrons drift eastward in about 1 to 10 h. This drift leads to the longitudinal uniformity of the radiation belts.

Let's investigate these charged particle motions quantitatively. The *Lorentz force law*

$$\mathbf{F} = q\,(\mathbf{v} \times \mathbf{B}) \qquad (4-4)$$

tells us the force **F** (in newtons) experienced by a particle of *charge q* (in coulombs, where an electron's charge is 1.60×10^{-19} C) moving with velocity **v** (in m/s) through a magnetic field of strength **B** (in units of teslas). Note that the magnetic field is represented by the *vector* **B** and that **F** is perpendicular to both **v** and **B**, in accordance with the right-hand rule (Figure 4–23A). Although the mass of the proton (1.673×10^{-27} kg) is much larger than the mass of the electron (9.109×10^{-31} kg), the charge of the proton is equal to the charge of the electron but of opposite sign. No work can be done on a charged particle by the magnetic field since $\mathbf{F} \cdot dr = (\mathbf{F} \cdot \mathbf{v})dt = 0$ (**F** is perpen-

dicular to **v**!); hence, the kinetic energy of a particle is not changed by a magnetic field (the total speed v remains constant). The particle moves in a circular orbit in a *uniform* magnetic field (Figure 4–23B). The Lorentz force (Equation 4–4) provides the centripetal force for the orbit of radius r, so that

$$mv^2/r = qvB$$
$$r = mv/qB \qquad (4-5)$$

For example, a proton with speed $v = 10^8$ m/s circles the magnetic field lines with a radius $r = 10$ km when $B = 10^{-4}$ T.

If the particle has some speed along a magnetic field line, it will follow a helical trajectory (Figure 4–23C). When the particle moves into a region of higher field strength (B increases), the circular orbit shrinks while the circular speed increases. Since the particle's total kinetic energy cannot change, its motion along the field line must slow down; eventually the forward motion reverses if one particle collides with another. Note that charged particles moving into a region of narrowing magnetic field lines are effectively accelerated to very high speeds. This process occurs in a number of astrophysical situations, such as aurorae, at the *mirror point.*

(d) The Aurorae

The inner radiation belt interacts with the Earth's upper atmosphere to produce the colorful *aurorae:* aurora borealis, or northern lights, in the high northern latitudes and aurora australis in the high southern

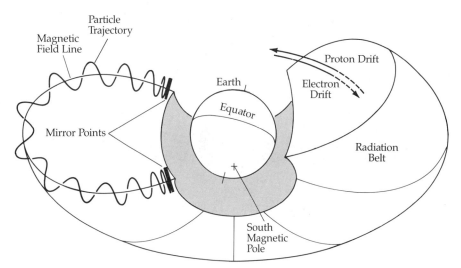

Figure 4–22

Particle motion in the Van Allen belts. Charged particles from the Sun spiral along the magnetic field lines and bounce between the northern and southern mirror points. Protons and electrons drift in opposite directions because of their opposite charges.

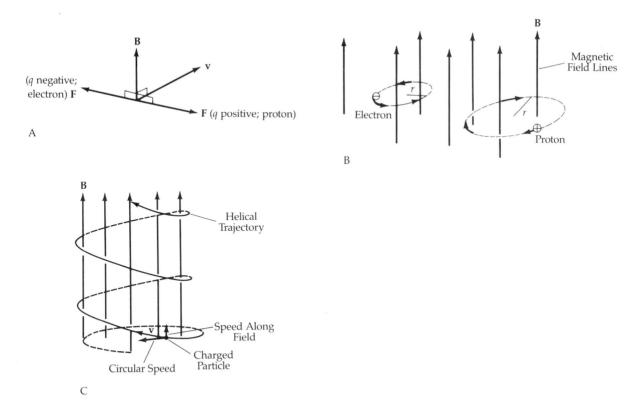

Figure 4–23

Motions of charged particles in a magnetic field. (A) Vector diagram for the direction of the Lorentz force. (B) Circular orbits of electrons and protons in a uniform magnetic field. (C) Helical orbit of a charged particle with some component of its velocity along uniform field lines.

latitudes. Diffuse luminous forms over large areas of the sky have been observed since ancient times in northern regions between 15 and 30° from the magnetic pole. Auroral displays may be flickering streaks or curtains of light, or steady lacework draperies that change intensity and color (pale pink, blue, and green) in the course of hours (Figure 4–24). Triangulation measurements of the heights of aurorae show that most occur between 80 and 160 km, with a few as high as 1000 km.

Auroral emissions result when low-energy electrons drop out of the inner radiation zone and collisionally excite and ionize atmospheric gases. As these excited gases, principally oxygen and nitrogen, return to their stable forms, visible light is emitted with characteristic colors. The solar wind plays an important role in aurorae, either by supplying the necessary electrons or by perturbing the magnetosphere so that the particles trapped in the radiation belts are

dumped into the atmosphere near the mirror points. That solar activity induces aurorae is inferred from the fact that it is so strongly correlated with the appearance of auroral displays.

Recent models picture aurorae as a result of the release of energy from the magnetotail by a process called *magnetic reconnection*. This takes place when regions of opposite magnetic fields come together, and the magnetic field lines can break and reconnect in new combinations. One site of such reconnections is the plasma sheet in the magnetotail; the point of reconnection usually lies about $100R_{\oplus}$ downstream from the Earth. The net magnetic field at this point is zero. Occasionally, the solar wind adds sufficient magnetic energy to the magnetosphere so that the field lines there overstretch and a new reconnection takes place only $15R_{\oplus}$ from the Earth. The field on the Earthside collapses, injecting electrons into the atmosphere to drive the auroral emissions. Meanwhile, a

Figure 4–24

The aurorae borealis. Note the drapelike structure in this time exposure. *(National Research Council of Canada)*

blob of plasma containing looping field lines is free of magnetic attachment from the Earth and moves downstream. The equilibrium magnetosphere is then re-established. The process of magnetic reconnection in a plasma applies to the environment of comets (Section 7–3) and other astrophysical situations.

4–7   The Evolution of the Earth–Moon System

We are now in a position to rely upon the physical properties of the Earth and Moon—especially radioactive dating of rocks—to infer their evolution. This comparison will highlight the Earth as the most evolved—and still evolving—terrestrial planet; the Moon, as a fossil world, preserves evidence of the early stages in terrestrial planetary evolution. (See Section 7–6 for a presentation about the origin of the Earth and Moon.)

(a) Lunar History

About 4.6 billion years ago, the Moon formed. Whether the Moon came together near the Earth or not, it probably formed by the accretion of chunks of material. These pieces continued to plunge into the Moon after most of its mass had gathered. During the first 200 million years after formation of the Moon, these projectiles from space bombarded the surface

and, along with radioactive decay, heated it enough to melt it. Less-dense materials floated to the surface of the melted shell; volatile materials were lost to space. The crust began to solidify from this melted shell about 4.4 billion years ago. From 4.1 to 4.4 billion years ago, the crust slowly cooled. The bombardment from space continued but began to taper off. This falling debris made many of the craters now found in the highland areas.

Below the surface, the Moon's material remained molten. About 3.9 to 4.1 billion years ago, a few huge chunks smashed the crust to produce basins which later became maria. For example, the Mare Orientale basin formed 4 billion years ago when an object about 25 km across smashed into the Moon. Only later did the basins fill with lava. Though the crust lost its original heat of accretion, short-lived radioactive elements that decay rapidly reheated sections of it. From 3.0 to 3.9 billion years ago, lava formed by the radioactive reheating punctured the thin crust beneath the basins, flowing into them to make the maria.

For the past 3 billion years, the crust has been inactive. However, small particles from space have incessantly plowed into the surface since it solidified. These sand-sized grains scoured the surface, smoothed it down, and pulverized it. Continued bombardment by larger bodies churned the fragmented surface. Impacts melted the soil, which swiftly cooled to form breccias and glass spheres.

Note that in this evolutionary history (Table 4–2), the highlands make up the oldest sections of the Moon's crust and the lowlands (maria) make up the youngest. Once the maria formed, the Moon's interior and crustal evolution halted because the interior lost much of its primodial heat in 1 billion years. This internal heat, from the original accretion and radioactive decay, drives the evolution of a terrestrial planet. In general, the rate of heat loss is proportional to a planet's surface area; the reservoir of heat is proportional to its mass. Hence the lifetime of a terrestrial planet's evolution will be roughly proportional to the total amount of energy available divided by the loss rate, or $t \propto R^3/R^2 \propto R$. So larger terrestrial planets should be the more evolved in the sense that the processes driven by internal heat will last a longer time. Also, the more massive planet will retain an atmosphere longer, resulting in wind and rain erosion of the surface. Let's examine the Earth with this point in mind.

(b) The Earth's History

Driven by its internal heat, evolutionary processes have moved the Earth through six stages of evolution. The first stage began 4.6 billion years ago, when the Earth accreted from small, rocky bodies in only a few million years. That left a pockmarked planet of more or less uniform composition since each piece was made of about the same combination of materials. The atmosphere at the time of accretion was rich in hydrogen and inert gases. Accretion and radioactive decay produced rapid internal heating. The low-mass gases escaped into space.

In the second stage, beginning about 4.5 billion years ago, radioactive heating (and perhaps gravitational contraction) melted the interior. It differentiated to form a core of dense materials and a crust of light ones. In the third and fourth stages, volcanic activity caused by interior heating created the second atmosphere, containing outgassed water, methane, ammonia, sulfur dioxide, and carbon dioxide. An infall of large objects continued, fracturing the crust. Ocean basins were formed, and the Earth's surface cooled enough for rain to fall and begin filling the basins.

In the fifth stage, about 3.7 billion years ago, the first continents appeared and plate tectonics began. Mountains grew, only to succumb to weathering by wind and rain. Slowly the atmosphere evolved, affected in part by the presence of life. Roughly 2.2 billion years ago, crustal cooling thickened the crust enough to allow plate activity as we see it today. By 600 million years ago, the planet had entered its sixth stage of evolution. The processes that had begun in the fifth stage continued at a lower rate until the Earth came to look much as it does today.

The evolution of terrestrial planets involves change to their interiors, surfaces, and atmospheres. These changes are driven mostly by the planets' internal heat: the more massive a planet, the more internal heat it generates (from radioactive decay), the longer it retains this heat, and the greater the degree to which it evolves.

Table 4–2 Evolution of the Moon: A Contemporary Model

Event	Time (billions of years ago)	Processes
Formation	4.6	Accretion of small chunks of material
Melted shell	4.6–4.4	Melting of outer layer by heat from infall of material and/or radioactive decay; volatile elements lost
Cratered highlands	4.4–4.1	Solidification of crust while debris still falls to form craters
Large basins	4.1–3.9	Reduced infall, but formation of basins by impact of a few large pieces; outflow of basalts from lava below solid crust
Maria flooding	3.9–3.0	Flooding of basins by lava produced by radioactive decay
Quiet crust	3.0–now	Bombardment by small particles to pulverize and erode surface

Surfaces are modified by several processes: impact of interplanetary debris, outflow of internal heat, volcanism, erosion by wind and water, and crustal movements (if the mantle is hot). The atmospheres change as a result of interaction with sunlight, escape into space, degassing (from a hot interior), and life (if it exists). In particular, life on Earth has transformed its second atmosphere (which came from outgassing) into the oxygen-rich atmosphere that we enjoy now.

Problems

1. (a) An Apollo spaceship heads for the Moon. At what point between the Earth and the Moon will the ship experience no net gravitational acceleration?
   (b) How long will it take the ship to circumnavigate the Moon in a circular orbit 50 km above the lunar surface?

2. Using the length of the sidereal month (27.322^d) and the periods of the Earth's revolution (365.26^d), the regression of the nodes (18.6 years), and the precession of the Moon's perigee (8.85 years), compute the lengths of
   (a) the synodic month
   (b) the nodical month
   (c) the anomalistic month

3. Assume that the Moon's orbit is circular and lies in the ecliptic plane. Find the difference in the solar attraction on the Moon's *orbit* at opposition and inferior conjunction, and compare this with the gravitational attraction from the Earth. Can you now understand why the lunar orbit is not a simple ellipse?

4. (a) If the Moon had a core of radius $R_m/10$ and the remainder of its interior had a uniform density $\rho = 3000$ kg/m^3, what must be the uniform density of this core?
   (b) Compare the Earth's tidal forces on the Moon at the perigee and apogee of the Moon's orbit; comment on your results.

5. The mean kinetic energy per molecule in a gas at temperature T is $mv^2/2 = 3kT/2$, where $k = 1.380 \times 10^{-23}$ J/K. A particle with a vertical speed v at the Earth's surface will rise to a height $h = v^2/2g$ before it falls back to the Earth.
   (a) Show that the *characteristic height* for a molecule of mass m at temperature T is $h = 3kT/2mg$.
   (b) At $T = 250$ K, compute the characteristic heights for nitrogen (N_2), oxygen (O_2), carbon dioxide (CO_2), and hydrogen (H_2). What does this tell you about the *compositional* structure of the Earth's atmosphere?
   (c) How does this calculation differ from that of the scale height done in the chapter?

6. (a) If a star emits the same intensity of radiation at all visible wavelengths, what will be its apparent color at the Earth's surface?
   (b) Explain why the Sun appears *flattened* (like an ellipse) at sunset.

7. Show that the circular period P (in seconds) for a charge in a uniform magnetic field B does *not* depend upon the radius of the orbit. Evaluate this period for a proton moving at speed $v = 10^7$ m/s in a magnetic field of 10^{-4} T.

8. Draw a diagram and use the Lorentz force law to explain why electrons in the radiation belts drift *eastward* owing to the decrease in the magnetic field strength with distance from the Earth.

9. Find the radius of the Earth's core relative to the total radius if the core density is 10,000 kg/m^3, the mantle density is 2500 kg/m^3, and the average density is 5500 kg/m^3.

10. In the Earth's exosphere, the temperature can reach 2000 K. Estimate the lifetime of water vapor here by comparing its mean velocity with the Earth's escape velocity.

11. The equation of hydrostatic equilibrium can be applied to planetary interiors because they do not expand or contract. Use this equation with suitable approximations to estimate the central pressures of the Earth and Moon. *Hints:* Use Equation 4–3 with $dP \approx \Delta P = P_c - P_s$, where P_c is the central pressure and P_s the surface pressure. Let $P_s \approx 0$. For $\rho(r)$, use the average density, and for r, the radius of the planet.

12. (a) Use the strength of the magnetic field at the Earth's surface to estimate the strength in the Van Allen belts.
    (b) Use this estimate to calculate the radius of curvature of a 50-MeV proton in the belts.

13. Demonstrate that a dipolar field decreasing as $1/r^3$ beats out an atmospheric density falloff that is exponential.

The Terrestrial Planets: Mercury, Venus, and Mars

We have so far presented the general properties of the Solar System (Chapters 1 and 2) and the basics of two special planets, the Earth and Moon (Chapters 3 and 4). We will use these two familiar worlds as the basis for the *comparative planetology* of the other terrestrial planets: Mercury, Venus, and Mars. We will try to infer the evolutionary histories of these bodies by drawing analogies to the Earth and Moon.

5–1 Modern Planetology

Because the human eye alone cannot discern details smaller than 1' of arc, it took the invention of the telescope in the seventeenth century to spark the investigation of the planets — *planetology*. In 1610, Galileo described telescopic observations of the four largest (Galilean) moons of Jupiter, the phases of Venus, the cratered surface of our Moon, the dark spots on the Sun, and the strange nonspherical shape of Saturn (its rings). The Earth's atmospheric seeing limits the visible resolution of telescopes to about 1" of arc, however, so that the surface markings and atmospheres of only those objects of much greater angular size can be studied in any detail. The maximum angular diameter of Mercury seen from the Earth is only 12", and that of Uranus is about 4"; in such cases, optical viewing is marginal at best.

New technologies and techniques characterize modern planetology. The optical window has been expanded to the near ultraviolet (photography) and near infrared (solid-state detectors). Radio telescopes and radar now peer through the radio window. High-altitude balloons, rockets, artificial satellites, and space probes rise above the Earth's atmosphere to give a broader and clearer view of the Solar System. Spacecraft sent to the Moon and planets have dramatically improved our picture of these bodies. The main point here is that a flyby spacecraft improves our resolving power enormously — by a factor of 10^6 in the case of the Voyager missions to Jupiter and Saturn.

5–2 Mercury

Mercury is the planet closest to the Sun and the smallest of the terrestrial planets. Each year, it appears about three times as a bright evening star near the sunset horizon and as a predawn morning star. Because of the swiftness of its celestial motion, it is named after the mythological god of flight. At times, Mercury rivals Venus in brightness, but it is usually obscured by the brilliance of the nearby Sun.

(a) Motions

Mercury revolves about the Sun in a highly eccentric ($e = 0.2056$) and inclined (7.00° to the ecliptic plane) orbit of semimajor axis 0.3871 AU, with a sidereal orbital period of 87.96^d. The greatest elongation of this planet viewed from the Earth ranges from 18 (perihelion) to 28° (aphelion); it averages 23°.

The sidereal rotation period of Mercury was thought to be either 24^h (Earthlike) or the *synchronous* value of 88^d. However, in the early 1960s, radar pulses reflected from the surface of Mercury were first detected, and by 1965 G. H. Pettengill and R. B. Dyce had shown the rotation period to be about 59^d direct, using Doppler radar techniques. A re-analysis of the old data and new visual data obtained from the Pic-du-Midi Observatory in France verified the radar period and resulted in a measured period of 58.65 ± 0.01^d, with the planet's equator essentially parallel to its orbital plane. Mercury is in a spin-orbit resonance with the Sun [Section 2–1(a); Figure 2–5]; its sidereal rotation period is 58.64^d, which is two-thirds of its sidereal orbital period. The earlier periods were probably in error because surface features on Mercury can be observed visually only near greatest elongation, which occurs at the *synodic* orbital period of 115.88^d — essentially twice the sidereal rotation period.

Radar astronomy permits us to deduce a planet's distance, size, rotation, and large-scale surface features by analyzing radar pulses (wavelengths ≈ 1 to 10 cm) reflected from the planetary surface. Whereas lunar surface features, such as craters and mountains, are easily resolved, a radar pulse directed at a terrestrial planet is usually wider in angle than the planet itself. Hence, we must resolve the return pulse into its *time-delayed* and *Doppler-shifted* components (Figure 5–1). The shortest time delay occurs at the sub-Earth point on the planet and tells us the planet's distance from the Earth. Successively longer time delays originate from rings concentric about the sub-Earth point at constant distances, with the longest delay defining the planet's *limb* (edge) and therefore its radius. As the planet rotates, the reflected radar waves are Doppler-shifted in frequency (Chapter 8), so that

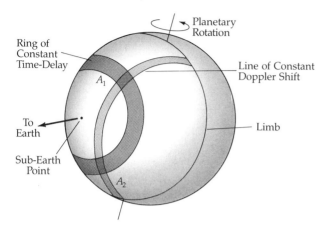

Figure 5-1

Radar mapping. Imagine a radar pulse reflecting from a planet's hemisphere facing the Earth. Different time delays form a series of concentric rings. They intersect lines of constant Doppler shift at two points (A_1 and A_2), so that a feature at either A_1 or A_2 has an ambiguity in position.

higher frequencies than the original return from the approaching limb and lower frequencies from the receding limb; these Doppler shifts reveal the rotation speed of the planet. The shifts are maximum at the limbs and zero at the sub-Earth point. Radar mapping of the planetary surface is limited only by uncertainties in signal strength, time delay, and frequency shift; in addition, a twofold positional ambiguity (Figure 5-1) exists, but this may be resolved as the Earth–planet direction changes.

We can see the relationship between Doppler shift and rotation period as follows. The shift in wavelength, the difference between that sent out and that received, is

$$\Delta\lambda/\lambda_0 = v/c$$

where λ_0 is the emitted wavelength, v the radial velocity, and c the speed of light. For a spherical planet, the maximum linear rotation rate occurs at the equator:

$$v_{eq} = 2\pi R/P$$

Because each limb contributes to the total shift,

$$\Delta\lambda/\lambda_0 = 2v_{eq}/c$$
$$= 4\pi R/Pc$$

from which we calculate the rotation rate from the planet's radius and the measured shift.

How long is the solar day on Mercury? We can derive a relationship between the solar day and the sidereal day in a manner very similar to the derivation of the relationship between the synodic and sidereal orbital periods of the planets. Suppose you were on Mercury with the sun directly overhead (Figure 5-2). Let T be the period of revolution of Mercury about the Sun (88 days), P the sidereal period of rotation (58.7 days), and S the synodic period of rotation (the solar day). After one Earth-day, Mercury will have rotated $360°/P$ (angle A) with respect to the stars but only $360°/S$ (angle B) with respect to the Sun. The difference between these two angles (angle C) is equal to the angular distance Mercury has moved in its orbit, $360°/T$ (angle D). So

$$360°/P - 360°/S = 360°/T$$

or

$$1/S = 1/P - 1/T$$

Putting in the numbers,

$$1/S = 1/58.7 - 1/88.0 = 0.00567$$
$$S = 176 \text{ days}$$

Mercury's solar day is just twice the length of its year.

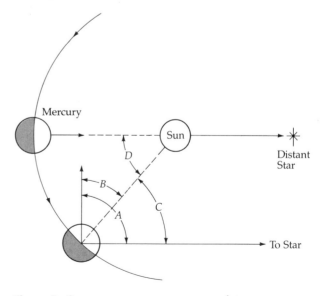

Figure 5-2

Geometry for the solar day on Mercury.

(b) Physical Characteristics

Mercury's radius is 2440 km. Its mass of $0.055 M_\oplus$ (3.3×10^{23} kg) is deduced from gravitational perturbations upon spacecraft (Mercury has no natural satellites). The average density is 5420 kg/m³, typical of a terrestrial planet but high for one of Mercury's size. Because Mercury's overall gravity is less than the Earth's (compressing it less) but its bulk density is about the same, it must contain a greater proportion of metals. We infer that Mercury's interior (Figure 5–3) has a rocky mantle and a large metallic (perhaps nickel–iron) core.

The surface albedo is very low (0.056 at visible wavelengths), indicative of rocky material even darker than the Moon's surface. Surface temperatures vary from about 700 K at the perihelion subsolar point to about 100 K on the dark side. The long, hot solar day and low escape velocity (4.2 km/s) make it unlikely that Mercury has much of an atmosphere. Gas molecules, even the more massive ones, would easily reach escape velocity, and so any atmosphere could not be expected to last long. The Mariner 10 space probe had on board an ultraviolet spectrometer to search for an atmosphere. This device detected a thin atmosphere of helium and hydrogen, but the surface atmospheric pressure was very small,

a little less than 10^{-15} atm. Here is the atmospheric escape calculation. For helium at noon,

$$
\begin{aligned}
V_{\text{rms}} &= (3kT/m)^{1/2} \\
&= [3(1.4 \times 10^{-23})(700)/(4)(1.7 \times 10^{-27})]^{1/2} \\
&= 2.1 \text{ km/s}
\end{aligned}
$$

so that

$$
\begin{aligned}
V_{\text{rms}}/V_{\text{esc}} &= 2.1/4.2 \\
&= 0.5
\end{aligned}
$$

and the lifetime is but a few hours.

No significant atmosphere means no insulation from space. That's why the range of noon-to-midnight temperatures on Mercury is so severe. Let's contrast them to the Moon's. Night on the Moon and Mercury is essentially the same: no insulating blanket, and so infrared radiation from the sunless side escapes directly into space during the long night. Thus both bodies have about the same midnight temperature: approximately 100 K. At noon, the surface temperature then depends on the surface albedo and distance from the Sun. The Moon and Mercury have about the same albedo, but Mercury is 2.5 times closer to the Sun than is the Moon, and so it is much hotter at noon because the Sun's flux is 6.25 times greater.

With the day so hot and the escape velocity so low, why does Mercury have any atmosphere at all? One answer is the solar wind. The influx of material from the solar wind, which contains some 10 percent helium, could possibly replenish the loss. Another possible source, at least for helium, is decay of radioactive elements in Mercury's interior.

(c) Surface Features

Television cameras on board Mariner 10 scanned 50 percent of Mercury's surface to increase our resolution of its details by 5000 times. The photographs revealed a surface like our Moon's. Key differences are fewer mid-sized craters; no mountain ranges; many shallow, scalloped cliffs, called *scarps*; fewer basins and large lava flows; and more relatively uncratered plains amid the heavily cratered regions.

These differences are important, and yet Mercury's surface clearly resembles the farside of the Moon. Mercury's highlands (Figure 5–4) are riddled with craters like the Moon's bleak highlands. Light-colored rays spring from some of the craters, an indication that these were formed by violent impacts

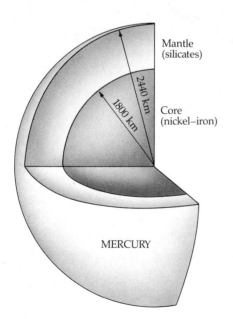

Figure 5–3

Model of Mercury's interior. Note the large fraction of the interior taken up by the metallic core.

Figure 5-4

Mercury's cratered highlands. Note the bright, rayed crater (lower left) and the double-ringed crater above it. The outer ring is about 170 km across. *(NASA)*

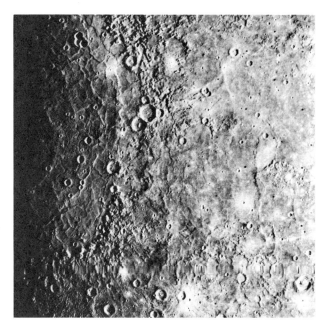

Figure 5-5

Caloris Basin. One half of this 1300-km-diameter, ringed basin is visible at the terminator, the line between night and day on the left. Wrinkled ridges and flooded lowlands are visible to the left. *(NASA)*

during Mercury's stormy past. Some craters are more than 200 km in diameter, comparable to the largest lunar craters. Mariner 10 found but one mare basin: the Caloris ("hot") Basin on the northwest part of the planet (Figure 5-5). It probably has an overall diameter of 1300 km. The basin is bounded by rings of mountains about 2 km high. In size and structure, the Caloris Basin strongly resembles the Moon's Orientale Basin.

All these similarities do not mean that the Moon and Mercury are identical. Their surfaces differ in at least three ways: (1) Mercury's surface has scarps hundreds of kilometers long, (2) even the most heavily cratered regions of Mercury are not completely saturated with craters but are interspersed with intercrater plains, and (3) Mercury has fewer small craters than the number expected if the craters had formed at the same rate as on the Moon. From these last two differences, we can infer either that the cratering material had a different range of sizes for Mercury than for the Moon or that some process modified the surface after the cratering.

Mercury's scarps (Figure 5-6) vary in length from 20 to 500 km and have heights from a few hundred meters up to 2 km. Individual scarps often traverse different types of terrain. If the regions photographed by Mariner 10 represent Mercury's overall surface, the characteristics of the scarps imply that Mercury's radius has shrunk by 1 to 3 km, from cooling of either its core or its crust, or both, thereby forming the thrust faults that created the scarps.

(d) Magnetic Field

Mariner 10 detected a weak planetary magnetic field with a strength of about 220 nT(1 nT = 10^{-9}T). Although small, this is sufficient to carve out a magnetosphere in the solar wind. Here the magnetic field deflects the charged particles (mostly protons) of the solar wind around the planet.

Mercury's field appears to be a dipole, more or

Scarp

Figure 5–6

A scarp in Mercury. This ridge (arrow) extends for hundreds of kilometers. (*NASA*)

less aligned with the planet's spin axis. In general, then, Mercury's magnetic field is similar to the Earth's, only weaker. The presence of a magnetic field as well as the planet's high density indicates that Mercury, like the Earth, has a metallic, largely iron–nickel core. This core is presumed to be relatively cold and solid now because a small planet loses heat quickly. Recall that the Earth's magnetic field supposedly arises from an internal dynamo whose churning motions are thought to be driven by the Earth's spinning. Because Mercury rotates much more slowly than the Earth, a planet-wide magnetic field was not really expected and its source is not clear. Perhaps Mercury's core is still partially molten.

(e) Evolution of the Surface

Because both the Moon and Mercury lack substantial atmospheres, weathering does not erode their surfaces. Both are tiny worlds with interiors cooler than the Earth's. Neither has much (if any) volcanic activity now, and neither has undergone the continual surface evolution the Earth experiences from the shifting of crustal plates.

The lack of atmospheres and the short time of crustal evolution are both related to the small masses of Mercury and the Moon. Their surface gravities are so low that most gases achieve escape velocities, and atmospheres are not retained for long. The small masses also imply that internal heating from radioactive decay would be less than that for the Earth and the flow of heat outward would be so fast that both bodies would cool off quickly. The Earth is hot in its interior, and the outward flow of heat sets up currents in the plastic mantle; these power the evolution of the Earth's crust. Both Mercury and the Moon lack this combination of hot interior and plastic mantle.

Using lunar analogies, we can set up a working model for the evolution of Mercury, with the following general stages: (1) heating of the surface (by impacts and/or radioactive decay) and formation of a solid crust, (2) heavy cratering, (3) formation of impact basins, (4) filling in of basins by lava flows, and (5) low-intensity cratering.

A lunar comparison suggests that the intense sculpting of Mercury's surface took place about 4 billion years ago, not long after the planet formed (stage 1). The earliest phase of cratering (stage 2) was wiped out by later volcanism. We come to this conclusion because the intercrater plains seem to have covered up any scars of the earliest accretion. At about the same time, the scarps developed from global shrinking of the crust, the interior, or both.

Basin formation (stage 3) must have come at the end of the heavy bombardment of the surface. A few large pieces crashed into the surface; one made the Caloris Basin. Not long after this, widespread lava flows (similar to those which made the lunar maria) created the broad, smooth plains, such as those adjacent to the Caloris Basin (stage 4). Since then, a few impacts punched out the fresh rayed craters (stage 5).

5–3 Venus

Venus is the second terrestrial planet from the Sun and the closest planet to the Earth. As a morning and evening star, Venus rises to a greatest elongation of 48°. Its maximum brilliance is exceeded only by those of the Sun and Moon. Named for the goddess of love, Venus closely resembles the Earth in size and mass but otherwise differs enormously.

(a) Motions

The nearly circular ($e = 0.0068$) orbit of Venus is inclined 3.39° to the ecliptic, with a semimajor axis of 0.7233 AU and a sidereal orbital period of 224.70 days. At the orbit of Venus, Mercury and the Earth are spectacularly brilliant planets.

Doppler-shift radar studies show that the planet rotates retrograde with a sidereal period of 243.01 days and with its equator inclined only 3° to its orbital plane (conventionally this inclination is written as $+177°$ or $-87°$ to indicate the retrograde behavior).

(b) Physical Characteristics

We can measure the Earth–Venus distance directly by radar; then from the planet's angular diameter, we can calculate its physical radius. Venus turns out to have a radius of 6052 km, only 5 percent smaller than the Earth's.

Like Mercury, Venus has no known natural moon, and so we can accurately determine its mass only when a spacecraft passes or orbits it. If the craft is in orbit, we simply use Kepler's third law. During a

flyby, we measure the acceleration of the spacecraft (from the Doppler shift of its radio signals) and use Newton's law of gravitation. Venus' mass comes out to be 4.86×10^{24} kg (82 percent of the Earth's mass). The resulting bulk density is 5200 kg/m³, almost the same as the Earth's. We infer that the interior of Venus closely resembles the Earth's interior (Figure 5–7): a rocky crust (which Venus landers confirmed), a mantle, and a metallic core. Because Venus has a lower density than the Earth, we imagine that it has a somewhat smaller core.

Interplanetary probes launched by both the United States and the Soviet Union indicate that the atmosphere of Venus contains about 96 percent carbon dioxide, 3 percent nitrogen, and traces of water vapor (0.1–0.4 percent), oxygen, hydrogen chloride, argon, hydrogen fluoride, hydrogen sulfide, sulfur dioxide, helium, and carbon monoxide.

The Russian Venera landers found the surface atmospheric pressure to be 95 atm and the sunlit surface temperature about 740 K. This high temperature probably results from the effective trapping of surface heat because carbon dioxide absorbs infrared radiation well (the so-called greenhouse effect). Atmospheric winds on Venus blow from the day to the night side and from the equator to the polar regions; the wind flow carries heat. Along with the very effective greenhouse effect, this atmospheric transport of heat keeps the surface temperatures fairly constant over Venus' surface. They vary 10 K or less.

The optical brightness of Venus is a result of the high albedo (0.76) of its clouds and thick atmosphere. The structure and motion of the clouds can be seen in ultraviolet photographs (Figure 5–8). The Pioneer

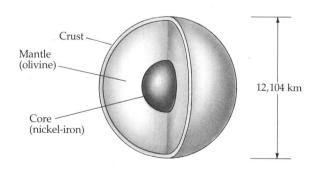

Figure 5–7

Interior of Venus. This model is based on the planet's bulk density and on the assumption that its interior is similar to that of the Earth.

Ultraviolet photography by Mariner 10 revealed that these cloud tops flow with the upper atmosphere (Figure 5–8) in patterns similar to the jet streams of the Earth. Ringing the equator, the clouds whiz around at roughly 300 km/h—fast enough to circle the planet in only four days. In addition to this planet-wide circulation, winds also blow from the equator to the poles in large cyclones 100 to 500 km in diameter. They culminate in two gigantic cloud vortices capping the polar regions.

The circulation pattern of Venus' atmosphere is in large part explained in terms of Hadley cells. Air rises over warm regions—the equator—and flows on top of cooler air at higher latitudes. The cooler air then replaces the rising warm air at the surface. On Earth, instabilities caused by the rapid rotation break up the Hadley circulation into three cells in each hemisphere. On Venus a single-cell model is a rea-

Figure 5–8

Venus atmospheric features. The "Y" feature, which is a combination of recurring clouds, has changed in these four photographs, taken one day apart. Note the cap of clouds at each pole. *(NASA)*

Venus probes found that clouds float in two broad layers (Figure 5–9). The upper cloud deck tops are at roughly 65 km; it has a thickness of some 5 km. The liquid droplets in the clouds have diameters of about 1 μm. Below the upper deck floats a thin haze layer. Below that, at 50 km, is the lower cloud deck, by far the densest layer. The cloud particles here are both liquid and solid, some 10 μm in diameter. Below an altitude of 50 km, the clouds gradually thin out; below 33 km, the atmosphere is clear of any particles down to the surface. Models of the clouds designed to match their observed infrared spectra imply that they contain a solution of 90 percent sulfuric acid mixed with water. Although the atmosphere and clouds of Venus do contain some water vapor, it would amount to a layer only 30 cm thick if it all condensed out uniformly on the surface.

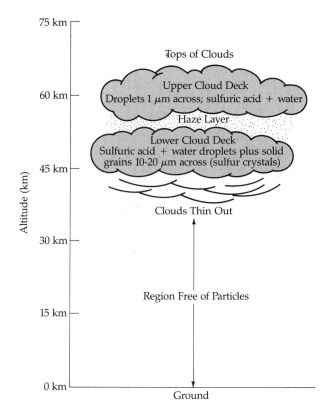

Figure 5–9

Structure of Venus' clouds, which come in two high layers with haze between them. Below the clouds, the atmosphere is clear.

sonable approximation in the lower atmosphere. The rising warm air can be understood in terms of the dependence of scale height H on temperature [Section 4–5(b)].

(c) Surface Features

We probe the surface of Venus by sending landers to photograph it or by bouncing off radar signals to map the terrain. This recent work has revealed high plateaus, gigantic volcanoes, impact craters, and long rift valleys. Overall, Venus looks fairly flat. Elevation differences are small, only 2 to 3 km, with the exception of a few highland regions. Here the land rises up to 12 km, compared with 4-km highland–lowland differences on the Moon and Mercury, 25-km differences on Mars, and 9 to 20-km differences on the Earth (Mt. Everest is 8.85 km high, and the Marianas trench is 11.03 km deep).

The southern and northern halves of the mapped face of Venus differ remarkably from each other. The northern region is mountainous with uncratered *upland plateaus*. In contrast, the southern part consists of relatively flat *cratered terrain*. The great (1000 by 1500 km) northern plateau is called Ishtar Terra. It is larger than the biggest upland plateau on the Earth, the Himalayan Plateau. Three mountain ranges border Ishtar on the west, north, and east. The eastern range, called Maxwell Montes (Figure 5–10), contains the highest elevations on Venus seen to date: 12 km. Radar shows a rough terrain, as expected from a lava flow, and a volcanic cone near Maxwell's center. The northern mountain range rises about 3 km above Ishtar; the western one reaches only 2 km above the plateau. These three mountain ranges may have folded and risen from moving plates in Venus' crust and so may be similar to mountains built from plate tectonics on the Earth.

The southern half of Venus' face consists of low, rolling plains punctuated by craters (Figure 5–11), both large craters (up to 800 km in diameter) and smaller ones (less than 1 km across). Craters smaller than this do not form on Venus because the dense atmosphere completely burns up incoming objects before they reach the ground. The craters of Venus resemble those of the Moon, Mercury, and Mars, and so they most likely have an impact origin. Venera 8 landed in the midst of the southern terrain. Its instruments indicated that the rock was more granitic than basaltic—one clue that at least some of the surface

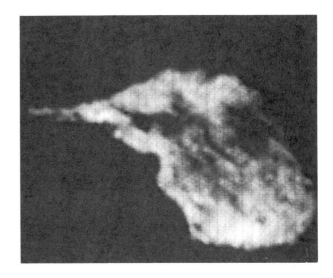

Figure 5–10

Radar map of Maxwell Montes. This high mountain appears to be volcanic; note the cone at the top and the irregular surface of the sides. *(Arecibo Observatory)*

here is old. Later lava flows could have poured out basaltic rock. Also, the craters here imply the surface must be old; otherwise volcanoes and mountain building would have obliterated the craters.

Beta Regio, which seems to contain at least two separate volcanoes, appears to be an enormous volcanic complex that formed from a great north–south fracture zone (Figure 5–12). The volcanoes here have gentle slopes; they are called *shield volcanoes*. (Mauna Loa and Mauna Kea in Hawaii are shield volcanoes, which are relatively flat. Often shield volcanoes have a collapsed central crater—a caldera—at their summits.) One volcano in Beta Regio has a diameter of 820 km, a height of 5 km, and a summit crater 60 by 90 km. In contrast, the island of Hawaii (a volcanic island) is 200 km across and 9 km high. (The largest volcano on Mars, Olympus Mons, is 550 km in diameter and 20 km high.)

There are hints that some of the volcanoes on Venus may be active now. The evidence, however, is very indirect. First, the levels of sulfur dioxide (SO_2) detected by the Pioneer Venus ultraviolet spectrometer have shown a steady decline with occasional bursts to higher amounts. Terrestrial volcanoes spew out sulfur dioxide (as well as hydrogen chloride and hydrogen fluoride, which are minor constituents of

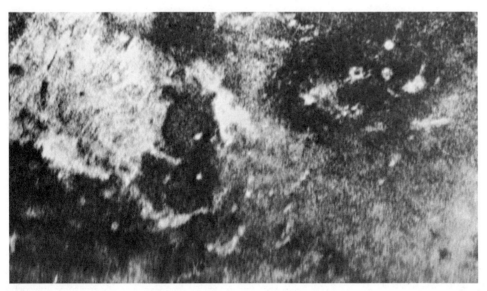

Figure 5–11

Surface of Venus. This radar map shows a part of the southern hemisphere. Note the craters in the upper right and the ripple structure at the upper left. *(Arecibo Observatory)*

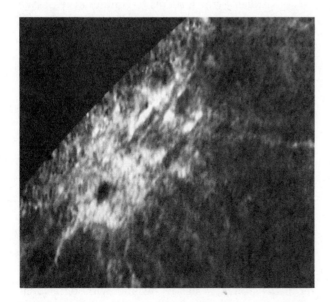

Figure 5–12

Beta Regio. This radar map has a resolution of 10 km and shows two large shield volcanoes in the upper right with flows radiating from them. *(Arecibo Observatory)*

Venus' atmosphere). One interpretation of these results is the episodic injection of sulfur dioxide by episodic volcanic eruptions. The haze above the clouds may also be supplied by volcanic injection, at levels equivalent to the largest such eruptions on the Earth (Krakatoa in 1883, for instance). Second, Pioneer Venus also recorded low-frequency radio bursts believed to be emitted by lightning strokes. On Earth, lightning discharges often flicker through the plumes of erupting volcanoes. The Venus bursts were clustered over three regions thought to be volcanic, including Beta Regio.

If the size and mass of a planet are known, geologists are able to relate the height of a volcano to the thickness of the crust below it. Basically, a larger volcano needs a thicker crust to support it. On Earth, the 9-km-high Hawaiian volcano Mauna Loa has a crust below it 57 km thick. Because the Earth and Venus are so similar in mass and size, the 10-km volcano on Venus should have an underlying crust of about the same thickness, 60 to 65 km.

On the side of Venus facing the Earth at inferior conjunctions, a huge canyon extends for more than 1300 km. It is about 150 km wide and 2 km deep. On the other side of Venus, an even larger canyon scars the landscape: it is 5 km deep, 320 km wide, and at least 1400 km long. These canyons appear to be rift valleys, which form along fault zones rather than by water erosion. Rift valleys appear on the Earth where the crust is spreading apart.

Overall, the surface of Venus is much flatter than the Earth's: only 18 percent of the mapped surface extends above 7 km, 11 percent above 10 km. In contrast, about 30 percent of the Earth's surface reaches above 10 km. Venus does not appear to have lunar-type basins, lowlands filled by lava flows. As indicated by the presence of craters, the lowlands of Venus must be older than the rest of the crust. The highlands are more evolved. That's just the opposite of what we find on Earth, where the lowlands (the ocean basins) make up the youngest part of the crust.

The Soviet Union has sent four spacecraft (Veneras 9, 10, 13, and 14) to the planet's surface. The pictures from Veneras 9 and 10 show slabby rocks about 60 cm long and 20 cm wide. A few rocks are vesicular, which implies a volcanic origin. Some rocks show jagged edges, indicating little erosion, and others show blunted, rounded edges, indicating much erosion. The rocks rest on loose, coarse-grained dirt. Lander measurements of gamma rays emitted by the radioactive potassium, uranium, and thorium in the rocks indicate that at one lander site they are basaltic, like those lining the Earth's ocean basins, but at the other site they are granitic, like the Earth's mountains. Both are igneous rocks formed from lava.

The photographs from Veneras 13 and 14 show quite different views (Figure 5–13). Those from Venera 13 reveal a rocky plain with clustered outcroppings; the rocks are basaltic, similar to those on Earth around continents. In contrast, those from Venera 14 show even layers of broken, rocky plates that are basically volcanic, like the basalts in the Earth's ocean floor near mid-oceanic ridges. So in both places (separated by 900 km), the rocks appear to be relatively young.

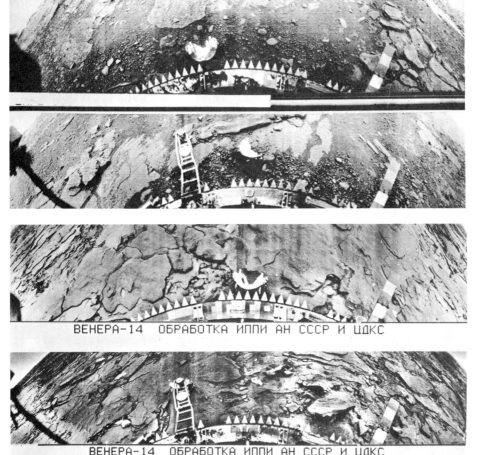

ВЕНЕРА-14 ОБРАБОТКА ИППИ АН СССР И ЦДКС

ВЕНЕРА-14 ОБРАБОТКА ИППИ АН СССР И ЦДКС

Figure 5–13

Closeups of Venus' surface. The upper two frames were taken by the Venera 13 lander. Note the slabs of rock, patches of dark soil, and pebbles close to the edge of the lander. The view is distorted by a wide-angle lens. The lower two frames were taken by Venera 14. Here the flat expanses of rock are unbroken by soil or pebbles.

(d) Magnetic Field

A nickel–iron core, liquid in part, implies by comparison with the Earth that Venus should have a planetary magnetic field. Because Venus rotates 243 times more slowly than the Earth, we expect its internal dynamo to be weaker and so the magnetic field to be less intense than the Earth's but still there. No probe to date has detected any magnetic field. If one exists, measurements imply it must be at least 10^{-4} times the Earth's magnetic field! That's much weaker than expected from a simple dynamo model. One possible explanation: we know that the Earth's magnetic field reverses its polarity; in the middle of a reversal, the magnetic field is essentially zero. That may be the situation with Venus now. (Recent reversals on the Earth have occurred roughly every million years or so.)

(e) Evolution of the Surface

The crust of Venus has experienced some crustal plate movement, as evidenced by the presence of rift valleys, where a few plates have pulled apart, and by mountain plateaus, where plates have collided. The widespread cratered terrain indicates that plate movement has not been a planet-wide process, however, as it has been on the Earth.

What do these observations imply about the geologic history of Venus? Its early history (earlier than 4 billion years ago) must have followed the Earth's early history because the two planets have similar masses, densities, and sizes. We infer that Venus formed about 4.6 billion years ago with the other terrestrial planets. As happened to the Earth, Venus's interior differentiated from internal heating. During the first 500 million years, a crust—part basalt and part granite—formed and solidified. About 4 to 3 billion years ago, large masses bombarded the surface and fractured the crust. Volcanoes erupted. Bombardment by smaller bodies from space cratered the surface. About 4 billion years ago, this intensive bombardment ended, and erosion has somewhat altered the ancient surface. Since then, plate movement helped to push up some highland regions. Rift valleys appeared. Huge volcanoes vented through cracks in the surface; their cones formed the shield volcanoes of today.

Venus seems to have evolved in a sequence similar to that of the Earth but more slowly and not as much in the later stages, when plate tectonics have dramatically altered the Earth's surface.

5–4 Mars

Mars shows a ruddy red-orange color, which links it to the Greek god of war. This fourth planet from the Sun runs the Earth a close race in its orbit, so that it exhibits the most pronounced retrograde motion at opposition. Opposition occurs with a synodic period of 779.9^d, or about 26 months, and at this time the planet has an apparent angular size of 18″ and may be viewed at full phase throughout the entire night.

(a) Motions

Mars' orbit of semimajor axis 1.5237 AU is inclined only 1.85° to the ecliptic. The orbital eccentricity of $e = 0.0934$ implies that at opposition the Earth–Mars distance may range from about 10^8 km (at which time Mars shows an angular diameter of 14″) to 5.5×10^7 km (25″). The thin atmosphere of Mars permits well-defined surface markings to be followed clearly; the Martian sidereal rotation period is $24^h37^m22.6^s$, and the rotation axis is inclined 25°12′ to the orbital plane. This near coincidence with the Earth's rotational properties ($23^h56^m4.09^s$ and 23°27′) implies that you might feel very much at home on Mars, except that the Martian seasons last about twice as long as the Earth's.

(b) Physical Characteristics

Mars has a radius of 3394 km, which is only 53 percent of the Earth's radius. Mars has two moons, and so we can use Newton's form of Kepler's third law and the orbits of these moons to infer Mars' mass. It is only 6.4×10^{23} kg, about 11 percent of the Earth's mass. The density is 3900 kg/m^3, only a bit higher than the Moon's (3300 kg/m^3) and much lower than the Earth's (5500 kg/m^3).

This comparatively low density implies that Mars' interior (Figure 5–14) must be different from the Earth's. In particular, its core must be smaller and probably consists of a mixture of iron and iron sulfide, which has a lower density than the materials in the Earth's core. The Martian mantle probably has the same density as the Earth's. The exact composi-

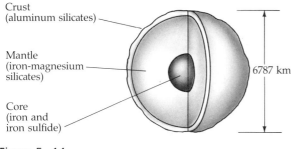

Crust
(aluminum silicates)

Mantle
(iron-magnesium
silicates)

Core
(iron and
iron sulfide)

6787 km

Figure 5-14

Interior of Mars. Note the relatively small size of the core.

tion of the mantle is not known; one model has a mantle with olivine (an iron–magnesium silicate), iron oxide, and some water (0.3 percent).

Astronomers have known for a long time that Mars has an atmosphere. The Viking landers found average surface pressures of roughly 0.02 atm. In the Earth's atmosphere, the pressure falls this low at an altitude of 40 km. This thin atmosphere consists of 95 percent carbon dioxide, 0.1 to 0.4 percent molecular oxygen, 2 to 3 percent molecular nitrogen, and about 1 to 2 percent argon—a composition very similar to the atmosphere of Venus.

The Viking orbiters measured the amount of water vapor in the atmosphere and found the greatest amounts in the high northern latitudes. Peak concentrations were about 0.01 mm of precipitable water (the thickness of the water layer if all the water were taken out of the atmosphere and spread over the surface). On Earth, the atmospheric water vapor is typically several centimeters of precipitable water, and of course the oceans are several kilometers thick. Mars is a much drier planet than the Earth. Liquid water cannot exist on Mars today because of the low surface pressure. Only in the deepest canyons, where the atmospheric pressure is higher, could water be liquid on the surface. However, it is common to have water ice on Mars, either on the surface or in the clouds. Some evidence suggests water in a permanent frost layer beneath the surface.

Although the atmosphere contains mostly carbon dioxide, its low density does not provide much of a greenhouse effect against temperature extremes. At the Martian equator, the difference between noon and midnight surface temperature amounts to almost 100 K when Mars is closest to the Sun. The summer tropical high temperature of 310 K is exceptional. For a period of two Martian months, the surface temperature remains below the freezing point of water both day and night over the entire surface. At the Viking 1 site, 23°N latitude, the air temperature near the ground ranges from −85 to −29°C. At the Viking 2 site, which is farther north, the temperature falls even lower and water condenses as frost on the surface. The layer of water ice that coated the rocks and soil is extremely thin, less than 1 mm.

(c) Surface Features from Afar

Because an astronomer must look through two atmospheres, the view of Mars is usually poor in clarity. A small telescope shows the main surface features: the white polar caps, light reddish orange regions, and darker areas. Spacecraft visits improved dramatically the resolution of our image of Mars.

The visible surface features on Mars are dark, apparently greenish gray areas, in contrast to the reddish orange color of the rest of the surface (Figure 5–15). This greenness turns out to be an illusion caused by the contrast of the light and dark areas. The dark regions are not really green; they are actually red, with greatest intensity in the red region of the spectrum.

The light orange and yellow-brown regions make up almost 70 percent of the Martian surface. They give Mars its striking reddish appearance. In 1934, the American astronomer Rupert Wildt suggested that these areas contain ferric oxide (Fe_2O_3). Iron oxides come in many forms on the Earth; all are characteristically brown, yellow, and orange. The Viking landers' measurements indicated a surface composition of about 19 percent ferric oxide. In addition, the landers measured about 44 percent silica, which leads to the conclusion that silicate minerals make up a major part of the surface.

The rusty sand, some of which is much finer than that on Earth's beaches, is blown up by fierce winds (speeds greater than 100 km/h) to create planet-wide dust storms. These storms occur most violently when Mars is closest to the Sun. Then the dust clouds, whipped up to heights of 50 km, shroud the entire planet, covering it in a yellow haze for about one month. It takes many months for the fine dust to completely settle to the surface. These global storms sandblast the surface and mix it up so much that the surface composition over the planet becomes essentially uniform. We now know that this wind-driven

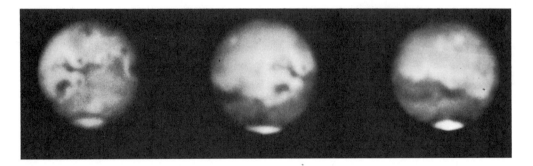

Figure 5–15

Mars in 1971. These photographs were made during the 1971 opposition, when the Earth and Mars were very close to each other. Note the polar cap and dark regions. *(R. Minton and S. Larson, Lunar and Planetary Laboratory, University of Arizona)*

dust caused most of the changes in Martian surface features seen in the past. For example, the wind storms blow dust into dunes or deposit it in streaks around mountains and craters.

In 1877, Schiaparelli recorded Martian surface features in great detail. He charted a number of dark, almost straight features, which he called *canali,* Italian for "channels." This word was translated into English as "canals," which implied to some people that they were artificial structures. These so-called canals ignited the curiosity of the American astronomer Percival Lowell (1855–1915). To pursue his interest in Mars, Lowell in 1894 founded an observatory near Flagstaff, Arizona, to take advantage of the excellent observing conditions there. Shortly afterward, he published Martian maps showing a mosaic of more than 500 canals. In a series of popular books, he argued that the canals were artificial waterways constructed by Martians to carry water from the polar caps to irrigate arid regions for farming.

We know now that the polar caps do indeed consist mostly of water ice, especially the residual cap left in the summer, which ranges in thickness from 1 m to 1 km (Figure 5–16). The outer reaches of the caps, prominent in winter, consist of carbon dioxide ice, which condenses at a lower temperature than water ice. (At Martian surface pressures, water ice condenses at about 190 K, carbon dioxide ice at 150 K.) The total amount of water in the polar caps is only some 10 m of precipitable liquid.

Lowell was wrong about the canals, however,

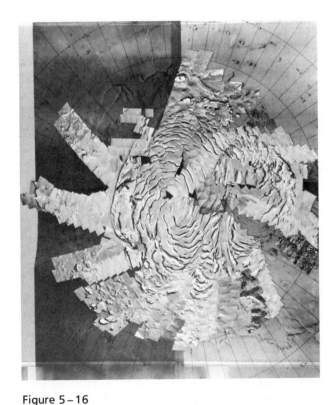

Figure 5–16

A summer polar cap. The residual cap of water ice in the summer shows the layered structure of the polar region. *(NASA)*

Planetary astronomers now believe that wind-blown dust deposits might have created temporary features that were seen as the largest and fuzziest of the canals. A comparison of Lowell's canal maps with orbiter photographs indicates that only one real feature (part of Valles Marineris) corresponds to any of the so-called canals.

(d) Surface Features Close Up

In 1969, data from Mariners 6 and 7 reinforced the view (first developed from Mariner 4 pictures of 1965) that parts of the surface of Mars resemble that of the Moon. These craft photographed abundant Martian craters visible even under the thinner regions of the polar caps. The Martian craters do look like impact craters but tend to be shallower than lunar ones. Their flat floors and low rims indicate that they have been strongly eroded, an important clue to conditions in the past.

In contrast, the photographs taken by Mariner 9 (1971) also showed spectacular features of a geologically once-active planet. The two Martian hemispheres have different topological characteristics: the southern is relatively flat, older, and heavily cratered; the northern is younger, with extensive lava flows, collapsed depressions, and huge volcanoes. Near the equator separating the two hemispheres lies a huge canyon, called Valles Marineris (Figure 5–17). This chasm is 5000 km long (about the length of the United States) and 500 km wide in places.

The Viking landers touched down on Mars in 1976. They photographed a bleak, dry surface with large rock boulders strewn about amid gravel, sand, and silt. The boulders are basaltic. Some contain small holes (Figure 5–18) from which gas has apparently escaped. On Earth, such basalts originate in frothy, gas-filled lava; the Martian rocks probably had a similar origin.

Both landers uncovered indirect evidence of once-flowing Martian surface water. The region around Viking 1 seems to be a flood plain where water sorted the smaller rocks into gravel, sand, and silt. The ground there also resembles the hardened soil of Earth's deserts. Such soil forms when underground water percolates upward and evaporates at the surface. Minerals left behind when this water evaporates harden the soil.

The Mariner 9 mission discovered and the Viking orbiters confirmed a number of sinuous channels that appear to have been cut in the surface by running water (Figure 5–19). The largest ones have lengths up to 1500 km and widths up to 100 km. (These channels are not the canals seen by Lowell and others; they are crooked and too small to be visible from the Earth.) They resemble the arroyos commonly found in the southwestern United States. An *arroyo* is a channel in which water flows only occasionally. The evidence is strong that the Martian channels were cut by flowing water: (1) the flow direction is downhill; (2) the flow patterns meander; (3) tributary structures indicate where several flows merged to form a larger one; (4) sandbars are cut by smaller flow channels, as is commonly found in arroyos on the Earth. The presence of Martian arroyos requires extensive running water for at least a short period of time. Since Mars does not have liquid surface water now, conditions for its presence must have occurred in the past and would have required a denser, warmer atmosphere.

By far the most striking Martian surface features are the shield volcanoes clustered on and near the Tharsis ridge. The largest is Olympus Mons, 550 to 600 km across at its base (Figure 5–20). The cone's surface shows a wavy texture that is the result of lava flows. The cone reaches 25 km above the surround-

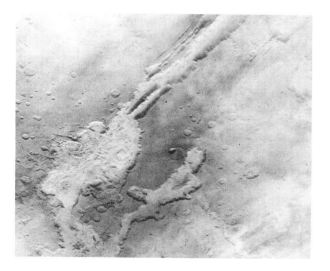

Figure 5–17

Valles Marineris. This view shows an area 1500 by 2000 km; the total length is 5000 km. *(NASA)*

Figure 5–18

Viking 1 site. The rocks are a few tens of centimeters in diameter. Note the dark rock with holes to the left of the scoop; it is a volcanic rock. The scoop has sampled the soil just to the left of this rock. *(NASA)*

ing plain, and its base would span the bases of the islands of Hawaii, which are made of several volcanoes. Olympus Mons soars more than 2.5 times the height of Mt. Everest above sea level. Indirect evidence from the lava flows implies that the Martian volcanoes are about 1 billion years old. The huge mass of Olympus Mons requires that the Martian crust beneath it be 120 to 130 km thick, about twice the thickness of the Earth's crust, despite the lower Martian surface gravity.

Olympus Mons crowns a string of volcanoes situated on the Tharsis ridge. Occasionally, thin ice clouds have been seen decorating the tops of the volcanoes there. These clouds might result from erratic spurts of outgassing. On Earth, volcanic activity spews forth gases (including water vapor) from the Earth's interior. Such outgassing by the gigantic Martian volcanoes in the past may have contributed significantly to the Martian atmosphere.

The Tharsis ridge is a hallmark of Mars' northern hemisphere, which differs dramatically from the southern one. The ridge rises about 10 km above the average surface height of the planet and contains numerous volcanic structures. Very few impact craters are visible. In contrast, the southern hemisphere is basically a desert pockmarked by old, eroded craters. The geologic inference from this difference is that, about 3 billion years ago, a huge mass of lava oozed out from under the surface in the northern hemisphere, creating the volcanic plains and the volcanoes over a long period of time. This flow wiped out the older deserts and craters. Other flows have since taken place in this region.

The southern hemisphere of Mars has a cratered terrain (Figure 5–21) that resembles the ancient highlands of the Moon or the intercrater plains of Mercury. The landscape contains impact craters that range in size from huge, lava-filled basins down to some only a few meters across. The Martian craters come in the same variations as lunar and Mercurian ones, some with central peaks that mark their impact origin. In general, the Martian craters are shallower than the ones on the Moon and Mercury because of wind erosion.

(e) Magnetic Field

Mars has an extremely weak planet-wide magnetic field, the magnitude of which is only 60 nT at the surface. That small a value presents a puzzle if the dynamo model correctly describes the origin of planetary magnetic fields. Mars rotates as fast as the Earth. Though the Martian core is smaller, it should contain a substantial amount of iron or nickel–iron. We have no direct evidence that the core is liquid, but the evidence for past volcanic activity implies a hot mantle and therefore a hot, probably liquid, core. So Mars should have a moderately strong field, but it does not. Perhaps, as for Venus, we are viewing Mars in the middle of a magnetic field reversal, but it is unlikely that we would catch both planets in the process of changing polarity.

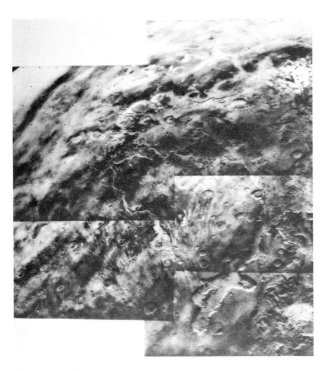

Figure 5–19

Arroyos and craters on Mars. *(NASA)*

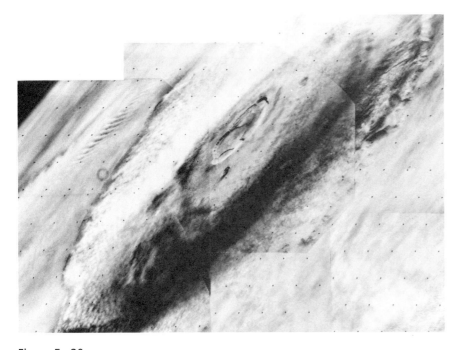

Figure 5–20

Olympus Mons. *(NASA)*

Figure 5 – 21

Southern cratered terrain. An arroyo 700 km long is surrounded by old, cratered terrain in Mars' southern hemisphere. *(NASA)*

(f) Evolution of the Surface

The evolution of Mars has reached a level midway between that of the Moon and that of the Earth. After the formation of Mars by accretion, impact craters covered the surface. Shortly thereafter, the planet differentiated to form a crust, mantle, and core. Regions of thicker crust rose to higher elevations. In the second phase, thin regions of the crust fractured and the Tharsis ridge uplifted, cracking the surface around it. During this time, a primitive atmosphere, denser and warmer than at present, held large amounts of water vapor from the volcanic outgassing. Rainfall may have eroded the surface in furrows and then percolated to a depth of a few kilometers. Decreasing temperatures formed ice at shallow depths. When heated (perhaps by volcanic activity), this ice melted, leading to the formation of collapse and flow features. Planet-wide water erosion carved the surface.

In the next phase, extensive volcanic activity occurred, especially in the northern hemisphere. The Tharsis region continued to uplift, generating more faults. Valles Marineris formed at this time. Finally, recent volcanism—most of it concentrated on the Tharsis ridge—broke the surface and spewed out great flows of lava. Since those great eruptions, it is

mainly wind erosion that has sculpted the Martian surface. A few small impact craters have probably formed from time to time.

5 – 5 Comparative Evolution of the Terrestrial Planets

Internal heat, generated by radioactive decay now or left over from the planet's formation, drives the evolutionary processes of Earth-like planets. As we argued in Chapter 4, the lifetime of these processes is roughly proportional to the planet's radius, so that we expect—and find—the Earth the most evolved (and still evolving!) of the terrestrial planets.

We can divide the evolutionary sequence of terrestrial planets into five main stages:

  I. Formation, heating of crust and interior
 II. Crust solidification, intense cratering
III. Basin formation and flooding
IV. Low-intensity cratering, atmosphere by outgassing
 V. Volcanoes, continents, crustal movement

We can compare and contrast the overall evolutionary status of the terrestrial planets in this scheme. Both Mercury and the Moon evolved to the end of stage III; they provide fossil records of the early history of all the terrestrial planets. In particular, they show that an intense bombardment heavily cratered the surfaces 4 billion years ago. On Mars, some of those impact craters are still visible, though highly modified. On the Earth and Venus, subsequent evolution has wiped out evidence of that era. Mars appears, in fact, to have reached an evolutionary status between that for the Moon and Mercury and that for Venus and the Earth; it reached the beginning of stage V. Venus has entered stage V, and the Earth is far into this stage.

Note that the differentiated interior structure of the terrestrial planets implies that all of them became hot enough in their interiors to melt the materials there or at least make them plastic. The original mixture then separated into layers on the basis of different densities.

Why do the Earth and Venus differ so dramatically now even though their masses are almost the same? In particular, why does Venus have an atmosphere that is almost all carbon dioxide and why did this planet never develop oceans on its surface? The answer probably relates to the fact that Venus is closer to the Sun than the Earth is. Consider the atmospheric carbon dioxide. If the carbon dioxide in the Earth's crust and oceans is added to that in its atmosphere, the total is about the same as that in the atmosphere of Venus. The carbon dioxide on Venus did not end up in the crust because of the lack of water. (Carbon dioxide in solution in the oceans reacts with silicates to form carbonates, and the shells of sea life also contain carbonates.) We infer that Venus is so close to the Sun that the greenhouse effect has always kept the surface temperature too high for liquid water. Also, any water vapor in Venus' atmosphere has probably escaped into space. Ultraviolet light from the Sun has enough energy to dissociate water into hydrogen and oxygen; the hydrogen is then lost as it has high enough speeds to escape rapidly to space.

Finally, a comment about cratering, which clearly contributed to the early evolution of all the terrestrial planets. As you will see in the next two chapters, nearly every solid surface in the solar system is cratered. The number of craters of various sizes

on a surface allows us, with some assumptions, to infer the range of sizes and infall rates of the impacting masses and the ages of the cratered surfaces.

Basically, the diameter of a crater relates to the kinetic energy of the impacting object, which is a stronger function of infall velocity than mass. Hence, the fact that on most planetary surfaces smaller craters outnumber larger ones implies that smaller objects were in greater abundance. The largest craters and basins were formed by objects more than 100 km in size. Note that, once formed, craters are modified by later impacts, lava flows, and erosion (for a planet with a substantial atmosphere).

The Moon serves as the baseline for estimating cratering rates as a function of time because we can date the surface areas. One interpretation of the data (Figure 5–22) envisions a sharp but smooth decline in the rate over the past 4.5 billion years. The rate now may be as much as a million times less than that during the era the planets formed—that is what is meant by the early intense bombardment of the sur-

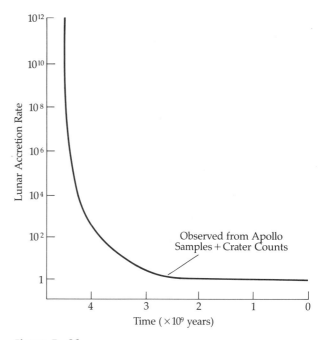

Figure 5–22

Schematic reconstruction of the relative rate of cratering of the Moon. Present rates are observed for the Moon; the rates from 4.0 to 4.5 billion years are very uncertain and may be a factor of 100 lower than shown here. *(Adapted from a figure by W. K. Hartmann)*

faces. Impact rates varied from planet to planet, but should be roughly the same (within a factor of 2) for at least the terrestrial planets. It is probable that the evolution of the cratering rate (Figure 5–22) was more or less the same (in a relative sense) for all the planets and satellites.

For the terrestrial planets, we can use the spectrum of crater sizes and surface densities to estimate different degrees of surface evolution. Our Moon ranks as the least modified (along with Deimos and Phobos, the moons of Mars; see Section 7–1), followed by Mercury, Mars, Venus, and the Earth.

Problems

1. (a) Calculate Mercury's surface temperature at noon at perihelion and at aphelion.
    (b) At what wavelength would Mercury radiate most strongly on the dayside? On the nightside?

2. At a frequency of 10 GHz, calculate the difference (due to rotation) between the Doppler shift of a radio signal bounced off one side of Mercury and that of a signal bounced off the other side.

3. Assume Mercury had a satellite whose composition is the same as Mercury's and whose mass is 1 percent of the mass of Mercury. How close must it orbit Mercury to remain bound against the tidal force of the Sun?

4. Calculate the length of the solar day on Venus. Watch out for the retrograde rotation!

5. (a) Calculate the maximum radar Doppler shift (at 10 GHz) due to rotation from Venus and Mars.
    (b) What accuracy in timing of radio signals must be obtained in order to detect minimum height differences on Venus?

6. At what distance from the center of Venus would you expect a magnetic field strength equivalent to that of

the Earth's Van Allen belts? Do the same calculations for Mars. What do you conclude?

7. Compare the appearance of the Earth–Moon system viewed from Mars with that viewed from Mercury.

8. Calculate the Roche limit for Mars (with the planetary and lunar densities equal) and compare your results with the orbits of Deimos and Phobos, the moons of Mars.

9. Estimate the lifetime of CO_2 in the atmospheres of Mars and Venus.

10. Compare the scale heights of CO_2 in the atmospheres of Mars and Venus. At what altitude in Venus' atmosphere does the pressure drop to 1 atm?

11. For a magnetic dipole, the field strength far from the dipole varies as $1/R^3$. At what distance from Mercury's center does that planet's magnetic field have the same strength as the Earth's at the Van Allen belts?

12. Use the equation of hydrostatic equilibrium to estimate the central pressures of Mercury, Venus, and Mars. (*Hint:* Call the surface pressure zero and use the average density.)

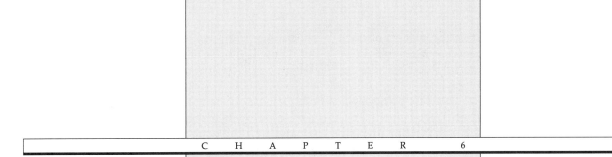

The Jovian Planets

Based on physical properties, the planets of the Solar System fall into two categories: terrestrial and Jovian. The previous chapter highlighted the rocky, earthlike worlds. This chapter turns to the gigantic, liquid worlds that make up the Jovian planets — places quite different from the terrestrial planets. The main point of the comparison is this: the Jovian planets are primitive worlds that have evolved less since their time of formation than the active terrestrial planets (Earth, Venus, and Mars).

6 – 1 Jupiter

Beyond Mars, we pass the *asteroid belt* (Section 7 – 2) near 3 AU and finally come to the lord of the Jovian planets, Jupiter, named after the king of the Olympian gods. Because of its enormous size and high albedo (0.51), Jupiter is a very bright planet in the Earth's night sky, especially at opposition.

(a) Motions

Jupiter's orbit about the Sun has a small eccentricity (0.0484) and is inclined only 1.31° to the plane of the ecliptic; at a semimajor axis of 5.2028 AU, the planet completes one sidereal orbit in 11.862 Earth years. The synodic orbital period of 398.88^d implies that Jupiter returns to opposition (at full phase) about one month later each year.

Because we can see only the dense atmosphere of Jupiter (Figure 6 – 1), the planet's rotation period is determined by following the rotation period of atmo-

Figure 6 – 1

Jupiter. Taken in 1969, these photographs show Jupiter through blue (1), green (2), and red filters (3). The dark spot is the shadow of Io. *(New Mexico State University Observatory)*

spheric features, such as the Great Red Spot, by measuring the Doppler shifts of light from the approaching and receding limbs, and by studying the rotation of the magnetic field structure. We find that Jupiter's rotation axis is inclined 3°7′ to its orbital axis, but the sidereal rotation period varies from 9^{h}50^m near the equator to 9^{h}55^m at higher latitudes. Hence, the gaseous Jovian atmosphere exhibits *differential rotation:* fastest at the equator and slowest at the poles. (The Sun also rotates differentially because it is a fluid.) The rotation structure of Jupiter's atmosphere is reminiscent of the Hadley cells and characteristic trade winds in the Earth's atmosphere [Section 3 – 2(a)]. Jupiter's extremely rapid rotation results in its large oblateness (0.062).

(b) Physical Characteristics

The equatorial radius (11.19$R_\oplus$) and mass (318$M_\oplus$) of Jupiter have been accurately determined by observing the orbits and occultations of its moons, by noting its gravitational perturbations upon the orbits of comets and asteroids, by measuring the angular diameter of its visible disk (47″ at opposition), and by Voyager flyby measurements. This gigantic prototype of the Jovian planets has a mean density of only 1330 kg/m³. This low density implies that the composition of Jupiter is similar to the solar mass abundances of about 75 percent hydrogen, 24 percent helium, and 1 percent all heavier elements (by mass).

The disk of Jupiter shows bands of white, blue, red, and yellow clouds (Figure 6 – 2). The colors result from the various chemical compounds formed there. These bands change their structure with time, but the relatively stable Great Red Spot has been constantly observed since 1831. This enormous atmospheric feature measures about 20,000 km by 50,000 km, and it changes shape, position, and intensity. Jupiter's cloudy atmosphere reflects light well and results in the planet's high albedo.

The alternating strips of light and dark regions that run parallel to the equator are called *zones* and *belts*. Infrared observations show that the zones have lower temperatures than the belts. Because the temperature falls at greater altitudes, the zones are higher up than the belts. These differences in temperature imply that the zones mark the tops of rising regions of high pressure and the belts descending regions of low

Figure 6–2

Jupiter, Voyager 1 and 2. The two top photographs, taken by Voyager 2, show changes in the atmosphere that occurred after the small center photograph had been taken by Voyager 1 four months earlier. *(NASA)*

pressure. (The Coriolis effect stretches these convective regions out parallel to the equator.) The convective atmospheric flow transports heat out to space from the planet's interior. The Voyager missions zoomed in on these complex streams and swirls of Jupiter's upper cloud layer. The stunning photographs of Figure 6–2 show the turbulent atmospheric flow. Earth-based telescopes have revealed complex changes in the belts and zones. Occasionally, dark blue, red, brown, and white ovals appear against the banded background. These small oval spots last as long as one or two years.

From a variety of observations, we know that the Red Spot is a few degrees cooler than, and extends about 8 km above, the surrounding zone; it is a rising region of high pressure. Also, it rotates counterclockwise like a vortex, just as expected from a high-pressure region in Jupiter's southern hemisphere. A model emerges in which the Red Spot turns like a huge wheel pushed by the surrounding atmospheric flow and in turn deflects nearby clouds to force them around it. So the Red Spot is a huge, long-lived atmo-

spheric eddy with a turbulent region flowing past it (Figure 6–3).

Infrared spectroscopy reveals the atmospheric composition above the clouds. In 1934, methane (CH_4) was discovered, the first molecule definitely identified. Later, ammonia (NH_3), molecular hydrogen (H_2), and atmospheric helium (He) were found. Voyager also observed acetylene (C_2H_2), ethane (C_2H_6), phosphine (PH_3), water (H_2O), and germane (GeH_4). Some of these molecules had been previously detected in spectra taken from the Earth, as had CO and HCN. A few molecules containing deuterium are also known to be present. An analysis of Voyager spectroscopic data implies that Jupiter's upper atmosphere contains about 78 percent hydrogen, 20 percent helium, and 2 percent all other elements, a composition essentially the same as the Sun's. Most of this material exists as molecules. The visible clouds at the tops of the zones are most likely ammonia ice crystals because the temperature here is about 130 K. Below them, according to one model, lies a layer of ammonia hydrosulfide (NH_4HS)

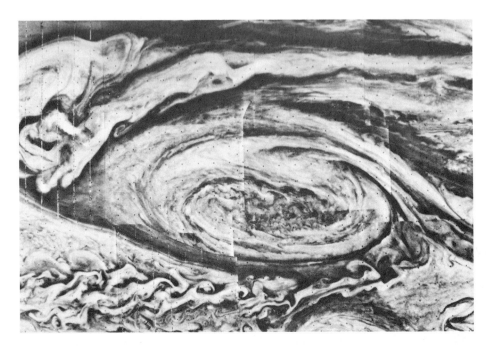

Figure 6–3

The Red Spot. This computer-enhanced view emphasizes the flows around the Red Spot. The smallest details are 30 km across. *(NASA)*

clouds. Below these float ammonia vapor and water ice clouds. This model describes three separate cloud layers making up the upper atmosphere. The top layer is ammonia, below that are ammonia and hydrogen sulfide, and the bottom layer is water ice. The colors of the Red Spot and other atmospheric features probably result from chemical reactions of major and trace molecules, with photoionization and dissociation driving the reactions.

We infer Jupiter's internal structure from physical models that include two key pieces of information: (1) Jupiter's low density and atmospheric composition imply a solar mix of material throughout and (2) Jupiter radiates into space more energy than it receives from the Sun (about twice as much) because its temperature is higher than that of a blackbody at the same distance from the Sun and having the same albedo. The internal heat is probably left over from Jupiter's formation.

Recent models of Jupiter come up with the following picture (Figure 6–4). The atmosphere covers the planet like a thin skin and consists mostly of molecular hydrogen. Going into the planet, the density, temperature, and pressure increase (as expected from hydrostatic equilibrium), and so the hydrogen exists in a liquid state. At a pressure of about 2 to 3 million atm, the hydrogen is squeezed so tightly that the molecules are separated into protons and electrons

that freely move around and can conduct electricity. This state is called *metallic hydrogen,* and although it has never been observed in a laboratory on the Earth, its existence is predicted from quantum physics. This strange state persists to within about 14,000 km of the planet's center. Here, perhaps, if Jupiter does have a solar composition, lies a solid(?) core of heavy elements.

Note that the equation of hydrostatic equilibrium (Equation 4–3) can be applied to the interior of a stable planet, which is neither expanding nor contracting. This condition then requires that the internal pressure increases as one goes deeper into a planet's interior (because a greater weight has to be supported). Also, we can use the equation of hydrostatic equilibrium to estimate roughly the central pressure of a planet. Take the equation in differential form

$$dP = -\rho(r)\,(GM/r^2)\,dr$$

where M is the mass within radius r:

$$M = (4/3)\pi r^3 \bar{\rho}$$

and $\bar{\rho}$ the average density within r. To simplify the calculation, assume that the density is constant throughout the planet and equals the average density, $\bar{\rho}$, and that the surface pressure is essentially zero compared to the central pressure. Then

$$dP = -\bar{\rho}G(4/3)\pi\bar{\rho}rdr$$

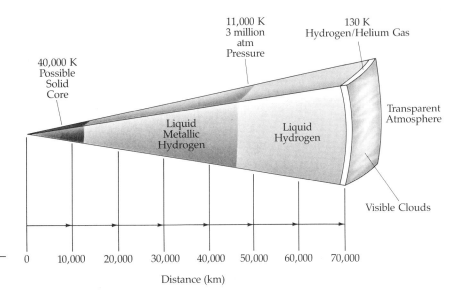

11,000 K
3 million
atm
Pressure

130 K
Hydrogen/Helium Gas

40,000 K
Possible
Solid
Core

Liquid
Metallic
Hydrogen

Liquid
Hydrogen

Transparent
Atmosphere

Visible Clouds

Figure 6-4

Jupiter's interior. Note the large zone of liquid metallic hydrogen.

0   10,000   20,000   30,000   40,000   50,000   60,000   70,000

Distance (km)

which we then integrate from the center to the surface:

$$\int_{P_c}^{0} dP = -\bar{\rho}^2(4/3)\pi G \int_{0}^{R} r\,dr$$

so

$$P_c = (2/3)\pi G\bar{\rho}^2 R^2$$
$$= (1.4 \times 10^{-10})\bar{\rho}^2 R^2$$

in SI units (N/m^2 for pressure). Then for Jupiter, we have a radius of about 70,000 km and an average density of roughly 1300 km/m³, so

$$P_c \simeq (1.4 \times 10^{-10})(1300)^2(7 \times 10^7)^2$$
$$= (1.4 \times 10^{-10})(8.3 \times 10^{21})$$
$$= 1.2 \times 10^{12} \ N/m^2$$
$$\simeq 1.2 \times 10^7 \ atm.$$

As mentioned above, Jupiter (along with Saturn and Neptune, Table 6-1) emits more infrared energy than the amount absorbed from incoming sunlight. The total internal excess power amounts to about

4×10^{17} W. How can we conclude that the source of this excess is energy from the time of Jupiter's formation? Consider an alternative: the heat may come from a very slow gravitational contraction as the central metallic layer grows at the expense of less dense material. How much shrinkage is needed? Roughly, the gravitational potential energy of a spherical mass is

$$PE \approx GM^2/R$$

so that

$$\text{Energy loss} = d(PE)/dt \approx (-GM^2/R^2)\,dR/dt$$

$$(4 \times 10^{17} \ W)(3 \times 10^7 \ s/year) \approx 10^{25} \ J/year \approx$$

$$[(7 \times 10^{-11})(1.9 \times 10^{27})^2/(7.1 \times 10^7)^2]\,(dR/dt)$$

which implies that

$$dR/dt \approx 10^{-4} \ m/year$$

This amount of shrinkage would be undetectable. However, the problem with this idea is that Jupiter's

Table 6-1   Internal Heat Sources of the Jovian Planets

Planet	T_B (K) (blackbody temperature)	Total Power	Internal Power W/m²	W
Jupiter	127 ± 3	$(2.0 \pm 0.2) \times$ solar input	7	4×10^{17}
Saturn	97 ± 3	$(2.8 \pm 0.4) \times$ solar input	3.5	2×10^{17}
Uranus	58 ± 3	$\approx 1 \times$ solar input	<0.1	10^{15}
Neptune	56 ± 3	$(2.5 \pm 0.5) \times$ solar input	0.4	3×10^{15}

liquid interior is incompressible, and so this shrinkage, though small, could not happen for long. In contrast, if the thermal conductivity of metallic hydrogen is not high, then Jupiter could easily retain its primordial internal energy for billions of years. The period of gravitational contraction then could have occurred early in the planet's history until it could no longer contract.

Most of Jupiter is hydrogen, and most of that hydrogen is liquid—quite a contrast to the Earth's interior and those of the other terrestrial planets. The core temperature may be about ten times hotter than the Earth's core—as high as 40,000 K. It is the flow of heat outward from the core that drives the convective circulation of the atmosphere. The rapid rotation of the planet creates a large Coriolis acceleration that produces the beautiful banded atmosphere.

(c) Magnetic Field

Jupiter exhibits radio emissions that have been linked to a *Jovian magnetic field* of about 10^{-4} T at the surface. This strong magnetic field arises from a dynamo mechanism in a rapidly rotating liquid core of metallic hydrogen. At wavelengths from 3 to 75 cm, the planet is observed to radiate nonthermally; this *decimeter*, or DIM, radiation (1 decimeter = 10^{-1} m) is *synchrotron radiation* from relativistic electrons spiraling at speeds very near the speed of light in Jovian *radiation belts* trapped by Jupiter's magnetic field. Radio interferometer measurements and the Voyager missions reveal radiation belts similar to the Earth's Van Allen belts (Figure 6–5), extending beyond three Jovian radii at the magnetic equator; the magnetic axis is inclined about 10° to Jupiter's rotation axis. This intense field creates around Jupiter a huge magnetosphere that holds off the solar wind.

In 1955, scientists at the Carnegie Institute observed sporadic radio bursts at decameter, or DAM, wavelengths (1 decameter = 10 m); these bursts correlated with the transit of Jupiter. In 1964, some DAM radiation was found associated with the position of the Jovian satellite Io relative to Jupiter's magnetic axis. The probability of radio bursts is greatly

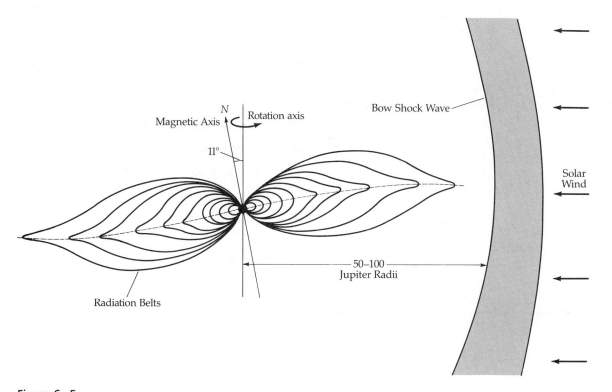

Figure 6–5

Jupiter's magnetosphere, based on spacecraft data. The size and shape vary depending on the strength of the solar wind.

enhanced when Io is in one of two positions relative to Jupiter and the Earth: either 90° or 240° from the direction of superior conjunction. Although the details are not year clear, this DAM radiation likely is generated in and above Jupiter's ionosphere on magnetic field lines that thread to Io [Section 7–1(b)] along a magnetic flux tube. Other DAM radiation occurs as sharp, intermittent bursts that last about 1 s. A strong burst generates approximately 10^{11} J; in comparison, an intense lightning bolt on Earth discharges about 10^5 J. These radio bursts are likely generated by superbolts of lightning. Some Voyager pictures show bright regions on the planet's nightside that may be lightning. Violent updrafts and turbulence in Jupiter's atmosphere probably generate this lightning in much the same way as thunderstorms do on the Earth.

Earth-based radio observations and satellite flybys reveal that Jupiter's magnetosphere has a complex structure. If it were visible to us optically, the magnetosphere seen head-on would subtend an angle of 2°—four times larger than the angular size of the Sun or Moon (Figure 6–6)! Jupiter's magnetosphere falls into three distinct zones. An *inner magnetosphere* marks the region where the magnetic field generated by currents within the planet dominates; it extends out to about $6R_J$. From there to $30R_J$ to $50R_J$ extends the *middle magnetosphere,* where equatorial azimuthal currents control the field configuration. Beyond $50R_J$, the geometry of the *outer magnetosphere* depends on the sunward or nightside orientation. The sunside fields act as a buffer zone that expands and contracts with variations in the solar wind's intensity. The nightside shows a long magnetic tail, about $400R_J$ in diameter and a few AU in length.

The planet's dipolar field controls the inner magnetosphere. It is mostly simply modeled by a dipole tilted about 10° with respect to the spin axis and offset from Jupiter's center. The magnetic field strength is 4.2×10^{-4} T at $R_J = 1$. As this field spins, it accelerates electrons to relativistic speeds—to tens of MeV. Those in the region from $1.3R_J$ to $3R_J$ generate the DIM synchrotron emission.

From our previous discussion [Section 4–6(c)] of the motion of charged particles in the Earth's magnetic field, you know that the particles will gyrate around the magnetic field lines with a radius

$$r = mv/qB$$

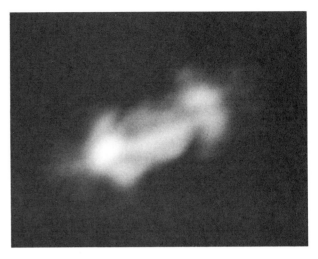

Figure 6–6

Radio map of Jupiter. This radio map at 1.5 GHz shows the synchrotron emission from the relativistic electrons in the radiation belts. *(J. A. Roberts, G. L. Berge, and R. C. Bignell, National Radio Astronomy Observatory)*

where m is the particle's mass, v its velocity, q its charge (in coulombs), and B the magnetic field strength (in teslas). The frequency of its motion is

$$
\begin{aligned}
f &= v/2\pi r \\
&= v/2\pi(mv/qB) \\
&= qB/2\pi m
\end{aligned}
$$

This frequency is often called the cyclotron frequency for it is the frequency with which charged particles travel around a cyclotron, which uses a uniform magnetic field to contain them. For electrons, the cyclotron frequency is

$$f_c = Be/2\pi m \approx (2.8 \times 10^4)B \qquad \text{(MHz)}$$

where B is in teslas. These accelerated particles emit electromagnetic radiation. Such radiation moving at nonrelativistic speeds is emitted in all directions. For electrons moving at relativistic speeds, the emission is concentrated in the forward direction in a beam of opening angle

$$\theta = (1 - v^2/c^2)^{1/2} \qquad \text{(radians)}$$

or

$$\theta \approx 56/E \qquad \text{(degrees)}$$

when E is in MeV. An observer located in the electron's plane of gyration would see one pulse per revolution at frequency f_c, and the emission has linear polarization. This polarized radiation is called *synchrotron radiation*.

The actual synchrotron emission of a group of electrons at various energies consists of a broad envelope characterized by the spread in f_c. If their velocities are not perpendicular to the magnetic field, the particle paths are helixes with a pitch angle of α (α is 90° when an electron moves perpendicular to the magnetic field). Then the peak in the synchrotron emission spectrum occurs at approximately

$$f_{max} \approx (4.8 \times 10^{-4})E^2B \sin \alpha \qquad \text{(MHz)}$$

where E is in MeV and B in teslas. The flat peak in the emission from Jupiter occurs at roughly 1 GHz (=1000 MHz). For a 90° pitch angle, this implies that in a field of 10^{-4} T, the electron energies are

$$E^2 \approx f_{max}/(4.8 \times 10^{-4})B \sin \alpha$$
$$= (1000)/(10^{-4})(1)(4.8 \times 10^{-4})$$
$$E \approx 140 \text{ MeV}$$

and the emission is beamed within a cone angle of 0.4°.

Voyager pictures of Jupiter's nightside show polar *aurorae* for the first time. The photographs indicate that the aurorae occur in at least three layers: at 700, 1400, and 2300 km above the cloud tops. We presume that these aurorae happen for the same reason as on the Earth: excitation of the upper atmosphere by energetic charged particles pouring in near the north and south magnetic poles. Some of these particles flow from Io and are trapped by Jupiter's magnetic field.

6–2 Saturn

Beyond Jupiter orbits the last of the seven planets known to the ancients: Saturn, named after the father of Jupiter. Saturn is the second largest Jovian planet in the Solar System, and it is girdled by a most magnificent system of rings (Figure 6–7). In our night sky, Saturn shines brightly because of its large size and the high albedo (0.50) of its atmosphere. The splendor of the Saturnian sky is marked by the bright bands of its rings and its many moons.

(a) Motions

At the orbital semimajor axis of 9.539 AU, Saturn's sidereal revolution period is 29.458 years in a moderately eccentric (0.0557) orbit inclined 2.49° to the ecliptic. From the Earth, Saturn's angular diameter at opposition is about 20″.

Like Jupiter, Saturn has a thick, cloud-filled atmosphere that rotates differentially. By observing the Doppler shifts across the planet and accurately timing atmospheric markings, we find the sidereal rota-

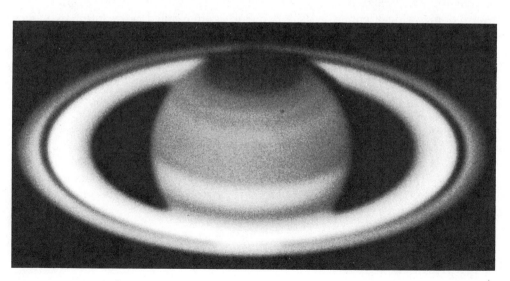

Figure 6–7

Saturn. Note the banded structure of the atmosphere. *(Palomar Observatory, California Institute of Technology)*

tion period to be 10^h14^m near the equator and 10^h38^m at high latitudes. Again, we have a differential rotation similar to Jupiter's. Saturn's equator is inclined 26°45′ to its orbital plane, so that alternate poles of the planet are tilted toward the Earth at intervals of about 15 years. The rotation causes the large oblateness (0.096) of Saturn; the polar and equatorial radii are in the ratio of about 9:10.

(b) Physical Characteristics

Saturn bears considerable resemblance to Jupiter. Saturn is slightly smaller ($9.0R_\oplus$) and less massive ($95M_\oplus$) than Jupiter. It has the lowest bulk density of any of the planets—only 680 kg/m³, even less than that of water!

The atmospheric structure of Saturn also resembles that of Jupiter with its belts running parallel to the equator (Figure 6–8). Disturbances in Saturn's belts are much rarer (only ten spots have been observed to date from the Earth) than on Jupiter. Voyager 1 discovered a reddish spot, but it is much smaller than the Great Red Spot on Jupiter; clouds only a few hundred kilometers across were detected at high latitudes. A general haze layer restricts our view of atmospheric features.

The atmosphere of Saturn probably has much the same composition as that of Jupiter. So far, methane (CH_4), ammonia (NH_3), ethane (C_2H_6), phosphine (PH_3), acetylene (C_2H_2), methylacetylene (C_3H_4), propane (C_3H_8), and molecular hydrogen (H_2) have been detected. The percentage of ammonia is less than that found on Jupiter; probably just as much exists, but at the lower temperature of Saturn (100 K) it has frozen and fallen out of the upper atmosphere as ammonia snow. Infrared spectroscopy has detected abundant molecular hydrogen and a substantial percentage of helium. However, Voyager spectrometers measured only 11 percent helium by mass, compared with 20 percent for Jupiter. Some of the helium may have condensed and settled out toward the interior.

Saturn's clouds appear far less colorful than those of Jupiter—the former being mostly a faint yellow and orange. Because temperatures on Saturn are lower than on Jupiter, the Saturnian clouds lie lower in the atmosphere. The Voyager photographs show much of the same complexity of cloud patterns seen on Jupiter, with wind speeds much higher (up to 500 m/s near Saturn's equator). Voyager 2 photo-

Figure 6–8

Saturn's atmosphere. This Voyager photograph shows jet streams and turbulence in the upper atmosphere. *(NASA)*

graphs show that the weather on Saturn can change enormously in a time equivalent to 1 week on the Earth. Large storm systems changed shape, but still remained visible—a hint that Saturn's storms, like Jupiter's, are longer-lived than Earth's storms.

The Voyager photographs also show that the upper atmospheric wind flow of Saturn is different from Jupiter's. Near the equator, the winds all blow eastward at speeds four times those found on Jupiter. At the higher latitudes, the pattern follows an east–west flow alternation, as found on Jupiter. The wind velocities for both planets fall off rapidly away from the equator, but Saturn's atmospheric bands do not mark jet stream flows as they do on Jupiter.

Saturn resembles Jupiter in another important respect: infrared observations show that Saturn emits more energy, as infrared radiation, than it receives from the Sun. The excess (about 2×10^{17} W) is about three times the energy Saturn receives from the Sun. In other words, the infrared blackbody temperature is about 97 K but we expect it to be only 82 K from the amount of incoming sunlight and Saturn's albedo. The excess heat from Saturn is somewhat of a puzzle. For Jupiter, the emission can be accounted for as that left over from a period of gravitational contraction during its formation; a similar model for Saturn fails to account for the infrared excess.

Saturn's interior (Figure 6–9) probably reflects Jupiter's composition. Theoretical estimates are about 74 percent hydrogen, 24 percent helium, and 2 percent heavier elements. This composition is also roughly the same as that of the Sun. Saturn may have a small, rocky core 20,000 km in diameter and having a mass of $20M_\oplus$ (25 percent of the total mass). Other models have the metallic hydrogen region extending right to the center. The level at which hydrogen becomes metallic is much deeper in Saturn than in Jupiter. The change of state occurs at a pressure of 3 million atm. This level is reached deeper in Saturn than in Jupiter because Saturn has a smaller mass and density and so the internal pressure does not rise as fast as it does in Jupiter.

(c) Magnetic Field

Saturn's magnetic field has a total moment only 1/35 that of Jupiter's, but that is still strong enough to generate a Jovian-type magnetosphere with Earth-type radiation belts. The magnetic dipolar moment aligns within 1° of Saturn's spin axis, in contrast to the clear tilt of Earth's and Jupiter's magnetic axes.

Saturn's magnetosphere stores far fewer particles than Jupiter's. Two reasons for this difference are (1) the lack of a local source of charged particles, as provided by Io's eruptions for Jupiter and (2) Saturn's visible rings, which effectively absorb charged particle and so sweep clean the inner magnetosphere. Outside the edge of the rings (Figure 6–10), the density of charged particles rises quickly; it hits a peak at about $5R_S$ to $10R_S$. Here, the charged particles are tightly coupled to the rapidly spinning magnetic field; this interaction generates a plasma sheet about $2R_S$ thick that extends out to $15R_S$. Beyond this, the magnetosphere lacks structure; its size varies with the solar wind. At high solar wind pressures, the magnetosphere can shrink to a radius of $20R_S$; at low pressures, it balloons to $30R_S$ and larger.

6–3 Uranus

Uranus (Figure 6–11) is named after the progenitor of the Titans and father of Saturn; it is the seventh planet from the Sun and the third Jovian planet. William Herschel discovered it in 1781; at first he thought it was a comet, but his observations implied a low-eccentricity elliptical — hence, planetary — orbit about the Sun. Uranus is just at the limit of naked-eye visibility from the Earth, with an angular diameter at opposition of only 3.6''.

(a) Motions

The orbit of Uranus has a semimajor axis of 19.182 AU, an eccentricity of 0.0472, and an inclination of only 0.77° to the ecliptic; the sidereal orbital period is 84.013 years. Chapter 2 discussed the bizarre rotational behavior of Uranus; with its equatorial plane inclined 98° to its orbital plane, Uranus rotates *retrograde* in 17^h. Since the rotation axis lies essentially in the ecliptic plane, we observe the following phenomena: if we see one pole of the planet now, the equatorial plane will be seen edge-on in 21 years and in 42 years the opposite pole will point toward the Earth.

(b) Physical Characteristics

Because it is so far from the Sun, Uranus' atmosphere must be very cold. Infrared observations put

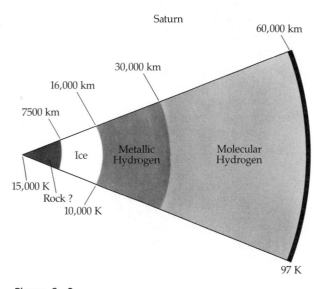

Figure 6–9

Interior of Saturn. Note that the liquid metallic hydrogen zone is smaller than Jupiter's.

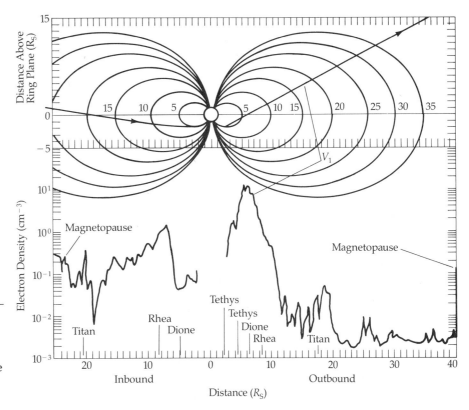

Figure 6-10

Saturn's magnetosphere, showing the path of Voyager 1 through the magnetosphere and the plasma density measured by it. Units are in Saturn radii. Note that the satellites can be sources of charged particles. *(NASA)*

the temperature at 58 K. In such a deep freeze, all the ammonia has frozen out of the atmosphere and cannot be detected spectroscopically. Methane and hydrogen do appear in the spectrum. Helium may also have been detected, but this result has not been confirmed.

Viewed through a telescope, Uranus has a distinctive pale green color, which comes from sunlight that penetrates deep into the planet's atmosphere; some red light is absorbed in the atmosphere, and much of the green is reflected back into space. This selective absorption and reflection occur because the atmosphere's spectrum is dominated by absorption bands of methane.

The low bulk density of Uranus, 1600 kg/m³, implies that it contains mostly lightweight elements. Uranus (Figure 6-12) is thought to consist of roughly 15 percent hydrogen and helium, 60 percent icy materials (H_2O, CH_4, and NH_3), and 25 percent earthy materials (silicates) by mass.

Measuring the rate of Uranus' rotation has been a frustration to astronomers for years, in part because the axial tilt limits the usefulness of Doppler studies Earth-based observations reported rotation rates of

Figure 6-11

Uranus and its moons. *(W. Liller, National Optical Astronomy Observatories)*

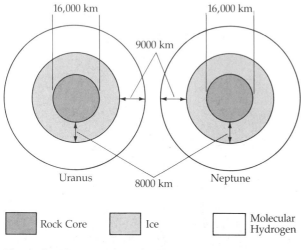

Figure 6–12

Interiors of Uranus and Neptune. Note that the internal structures are essentially the same.

10.8, 15, 15.6, 23, and 24 hours! Recent infrared observations imply a period close to either 15 or 17 h. The technique here is a direct one (Figure 6–13): place the slit of a spectroscope at right angles to the rotation axis. Then absorption lines will be tilted as part is Doppler redshifted and part is blueshifted. The amount of tilt of the lines corresponds to the rotation rate. Unfortunately, such observations are very difficult to make with precision for Uranus.

(c) Voyager Results

The Voyager 2 flyby of Uranus in January 1986 confirmed some of our ideas about this planet and provided some new ones. The planet's rotation rate is now known to be close to 17 h from observations of a magnetic field. Curiously, this field is tilted 55° to the spin axis, and the north magnetic pole is closest to the south geographic pole. (This peculiarity has led some astronomers to speculate that we are observing Uranus in the middle of a magnetic pole reversal.) The spacecraft measured high-altitude winds at speeds of 300 to 400 m/s and an atmospheric content of helium of 10 to 15 percent (by number)—consistent with the values for Jupiter and Saturn. Voyager also noted emissions from the daylight side of the atmosphere, which are very puzzling because they are not auroral in nature. They seem to occur from molecules excited by low-energy electrons, but the source of the electrons is unclear. Overall, these

observations support a model of a planet with a rock and liquid core surrounded by an ocean of water and dissolved ammonia.

Computer-enhanced photographs showed that the ammonia clouds lie low in Uranus' atmosphere below deep layers of haze. Images taken through special filters revealed a banded structure to the clouds; those near the equator rotated once in 17 h, near the pole in 15 h—a resolution to the ground-based rotation rate measurements of both periods. Atmospheric winds generally blow the clouds in the same direction that the planet rotates. Strong plumes appeared in the upper atmosphere, probably generated by violent convection lower down.

Charged particle detectors onboard the spacecraft confirmed that a substantial magnetosphere surrounds the planet. Its presence was first suspected from synchrotron emission measured on the incoming flightpath. Overall, the total magnetic field

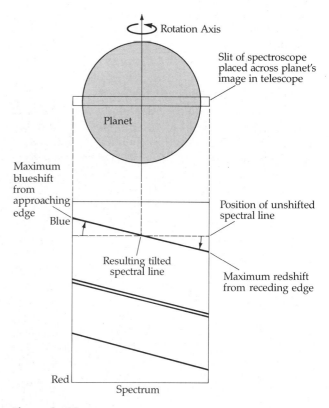

Figure 6–13

Doppler shifts and rotation. The variation of the Doppler shift across the disk of a rotating planet creates a tilt in the lines in an optical spectrum.

strength is about 0.1 that of Saturn, but the magnetosphere contains more energetic particles, many with energies greater than 1 MeV. As expected from a planet with a strong magnetosphere, aurorae actually were observed on the dark side of Uranus.

6–4 Neptune

The last of the Jovian planets, and the eighth planet from the Sun, is Neptune. This near-twin of Uranus is named for the god of the sea. Between 1790 and 1840, the orbit of Uranus exhibited perturbations from an unknown source, and the existence of a more distant planet was suspected. J. C. Adams (in 1843) and U. J. Leverrier (in 1846) independently used Newtonian celestial mechanics to deduce the mass and orbit of this eighth planet from the observed perturbations on Uranus. In 1846, Johann G. Galle at the Berlin Observatory found Neptune within 1° of the predicted position! Neptune may have been first seen by none other than Galileo, however, 234 years earlier. Calculations of Neptune's orbit show that it should have been very close to Jupiter in January 1613. Galileo's journals have entries showing that he observed an object in the vicinity of Jupiter near Neptune's predicted position on both December 27, 1612, and January 28, 1613, when he detected a small motion of Neptune with respect to a nearby star. Inexplicably, Galileo never followed up on this discovery and so failed to recognize the object as a new planet.

(a) Motions

Neptune moves around its low-eccentricity (0.0086), low-inclination (1.77°) orbit with a semimajor axis of 30.06 AU in a sidereal period of 164.79 years. Since its discovery, Neptune has traversed only three-quarters of its orbit. Its average distance from the Sun is less than Pluto's, but Pluto has such an eccentric orbit that it can—and has—come within Neptune's orbit. That occurred in January 1979, and until March 1999, Neptune will be the outermost planet in the Solar System!

Neptune's rotation period has been hard to pin down. Recent images in the near infrared (Figure 6–14) show atmospheric features that rotate in 17 h 50 min (±5 min), in good agreement with infrared photometry of rotation variations with a period of 17 h 43 min.

(b) Physical Characteristics

In many ways, Neptune is the twin of Uranus. Like Uranus, Neptune has a light green color from selective methane absorption. The upper atmosphere

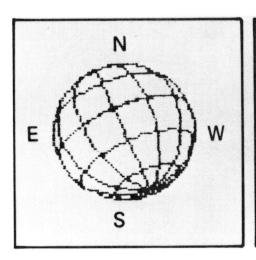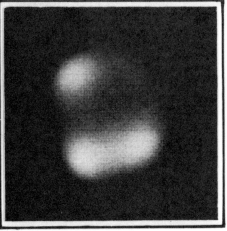

Figure 6–14

Infrared image of Neptune, taken at a wavelength of 890 nm. Note the three bright cloud features (each about the size of the Earth!) in the atmosphere. *(R. J. Terrile and B. A. Smith, Las Campanas Observatory, Carnegie Institution, and the Jet Propulsion Lab, NASA)*

displays faint cloud bands. This cold (≈ 55 K) atmosphere probably contains water ice and ammonia ice mixed with gaseous methane, hydrogen, and helium. One difference: ethane (C_2H_6) has been detected in Neptune's atmosphere but not in Uranus'. The internal structure of Neptune (Figure 6–12) probably resembles closely that of Uranus because the bulk densities and masses are similar.

Recent infrared observations show that Neptune's temperature is about 56 K; the expected blackbody value is 44 K if Neptune were heated only by the Sun. So Neptune, unlike Uranus, has internal heat. It gives off 3×10^{15} W, 2.5 times as much energy as it receives from the Sun — heat most likely left over from its formation.

Neptune may have weather in the sense of changing meteorological conditions in its upper cloud layers. Infrared observations indicate that its atmospheric reflectivity from 1 to 4 μm increased substantially over a one-year period. One interpretation of this is the formation of extensive cloud cover that then partially dissipated.

6–5 Pluto and Charon

Pluto, the ninth planet from the Sun (sometimes!), is named after the god of the underworld (Hades). From the Earth, Pluto presents only a faint stellar image at the telescope; from Pluto, the rest of the Solar System is distant and close to the Sun, which appears only as a very bright star in the sky.

After the discovery of Neptune, small unexplained perturbations appeared in the orbit of Uranus. Because the predicted position of a ninth planet that would cause these perturbations was very uncertain, initial attempts to find the planet were unsuccessful. It was not until March 1930 that Clyde W. Tombaugh found Pluto near a position predicted by Percival Lowell. Today we know that this discovery was a fluke because Pluto's low mass could not have caused the apparent perturbations of Uranus. If Pluto had not been near the ecliptic in 1930, it would never have been seen by Tombaugh. (Pluto is in this chapter mostly because of its position just outside the orbits of the Jovian planets. Pluto is much smaller than the Jovian planets, but it may have a similar density. In fact, it resembles the icy moons of the Jovian planets — Chapter 7.)

Pluto's average distance from the Sun is 39.44 AU. Since it has a highly eccentric orbit ($e = 0.25$), it ranges from 29.7 to 49.3 AU from the Sun and so is never closer to the Earth than 28.7 AU at opposition. Because of its great distance and small diameter, Pluto presents a difficult object to observe well. Attempts to measure its diameter have been frustrating. One near occultation indicated that Pluto is less than 6800 km in diameter; speckle interferometric observations with the Hale 5-m telescope indicate a diameter of 3000 to 3600 km.

Infrared spectral observations show that methane ice coats some of Pluto's surface. The methane ice there means that the surface temperature is no more than 40 K. This discovery also leads to a means of estimating Pluto's size by using its brightness, which depends on its distance from the Earth, its diameter, what fraction of its surface is covered with ice, and the albedo of ice and rock. If the surface is completely covered by highly reflective ice, then Pluto's diameter is roughly 3000 km — a lower limit.

Observations of Pluto's brightness have revealed cyclic variation every 6.4 days. It is the only evidence of rotation, and 6.4 days is generally accepted as

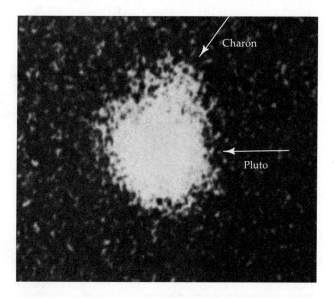

Figure 6–15

Pluto and Charon. The two bodies are so close together that their images are merged; Charon is a bump (arrow) off Pluto. (*J. Christy, U.S. Naval Observatory*)

Pluto's rotation period. Pluto's axis of rotation lies close to the ecliptic, like that of Uranus.

In June 1978, James Christy of the U.S. Naval Observatory in Flagstaff, Arizona, noticed what appeared to be a bump on Pluto's image in a photograph (Figure 6–15). Checking older photographs, Christy found seven showing the same bump, always oriented approximately north–south. He proposed that the bump was the faint image of Pluto's moon partially merged with the image of the planet. Christy named this moon Charon, after the mythological boatman who ferried souls across the river Styx to the underworld god Pluto.

Knowing the orbital properties of the Pluto–Charon system, we can find Pluto's mass by Kepler's third law. The orbital period is 6.4 days (the same as Pluto's rotation period; the two periods are tidally locked). The separation is 17,500 km (Figure 6–16). Let's compare the Earth–Moon system with the Pluto–Charon system. For the Earth and Moon,

$$M_E + M_M = (4\pi^2/G)(a_{EM}{}^3/P_{EM}{}^2)$$

and for Pluto and Charon,

$$M_P + M_C = (4\pi^2/G)(a_{PC}{}^3/P_{PC}{}^2)$$

Divide the second equation by the first to get

$$(M_P + M_C)/(M_E + M_M) = (a_{PC}/a_{EM})^3(P_{EM}/P_{PC})^2$$

Assume that the masses of the moons are smaller than their parent planets to approximate $M_E + M_M$ by M_E and $M_P + M_C$ by M_P. Then

$$\begin{aligned} M_P/M_E &= [(1.75 \times 10^4 \text{ km})/(3.8 \times 10^5 \text{ km})]^3 \\ &\quad [(27.3 \text{ days})/(6.4 \text{ days})]^2 \\ &= (1.0 \times 10^{-4})(18.2) \\ &= 1.8 \times 10^{-3} \end{aligned}$$

The Earth has a mass of 6×10^{24} kg, and so

$$\begin{aligned} M_P &= (6.0 \times 10^{24} \text{ kg})(1.8 \times 10^{-3}) \\ &= 1.1 \times 10^{22} \text{ kg} \end{aligned}$$

With Pluto's mass and diameter, we can figure out its density (within the range of uncertainty of the radius). The result is 500 to 800 kg/m^3, which implies that Pluto consists mainly of ices and other frozen gases.

We note that using Kepler's third law with the usual approximation does *not* give an *accurate* value for Pluto's mass because Charon has a mass about 10 percent that of Pluto. We find this mass ratio from

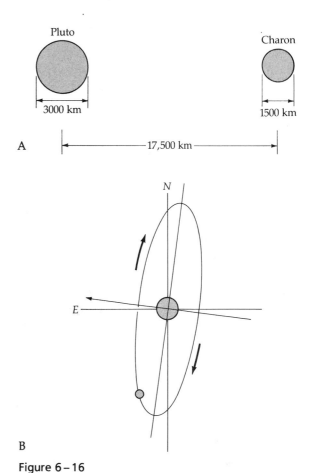

Figure 6–16

Pluto and Charon. (A) Relative sizes. (B) Charon's orbit around Pluto as viewed from the Earth in 1980.

observations of the center of mass of the Pluto–Charon system; Kepler's third law provides the sum of the masses. If Charon then has a density roughly that of Pluto (≈ 800 kg/m^3), its diameter is roughly 1300 km, about half that of Pluto.

The search at Lowell Observatory for other planets beyond Neptune ended in 1945 without yielding further positive results. A planet with the same characteristics as Pluto but placed at a greater distance from the Sun has little chance of discovery. Some astronomers have postulated a tenth planet orbiting retrograde beyond Pluto (from perturbations observed for comets), but nothing has been observed to date. Tombaugh's search, which covered 13 years, would have uncovered a planet like Neptune as far as 100 AU from the Sun.

Problems

1. Determine the orbital periods of particles at the inner and outer edges of Saturn's rings. At what distance from the center of Saturn will a particle orbit the planet in 10^h14^m? Show that the inner particles of the rings rise in the *west* and set in the *east* of Saturn's sky and the outer particles rise in the *east* and set in the *west*. Is this result paradoxical? Explain.

2. Show how the orbits of Uranus' moons appear from the Earth over a period of 100 years.

3. Show that the moons of Neptune obey Kepler's third (harmonic) law and deduce the mass of Neptune. (*Hint:* Use appropriate units or ratios.)

4. If Pluto's radius is 1500 km, what must its *mass* be to give the planet the same density as an icy moon of Saturn?

5. Jupiter has a strong magnetic field, about 10^{-5} T at a distance of 25×10^3 km from the surface (Pioneer 11). Estimate the size of Jupiter's magnetosphere and compare it with that of the Earth. Assume that the field is a dipole and that the solar wind pressure falls off as $1/R^2$ with distance R from the Sun. (*Hint:* Magnetic field pressure is proportional to the *square* of field intensity.)

6. Infrared observations indicate that Saturn gives off 2.8 times the energy it receives from the Sun for a total internal power loss of 2×10^{17} W. Assume that gravitational contraction releases this thermal energy. How much must Saturn shrink per year to account for this output?

7. Assume that Saturn's internal heat is left over from primordial contraction. Calculate the *maximum* bulk thermal conductivity the planet would need to retain enough of its internal energy to account for its present luminosity. Theoretical calculations indicate that Saturn's maximum luminosity was about 10^{20} W 4.5 billion years ago. The *thermal conductivity, κ*, is the flow of heat energy per unit time per unit area per unit temperature gradient (units: $J/s \cdot m \cdot K$), so

$$\kappa = -H/A(\Delta T/\Delta x)$$

where H is the flow of heat energy (J/s), $\Delta T/\Delta x$ is the temperature gradient (K/m), and A is the surface area (m^2).

8. Spectroscopic observations suggest that Pluto is covered with icy frost and thus has a high albedo (0.5). The brightness of Pluto at opposition (38 AU from the Earth) is 2×10^{-17} as bright as the Sun (1 AU from the Earth). From these two observations, calculate the radius of Pluto.

9. Imagine that you are viewing an eclipse of Charon by Pluto.
    (a) How could you use your observations to infer a diameter for Pluto?
    (b) By what percentage would the total brightness of the system dim at mid-eclipse?

10. Estimate the lifetime of methane in the atmospheres of Jupiter and Uranus.

11. Use the equation of hydrostatic equilibrium to estimate the central pressures of Saturn and Uranus.

12. Calculate the blackbody equilibrium temperatures of Uranus and Neptune and compare your values with the measured temperatures given in the text.

Small Bodies and the Origin of the Solar System

The previous four chapters put forth our current understanding of the planets. We have focused on the evolution of these large bodies of the Solar System but so far have only hinted at their formation. Key clues to the origin of the Solar System are locked in its small bodies: moons, rings, asteroids, meteoroids, comets, and interplanetary dust. This chapter treats the properties of this interplanetary debris and connects them to a contemporary model of the Solar System's formation, which took place 4.6 billion years ago.

7 – 1 Moons and Rings

Only three moons orbit all the terrestrial planets (our Moon and Deimos and Phobos of Mars). In contrast, the Jovian planets carry at least 50 moons (not counting Charon) as well as many rings, which contain a multitude of tiny moons. This section examines these objects, which range from small rocks to planetary-sized bodies.

(a) The Moons of Mars

Two moons encircle the planet Mars; appropriately they are named Phobos and Deimos ("fear" and "panic") after the mythological companions of the god Mars. Asaph Hall (1829–1907) at the U.S. Naval Observatory discovered the two moons in 1877, both of which lie close to Mars and orbit the planet rapidly (Figure 7–1). Deimos, the outer moon, circles Mars in 30.3 h; Phobos, the inner moon, takes a mere 7.67 h. In fact, Phobos is one of two moons (another is Jupiter's innermost satellite, 1979 J1, discovered by Voyager) that orbit their parent planets faster than the planets spin. So Phobos rises in the west and sets in the east as seen from the Martian surface! Like the Earth's Moon, the Martian moons keep the same face to the planet in synchronous rotation.

Spacecraft observations have found that Deimos and Phobos have ellipsoidal shapes with three axes. Phobos, the larger, has axes about 27, 21, and 19 km long; Deimos' axes are only 15, 12, and 11 km. Photographs also show that Phobos (Figure 7–2) and Deimos have cratered surfaces. The sizes and numbers of these craters indicate that the surfaces of these satellites are at least 4 billion years old and little modified since the cratering.

What is the origin of these miniature moons? One

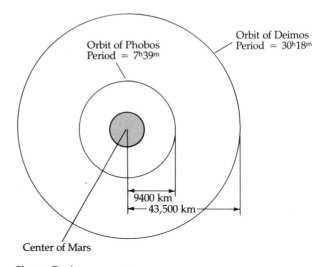

Figure 7 – 1

Orbits of Deimos and Phobos.

hint comes from the albedos: 0.022 for Deimos and 0.018 for Phobos in the visible — much less than our Moon's reflectivity. These dark surfaces resemble a certain class of meteorite (carbonaceous chondrites, Section 7–4) and asteroids (such as Ceres, Section 7–2). Also, recall that Mars is the closest of the terrestrial planets to the asteroid belt (the average distance between Mars and the asteroids is 1.3 AU). Hence, Mars may have captured both moons from high-eccentricity asteroids that passed close by it. Theoretical calculations indicate that such captures are possible. (And you will see that Jupiter has a set of asteroidal moons.)

(b) The Moons and Rings of Jupiter

Jupiter possesses an entourage of at least 16 moons. The brightest and largest were first discovered with a telescope by Galileo. Their orbits lie within 3° of Jupiter's equatorial plane, close to our line of sight. These huge moons orbit within 2×10^6 km of Jupiter in the following order: Io, Europa, Ganymede, and Callisto. As our Moon does to the Earth, each keeps one face toward Jupiter. The Galilean moons are relatively large (Table 7–1): Ganymede and Callisto are both larger than Mercury, and Io is somewhat larger and Europa somewhat smaller than our Moon.

Each Galilean moon is a world of its own, different from the others (Figure 7–3). These differences

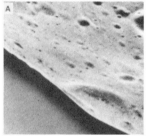

Figure 7-2

Phobos. These closeup photographs show craters as small as 10 m in diameter. Their flat bottoms indicate a surface soil hundreds of meters deep. *(NASA)*

are hinted at by the bulk densities: Io, 3500 kg/m³; Europa, 3000 kg/m³; Ganymede, 1900 kg/m³; and Callisto, 1800 kg/m³. This list is in order of increasing distance from Jupiter. Note the pattern: density decreases with increasing distance from Jupiter. Such density differences show that the compositions of Io and Europa resemble that of our Moon—mostly rock, with perhaps a little icy material. In contrast, Ganymede and Callisto must contain substantial amounts of water ice or other low-density icy mate-rials and proportionally much less rock than the inner moons.

Io

One of the most fascinating objects in the Solar System, Io is a world in its own right—three-fourths the size of Mercury! Remarkably Io has a thin atmo-sphere. (Only two other satellites, Saturn's Titan and Neptune's Triton, are known to have atmospheres.)

Table 7-1   Properties of the Galilean Satellites

Name	Diameter (km)	Average Distance from Jupiter (km)	Orbital Period (days)	Bulk Density (kg/m³)	Mass (moon = 1)
Io	3638	4.22×10^5	1.77	3530	1.21
Europa	3126	6.71×10^5	3.55	3030	0.66
Ganymede	5276	1.07×10^6	7.16	1930	2.03
Callisto	4848	1.883×10^6	16.69	1790	1.45

Figure 7-3

The Galilean moons of Jupiter in their correct relative sizes: Io (top left), Europa (top right), Ganymede (bottom left), and Callisto (bottom right). *(NASA)*

At the surface, the atmospheric pressure on Io is about 10^{-10} atm.

Io's atmosphere has a peculiar property: it gives off a yellow glow from emission by sodium atoms. This sodium glow surrounds Io like a yellow halo out to a distance of about 30,000 km (Figure 7-4) and then extends about 200,000 km along Io's orbit, forming a partial ring of gas around Jupiter. Volcanic eruptions, at least in part, produce Io's sodium cloud. We know from Voyager photographs that Io has at least 11 active volcanoes. In fact, it is the most volcanically active body in the Solar System; erupting volcanoes and fuming lava lakes imply that the interior is hot. The volcanoes eject plumes of gas and dust to heights of 250 km at velocities of up to 1000 m/s. (In contrast, the Earth's large volcanoes spit out material at about 50 m/s.) On a nearly airless body like Io, the volcanic gas and dust crest like a fountain plume in several minutes and then spread and fall in a dome shape (Figure 7-5).

Io's volcanoes have a different shape from those commonly found on the Earth, Venus, and Mars. Few appear as cones or shields. Instead, they resemble collapsed volcanic craters from which lava simply pours from a crater vent and spreads outward for hundreds of kilometers. Thus multicolored lava lakes surround many of Io's volcanoes; the temperatures in these lava lakes are about 330 K. The red, black, yellow, orange, and white coloration, most of which can be attributed to sulphur and sulphur compounds, makes Io spectacularly colorful. Because volcanic activity continually alters Io's surface, it must be very young. No impact craters appear on Io; volcanic flows have covered them up. Its surface is the youngest in the Solar System, probably less than 1 million years old.

Why is Io's interior hot? The other Galilean moons force Io gravitationally into an eccentric orbit, and so its distance from Jupiter changes significantly. These distance variations cause large and variable tidal forces from Jupiter to act on Io, whose interior heats up from the continual stress from the tidal forces.

Europa

The surface features of Europa consist of bright areas of water ice among darker orange-brown areas. Europa's surface is criss-crossed by stripes and bands that may be filled fractures in the icy crust.

The most impressive features on Europa are dark markings that criss-cross its face, making it look like a cracked eggshell (Figure 7-6). Some of these cracks extend for thousands of kilometers, splitting to widths of 50 to 200 km but reaching depths of only 100 m or so. Europa's surface is almost devoid of impact craters and so cannot be a primitive one; it must have evolved since its formation. The crust must have been warm and soft sometime after for-

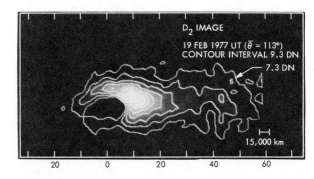

Figure 7-4

Io's sodium cloud. The contour lines represent the emission from the sodium surrounding Io. *(Jet Propulsion Lab, NASA)*

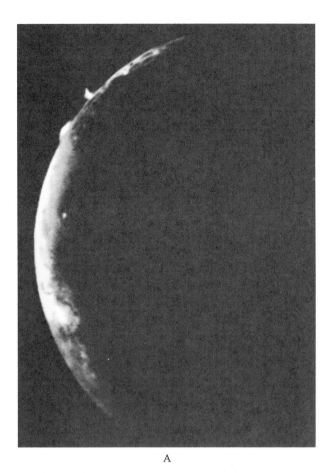

A

B

Figure 7 – 5

Volcanoes on Io. (A) Two eruption plumes are visible at the top. Their height is about 100 km. (B) The surface of Io with an erupting volcano. *(NASA)*

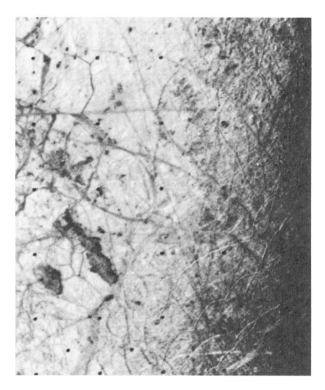

Figure 7 – 6

Europa's surface. This closeup view shows an area 600 by 800 km. The bright ridges are 5 to 10 km wide, the dark bands 20 to 40 km wide. *(NASA)*

mation to wipe out evidence of the early, intense bombardment.

Europa's cracked surface indicates that its solid, icy crust is thin and its interior hot and primarily molten. How did Europa get this way? One tentative model proposes that its crust long ago may have been a slush kept partially melted by a hot interior. As Europa cooled, its crust turned to smooth, glassy ice that later cracked.

Ganymede

The largest moon of Jupiter (radius 2640 km), Ganymede ranks overall as the largest moon in the Solar System. (Titan of Saturn is second.) Its surface looks vaguely like our Moon's, with dark, maria-like regions. It also has huge fault lines along its surface, as Europa does.

Ganymede has two basic types of terrain (Figure 7 – 7): cratered and grooved. Craters up to 150 km in diameter densely mark the cratered terrain. Their

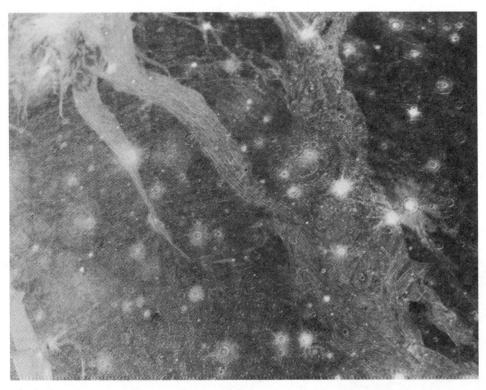

Figure 7–7

Ganymede's surface. This photograph shows features as small as 5 km. Note the fresh impact craters with bright rays. The dark regions are the oldest part of the surface. *(NASA)*

abundance indicates that the cratered terrain is some 4 billion years old. Compared with those on the Moon and Mercury, the craters are shallow for their size, and some have convex rather than concave floors. The craters of Ganymede also differ from those of the Moon and Mercury in that they have central pits rather than central peaks. Many craters on Ganymede have very bright rays extending from them (Figure 7–7), attesting to their formation by impacts on an icy surface.

The grooved terrain separates the cratered terrain into polygon-shaped segments. The grooved terrain consists of a mosaic of light ridges and darker grooves where the ground has slid, sheared, and torn apart. Long cracks, where the surface has moved sideways for hundreds of kilometers, also abound.

No large mountainous regions or large basins exist on Ganymede; nowhere does any relief amount to more than about 1 km. In some regions, small, bright patches with just a hint of surrounding walls look as if a crater has sunk into a soft surface. These features suggest that the crust of Ganymede is somewhat plastic. Crater rims and mountains slowly sink back into the surface; crater floors gradually fill in. This plastic flow probably is due to the large amount of water ice in Ganymede's crust. Ganymede's bulk density implies that its interior contains about half water and half rock. Occasional stresses on the water–rock crust have created the fracture patterns. Some ridges and grooves overlie others, an indication that there have been many episodes of crustal deformation.

Callisto

Farthest out of the Galilean moons, Callisto (Figure 7–8) has a surface that most resembles our Moon's and Mercury's. It is riddled with craters of a wide range of sizes. Some have bright ice rays; others are filled with ice. Callisto's craters are shallow, less

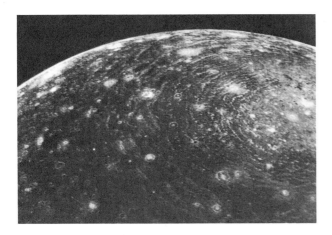

Figure 7–8

Callisto's surface. The outer ring of the large impact basin is about 2600 km across. Note the fresh impact craters on the dark, icy surface. *(NASA)*

than several hundred meters deep, because the surface is a mixture of ice and rock. The surface slowly flows, flattening out any relief.

Callisto has one huge and beautiful multiringed feature (Figure 7–8). The central floor of this feature is 600 km in diameter, and it is surrounded by 20 to 30 mountainous rings having diameters up to 3000 km. The rings look like a series of frozen waves that may have been formed in a stupendous collision that melted subsurface ice, then caused the water to spread in waves that quickly froze in the 100-K surface temperature. The ripple marks are preserved as rings: frozen blast waves. The central floor of this ringed feature has fewer craters than the rest of the terrain. This difference indicates that the impact forming the rings occurred after much of the surface cratering, probably 3.5 to 4 billion years ago.

Asteroidal Moons

Jupiter's other moons are asteroid-like bodies, and we expect that they are indeed captured asteroids. (Recall that Jupiter lies just outside the asteroid belt.) There are two groups of four moons each, one group at a distance of about 12×10^6 km that orbit counterclockwise (direct) and another at about 23×10^6 km that orbit clockwise (retrograde).

We have observed closely one other asteroidal moon—Amalthea. Only 181,000 km out from Jupiter, it orbits once every 12 h. It is elongated, 270 km by 155 km; the surface is cratered and has a dark red color. This moon's irregular shape, small size, and dark, cratered surface imply its asteroid-like character (more about asteroids in Section 7–2). It seems to resemble Deimos and Phobos (except that it is larger).

Rings

Voyager 1 discovered Jupiter's ring system. The rings are so thin (less than 30 km thick) that they are essentially transparent. They are most visible when viewed edge on; then the particles scatter light well, which implies that the particles must be small, about 3 μm in diameter. We do not yet know what they are made of, but on the basis of their infrared properties, it is likely they are a rocky material.

Dramatic pictures of the backlit rings (Figure 7–9) show them to have a definite structure. The outer, brightest part is 800 km wide and lies about 128,500 km from Jupiter's center. Within it is a broader ring 6000 km wide, and within that ring lies a faint sheet of material that extends from 119,000 km out from Jupiter's center down to the cloud tops. Note that the main ring extends from $1.72R_J$ to $1.81R_J$, which places the entire ring system well inside the Roche limit (Equation 3–9) for a fluid moon.

Recent computer processing of Voyager wide-

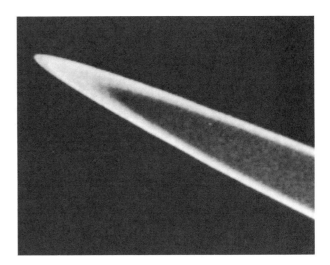

Figure 7–9

Jupiter's rings. This backlit view shows that the outer edge of the rings is thicker than the inner edge and that there is a sheet of particles between the two. *(NASA)*

angle photographs of the rings shows that a faint, outer ring surrounds the system. Called the gossamer ring, it lies mostly within the orbit of Amalthea, though some of its material extends out to 210,000 km. The ring's thickness is less than 200 km, and it is made of micrometer-size particles. These particles collide with the plasma in the magnetosphere and are swept out by these collisions in only about 1000 years. Hence, unless the gossamer ring is very young, its particles must be replenished from moons that lie close by.

(c) The Moons and Rings of Saturn

Saturn's band of moons totals at least 17. With two exceptions (Phoebe and Iapetus), all the moons stick close to Saturn's equatorial plane. Masses for some of the moons were determined from their gravitational attraction on spacecraft. The densities range from 1200 kg/m³ for Tethys to 1400 kg/m³ for Dione, similar to the densities of the outer Galilean moons of Jupiter.

The moons of Saturn fall into three groups: Titan by itself, the six large, icy moons (Mimas, Enceladus, Tethys, Dione, Rhea, and Iapetus, in order outward from Saturn), and the ten small moons (Phoebe, Hyperion, and the rest). Overall, their densities are less than 2000 kg/m³, which implies that they are mostly ice (60 to 70 percent) with some rock (30 to 40 percent). In contrast to the Galilean moons, no trend of densities follows with the distance from Saturn. Like Jupiter's moons, all of Saturn's moons except one (Phoebe) keep the same face toward the planet; they orbit in synchronous rotation.

Most of the moons are cratered. Some cratered terrain has been modified on the larger moons, which implies internal heating to melt parts of the icy surfaces. In contrast, the small moons, which are also cratered, show no changes—they still have their original surfaces. They may be pieces of an originally larger body. Note that the craters we see here imply that intense bombardment of planetary surfaces 4 billion years ago took place *throughout* the Solar System.

Titan

Titan, the largest moon, has a mass of 1.37×10^{23} kg and a radius of 2560 km. Its density is 1900 kg/m³, which implies a composition of half ice and half rock. Titan was the first moon found to have an atmosphere. The ultraviolet spectrometer on Voyager showed that this atmosphere consists mostly of nitrogen (99 percent) with about 1 percent methane. Several hydrocarbons have also been detected, including ethane, acetylene, ethylene, and hydrogen cyanide. The atmosphere's surface pressure is about 1.5 atm. The surface temperature is roughly 94 K.

Color photographs show a stratospheric layer of orange smog that varies to a blue color along Titan's edge. This variation indicates that the atmosphere varies in composition. No surface features were seen (Figure 7–10A). Voyager's pressure and temperature data, along with the spectroscopic detections of nitrogen and hydrocarbons, have led to models of a surface covered with a frigid ocean of nitrogen, methane, and ethane up to 1 km deep.

Other Moons

After Titan, Saturn's four largest moons are Iapetus, Rhea, Dione, and Tethys, with diameters ranging from 1020 to 1530 km. They appear heavily cratered (Figure 7–10B). In a few cases, wispy white streaks form rayed patterns around impact craters. These streaks are probably deposits of frozen ice, but whether from material emanating from the interior or from debris deposited by colliding bodies is unknown. Iapetus (Figure 7–10C) has the most extremes of surface cover. The hemisphere leading in its orbit is only 1/15 as bright as that following. The leading surface seems covered with dark debris picked up during its journey around the planet. Only Enceladus does *not* have a surface thick with craters, a sure sign that this satellite has suffered recent modification of the surface. A hot interior can melt the icy surface; one photograph shows a possible volcanic plume, which would clearly indicate a hot interior now.

The rest of the moons are all small bodies, a few hundred kilometers or less in diameter. The largest is Hyperion, 300 km in diameter. This moon has a strange shape, like that of a thick hamburger, and a cratered surface. The other moons are also cratered but much smaller, less than 30 km in diameter (Figure 7–11). We presume that all these bodies are basically ice, as are the larger moons.

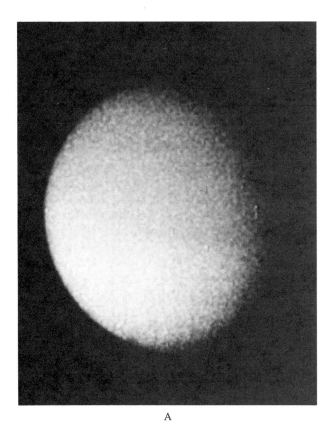

A

B

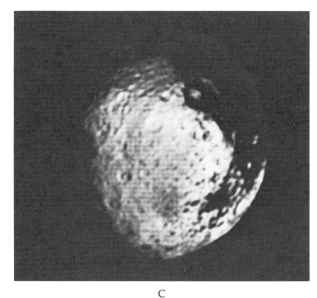

C

Figure 7–10

The moons of Saturn. (A) Titan, showing its hazy atmosphere. (B) Dione, showing its icy, cratered surface. (C) Iapetus, showing a black layer on one side covering its icy crust. *(NASA)*

Ring System

In 1659, Christian Huygens observed that Saturn "is surrounded by a thin, flat ring" that does not touch the body of the planet. Further observations by Cassini uncovered a gap in Huygens' single ring; this gap is known as *Cassini's division.* The rings lie in the planet's equatorial plane and therefore are tipped about 26° to the orbital plane; because of their tilt, they change their appearance as viewed from the Earth during the course of Saturn's revolution about the Sun. The near disappearances of the edge-on rings indicate that they are very thin, no more than 2 to 5 km thick. Although thin, the rings are wide; the three main ones visible from the Earth reach from 71,000 to 140,000 km from Saturn's center (Figure 7–12).

Figure 7–11

The small moons of Saturn. This composite photograph shows the heavily cratered surfaces of these irregularly shaped moons. *(NASA)*

The Voyager photographs reveal spectacular detail in the ring system. Although the A ring is relatively smooth, the B and C rings break up into numerous small ringlets (Figure 7–13). Many hundreds, perhaps a thousand, light and dark ringlets surround the planet, with widths as small as 2 km, the best resolution of the Voyager cameras. Some (in the C ring) appear elliptical rather than circular. Even the Cassini division, apparently empty as seen from Earth, was found to be filled with at least 20 ringlets.

Dark, spoke-like features occur in the B ring (Figure 7–14). Typically, the spokes are about 10,000 km long and 1000 km wide. They consist of very small particles, much smaller than the average particle in the rings. Because the inner particles orbit faster than the outer ones, the spokes last only a few hours. They may be small, darker particles with electric charges lifted out of the main ring plane by Saturn's magnetic

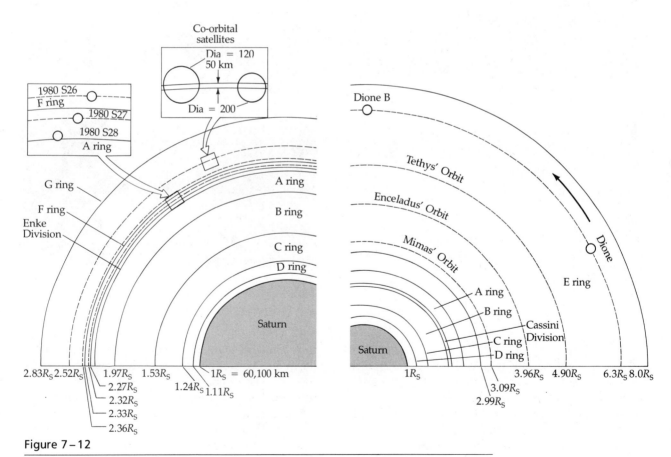

Figure 7–12

Saturn's rings. These diagrams show the ring system and some orbits of satellites as viewed from above Saturn's north pole. Units are Saturn radii.

Figure 7–13

Ringlets. Saturn's rings contain many smaller rings within them. *(NASA)*

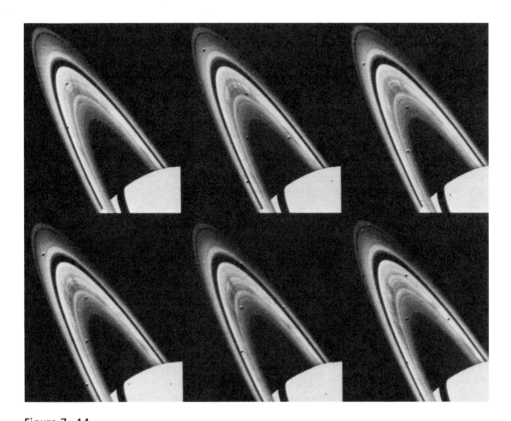

Figure 7–14

Spoke-like features in the rings. This sequence was taken at intervals of 15 min. *(NASA)*

field. Note that these spokes do *not* obey Kepler's laws because different orbital periods would break up the spokes in one or two revolutions. Hence, the particles producing the spokes cannot be attached to the ring particles.

Pioneer 11 discovered a new ring out beyond the previously known ones. Called the F ring, it lies 3500 km outside the edge of the rings visible from the Earth. The F ring appears to be 320 km wide and a mere 3 to 4 km thick. Voyager 1 photographs resolve this ring into a complex system of knots and a braided structure of at least three strands; Voyager 2 photographs taken nine months later show that the braiding had disappeared. Apparently, the braiding was a dynamically unstable situation. Another such very narrow ring (the G ring) is 10,000 km farther out. Two other extremely faint rings are known. The E ring extends out beyond the F ring to at least 6.5 Saturn radii (400,000 km). A ring inside the C ring, called the D ring, extends at least halfway to the surface of Saturn.

The gravitational effects of the satellite just outside the A ring (1980 S28) and the two that straddle the F ring (1980 S27 and S26) play important roles in the dynamics of the rings. The F-ring moons, in particular, are called the *shepherd satellites* because they keep the ring particles in a narrow range of orbits. The inner moon accelerates the inner ring particles as it passes them (as expected from Kepler's third law, it has a shorter orbital period). They spiral outward to larger orbits just as the tidal force of the Earth on the Moon forces the Moon into a larger orbit [Section 3–4(b)]. In a physically similar way, the slower-moving outer moon decelerates outer-ring particles as they pass by, so that they spiral inward. The balance of these interactions constrains the particles' motions and preserves the narrowness of the F ring. Likewise, the A-ring shepherd causes the sharp outer edge of the A ring.

The rotation rate of the rings varies according to Doppler-shift data. The velocities range from 16 km/s at the outer boundary of the A ring to 20 km/s at the inner boundary of the B ring. The measured velocities agree with those expected from Kepler's third law for individual masses placed at the ring distances from Saturn; this agreement indicates that separate particles make up the rings.

Infrared observations of Saturn's rings show that they are made of particles of water or ice or of rocky particles coated with water ice. The ice does not evap-orate because the surface temperature of the particles is only about 70 K. At this equilibrium temperature, the rings' icy material has a very low vapor pressure, and so the ice stays in solid form. Radio signals from Voyager reflected by the rings indicate that the particles are about 1 m in diameter, but a range of sizes, from centimeters to tens of meters, probably exists. Although covering a large area of space, the rings have a total mass estimated to be only 10^{16} kg, about 10^{-10} the mass of Saturn.

(d) The Moons and Rings of Uranus

The five major moons of Uranus (Figure 6–11) move in the planet's equatorial plane and revolve in the same direction as the planet rotates. Because the moons lie in the same plane as Uranus' equator, their orbits as seen from the Earth are alternately edge-on and fully open every 21 years; in 1966 they appeared on edge, but in 1987 they will appear as circles.

The five moons are called Miranda, Ariel, Umbriel, Titania, and Oberon. Miranda is the smallest (less than 320 km in diameter) and closest to Uranus. The others range in diameter from 1110 km (Umbriel) to 1630 km (Oberon). Their surfaces appear to be made of a dirty ice, very much like that of Saturn's Hyperion. Recently, the masses of these moons have been estimated, and the inferred bulk densities range from 1300 to 2700 kg/m^3, which implies that these are bodies made of rock and ice.

Voyager 2 transformed our view of Uranus' moons much as it did our view of the moons of Jupiter and Saturn. First, the spacecraft discovered ten more moons, for a total of at least 15. The first new moon discovered by Voyager is called 1985U1; it has a diameter of only 100 km and orbits closer to Uranus than do the five large moons. Six other moons, named 1986U1 through 1986U6, orbit between Uranus and 1985U1. These inner moons have diameters between 30 and 50 km. Second, it provided closeup views of the largest moons (Figure 7–15A through E). Miranda (Figure 7–15A) has the most complex surface, with ten types of terrain that appear tectonically shuffled (perhaps by tidal forces of Uranus). Oberon (Figure 7–15B) is densely covered with impact craters and has at least one mountain peak—probably volcanic—about 5 km high. Titania (Figure 7–15C) has a surface plastered with impact craters and strewn with valleys 50 to 100 km wide and hundreds of kilometers long (one of them

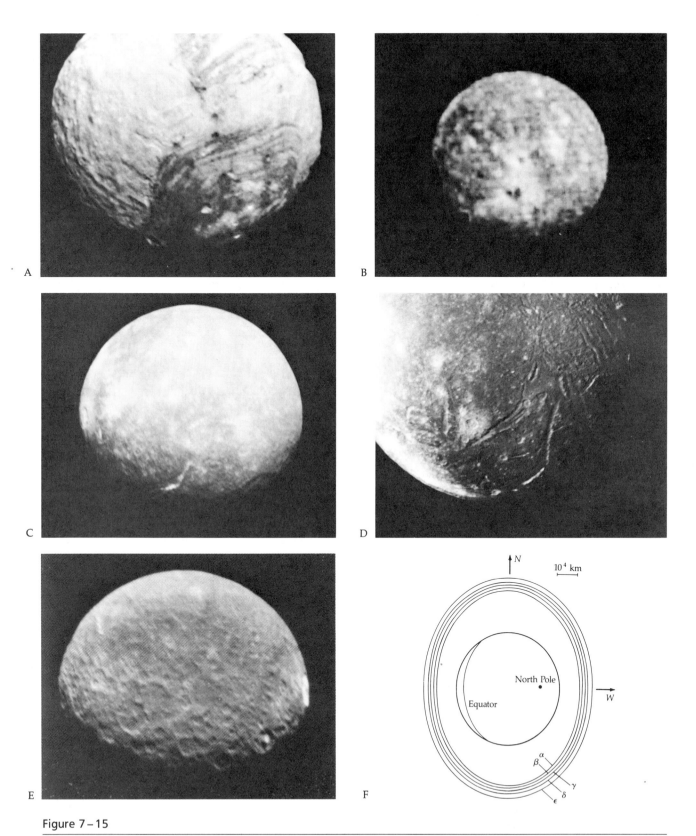

Figure 7–15

(A) Miranda. *(NASA)* (B) Oberon. *(NASA)* (C) Titania. *(NASA)* (D) Ariel. *(NASA)* (E) Umbriel. *(NASA)* (F) Rings of Uranus. The five major rings of the nine total are shown here.

cuts across the entire surface). Ariel also has impact craters and large fractures and valleys (Figure 7–15D). Finally, Umbriel shows the least dramatic surface, with overlapping impact craters but no special dramatic features (Figure 7–15E). We again find evidence here for the era of torrential impacts early in the history of the Solar System.

Uranus has rings discovered accidentally in March 1977 by three groups of astronomers — from Cornell, Lowell, and Perth Observatories and the Indian Institute of Astrophysics. They watched Uranus occult a faint star and were surprised to see the star momentarily dimmed a few times before and again after Uranus had covered it. Later observations of Uranus passing in front of another star in 1978 showed nine rings. Voyager 2 found two more for a total of at least 11.

These observations paint a picture of a ring system dramatically different from that of Saturn. The 11 rings circle the planet in roughly three groups (Figure 7–15F): rings 6, 5, and 4 at about 42,500 km; α and β at 45,000 km; η, γ, and δ at 48,000 km; and the ϵ ring and two others at 51,000 km from the center of Uranus. The narrowest rings have widths of only about 5 km; the ϵ ring is 100 km wide.

Voyager images indicate that one new ring, called 1986U1R, lies midway between the δ and ϵ rings and that the ϵ ring splits into at least two ringlets. A continuous distribution of dust particles fills the entire ring plane. And the outermost ϵ ring is flanked by two shepherd satellites (called 1986U7 and 1986U8; each is only about 15 km in diameter), which serve to keep the ring stable and intact.

We did not detect the rings of Uranus until recently because they are dark and not very wide. The material they are made of is almost black (visual albedo less than 0.03)! In contrast, because they are covered with water ice, the particles in the rings of Saturn reflect more than 80 percent of the light hitting them. Voyager 2 radar observations showed that the particles in the rings of Uranus are dark ice with a size of about 1 m and an orbital period of 8 h.

(e) The Moons and Rings(?) of Neptune

Neptune has two known moons, Triton and Nereid (Figure 7–16). Triton has a diameter of about 5200 km, which makes it one of the largest satellites in the Solar System. Nereid has a diameter of 900 km or so.

Figure 7–16

Neptune and Triton (arrow). *(Lick Observatory)*

Triton revolves with a period of about 5 days in a retrograde (east-to-west) orbit that is inclined 20° to the plane of Neptune's equator. No other planet has a close moon moving retrograde and so steeply inclined. This strange orbit may be related to the speculation that Pluto and Charon are escaped moons of Neptune. Nereid, the outer moon, has an orbital eccentricity of 0.75 — two times larger than that of any other solar system satellite; its distance from Neptune ranges from 1 million to 10 million km. Nereid's orbital velocity of 3 km/s at closest approach to Neptune is only 0.2 km/s shy of escape velocity. Triton's and Nereid's unusual orbital characteristics can be explained if the system was disrupted by an outside body that had more mass than the present satellites.

Let's examine one scenario. Assume that Neptune's moons (four major moons in all) originally had circular, equatorial orbits modeled after the Galilean moons of Jupiter. A hypothetical disrupting body, with a range of assumed masses, runs through the satellite system. Theoretical calculations have focused on encounters that maximize the frequency of interesting results for a mass between $0.2M_N$ and $2M_N$. One trajectory has especially curious results (Figure 7–17). Here a planet having a mass of $3M_\oplus$ comes in at a 30° inclination angle and plunges through the system of four satellites. Satellite 1 gets captured by the perturbing planet and pulled off Neptune. Satellite 2 also escapes and ends up in a Pluto-like orbit. Satellite 3 has its orbit flipped to

in May 1981 and observed no rings. However, unpublished data on the occultation that occurred in 1968 have recently been uncovered, showing a dip in brightness 3 min after the star passed behind the planet and lasting for 2.5 min. If this dip was caused by a ring, then the radius of the ring is 28,600 km ($1.14R_N$) and its width is 4300 km.

7-2 Asteroids

Chapter 2 presented the general orbital properties of asteroids. Here we will focus on their physical characteristics because these inform us about the conditions of solid matter early in the Solar System's history. Basically, an asteroid is an irregular, rocky body both smaller and less massive than a planet. (Only about 200 asteroids have diameters greater than 100 km; some 10^6 are thought to be in the asteroid belt.) Ceres, the largest known asteroid, has a diameter of only about 1000 km. Asteroid sizes can be measured directly when an asteroid occults a star. We can also measure an asteroid's reflectivity of visible light and its infrared emission (typically at 10 μm) when it is at a known distance from the Sun. These two measures in an equilibrium state allow an indirect estimate of the asteroid's size.

The amount of solar energy falling on an asteroid is

$$(L_\odot/4\pi D^2)\pi R^2$$

where $L_\odot$ is the luminosity of the sun, D the distance from the Sun, and R the radius of the asteroid. A fraction A (the albedo) is reflected back into space. If the Earth is a distance d from the asteroid, the flux of reflected light at the Earth is

$$F_{vis} = (L_\odot/4\pi D^2)\pi R^2(A/4\pi d^2)$$

We can measure this flux and hence determine the quantity R^2A. The fraction of energy absorbed, $1 - A$, heats up the asteroid and is re-emitted into space as infrared radiation. We can observe this infrared flux at the Earth. The ratio of visible to infrared flux is

$$F_{vis}/F_{IR} = A/(1 - A)$$

From this measurement, we can determine A and then use this value with the previous determination of R^2A to calculate R.

Recent observations of the surface reflectivity of

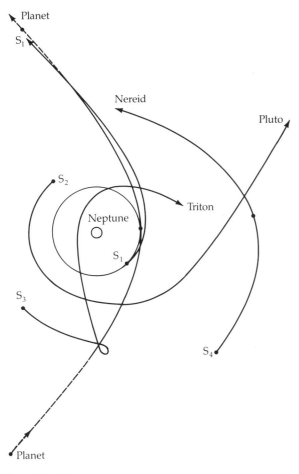

Figure 7-17

A possible scenario for the disruption of the Neptune system by a passing planet.

retrograde and becomes Triton. Satellite 4, the outermost, falls into a Nereid-like orbit. The disrupting planet ends up in an elliptical orbit with semimajor axis less than 100 AU and eccentricity less than 0.6 —another planet beyond Neptune's orbit.

Triton has an atmosphere; infrared observations shown an absorption band at 2.3 μm, which is attributed to gaseous methane. The pressure due to methane at Triton's surface is roughly 10^{-7} atm. Other observations indicate that the atmosphere may also contain molecular nitrogen. If so, the nitrogen is at such a high pressure that pools of liquid nitrogen may exist on the surface, as well as patches of methane ice.

Neptune may also have a ring system very close to itself. An international collaboration of astronomers observed two stellar occultations by Neptune

the larger asteroids have given some new hints about their compositions. Asteroids have a wide range of reflectivities, from Nysa (diameter 82 km), whose albedo is 35 percent, to Cybele (diameter 309 km), which reflects only 2 percent of the visible light that hits it. Nysa's surface reflects sunlight about as well as the icy satellites of Jupiter and Saturn; Cybele has an albedo similar to that of the rings of Uranus.

The reflectivities indicate that most asteroids fall into two major compositional classes. Some are relatively bright with albedos of about 15 percent, and others are much darker, with albedos of 2 to 5 percent, indicating that they contain a substantial percentage of dark compounds, such as carbon or the black mineral magnetite (Fe_3O_4). These dark asteroids resemble a class of meteorites (Section 7–4), the carbonaceous chondrites, which are dark because they contain carbon compounds (roughly 1 to 5 percent carbon). The lighter class is dubbed *S-type* asteroids, the darker ones *C-type* asteroids. The S type, in addition to having higher albedos, also show spectral absorption bands indicative of silicate materials. A third class, called *M-type* asteroids, has characteristics suggestive of metallic substances. They have albedos of about 10 percent. Only 5 percent of all asteroids belong to this last class.

Recent investigations indicate that, based on albedos, compositions in the asteroid belt vary with distance from the Sun. Near the orbit of Mars, almost all asteroids have S-type characteristics. Farther out are fewer high-albedo ones and more dark ones. At the outer edge of the belt, 3 AU from the Sun, 80 percent of the asteroids are C type.

We already noted [Section 2–1(c)] that many asteroids have irregular shapes as deduced from their light fluctuations. Some of the smaller natural satellites, like Mars' Phobos and Deimos, are also irregular. These objects are all small enough that their internal pressures are too low to overcome the strength of the material of which they are composed. Ceres, with a diameter of about 1000 km, is beyond this limit and so is spherical.

Do asteroids have moons? That depends on the stability of a binary (double) asteroid system to tidal forces. Asteroids have very small masses; Ceres, for example, if it were all rock, would have a mass of only about 10^{21} kg. A typical asteroid is about 10 times smaller and has a mass of only 10^{-3} that of Ceres. This small mass means that the gravitational force between the bodies would be so weak that the tidal force of the Sun or Jupiter could easily disrupt the binary system. Any existing system is likely to have a very small separation between asteroid and moon, and so it would be hard to see the two separate objects telescopically.

Occasionally, as seen from the Earth, an asteroid will occult a star. The asteroid will block out the star's light for a few seconds; a moon would cut off the light for even less time. A few such moon-occultation observations have been reported. The most impressive occurred in 1978 with the asteroid Herculina (diameter 220 km); the suspected moon was 975 km away and had a size of only 50 km. Only one other observation looks as good as that for Herculina (for the asteroid Pallas); about 20 others hint at companions. (These observations may not yet be valid evidence for asteroidal companions. None, for instance, has yet been repeated.)

If Herculina does have a moon, the system's lifetime, before tidal forces disrupt it, is only about 10^7 years. This short time implies that asteroidal moons are probably formed when collisions between asteroids break them up; some of the debris from the breakup may remain bound as moons for a short period of time.

7–3 Comets

When first sighted telescopically, a comet typically appears as a small, hazy dot. This bright head of the comet is called the *coma* (Figure 7–18A). Sometimes the coma contains a small, starlike point called the *nucleus* (Figure 7–18B). Cometary nuclei are very small, perhaps no larger than 1 to 2 km across, certainly less than 50 km; none has ever been viewed as more than a point of light. As a comet moves toward perihelion, it grows brighter and sprouts a *tail* (Figure 7–18C). A comet's tail may stretch for millions of kilometers and always points away from the Sun. Surrounding the entire comet when it is in the inner Solar System is a gigantic halo of hydrogen gas, spanning millions of kilometers. This halo, detectable from spacecraft only in the ultraviolet, results from the dissociation by sunlight (photodissociation) of the hydroxyl radical (OH^+) in the coma.

Comets may have two types of tails: ionized gas and dust (Figure 7–18B). The physical difference between the two shows up in their spectra. The spec-

A

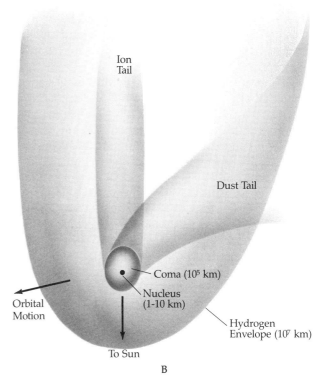

B

C

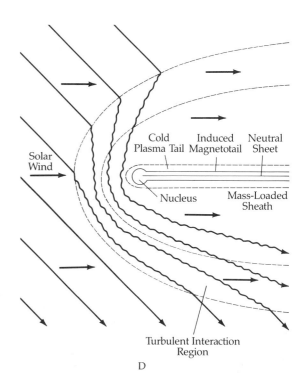

D

Figure 7-18

(A) Head of Halley's Comet, November 16, 1985. *(B. Laubscher, Capilla Peak Observatory, University of New Mexico)* (B) Main parts of a comet. (C) Head and tail of Comet Kohoutek in January 1974. *(Hale Observatory, California Institute of Technology)* (D) Schematic diagram of the structure of the interaction of Comet Giacobini-Zinner with the solar wind. *(NASA)*

trum of the ion tail has emission lines. The dust tail's spectrum is that of sunlight reflected from dust expelled from the coma. The pressure from sunlight detaches dust from the coma and pushes it out to form a tail.

In the spectrum of the ion tail, the most conspicuous spectral lines are those produced by carbon monoxide (CO), carbon dioxide (CO_2), nitrogen (N_2), and radicals of ammonia (NH_3) and methane (CH_4). Puffs of gas sometimes shoot through the tail. Radiation pressure from sunlight cannot account for the fact that these ion tails point straight away from the Sun. The German astronomer Ludwig Biermann suggested in 1951 that the solar wind has a major effect on the ionized tails. Measurements of the solar wind confirmed that the magnetic fields carried by the wind's particles can indeed drag ions from the comet's coma.

The ICE (International Cometary Explorer) intercept of Comet Giacobini-Zinner in September 1985 revealed the true complexity of the interaction between comets and the solar wind. ICE's passage through the comet's tail verified that the interaction produces many energetic ions around the comet. These ions are picked up by the wind; the additional mass slows down the flow behind the comet and produces the long, filamentary ion tail. Giacobini-Zinner's tail was 25,000 km thick where it was crossed by ICE (7800 km downstream from the nucleus) and contained an induced bipolar magnetotail split by a neutral sheet between the opposite current flows (Figure 7–18D). Electron densities in the heart of the tail exceed $10^9/m^3$. The magnetic field here is presumed to come from the field carried by the solar wind. Hence, we finally have good information about the magnetic field–plasma interactions that fix the form of a visible comet.

At great distances from the Sun, the coma shows a reflected solar spectrum, and so the cometary head must also contain solid particles that reflect sunlight. At about 1 AU from the Sun, the head exhibits molecular emission bands of carbon (C_2), cyanogen (CN), oxygen (O_2), hydroxyl (OH), and hydrides of nitrogen (NH and NH_2). As the comet speeds nearer to the Sun, emission lines of silicon (Si), calcium (Ca), sodium (Na), potassium (K), and nickel (Ni) appear. Table 7–2 summarizes the materials that have been observed to date in the heads and tails of comets. Note these two points about the composition. First,

Table 7–2   Observed Composition of Comets

Head	Tail
H, C, C_2, C_3, CH, CN, HCN, CH_3, CN, NH, NH_2, O, OH, H_2O, Na, K, Ca, V, Cr, Mn, Fe, Co, Ni, Cu plus dust particles with silicates	CH^+, CO^+, CO_2^+, N_2^+, OH^+, H_2O^+, Ca^+ plus dust particles with silicates

Source: F. Whipple and W. Huebner, *Annual Review of Astronomy and Astrophysics* 14:143 (1976).

the head contains some of the same molecules found in interstellar space, such as hydrogen cyanide (HCN) and methyl cyanide (CH_3CN). Second, infrared spectra of some comets display the 10-μm and 18-μm bands characteristic of silicate dust.

The observed brightness B of a comet depends upon its distance from the Sun R (which determines its fluorescence and the amount of sunlight reflected) and its distance from the Earth r (which determines the flux of radiation we receive):

$$B \propto R^{-n} r^{-2} \qquad (7-1)$$

Far from the Sun, no fluorescence occurs and so $n = 2$. Near the Sun, however, $n \approx 4$ and may range from 2 to 6, depending upon the particular comet; for Halley's Comet, $n \approx 5$ for distances closer than 6 AU. How much a comet glows (primarily in its coma) depends on the composition and amount of gases released by the nucleus.

For all their stunning length against the sky, comets have very small masses. Comet masses can be estimated only roughly because they are so small that they do not much affect the orbits of other bodies. Halley's Comet, one of the largest, has an estimated mass of only about 10^{16} kg, and it loses about 10^{11} kg during each perihelion passage. In 1910, the tail of Halley's Comet stretched about 100°, reaching from horizon to horizon; the lack of any noticeable effects as the Earth traveled through the tail indicates that the gas was quite rarefied, less than about 10^{-17} kg/m^3. The density of the coma is estimated at 10^{-7} kg/m^3, much higher than the tail's density but still a good vacuum.

The mass expelled from a comet, mostly as gas,

comes from the nucleus. Fred L. Whipple has developed a *dirty-iceberg cometary model* in which the comet nuclei are compact, solid bodies made of frozen gases (ices) of water, carbon dioxide, ammonia, and methane embedded with rocky material. Beyond Jupiter, low temperatures allow the ice–gravel conglomerate to persist unchanged for long periods of time. As the comet nears the Sun, the icy material vaporizes. This released material enlarges the coma and creates the tail. As the ice evaporates, a thin coating of rocky material remains to form a solid but fragile crust on the nucleus. The heating of the subsurface material creates gas jets that blow off puffs of gas and act as small rockets that slightly change the comet's orbit.

As a comet rounds the Sun, the semisolid nucleus can usually withstand the solar heat. Comet Ikeya-Seki, the great sun-grazing comet of 1965, passed within 470,000 km of the solar surface and survived. Other comets passing close to the Sun are not so lucky; Comet West (1976) split into at least four pieces after its perihelion passage (Figure 7–19).

Periodic comets lose a little material each time they pass the Sun and eventually must be completely outgassed. How, then, can we explain their abundance? All comets are gravitationally attached to the Sun. Most comets have long orbital periods. From the observed orbits, we find that the average value of their semimajor axes is about 50,000 AU and the corresponding orbital period is about 10^7 years. The orbits are highly elliptical; in accord with Kepler's second law, the comets travel very slowly at aphelion, only a few kilometers per day, and so such comets spend most of their time coasting far from the Sun.

A cometary cloud, proposed by the Dutch astronomer Jan Oort in 1950, is sometimes known as *Oort's cloud.* It makes up the Solar System's cometary reservoir. According to Oort's model, most comets never come very near the Sun and we never see them. Occasionally, however, the gravitational action of passing stars pushes (or pulls) a comet into an orbit that does bring it closer. Comets are eventually lost either by vaporization of the nucleus or by Jupiter's perturbation of their orbits. So the supply of Sun-approaching comets must be replenished with new comets from the cloud. To ensure sufficient input to make up the losses, Oort's picture requires at least 10^{11} (and perhaps as many as 10^{14}) comets to be clustered in the cloud.

A

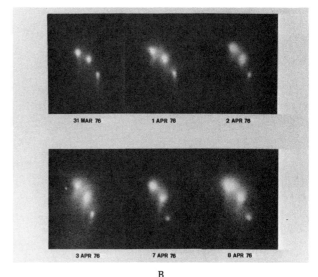

B

Figure 7–19

Comet West. (A) A few days before its breakup, Comet West had a knotty tail. (*Royal Observatory, Edinburgh*) (B) The breakup of the nucleus of Comet West in April 1976. (*C. F. Knuckles and A. S. Murrell, New Mexico State University Observatory*)

7–4 Meteoroids and Meteorites

When a meteoroid—interplanetary debris—enters the Earth's atmosphere, it generates the streak of light in the sky called a *meteor*. If any material survives the plunge through air, it strikes the ground as a *meteorite*. Meteoroids hit the atmosphere with speeds ranging from 12 to 72 km/s, and more meteors are seen after midnight than before (Figure 7–20). Meteors belong to the Solar System, so that their speed at the Earth's orbit cannot exceed the solar escape speed there of 42 km/s. Before midnight, only those meteoroids moving faster than the Earth (30 km/s) can catch up with it from behind; the relative speed of the fastest such meteor is 12 km/s. After midnight, all meteoroids except those moving faster than the Earth along its orbit will be seen; now velocities add to give a maximum relative speed of 72 km/s.

Because a meteor's trail provides a brief record of the body's disintegration, astronomers have been able to determine the orbit and general physical characteristics of meteoroids. Most meteoroids are fragile particles that crumble quickly when they come in contact with air (Figure 7–21). A block of meteoroid material having a volume of 1 m^3 would crumble under its own weight, for it is no stronger than cigarette ash.

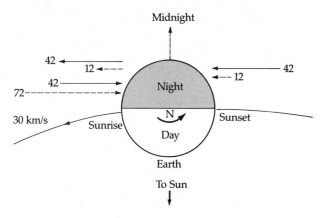

Figure 7–20

Meteor speeds. The solar escape velocity at the Earth's orbit is 42 km/s; meteoroids cannot move faster than this speed. In early evening, meteors catch up to the Earth. After midnight, the Earth (orbital speed 30 km/s) catches up to all but the fastest meteoroids moving along its orbit.

What is the source of this low-density meteoroid material? Comets. During a comet's successive passages by the Sun, solar heating causes a continual loss of icy material from the cometary nucleus. The dust and solid particles interspersed in the ice flake off and scatter in an array around the comet. This solid debris is very fragile and has a low density. The older the comet and the greater its number of passages by the Sun, the greater the total loss of icy material and release of meteoroid material. About 99 percent of all meteors are of cometary origin.

The remainder are probably associated with asteroids. Only those meteors over a certain size can survive vaporization in the atmosphere and hit the ground as meteorites. In terms of physical and chemical composition, astronomers divide meteorites into three broad classifications: irons, stones, and stony irons. The *irons,* which are generally about 90 percent iron and 9 percent nickel with a trace of other elements, are the most common finds. The *stones* are composed of light silicate materials similar to the Earth's crustal rocks. Though actually the most common kind of meteorite seen to fall, they are difficult to distinguish from an ordinary terrestrial stone and so make up a small fraction of the finds (Figure 7–22). When examined under a microscope, many stones are seen to contain silicate spheres called *chondrules* embedded in a smooth matrix. These stones are known as *chondrites.* The *stony irons* represent a cross between the irons and the stones and commonly exhibit small stone pieces set in iron. Irons are the densest meteorites, with densities ranging from 7500 to 8000 kg/m^3. Stones are the least dense, averaging from 3000 to 3500 kg/m^3. Stony irons, because they are a mixture of stones and irons, have an intermediate density of 5500 to 6000 kg/m^3.

One of the most curious kinds of chrondrites is the *carbonaceous chondrite.* The chondrules in these meteorites are embedded in material that contains much more carbon than other stony chondrites, typically from 1 to 4 percent carbon by mass. Their carbon content gives these meteorites a dark appearance. Carbonaceous chondrites also contain water (ranging from 3 to 20 percent) and volatile materials. In addition, the relative abundances of condensable elements in carbonaceous chondrites are closer to those found in the Sun's photosphere than to those found in the crust of the Earth. That is, if some gas were extracted from the Sun and cooled to below the freezing point of water, the condensed elements

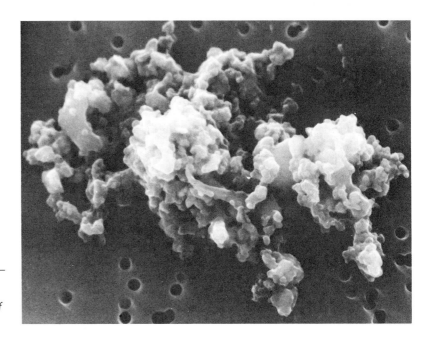

Figure 7-21

A meteoroid. This particle of interplanetary dust is about 10 μm across. Note its flaky structure. *(D. Brownlee, University of Washington)*

would have relative chemical abundances quite different from the Earth's but very similar to those of carbonaceous chondrites. This similarity suggests that carbonaceous chondrites formed out of the same primordial material as the Sun and have suffered no major bulk heating or changes since that time.

An important clue about the origin of iron mete-

orites comes from etching them with acid on a polished surface. Large crystalline patterns called *Widmanstätten figures* become visible (Figure 7-23). Terrestrial iron does not show such patterns when etched. Widmanstätten figures cleanly distinguish a meteorite from terrestrial iron and also give a clue to

Figure 7-22

A stony meteorite. A piece from the Allende, Mexico, fall. Note the light-colored inclusions. *(M. Zeilik, Institute of Meteoritics, University of New Mexico)*

Figure 7-23

An iron meteorite. The crystalline structure is visible in the flat, polished section of this meteorite, which fell in Dora, New Mexico. *(M. Zeilik, Institute of Meteoritics, University of New Mexico)*

the history of the meteorite material. A nickel–iron mixture, when cooled slowly under low pressures from a melting temperature of about 1600 K, forms large crystals. The cooling must be very gradual (about 1 K every 10^6 years). Metals conduct heat well, however, and in the cold of space a molten mass of nickel and iron would cool rapidly and not form large crystals. Nickel–iron meteorites could grow Widmanstätten figures only if they had protection from the cold, and so it's likely that nickel–iron meteorite material solidified inside small bodies, termed *parent meteorite bodies.* To allow a cooling of only 1 K in 10^6 years, these bodies must have been at least 10 km in diameter.

Such bodies probably formed with the formation of the Solar System. Parent meteorite bodies are envisioned as having been only a few hundred kilometers across. Once formed, they could be heated by the radioactive decay of such short-lived isotopes as ^{26}Al. When heated to melting, a parent meteorite body differentiates; the densest material falls to the center and the least dense comes to the surface. So the object ends up with a core of metals and a cover of rocky material, which cools to form a crust. This crust insulates the molten metals and allows them to cool slowly and form large crystals. Much later, the parent meteorite bodies collide and fragment. Pieces from the outer crust make stony meteorites, pieces from farther down become stony-iron meteorites, and the core produces iron meteorites.

This scenario requires that the parent meteor bodies were among the first solid objects to form in the young Solar System. Radiometric dating of meteorites backs up this idea: the ages all fall close to 4.6 billion years. So meteorites provide us with the most direct evidence of primitive solids—chemically, isotopically, and texturally. In addition, they give insights into the evolution of their parent bodies.

Not all meteorites originate from asteroids. Some (very few) come from comets—perhaps selected carbonaceous chondrites. However, we have four meteorites that very likely came from the Moon! Since 1981, scientists from the United States and Japan have recovered thousands of meteorites buried in old ice in Antarctica. Four of these—each only a few centimeters across—are breccias with light-colored inclusions. These meteorites have a texture similar to that of lunar breccias; their chemical composition is similar to that of lunar rocks (especially in noble gases, potassium, magnesium, and iron), and their

isotopic abundances of oxygen also appear similar to the Moon's. Hence, some lunar impacts provided enough energy to boost pieces above lunar escape velocity (2.4 km/s), and these pieces then traveled to the Earth. Those trapped in the antarctic ice have been well preserved for examination. Surprisingly, these rocks were *not* melted during their ejection from the Moon!

7–5 Interplanetary Gas and Dust

Interplanetary gas comes from various sources. Some has escaped from planetary atmospheres, and some has been released in the demise of comets. Most, however, comes from the Sun. The solar wind, essentially an expanding extension of the solar corona (Chapter 10), rushes past the Earth at about 500 km/s. The ionized gas does not stop there; it sweeps onward beyond Pluto's orbit until it slows down and dissipates in interstellar space.

However, the solar wind has only a minor effect on the heavier dust particles in the Solar System. These particles, which orbit approximately in the plane of the ecliptic, produce a phenomenon known as the *zodiacal light,* a faint cloud of light that extends in a roughly triangular shape above the horizon before sunrise and after sunset. The zodiacal light shines so faintly that the lights of even a small town can completely obscure it.

The spectrum of the zodiacal light resembles that of reflected sunlight, a fact that indicates that dust rather than gas is the source of the reflected light. Satellite observations show that the zodiacal dust particles in the Earth's neighborhood have a concentration of about 10^{-8} particles/m^3 and that they are composed mostly of silicates, iron, and nickel. When this cosmic dust falls to the Earth, it is found as micrometeorites. It adds perhaps a few million tons to the Earth's mass each year. The total mass of the zodiacal cloud is about 10^{16} kg.

Although dust is constantly being fed into the Solar System by disintegrating comets (and perhaps asteroid collisions), two physical processes efficiently remove it: radiation pressure and the Poynting-Robertson effect. *Radiation pressure* occurs because electromagnetic radiation carries momentum at the speed of light. The radiant energy flux E (J/$m^2 \cdot$ s) corresponds to a *momentum flux $P = E/c$* (J/m^3), where c is the speed of light (Chapter 8). The momentum flux is the light pressure on the particle. When solar radiation interacts with a dust particle of effective area or cross-section A, the rate of change of the particle's momentum is the *radiation force*:

$$F_R = PA = AE/c$$

From the discussion of planetary temperatures, $E = (R_\odot/d)^2 \sigma T_\odot^4$, where d is the distance from the Sun. If we assume that $A = \pi r^2$, where r is the particle's radius, then the radiation force pushing the particle *away* from the Sun is

$$F_R = (\pi \sigma r^2 R_\odot^2 T_\odot^4/c)/d^2 \qquad (7-2)$$

but the Sun's gravitational attractive force on the particle is

$$F_G = GM_\odot(4\pi r^3 \rho/3)/d^2 \qquad (7-3)$$

where ρ is the density of the particle. When we form the *ratio* of Equations 7–2 and 7–3 and substitute the appropriate numbers, we find

$$F_R/F_G = 3\sigma R_\odot^2 T_\odot^4/4cGM_\odot \rho r \qquad (7-4)$$
$$= 5.78 \times 10^{-5}/\rho r$$

where the ratio is dimensionless if we express r in meters and ρ in kilograms per cubic meter. Note that the particle's distance d from the Sun has disappeared. We see that $F_R = F_G$ for reasonable densities ($\rho \approx 1000$ to 6000 kg/m³) when $r \approx 0.1$ to 1 μm, so that dust particles smaller than 1 μm are blown out of the Solar System. The solar wind (Chapter 10) aids radiation in this cleansing operation.

For larger particles, $F_G \gg F_R$ the *Poynting-Robertson effect* becomes important. Just as the Earth's orbital motion leads to the aberration of starlight, so also will a particle's Keplerian orbit cause solar radiation to appear to be coming from slightly in front of the particle. If v is the speed of the particle in circular solar orbit, the angle between the incoming radiation and the radius vector to the Sun is clearly $\theta \approx v/c$ so that the component $(v/c)F_R$ of the radiation force is impeding the particle's motion. So a particle will spiral into the Sun. Calculation shows that a particle originally orbiting at distance d (AU) will fall into the Sun in a time

$$t = (7 \times 10^5)\rho r d^2 \text{ years}$$

For example, a particle of size $r = 1$ μm and density $\rho = 4300$ kg/m³ takes only 3×10^3 years to spiral into the Sun from a distance of 1 AU and about $5 \times$

10^6 years from 40 AU. In this fashion, the Solar System is purged of its small dust, but that material is replenished by the breakup of larger bodies, such as comets.

7–6 The Formation of the Solar System

We end this chapter with a composite sketch of the possible development of our Solar System based on the best models we have today. These models are basically *nebular models*, where a cloud of interstellar gas and dust contracts to form the Sun and planets. The overall features of this picture are probably reasonably correct even though the details are still vague and uncertain. We will focus on two major aspects of these models: dynamics and chemistry.

(a) Dynamics

The Solar System displays a regular structure in terms of its dynamic properties. Viewed from above the Sun's north pole, the Solar System shows the following regularities:

1. The planets revolve counterclockwise around the Sun; the Sun rotates in the same direction.
2. With the exceptions of Mercury and Pluto, the major planets have orbital planes that are only slightly inclined to the plane of the ecliptic; the orbits are nearly *coplanar*.
3. With the exceptions of Mercury and Pluto, the planets move in orbits that are very nearly circular.
4. With the exceptions of Venus and Uranus, the planets rotate counterclockwise, in the same direction as their orbital motions.
5. The planets' orbital distances from the Sun follow a regular spacing; roughly, each planet lies twice as far out as the previous one.
6. Most satellites revolve in the same direction as their parent planets rotate and lie close to their planets' equatorial planes.
7. Some satellites' orbital distances follow a regular spacing rule.
8. The planets together contain much more angular momentum than the Sun.
9. Long-period comets have orbits that come in from all directions and angles, in contrast to the

coplanar orbits of the planets, satellites, asteroids, and short-period comets.

10. Three of the Jovian planets are known to have rings.

The essential feature of nebular models is that the Sun and then the planets form from a cloud of interstellar material. The Sun's formation takes place in the center of a flattened cloud. The planets grow from the disk of the cloud. So the problem has two basic parts: (1) how to make a flat solar system and (2) how to get the planets to grow out of the cloud.

To tackle the first part, consider the conservation of angular momentum. The basic point is this: once a body starts spinning, it will keep on spinning as long as no torque affects it. The amount of spin angular momentum depends on how much mass the body has and how much it is spread out. If by itself the body changes size — for instance, if it contracts gravitationally — it will naturally spin faster to keep its spin angular momentum the same. And it will flatten as it falls in along the rotation axis, where the spin angular momentum per unit mass is the least. As a natural result of contraction with spin, we get the planets' orbits aligned in a thin disk and the Sun rotating in the same direction as the planets revolve.

With this neat solution to key dynamic features of the Solar System comes a serious objection: the present distribution of angular momentum. Although the sun holds 99 percent of the system's mass, it contains less than 1 percent of the angular momentum. The Jovian planets have the most, 99 percent of the total. To adopt a nebular model requires a process to account for the present distribution of angular momentum. One idea that has been worked on in some detail involves the interaction of magnetic fields and charged particles to rearrange the distribution of angular momentum. The basic solution requires that the spin of the central part of the nebula be decreased and transferred to the outer regions.

Charged particles and magnetic fields interact in such a way that the particles spiral along the magnetic lines of force. As the Sun forms, it heats up the interior regions of the nebula. Here the gas is ionized and the magnetic field lines trap the charged particles. As the Sun rotates, it carries its magnetic field lines with it; these drag along the charged particles, which in turn interact with and drag along the rest of the gas and dust. So the magnetic field spins around the material in the nebula near the Sun. At the same time, the inertia of the nebula resists the rotation. This drag on the magnetic field lines stretches them into a spiral shape (Figure 7–24). The magnetic field links the material in the nebula to the Sun's rotation, and so the nebular material gains rotation (and angular momentum) and in the process causes a drag on the Sun's rotation, which slows it down.

(b) Chemistry

To form a planet requires a multistep process. First, solid grains condense out of the solar nebula's gas. Second, these particles accrete into large bodies called *planetesimals,* which then collide and accrete to make *protoplanets,* which evolved into the planets of today. The condensation of grains determined the chemical composition of planets by a process called the *condensation sequence.*

The basic idea of the condensation sequence is this: the nebula's center must have been at a temperature of a few thousand kelvins. Here solid grains, even iron compounds and silicates, could not condense. Elsewhere, which materials would condense as new grains depended on the temperature. Just below 2000 K, grains made of terrestrial materials would condense; below 273 K, grains of both terrestrial and icy materials could form. At different temperatures, the gases available and the solids present react chemically to produce a variety of compounds. The densities and compositions of the planets can be

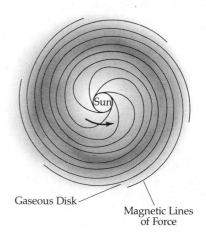

Figure 7–24

Magnetic fields in the solar nebula. A possible configuration of the twisting of magnetic field lines from the Sun trapped by ionized gas in the solar nebula.

well explained with the condensation sequence *if* the temperature of the nebula drops *rapidly* from the center outward. Then, at different distances from the Sun, different temperatures allowed different chemical compounds to condense and form grains that eventually made up the protoplanets (Figure 7–25). If a material could not condense because the temperature was too high, it would not end up in the protoplanet.

In general, the condensation sequence requires a certain *minimum* temperature to be reached to account for the known chemical composition of the planets. Roughly, these temperatures are 1400 K for Mercury, 900 K for Venus, 600 K for the Earth, 400 K for Mars, and 200 K for Jupiter. Note that, at a given distance from the Sun, the temperature varies with time, and so these values are the minimums reached in the interval of planetary formation.

(c) Accretion

Once the grains condense, they accrete into larger masses. Accretion processes fall into two distinct physical categories: (1) growth by collision and sticking due to the *geometric* cross section and (2) growth by collision due to gravitational attraction, with a gravitational cross section.

A grain's geometric cross section is simply πR^2 for a spherical grain of radius R. We can define a gravitational cross section (or impact parameter) as follows.

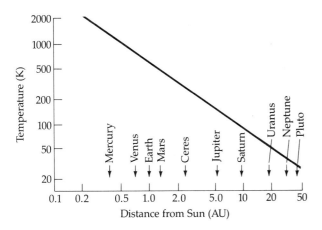

Figure 7–25

Temperature and condensation sequence. To result in the differences in composition from one planet to another, the temperature in the solar nebula followed the trend shown here.

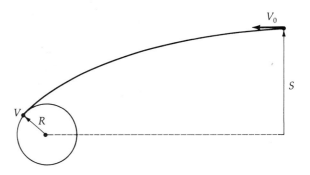

Figure 7–26

The geometry for particle impacts.

Consider a test particle at velocity V_0 approaching a grain of radius R. When the test particle hits the grain, it has velocity V. Let S be the maximum transverse distance from the center of the grain that a test particle can have and still strike the grain (Figure 7–26). The test particle starts out at position $S \gg R$ so that its potential energy at S is zero. Conservation of angular momentum says that

$$VR = V_0 S$$

Conservation of energy requires that

$$mV_0^2/2 = mV^2/2 - GmM/R$$

where m is the test particle's mass and M the grain's mass. Then

$$V^2/2 = V_0^2/2 + GM/R$$
$$V = (V_0^2 + 2GM/R)^{1/2}$$

Substitute this expression for V back into the angular momentum conservation equation:

$$V_0 S = (V_0^2 + 2GM/R)^{1/2} R$$

so that

$$S = (R/V_0)(V_0^2 + 2GM/R)^{1/2}$$
$$= (R^2 + 2GMR/V_0^2)^{1/2}$$

This is the equation for the gravitational impact parameter; note that it is a sum of a geometric term and a gravitational term.

Suppose particles grow by adding material of the *same* density so that the particle's density is constant. Then

$$dM/dt = d/dt[(4/3)\pi\rho R^3]$$
$$= 4\pi\rho R^2(dR/dt)$$

The particles grow in size to increase their mass. Assume that gravity dominates the growth. Now consider how a particle grows as it moves through a group of other particles. The collision rate depends on the velocity (greater velocity, more collisions in a given time), cross section (the larger the cross section, the larger the number of collisions in a given time), and the number density of the other particles (greater density means more collisions in a given time). So

$$dM/dt = \text{velocity} \times \text{particle density}$$
$$\times \text{cross section}$$
$$= V_0 \rho_0 \pi S^2$$
$$= V_0 \rho_0 \pi (R^2 + 2GMR/V_0^2)$$
$$= V_0 \rho_0 \pi R^2 + 2\pi \rho_0 GMR/V_0$$

Ignore the first (geometric) term for the case where the gravitational cross section dominates. Then

$$dM/dt \approx (2\pi G \rho_0 / V_0) MR$$
$$= (2\pi G \rho_0^2 / V_0)[(4/3)\pi R^3]R$$
$$dM/dt \propto R^4$$

As the particles grow, they add material at a very fast and accelerating rate.

At what particle size does the gravitational cross section dominate? Let's say that happens when $S^2 = 2R^2$, *twice* the geometric one. Then

$$S = \sqrt{2}R = (R^2 + 2GMR/V_0^2)^{1/2}$$
$$2R^2 = R^2 + 2GMR/V_0^2$$

so

$$R^2 = 2GMR/V_0^2$$

and

$$R = 2GM/V_0^2$$
$$= (2G/V_0^2)(4/3)(\pi \rho R^3)$$

or

$$R = (3V_0^2 / 8\pi G \rho)^{1/2}$$

Note that V_0 is the *relative* velocity of particles and ρ the particles' density. For $V_0 = 1$ km/s and $\rho \approx 3000$ kg/m³, we have

$$R \approx 1000 \text{ km}$$

as the transition size. An object larger than this is probably a planetesimal.

(d) The Formation of Jupiter and Saturn

To make matters somewhat confusing, Jupiter and Saturn may have formed in a way other than that described by the planetesimal accretion model. In analogy with star birth, Jupiter and Saturn may have condensed gravitationally from single, large blobs of material in the nebula rather than by accretion of planetesimals. If so, the evolutions of a proto-Jupiter and a proto-Saturn relate to the evolution of proto-stars (Chapter 19). In fact, the chemical compositions of Jupiter and Saturn match those of stars fairly closely. The main difference arises from the fact that Jupiter and Saturn don't have enough mass to get hot enough to ignite fusion reactions. The heat they do gain comes from the conversion of gravitational potential energy to heat during gravitational contraction.

Harold Graboske and his colleagues have made theoretical calculations of the evolution of Jupiter from a hot beginning, after the proto-Jupiter had come together. They assume a solar mixture of material (74 percent hydrogen, 24 percent helium, and 2 percent everything else by mass) and start the calculations with a proto-Jupiter 16 times Jupiter's present size, a central temperature of 16,000 K, a surface temperature of 1000 K and a luminosity of almost 10^{-2} the Sun's present luminosity. Gravity quickly collapses the proto-Jupiter at first. The shrinking slows down when the planet's interior is liquid and liquids are difficult to compress. In the next 4.5 billion years, Jupiter contracts to its present size and its central temperature drops to 30,000 K as it loses some of its heat of formation to space at a rate of $(1.8 \times 10^{-9}) L_\oplus$.

These models indicate that, soon after its formation, the proto-Jupiter went through a brief phase of high luminosity. This early high luminosity may explain, in the context of a condensation sequence, why the Galilean satellites decrease in density going outward from Jupiter. This density decrease implies, for example, that Callisto contains proportionally more icy materials than Io. At Io's closer distance, less ice condensed than at Callisto's distance. If Jupiter were hot at the time of the satellites' formation, the inner ones would not have accreted as much icy materials as the outer ones. So the Galilean moons may mimic the condensation and accretion of the terrestrial planets.

James Pollack and his colleagues have developed similar models for the early evolution of Saturn. The sequence resembles that for Jupiter. Their models begin 4.5 billion years ago with a planet ten times its present size, with solar composition and no rocky core. Gravity quickly shrinks the planet. In about 1 million years, Saturn is about twice its present size and has a central temperature of 21,000 K. This high temperature raises the internal pressure and slows down the gravitational contraction. It then takes 4.5 billion years for Saturn to contract to its present size.

(e) Asteroids, Meteorites, and Comets

Contemporary research on asteroids implies that they are planetesimals that just did not accrete to make a planet. Recall the compositional variation across the asteroid belt. On the inner edge, it contains mostly S-type asteroids; at the outer edge, mostly C types. These albedo differences fit in nicely with the condensation sequence if the C types contain more carbon than the S types. At the inside of the belt, the temperatures were low enough for silicates to condense but too high for carbon-bearing materials to do so. Farther out, both types of materials condensed to end up in planetesimals. Why didn't they form a planet? Probably because of the tidal influence of the proto-Jupiter. Solar tidal forces also helped to change the orbits of the planetesimals from circular to elliptical. Some crashed into others, shattering them into smaller pieces. Some of these pieces caromed into the inner part of the Solar System and eventually rained onto the surfaces of Mercury, Venus, the Moon, the Earth, and Mars, forming craters—the epoch of planetary bombardment 4 billion years ago.

How do meteorites fit into the nebular model? The characteristics of chondritic meteorites support the condensation picture. Their chemical composition (similar to that of the Sun) and unmixed structure suggest that they are the original condensed material of the nebula. The turbulence in the nebula may have created shock waves that swiftly melted the grains. After its passage, the drops cooled and solidified to form chondrules. The glassy spheres that resulted accumulated in planetesimals. Chondrules suggest that the bulk of their condensation took place at temperatures around 600 K. (According to the condensation sequence, this range produces materials like those that make up the Earth.) About 1 million years after formation, radioactive decay reheated some planetesimals, melting them to some extent and allowing them to differentiate into iron cores and stony mantles. The planetesimals that were not gathered into a protoplanet possibly became the parent meteorite bodies. These bodies later collided and fragmented.

What happened to other planetesimals? Some may have collided at high speeds with others and disintegrated into small pieces. A few may have passed close enough to a protoplanet to be captured in an orbit as a satellite. Others may have experienced near misses. Their orbits might have changed enough to throw them out of the Solar System.

Near the Jovian planets, planetesimals would be mostly icy materials; these may have formed the nuclei of comets. These icy bodies may have then been gravitationally directed into the Oort cloud. We expect from the condensation sequence that bodies formed near the Jovian planets would have an icy composition. Estimates for the actual compositions of Uranus and Neptune are 10 to 15 percent solar materials, 20 to 25 percent terrestrial materials, and 60 to 70 percent icy materials. Comets are believed to have almost the same relative percentage (perhaps a bit lower for the solar and terrestrial materials). The inference: these two planets grew by the accretion of icy planetesimals. The remaining planetesimals have their orbits perturbed by the Jovian planets and ejected into orbits (20,000 to 50,000 AU) typical for the Oort cloud. Uranus and Neptune have the most efficient gravitational effects and account for some 75 percent of the objects in the comet cloud. Hence, the Oort cloud is a by-product of planetary accretion in the outer Solar System.

(f) Formation of the Earth and Moon

Because the Moon is a simpler, more primitive body than the Earth, we have a better idea of its history—but not of its formation! Attempts to explain the Moon's formation have usually fallen into one of three categories: capture, fission, or binary accretion. In the capture model, the Moon forms by accretion some distance from the Earth and later is captured gravitationally by it. In the simple fission model, a rapidly rotating, plastic Earth throws off a large part of its mantle (right after its core is formed); this piece cools to become the Moon. The binary accretion model views the formation of the Earth and Moon as taking place more or less simultaneously from the same part of the solar nebula. Each model has different strengths and weaknesses, and planetary scientists now feel that each emphasizes some aspect of the many processes that actually took place, rather than presenting an either–or statement about the formation sequence.

The capture model has largely fallen out of favor. It focuses on the differences in the chemical composition of the Earth and Moon (the Moon's lack of water and iron and its higher abundance of uranium and rare-earth elements). But both bodies have the same isotopic abundances of oxygen (as do some types of meteorites) so that they probably formed nearby in the solar nebula. Also, the dynamics of capture require very special conditions that are hard to arrange — in particular, a massive third body is needed nearby during the capture to ensure the conservation of momentum.

In contrast, the fission model relies on similarities between the Earth and the Moon (specifically, the chemical composition of the Earth's mantle). But substantial differences exist for both volatiles and major elements. In particular, the Mg/Si ratio for the Moon is outside the range of possible values for the Earth's mantle; refractory elements (such as aluminum and calcium) are enriched by a factor of two in the Moon; volatile elements are depleted; and the Ni/Fe ratio is too low to be derived from the Earth's mantle.

Binary accretion models handle many of these problems. The Earth and Moon grow by accretion in the same region of the solar nebula from nearby material. Some matter is ejected off the accreting body by extraordinary impacts, adding debris to the area. The Earth accretes most of the iron, which falls into the core that forms early in the sequence. Large impacts then blast off mantle-like material. Once the Moon forms, radioactive decay (from Al-26) heats it enough to deplete the volatiles. The end result is that the Earth and Moon share many chemical similarities with some notable differences.

Recently a hybrid model has been developed for the origin of the Earth and the Moon. It envisions the impact of a Mars-sized body into the young Earth. The energy of this tremendous impact injects a large fraction of the mass of the Earth and the impacting body into a disk of material around the Earth from which the Moon accreted. In order to account for the angular momentum of the Earth-Moon system, the incoming body would have to strike the Earth tangentially at about 10 km/s. This is called the *giant impact model*, and it can be seen as a variation on the binary accretion picture. This idea has not encountered any serious objections yet and still needs to have many details worked out. It looks very promising.

We now turn to the Earth specifically. After the era of accretion (which may have lasted but a few million years), the Earth was a pockmarked planet (a surface cratered like the lunar highlands) of roughly uniform composition; its atmosphere was mainly hydrogen. Radioactive heating then melted the interior and the core formed. This heating promoted de-gassing from the interior to create a second atmosphere rich in water, carbon dioxide, methane, and ammonia. When the Earth's surface had cooled enough, intense rains fell to begin the formation of the oceans. Perhaps a billion years after formation, plate tectonics began to modify the crust, which cooled and thickened. The ocean basins were mostly filled. Life developed and modified the atmosphere into the one we have today. Some 200 million years ago, the continents broke up and were driven by plate tectonics into the positions we see now.

Problems

1. What effects do radiation pressure and the Poynting-Robertson effect have upon an artificial space probe of mean density 1000 kg/m^3 and radius 1 m? Justify your answer quantitatively and state your assumptions.

2. At the origin of the Solar System, the Sun's tidal forces and Roche limit might have played an important role. To avoid tidal disruption, what is the minimum density a protoplanet must have at the distance *d* (in AU) from the Sun? Comment upon your result.

3. According to theoretical models, 4 billion years ago Jupiter may have had a luminosity as high as $10^{-2} L_\odot$ and a surface temperature of 1000 K.
   (a) What was Jupiter's radius then if it radiated as a blackbody?
   (b) Under these conditions, how hot would a blackbody get at Io's distance from Jupiter?

4. (a) Estimate the present rate at which the Earth is accreting interplanetary dust.
   (b) Calculate how long it will take the Earth at this rate to increase its mass by 50 percent.

5. Calculate how fast the Sun would rotate if all the angular momentum of the planets were added to it.

6. Compare the gravitational impact parameters for the Earth (surface) and for Jupiter (top of atmosphere) for bodies falling in at their respective escape velocities.

7. Compare the escape velocities from Titan, Dione, and Hyperion. Estimate the lifetime of methane on each.

8. (a) If an asteroid's infrared flux (at 10 μm) is 1/4 its visible flux, what is its albedo?
   (b) What is its radius?

9. Assume the asteroid Herculina has a moon of radius 100 km, orbiting 1000 km from Herculina. Make both bodies rocky.
   (a) Calculate the moon's orbital periods.
   (b) Compare the tidal force of the Sun pulling the bodies apart with the mutual gravitational forces holding them together.

10. On a clear night, if you are patient and away from city lights, you can see about 10 meteors per hour. The meteoroid that creates a meteor has a mass of about 1 g and burns about 100 km above the Earth's surface. *Estimate* the amount of mass of meteoroids that enters the Earth's atmosphere per year. If this influx has been constant, how much mass has been added to the Earth since it was formed?

11. Comet Ikeya-Seki (the great sun-grazing comet of 1965) had an elliptical orbit with a period of about 700 years. As it rounded the Sun, the comet had a perihelion distance of 0.008 AU.
    (a) How far away from the Sun (in AU) will the comet be at aphelion?
    (b) If its perihelion velocity was 500 km/s, what will its orbital velocity be at aphelion?
    (c) Make some reasonable assumption about the average density of the comet's nucleus. Did the comet pass close enough to the Sun to be ripped apart by tidal gravitational forces? (*Hint:* Work out the Roche limit for the Sun.)

12. The crushing strength (pressure for material deformation) for iron meteorites is about 4×10^8 newtons/m². At what size would an iron asteroid have this central pressure? What would happen to it if it were this size or larger? (*Hint:* Use the equation of hydrostatic equilibrium.)

13. Assume that the particles in Saturn's rings have circular orbits ranging from 87×10^3 to 137×10^3 km from the center of Saturn.
    (a) Calculate the orbital velocities at the inner and outer edges of the rings, assuming that they are made of individual particles obeying Kepler's laws.
    (b) What resolution (what accuracy in wavelength measurement) would a spectrograph have to have in order to distinguish between the inner and outer edge speeds of the rings?

The Stars

Electromagnetic
Radiation
and
Matter

We turn now from planets to stars. The nearest star is our own Sun, which shows a visible disk. The other stars are mere points of light scattered across the night sky. How can we study such objects? Stars betray their presence by the light they emit, and only by detecting and deciphering these radiations can we infer the properties of stars.

This chapter deals with the characteristics of *electromagnetic radiation* (most familiar in the form of visible *light*), the atomic structure of matter, and the all-important interactions between matter and radiation. Astronomers handle light in two ways: by measuring the total intensity (over a certain range of wavelengths) and by dispersing the light into a *spectrum* and examining it in detail. Both techniques resolve the astronomer's problem of how to infer the physical properties of the faraway stars.

8–1 Electromagnetic Radiation

Keeping in mind our ultimate goal of understanding stars, let's begin by investigating the nature of light as electromagnetic waves.

(a) The Wave Nature of Light

What is a *wave?* You are familiar with water waves, which are undulatory disturbances traveling along the surface of the liquid. Place corks along the line of the wave's motion, and you may characterize such a wave (a transverse, or up–down, wave) by the height h of the corks above the average surface level (Figure 8–1) in the mathematical form

$$h = h_0 \sin[(2\pi/\lambda)(x - vt)] \qquad (8\text{–}1)$$

Note that in a *transverse wave* the disturbance of the medium through which the wave travels is perpen-

dicular to the direction of motion of the wave. In contrast, a *longitudinal wave* involves a disturbance that lies along the direction of motion, such as the compression of a spring.

Equation 8–1 represents a sinusoidal wave of *amplitude* h_0 progressing with time t along the positive x axis at the speed v; the distance between successive wave crests is called the *wavelength* λ. At a given time (say, $t = 0$), the corks betray the oscillatory pattern

$$h = h_0 \sin(2\pi x/\lambda) \qquad (8\text{–}2)$$

If you observe the cork at $x = 0$, it rises and falls periodically. Note that a given cork completes one oscillation in the time λ/v s; in 1 s, v/λ wave crests will have passed a given point, and so the *frequency* (v) of oscillation is just $v = v/\lambda$ per second. Therefore, we may completely characterize such a wave by the fundamental relationship

$$\lambda v = v \qquad (8\text{–}3)$$

For example, water waves of wavelength $\lambda = 5$ cm moving at a speed $v = 10$ cm/s will pass a given point at a frequency $\lambda = 2/s$ (twice per second). The SI unit for frequency is the *hertz* (*Hz*), with 1 Hz equaling one cycle per second.

From Equation 8–3 we see that

$$\lambda v = c \qquad (8\text{–}4)$$

is the fundamental relationship between the wavelength and frequency of an electromagnetic wave (in vacuum), where c is 299,793 km/s—the speed of light.

Most waves require a material medium in which to be transmitted: water waves travel along the surface of water, sound waves move through air, and earthquake waves, both compressional and transverse, propagate through the solid Earth. Electro-

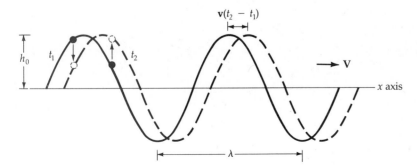

Figure 8–1

A traveling wave. A water wave moving with speed v to the right is shown at two times, t_1 and t_2. Floating corks mark the oscillating surface.

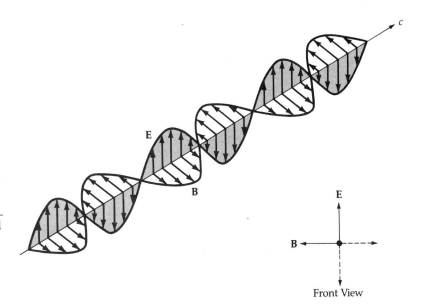

Figure 8–2

An electromagnetic wave. The electric field **E** and magnetic field **B** propagate at speed c through a vacuum. The two fields oscillate perpendicular to each other. The front view shows this for a linearly polarized wave.

magnetic waves, however, may propagate through a pure vacuum at speed c. What is it, then, that is propagating? The space around an electric charge may be characterized by an *electric field vector* **E**, which manifests itself as a force on a test charge placed nearby. If an electromagnetic wave encounters such a test charge, that charge will oscillate, so that we may ascribe the sinusoidal electric field

$$\mathbf{E} = \mathbf{E}_0 \sin[(2\pi/\lambda)(x - ct)] \qquad (8-5)$$

to an electromagnetic wave traveling in the positive x direction along the x axis. Maxwell's equations say that a time-varying electric field produces a perpendicular time-varying *magnetic field* **B**, however, and so an electromagnetic wave is a self-propagating disturbance of electric and magnetic fields in a vacuum (Figure 8–2). If the electric field always oscillates in a single plane, the wave is *plane-polarized*; otherwise, it is *elliptically* polarized (a special case is *circular* polarization).

Polarization plays an important role in astronomical observations, and so we deal with it in some detail here. Because light is a transverse wave, the orientations of the wave amplitude perpendicular to the direction of motion are the possible directions of polarization. Most light sources are unpolarized, which means that the electric field vector vibrates in a random orientation in the plane perpendicular to the propagation direction.

If we consider that the electric field has two components in this plane (E_x and E_y), then the phase differences between them determine the kind of polarization the wave has. If they are in phase, the wave is plane-polarized; if they are not in phase, the wave has an elliptical polarization; if they are 90° out of phase, the wave has a circular polarization.

The Electromagnetic Spectrum

Light is that class of electromagnetic radiation occupying the wavelength range $\lambda \approx 390$ nm (violet) to 720 nm (far red)—the *visible spectrum*. A wavelength λ is a *length* with units of centimeters or meters (or kilometers or miles); three convenient units of length in astronomy are the micrometer (sometimes called the micron), $1\ \mu m = 10^{-6}$ m; the nanometer, 1 nm $= 10^{-9}$ m; and the angstrom, $1\ \text{Å} = 10^{-10}$ m $= 10^{-4}\ \mu m$ (Appendix 7). Electromagnetic waves of different wavelengths are detected in different ways (Table 8–1), so that we give characteristic names to various parts of the electromagnetic spectrum: gamma rays ($\lambda \leq 10^{-2}$ Å), X-rays (10^{-3} to 10 nm), ultraviolet (10 to 300 nm), visible light (400 to 800 nm), infrared (1 to $10^3\ \mu m$), microwaves (1 mm to 10 cm), and radio waves ($\lambda \geq 1$ cm). Visible light occupies less than one decade of the 21 decades of wavelength shown in Table 8–1, and yet visible light is of paramount importance to astronomy because it easily penetrates the Earth's atmosphere and it is readily perceived by the human eye.

Table 8–1　The Electromagnetic Spectrum

λ	ν or $h\nu$	Type of e-m radiation	Detectors
10^{-5} Å	1240 MeV	Gamma rays	Geiger counters
10^{-4} Å			Scintillators
10^{-3} Å	12.4 MeV		Nuclear emulsions
10^{-2} Å			Proportional counters
10^{-1} Å		X-rays	Spark chambers
1 Å = 10^{-8} cm	12.4 keV		Cerenkov-light detectors
1 nm			
10 nm	124 eV	Ultraviolet	Photographic and photoelectric detectors
100 nm			Image-forming telescopes
1000 nm = 1 μm	1.24 eV	Visible	and spectrographs and
10 μm			spectrometers
100 μm	0.012 eV	Infrared	Photoconductive detectors
1000 μm = 1 mm			Thermal detection radiometers
10 mm = 1 cm	30,000 MHz	Radar	
10 cm		UHF	
100 cm = 1 m	300 MHz	FM	RADIO / Radio / Radio receivers — Telescopes
10 m		Short wave	
100 m	3 MHz		
1000 m = 1 km	300 kHz	Broadcast	
10 km		Long wave	
100 km	3 kHz		
1000 km			

Open window		eV = electron volt
		Hz = hertz = cycles per second
Partial window		M for mega = million = 10^6
		k for kilo = thousand = 10^3
		Å = angstrom
Opaque		μm = micrometer
		nm = nanometer

Reflection and Refraction

Now to consider some of the properties of light that make it possible for us to manipulate it—the goal of *optics*. A ray (or beam or pencil) of light falling upon a mirror is *reflected* in accordance with the rule (Figure 8–3) "the angle of reflection *r* equals the angle of incidence *i*." Note that these angles are defined with respect to the *normal* (perpendicular) to the reflecting surface. Reflecting telescopes utilize the law of reflection to collect, direct, and focus light.

When light is reflected, both the incident and reflected waves travel in the same medium at the same speed, but what happens when light passes from one medium into another? The speed of light v in a medium is generally different from the speed of light c in a vacuum. We may characterize a given medium by its *index of refraction*: $n = c/v$. The index of refraction of air is $n = 1.0003$, which is practically the same as that of a vacuum ($n = 1.0000$). Crown glass, however, has $n = 1.5$, and so in this medium $v = 2c/3 \approx 2 \times 10^5$ km/s. Light passing from a medium with index of refraction n_1 into a medium with index n_2 is bent, or *refracted*, in accordance with *Snell's law* (Figure 8–4):

$$n_1 \sin i = n_2 \sin r \qquad (8-6)$$

where *i* is the angle of incidence in medium 1 and *r* is the angle of refraction in medium 2 (both with respect to the normal to the interface between the media). A refracting telescope uses refraction to make images.

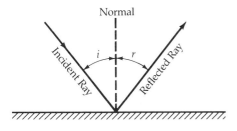

Figure 8–3

Reflection of light. The angle of incidence i equals the angle of reflection r.

In general, the index of refraction of a medium depends upon wavelength—$n = n(\lambda)$. That is, light of different wavelengths (or *colors*) is refracted through different angles $r(\lambda)$ when the incident beams all have the same angle of incidence i. So an incident mixed (white) beam is dispersed into separate beams of pure colors, as Newton demonstrated. The phenomenon of *dispersion* enables us to decompose light into its component colors—in the form of a spectrum—and is the basis of spectroscopy.

Diffraction and Interference

When water waves encounter an island, they *diffract* around the sides of the island to converge, and when these converging waves meet one another, they *interfere*. In the case of sound waves, we are also familiar with these phenomena of *diffraction* (sound

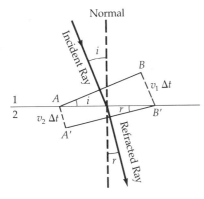

Figure 8–4

Refraction of light. The line AB represents a plane wavefront incident on the interface between medium 1 and medium 2 with angle of incidence i. Point A travels to A' in time Δt, a distance of $v_2\Delta t$. Point B travels $v_1\Delta t$. The ray of light is bent across $A'B'$.

bends around sharp corners) and *interference* (recall the silent spots in a large auditorium). Light exhibits these same phenomena, which are understandable in terms of a wave model.

Consider the situations where light waves meet an opaque wall having one or two apertures, with a viewing screen placed beyond (Figure 8–5). The sharp images of the apertures, which we expect to see on the viewing screen, are distorted (a) by the diffraction and spreading out of the light waves and (b) by the constructive and destructive interference of different light waves. We may understand both processes by considering how the intensity patterns at the viewing screen are formed. The intensity I of light is proportional to the square of the electric field:

$$I \propto |\mathbf{E}|^2 \qquad (8\text{–}7)$$

At an arbitrary point on the viewing screen (Figure 8–5B), the electric field of the wave is given by Equation 8–5:

$$E = E_0[\sin a + \sin(a + b)] \qquad (8\text{–}8)$$

where a refers to the wave from one aperture and $a + b$ to the wave from the other aperture. The *phase shift b* corresponds to the difference in path lengths from the apertures to the point on the viewing screen. From Equation 8–8, the intensity is

$$I \propto E_0^2[\sin^2 a + \sin^2(a + b) \\ + 2 \sin a \sin(a + b)] \qquad (8\text{–}9)$$

The first two terms on the right are just the intensities attributable to the separate apertures; the third term is responsible for the oscillatory interference pattern on the screen (as b changes with position).

Every point within the single aperture (Figure 8–5A) contributes to the *diffraction* pattern at the viewing screen. Application of Equations 8–7 and 8–8 shows that the *angular width θ* of the principal diffraction image is

$$\theta \approx \lambda/d \qquad (8\text{–}10)$$

where θ is in radians, λ is the wavelength of the light, and d is the size of the aperture. No optical image can be smaller than the *diffraction limit* of Equation 8–10, and so we say that this is the optimum *angular resolution* of the system. For example, the resolution of a telescope of aperture $d = 1$ m viewing light of wavelength 500 nm is

$$\theta \approx (500 \text{ nm})/(10^9 \text{ nm}) = 5 \times 10^{-7} \text{ rad} \\ = 0.1 \text{ arc-second}$$

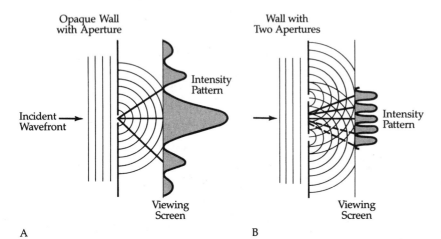

Figure 8–5

Diffraction and interference.
(A) Diffraction of light waves with a
single slit. (B) Interference of light
waves with two slits.

This theoretical resolving power may not be achieved by a ground-based telescope because of turbulence in the Earth's atmosphere.

The Doppler Effect

So far we have considered waves where both the source and the observer are relatively at rest. When either is in motion along the line connecting them, both the wavelength and the frequency of the wave are altered by the *Doppler effect,* named for the Austrian physicist C. J. Doppler (1803–1853). You are familiar with the Doppler phenomenon in water and air waves; for example, as a police car with its siren blaring approaches, the pitch (frequency) of the sound is high, but the frequency drops noticeably to a lower tone as the car passes you and then recedes.

Most familiar wave motions are associated with a material medium, which may complicate the picture by itself being in motion. Let's consider electromagnetic waves propagating in a vacuum. Armand Fizeau correctly explained the classical Doppler effect of light in 1848, and Albert Einstein gave the relativistic explanation in 1905.

Imagine a light source E (Figure 8–6) receding at speed v from an observer O while emitting radiation of wavelength λ_0 and frequency ν_0. In the time $t = 1/\nu_0$, one wavelength (λ_0) emerges from the source but, as seen by the observer, that wave has the length

$$\lambda = (c + v)t = c[1 + (v/c)]/\nu_0$$
$$= \lambda_0[1 + (v/c)] \qquad \text{(8–11)}$$

since the source has traveled the distance vt to the right. Note that the fundamental equation (8–4) has

been used in the last equality. The observed frequency is

$$\nu = c/\lambda = \nu_0/[1 + (v/c)] \qquad \text{(8–12)}$$

so that $\lambda > \lambda_0$ and $\nu < \nu_0$ and the light is *redshifted* (that is, shifted to longer wavelengths, or toward the red). When the source *approaches* the observer, v changes sign to give $\lambda < \lambda_0$ and $\nu > \nu_0$ and the radiation is *blueshifted.*

From Equation 8–11, we may extract the useful *wavelength shift:*

$$\Delta\lambda/\lambda_0 = (\lambda - \lambda_0)/\lambda_0 = v/c \qquad \text{(8–13)}$$

Remember that the sign of v is positive for recession and negative for approach.

When v approaches c, Einstein's theory of special relativity must be used. The two basic postulates of this theory are (1) the speed of light is independent of the motion of either source or observer and (2) only relative motions are observable. So v is the *relative*

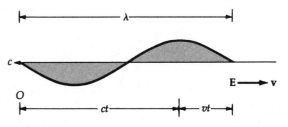

Figure 8–6

The Doppler effect. An observer at O sees a wave emitted by a source E moving at speed v to the right.

speed of the source and the observer, and no relative speed greater than the speed of light is possible ($v \leq c$). Equations 8–11 and 8–12 then necessarily take the symmetrical forms

$$\lambda = \lambda_0[(1 + v/c)/(1 - v/c)]^{1/2} \quad \text{(8-14a)}$$

$$v = v_0[(1 - v/c)/(1 + v/c]^{1/2} \quad \text{(8-14b)}$$

The nonrelativistic shift in wavelength (Equation 8–13) follows from Equation 8–14a in the limit when $v \ll c$.

(b) The Quantum Nature of Light: Photons

At the beginning of the eighteenth century, Newton proposed a particle theory of light, but it was not until the end of the nineteenth century that the particle-like manifestations of light began to appear in experiments.

Light is neither a particle nor a wave, but it can manifest itself as either one or the other! This apparently paradoxical behavior explains such phenomena as the photoelectric effect, Compton scattering, and blackbody radiation [Section 8–6(a)]. In addition, the interaction of light with atoms and molecules is understandable only if electromagnetic energy propagates in the form of discrete bundles, which we call *photons* or *quanta*. The energy of a light quantum (E) is proportional to the frequency characterizing the light wave:

$$E = hv \quad \text{(8-15)}$$

where $h = 6.626 \times 10^{-34}$ J · s is known as *Planck's constant.* We may crudely picture a classical light wave of wavelength λ and frequency v as composed of many quanta, each with the energy given in Equation 8–15.

(c) Intensity Versus Flux

In detecting the energy (or counting the photons) coming from a distant light source, we must be careful to distinguish between intensity and flux. *Intensity* depends upon direction, in the sense that the intensity I of a source is the amount of energy emitted per unit time Δt, per unit area of the source ΔA, per unit frequency interval Δv, per unit solid angle $\Delta \Omega$ in a given direction. The *solid angle* (Figure 8–7A) of the beam is related to the area Δa intercepted by the beam at the spherical surface of radius r by

$$\Delta \Omega = \Delta a / r^2 \quad \text{(8-16)}$$

The unit of solid angle is the *steradian* (ster), with the entire spherical surface subtending 4π ster — since $\Delta a = 4\pi r^2$ for the surface area of a sphere. (A steradian is essentially 1 rad^2; since 1 rad = 57.3°, 1 ster = 3283 square degrees of arc.) Common units of intensity are W/m^2 · Hz · ster. For example, the total energy flow from a spherical star of surface area A is just

$$E = 4\pi A \int_0^\infty I(v) \, dv \quad \text{W}$$

where $I(v)$ is called the *monochromatic intensity*, the intensity at a specific frequency.

Flux F relates directly to what we measure. The flux of energy through a surface (or into a detector) is the amount of energy per unit time passing through a unit area of the surface per unit frequency interval: $F(v) = \text{energy}/\Delta A \cdot \Delta t \cdot \Delta v$ (Figure 8–7B). Hence, the unit of monochromatic flux $F(v)$ is W/s · Hz. For example, if a star emits energy at the *rate E* (in watts), then the flux of energy through a concentric spherical surface of radius R is

$$F = E/4\pi R^2 \quad \text{W/m}^2 \quad \text{(8-17)}$$

Figure 8–7

Intensity and flux. (A) The intensity of light depends upon direction θ and solid angle Ω. (B) Flux depends only upon the energy passing through an area ΔA per unit time.

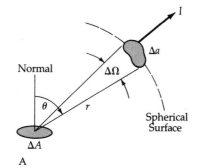

A

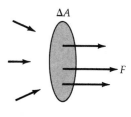

B

Equation 8–17 is a form of the well-known inverse-square law for the diminution of radiant flux with distance.

Note that a one-dimensional beam has flux but no intensity (since $\Delta\Omega = 0$). We may relate intensity and flux by noting that the energy passing through a certain area is composed of beams entering at different angles to the normal to the surface; to a beam entering at the angle θ, the surface area appears reduced by a factor $\cos\theta$ so that

$$F(\nu) = \int I(\nu, \theta)\cos\theta\, d\Omega$$

8–2 Atomic Structure

If we divide matter into smaller and smaller pieces, we find that at scales near 10^{-10} m, all forms of matter are composed of *atoms* (and molecules, which are bound aggregates of atoms). An elemental substance, such as gold, consists solely of atoms of the *element* gold; there are but 92 natural elements (hydrogen to uranium; Appendix 5), but this list has been extended to 109 with the addition of the synthetic transuranic elements.

(a) Atomic Building Blocks

What is an atom, and how may we characterize the elements? An atom is composed of a small *nucleus*, which contains most of the atom's mass in a size of the order of 10^{-14} m, surrounded by a diffuse cloud of *electrons* extending out to about 10^{-10} m. The atomic nucleus consists of *protons* and *neutrons* bound together by the strong interaction. These three elementary particles — the proton, the neutron, and the electron — are the fundamental building blocks of an atom. Table 8–2 lists the important parameters of each. Both the proton and the neutron are about 2000 times more massive than the electron. The neu-

tron is uncharged, and the proton (positive) and the electron (negative) have electric charges of opposite sign and equal magnitude $e = 1.602 \times 10^{-19}$ coulomb (C). A normal neutral atom consists of an equal number of protons and electrons, with approximately the same number of neutrons as protons (for the lighter elements); so an element is determined by the number of protons Z (called the *atomic number*) in the atomic nucleus. Heavier elements have a higher proportion of neutrons. For example, hydrogen ($Z = 1$) is one electron orbiting one proton; uranium ($Z = 92$) has 92 protons and 92 electrons, with about 146 neutrons in its nucleus.

A given element may exist in several different forms, called *isotopes*. All isotopes of an element have the same atomic number but differing numbers of neutrons (N). We term the different isotopic nuclei *nuclides*. To characterize a nuclide, the notation $^{Z+N}_{Z}X$ or $^{A}_{Z}X$ is used, where X is the symbol of the element having Z protons and $A = Z + N$ is the *atomic mass* (number of protons plus neutrons). For example, three isotopes of hydrogen are known: ordinary hydrogen, $^{1}_{1}$H; deuterium, $^{2}_{1}$H; and tritium, $^{3}_{1}$H. Note that X and the subscript Z are redundant, conveying the same information. The mass of an atom is conveniently given in terms of atomic mass units (amu); the standard is $^{12}_{6}$C, which has a mass of exactly 12 amu. Since 1 amu is essentially the mass of a proton, an atom's mass is A amu. When atomic masses are tabulated (Appendix 5), they frequently are not integers because it is the average mass of the naturally occurring isotopes that is being listed.

(b) The Bohr Atom

What dynamical configuration of electrons about a nucleus leads to a stable atom? Because the attractive electric force between a proton and an electron (Coulomb's law) obeys an inverse-square law, scientists envisaged electrons orbiting the nucleus. Electrons are *charged* particles, however, and charged particles radiate energy when they are accelerated (in circular orbit). So the very property that permits a bound atom should cause it to collapse almost at once because the electron's orbit would decay from the loss of energy! In 1913, Niels Bohr (1885–1962) advanced a simple theory that offered a way out of this problem; it resulted in the modern theory of *quantum mechanics* in the 1920s. Let's discuss Bohr's atomic theory and apply it to the hydrogen atom (which it accounts for well).

TABLE 8–2   Parameters of Atomic Particles

Name	Mass (kg)	Electron Mass	Electric Charge
Proton	1.6725×10^{-27}	1836	$+e$
Neutron	1.6748×10^{-27}	1838	0
Electron	9.1091×10^{-31}	1	$-e$

Quantized Orbits

In 1911, Sir Ernest Rutherford (1871–1937) proposed the nuclear model of the atom. Bohr spent a year with Rutherford and then stated two astounding postulates that breathed life into atomic theory. Of the infinite number of possible electron orbits in Rutherford's model, Bohr first postulated that *only a discrete number of orbits are allowed to the electron and when in those orbits, the electron cannot radiate.* The permitted orbits are those in which the orbital angular momentum of the electron is an *integral* multiple of $h/2\pi$, where h is Planck's constant.

Let's apply Bohr's postulate of quantized orbits to an electron in a circular orbit of radius r about a nucleus of charge Ze. The centripetal force maintaining the orbit (where m is the electron's mass),

$$mv^2/r$$

is provided by the coulombic attraction between the electron and the nucleus:

$$(Ze)e/r^2$$

but Bohr's postulate implies the additional constraint

$$mvr = n(h/2\pi) \qquad n = 1, 2, 3, \ldots \quad \textbf{(8–18)}$$

Combining these equations, we find

$$r = Ze^2/mv^2 = nh/2\pi mv$$

or

$$r = n^2(h^2/4\pi^2 me^2 Z) \qquad \textbf{(8–19)}$$

Therefore, the permitted discrete orbits occur at geometrically increasing (n^2) distances, with the smallest Bohr orbit occurring when the *principal quantum number n* equals unity.

We may now find the *total energy E* of these orbits. If E is negative, the system is bound and we have an atom. We know that E = kinetic energy + potential energy:

$$E = (mv^2/2) - (Ze^2/r) \qquad \textbf{(8–20)}$$

By using Equations 8–18 and 8–19 to evaluate Equation 8–20, we find

$$E(n) = -(2\pi^2 me^4 Z^2)/n^2 h^2 \qquad \textbf{(8–21)}$$

(That these energies are negative indicates that the orbits are *bound*.) The smallest Bohr orbit ($n = 1$) is the most strongly bound, and all orbits are bound until $n \to \infty$ where $E \to 0$. For $E > 0$, a *continuum* of unbound orbits is available to the electron.

Quantized Radiation

Bohr's second postulate concerns the absorption and emission of radiation by an atom—the fundamental interaction between matter and radiation. According to the first postulate, the electron cannot radiate while in one of the allowed discrete orbits, and so the second postulate states that (a) *radiation in the form of a single discrete quantum is emitted or absorbed as the electron jumps from one orbit to another* and (b) *the energy of this radiation equals the energy difference between the orbits.*

When an electron makes a *transition* from a higher orbit (n_a) to a lower orbit (n_b), a photon is emitted. The energies of this process may be symbolized by

$$E(n_a) = E(n_b) + h\nu \qquad (emission) \quad \textbf{(8–22a)}$$

where $n_a > n_b$. For the electron to make a transition from the lower orbit to the upper orbit, the atom must absorb a photon of exactly the correct energy:

$$E(n_b) + h\nu = E(n_a) \qquad (absorption) \quad \textbf{(8–22b)}$$

In both cases, the frequency of the proton involved is (Equations 8–15 and 8–21)

$$\begin{aligned} \nu_{ab} &= [E(n_a) - E(n_b)]/h \\ &= (2\pi^2 me^4/h^3)Z^2[(1/n_b{}^2) - (1/n_a{}^2)] \end{aligned} \quad \textbf{(8–23)}$$

Note that only one quantum is emitted or absorbed, even if $n_a > n_b + 1$; the electron may jump over several intermediate orbits. On the other hand, an electron may *cascade* to the lowest orbit, emitting several photons of different energy as it jumps through a series of adjacent orbits. The lowest orbit, called the *ground state*, corresponds to $n = 1$.

(c) The Bohr Model of the Hydrogen Atom

Now to apply Bohr's picture to the simplest atom, hydrogen ($Z = 1$), with its single electron. The electron's permitted orbital energies are, from Equation 8–21,

$$E(n) = -(2\pi^2 me^4/h^2)(1/n^2) \equiv -R'(1/n^2) \quad \textbf{(8–24)}$$

where $R' = 2.18 \times 10^{-18}$ J incorporates all the other constants.

It is convenient to express Equation 8–23 in the terms of the *wave number* (reciprocal wavelength):

$$\begin{aligned} 1/\lambda_{ab} &= \nu_{ab}/c = (R'/ch)(1/n_b{}^2 - 1/n_a{}^2) \\ &= R(1/n_b{}^2 - 1/n_a{}^2) \end{aligned} \quad \textbf{(8–25)}$$

The *Rydberg constant R* has the value $10.96776 \, \mu m^{-1}$ for hydrogen.

Equation 8–25 says that there is a *series* of wavelengths for *each* value of n_b when we consider a sequence of increasing values of n_a beginning with $n_a = n_b + 1$ (Figure 8–8). Transitions of the *Lyman* series (in the ultraviolet) all have the ground state ($n_b = 1$) as their lowest orbit, with $n_a \geq 2$. The visible spectral lines of the *Balmer* series have $n_b = 2$ and $n_a \geq 3$. Other important series are the Paschen, Brackett, and Pfund series, with $n_b = 3$, 4, and 5, respectively. All of these series are named for the physicists who first observed the spectral lines corresponding to the indicated transitions.

The first hydrogen lines discovered were the Balmer series, and they are designated H_α for $n_a = 3$, H_β for $n_a = 4$, H_γ for $n_a = 5$, and so forth. Use Equation 8–25 to compute the wavelength of the H_α line, which corresponds to $n_b = 2$ and $n_a = 3$:

$$1/\lambda_{H_\alpha} = 10.96776(1/4 - 1/9) \, \mu m^{-1}$$
$$= 1.52330 \, \mu m^{-1}$$

$$\lambda_{H_\alpha} = 656.3 \text{ nm}$$

Similarly, the Lyman α line ($n_b = 1$, $n_a = 2$) has the wavelength $\lambda_{L_\alpha} = 121.6$ nm (in the ultraviolet).

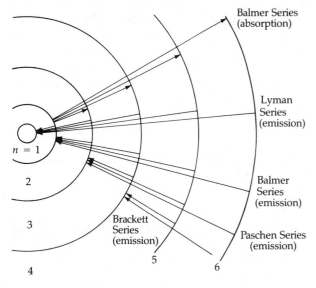

Figure 8–8

The Bohr model for hydrogen. The orbitals are the stable energy levels of the electron.

(d) Energy-Level Diagrams

We must warn you that Bohr's simple theory is an approximation to the actual dynamics of atomic phenomena. The theory encounters insurmountable difficulties when extended to atoms more complicated than hydrogen, and the full mathematical machinery of *quantum mechanics* is necessary to understand atoms in detail.

Atomic particles (such as electrons) also exhibit a wave nature, just as photons do; hence, an inherent uncertainty exists in both the position and the velocity of an electron (the *Heisenberg uncertainty principle*). The electron in the hydrogen atom may be protrayed as a *cloud* surrounding the proton, with the most probable position of the electron being one of the Bohr orbits. A multielectron atom has several such clouds, with the electrons tending to occupy fuzzy *shells* about the nucleus. The simplest shells are spherical, but in general more complicated shapes occur.

We will sidestep the sophistications of quantum mechanics in this book by abandoning spatial models of atoms. Instead, we represent atoms abstractly by means of an *energy-level diagram* (Figure 8–9). Such a diagram is directly related to atomic transitions, and so it can be constructed even for complicated atoms. As an example, consider the energy level diagram for hydrogen. The permitted energies of a bound electron in the hydrogen atom (Equation 8–24) are negative. Because we observe the positive-energy photons corresponding to electronic transitions, let's normalize to a positive energy scale by subtracting the ground-state energy $E(1)$ from all energies $E(n)$ to obtain

$$E(n) = R'[1 - (1/n^2)]$$

Note that $E(1)$ now equals zero. Finally, change our energy units to *electron volts*, where

$$1 \text{ eV} = 1.602 \times 10^{-19} \text{ J}$$

The electron volt is the energy acquired by an electron (or any particle of charge e) when accelerated through a voltage difference of 1 V; this unit is convenient in atomic and particle physics. Note that

$$E(\infty) = R' = 2.18 \times 10^{-18} \text{ J} = 13.6 \text{ eV}$$

The energy levels for several values of n are shown in Figure 8–9 [$E(1) = 0$, $E(2) = 10.2$ eV,

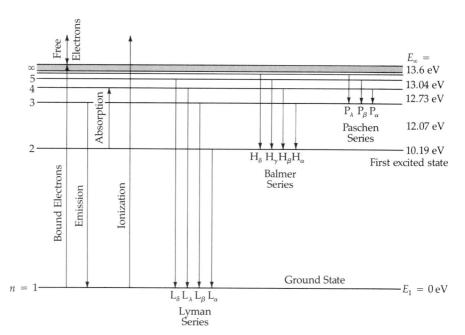

Figure 8–9

Hydrogen energy-level diagram. Transitions for the first three hydrogen series are shown in emission. On the left are shown the general types of transitions for absorption, emission, and ionization.

E(3) = 12.1 eV, and so on]. When the atom is in any level above the ground state, it is in an *excited state,* and the energy of such a level is called its *excitation potential.* To reach a higher level, the atom must be *excited*—an excessive energy of excitation leads to *ionization*—and in dropping back toward the ground state, the atom is *de-excited.*

Excitation

An atom may be excited to a higher energy level in two ways: radiatively or collisionally. *Radiative excitation* occurs when a photon is *absorbed* by the atom; the photon's energy must correspond exactly to the energy difference between two energy levels of the atom. This process produces *absorption lines* superimposed on a background continuous spectrum (Section 8–6).

An atom generally remains in an excited state for only an extremely short time (about 10^{-8} s) before re-emitting a photon. How then can an absorption line be produced? Recall that the electron may cascade through several energy levels on its way to the ground state, and so several lower-energy photons may be emitted for each photon absorbed. The input wavelength is converted to longer wavelengths, depleting the spectrum at the input wavelength. Also (Figure 8–10), the absorbed photons come predomi-

nantly from one direction, the direction of their source, whereas the emitted photons can travel away in *any* direction. Hence, fewer photons at the absorption wavelength reach an observer than photons at other wavelengths. An absorption line is darker than an unabsorbed continuum but not completely black because some photons of the critical wavelength still reach the observer.

Collisional excitation takes place when a free particle (an electron or another atom) collides with an atom, giving part of its kinetic energy to the atom. Such an inelastic collision does not involve any photons. A particle approaching the atom with speed v_i and leaving with speed v_f has deposited the energy $E = m(v_i^2 - v_f^2)/2$ with the atom; if E corresponds to the energy of an electronic transition, the atom is collisionally excited to a higher state. Such an excited atom returns to its ground state by emitting photons, producing an *emission line spectrum* in the process.

De-excitation

Atoms are always interacting with the electromagnetic field. This interaction causes an excited atom to jump spontaneously to a lower energy level (de-excite) in a characteristic time of order 10^{-8} s. Because a photon is emitted, we call the process *radiative de-excitation.* Another form of de-excitation,

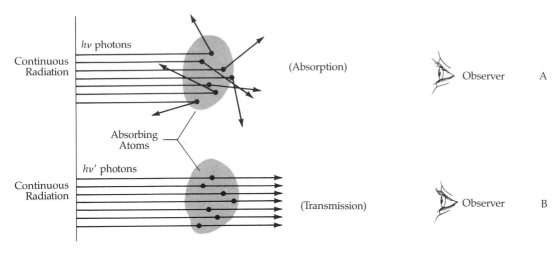

Figure 8–10

Absorption of light. (A) If *E* is the excitation energy of an atom, then only photons with this energy (= *hν*) can be absorbed. (B) Photons with energy *E'* cannot be absorbed and reach an observer through the cloud.

where the phenomenon is not announced by a photon, is *collisional de-excitation;* this is the exact inverse of collisional excitation, for the colliding particle *gains* kinetic energy in the exchange (a superelastic collision). In astrophysical situations, these two modes of de-excitation compete with one another.

Although most spontaneous downward transitions occur on short time scales, certain transitions — because of quantum mechanical rules — take place much more slowly. These transitions are called *forbidden transitions* and result in *forbidden lines.* They usually involve jumps from excited metastable states to the ground state of an atom. Under most conditions, collisions de-excite the metastable state before it loses energy by a radiative process. Hence, forbidden lines usually are produced when gas densities are low, so that the chances of collision are small during the time interval between excitation and radiative de-excitation.

Ionization

With sufficient energy (either radiative or collisional) to liberate an electron from a neutral atom, the atom is *ionized*. Schematically, this reaction is

$$X + energy \rightarrow X^+ + e^-$$

where X represents the atom. To denote neutral hydrogen (with *no* electrons removed), we write H or H

I, where the Roman numeral I signifies the neutral state; similarly, for neutral helium, He or He I. The singly ionized atom (with *one* electron removed) is in its first ionization state, such as $H^+ = H\ II$ or $He^+ = He\ II$. Hydrogen possesses but one electron and so may not be ionized beyond H II, but we can have $O^{++} = O\ III$ (doubly ionized oxygen) and $O^{+++} = O\ IV$ (oxygen with three electrons removed). For high stages of ionization (such as Fe XIV), the Roman numeral system is more common in astronomy than the multiple-sign convention.

The energy required to ionize an atom depends upon the ionization state of the atom, the particular electron to be liberated, and the excitation levels of that electron. For simplicity, consider the hydrogen atom with its sole electron. An electron in the ground state ($n = 1$; Figure 8–9) is removed from the atom when we supply it the energy $E \geq 13.6$ eV (the *ionization potential*); note that a *continuum* of energy states is available to the free electron above $E(n = \infty)$. If the electron is in the first excited state ($n = 2$), it will be freed when given energy $E \geq E(\infty) - E(2) = 13.6 - 10.2 = 3.4$ eV. In general (for hydrogen), the ionization potential for an electron in excitation level n is

$$IP(n) = E(\infty) - E(n) = 13.6/n^2 \text{ eV} \quad \textbf{(8–26)}$$

The kinetic energy available to the departing electron is the difference between the energy provided and the ionization potential ($E - IP$).

Radiative ionization (by photons) leads to spectral absorption continua because an infinite number of states are available above the ionization level. For example, hydrogen atoms in the ground state absorb discrete-wavelength photons to produce the Lyman absorption series, but this series ends at the *series limit* $\lambda = 91.2$ nm. For wavelengths shorter than the series limit, we observe the Lyman *absorption continuum,* corresponding to photons that can ionize hydrogen from its ground state. Similarly, the Balmer, Paschen, Brackett, and Pfund series limits and their associated absorption continua for ionizations arise from levels $n = 2$, 3, 4, and so forth.

Free electrons can *recombine* with ions by emitting a photon of the appropriate energy. Because this process is just the reverse of ionization, the various hydrogen emission series may end in an *emission continuum* if the conditions are right.

8–3 The Spectra of Atoms, Ions, and Molecules

(a) Atomic Spectra

In multielectron atoms, quantum mechanics and the *Pauli exclusion principle* dictate that only two electrons may occupy the innermost shell, eight the next, 18 the third, and so on. When a shell contains the maximum allowed number of electrons, it is filled; in this case, the atom is extremely stable and

hard to excite (helium, neon, argon, . . .). Innermost shells tend to be the first to be filled with electrons, and any excess electrons (*valence electrons*) are available for chemical interactions. For example, calcium (with 20 electrons) acts as if there are only two electrons in the outermost shell since the two inner shells (2 + 8) are closed and eight of the electrons in the third shell form two *subshells* (2 + 6).

Filled shells and subshells are tightly bound and shield the nucleus from the outer electrons, and so these electrons are only lightly bound (easily excited and ionized). The spectra of atoms with one outer electron, such as lithium and sodium, are similar to that of hydrogen but do show the effects of the inner closed shells. Atoms with more than one outer electron have increasingly complex spectra. To illustrate the various possibilities in the *periodic table*, Table 8–3 lists, for several representative atoms, (1) the excitation potential of the first excited state, (2) the ionization potential from the ground state, and (3) the wavelengths corresponding to these transitions.

(b) Ionic Spectra

Ions possessing at least one bound electron behave spectrally just like neutral atoms: they may be excited, de-excited, and ionized further. Except for overall wavelength modifications imposed by the greater charge of the nucleus, the spectrum of an ion closely resembles that of a neutral atom with the same number of outer electrons.

Consider ions with one electron remaining, such

TABLE 8–3   Excitation and Ionization Potentials for Selected Atoms

	Excitation* Potential	λ (nm)	Ionization† Potential	$\lambda_{series\ limit}$(nm)
Hydrogen (1 electron)	10.2	121.6	13.6	91.2
Helium (1 closed shell)	20.9	58.4	24.5	48.8
Lithium (1 filled shell, 1 outer electron)	1.8	670.8	5.4	225.0
Neon (2 filled shells)	16.6	73.5	21.5	57.6
Sodium (2 filled shells, 1 outer electron)	2.1	589.0	5.1	243.0
Magnesium (2 filled shells, 2 outer electrons)	2.7	457.1	7.6	163.0
Calcium (2 filled shells, 2 filled subshells, 2 outer electrons)	1.9	657.3	6.1	203.0

* From ground state to first excited state.
† From ground state of neutral atom.

as He II, Li III, O VIII, and even Fe XXVI. The Bohr wave number relationship for these cases is

$$1/\lambda_{ab} = RZ^2(1/n_b^2 - 1/n_a^2) \qquad (8-27)$$

Here Z equals the value of the ionization state. By analogy, each such ion exhibits Lyman, Balmer, and the other series, but the wavelengths differ from those of the hydrogen spectral lines by a factor of Z^{-2}. So the He II Lyman α line lies at 30.4 nm instead of 121.6 nm (since $Z = 2$).

(c) Molecular Spectra

Molecules are formed when atoms bind together (the *chemical bond*). Quantum mechanics applies to such a union, and three types of discrete energy levels are exhibited by molecules. (1) Electronic energy states are in the combined electron cloud surounding the nuclei. *Electronic transitions* similar to those in an atom can take place between these states, leading to excitation, de-excitation, and ionization of the molecule (such as $H_2 \rightarrow H_2^+ + e^-$). (2) The internuclear distances are quantized in discrete *vibrational energy states*, with consequent vibrational transitions. When the separation becomes so great that the atoms are no longer bound together, we say that the molecule has *dissociated*. (3) A molecule may rotate about various axes in space, resulting in discrete *rotational energy states*.

We will illustrate the concept of rotational molecular spectra with the example of the carbon monoxide (CO) molecule — one of the most common in interstellar space. Consider a two-mass rotator with two masses M and m separated by a distance r, spinning at angular velocity ω with moment of inertia I. The rotational energy is

$$E = I\omega^2/2 = \mu r^2\omega^2/2$$

where μ is the reduced mass of the system:

$$\mu = mM/(m + M)$$

Now, the rotational states of molecules are quantized so that the angular momentum can have only discrete values given by

$$I\omega = (h/2\pi)J$$

where J is the total angular momentum quantum number. Then the energy states corresponding to the possible states of J are

$$E = (h/2\pi)^2 J(J + 1)/2\mu r^2$$

Hence, for a diatomic molecule, the energy levels are equally spaced since J must vary by integral increments. For CO, the important transitions are from $J = 1$ to 0 and from 2 to 1. The former corresponds to a frequency of 115.2712 GHz, the latter to 230.5424 GHz.

The three classes of molecular transitions lead to numerous spectral lines superimposed upon one another. Since the vibrational and rotational transitions involve small differences in energy (usually much less than 1 eV), their spectral lines are closely spaced in wavelength and appear as *bands*. Molecular spectra are much more complex than atomic spectra, but they are easily recognized by their banded structure.

8–4 Spectral-Line Intensities

Let's now discuss the *strengths* of emission and absorption spectral lines (Figure 8–11). The intensity of an emission line is proportional to the number of photons emitted in that particular transition. Similarly, the strength of an absorption line relative to the

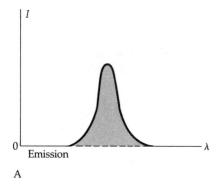

A   Emission

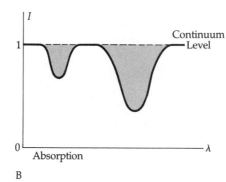

B   Absorption

Figure 8–11

Spectral line profiles. (A) An emission line. (B) An absorption line.

adjacent continuum depends upon the number of photons absorbed. Because the majority of astronomical spectra are absorption spectra, we will consider the case for absorption; very similar arguments follow for emission spectra.

An absorption line is never infinitely sharp. It exhibits a *profile*, the *intensity* (residual radiation in terms of the continuum as unity) of which varies with wavelength (Figure 8–11). A spectral line itself has no area, but if we represent the distribution of energy with wavelength on a two-dimensional diagram (like that of Figure 8–11), we may measure an area in the λ, I plane. The total strength of a line is proportional to its area, which may be represented by the line's *equivalent width* (Figure 8–12). We replace the area of the line profile by a rectangle of equal area, with one dimension of the rectangle being the height of the continuum and the other dimension being the equivalent width (in angstroms or milliangstroms in astronomy). Note that the equivalent width increases with the strength of the line. When the center of the line profile reaches zero intensity, we say that the line has become *saturated*; any further increase in the strength comes only from the wings of the line.

(a) Excitation Equilibrium: Boltzmann's Equation

The strength of the spectral line depends directly upon the number of atoms in the energy state from which the transition occurs. So we would like to know the fraction of all the atoms of a given element that are excited to that energy state. Recall that excitation and de-excitation may occur collisionally and/or radiatively (a radiative de-excitation is also termed a *spontaneous transition*). Both processes depend upon temperature since the mean kinetic energy of a gas particle goes as

$$\overline{mv^2/2} = 3kT/2 \qquad (8-28)$$

where m is the particle's mass, v its speed, $k = 1.380 \times 10^{-23}$ J/K is *Boltzmann's constant*, and T is the gas temperature in kelvins. The number of photons of a given energy increases rapidly with increasing temperature. So absorption lines originating from excited levels tend to be stronger in hot gases than in cool gases.

For simplicity, consider the situation where *thermal equilibrium* prevails and the average number of atoms in a given state remains unchanged in time

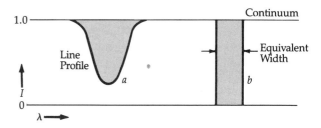

Figure 8–12

Equivalent width. The area of rectangle b is identical to that of line profile a. The continuum has been arbitrarily set to a level of 1.0.

(*steady state*). Each excitation is, on the average, balanced by a de-excitation. In this case, statistical mechanics shows that the *number density* (number per unit volume) of atoms in state B is related to the number density in state A (B > A) by *Boltzmann's equation* [named after the Austrian physicist Ludwig Boltzmann (1844–1906) who discovered the relationship]:

$$N_B/N_A = (g_B/g_A) \exp[(E_A - E_B)/kT] \qquad (8-29)$$

where N is the number density in the level, g is the multiplicity of the level (an intrinsic characteristic), and E is the energy of the level. The term exp ( ) denotes exponentiation; that is, raise $e = 2.71828 \ldots$ (the base of the natural logarithms) to the power ( ) (Mathematical Appendix). Because $E_B > E_A$, the bracketed quantity in Equation 8–29 is always negative, so that the ratio N_B/N_A increases as the temperature increases ($N_B/N_A \rightarrow g_B/g_A$ as $T \rightarrow \infty$). For a given temperature, the *excitation ratio* N_B/N_A increases as the excitation potential $E_B - E_A$ decreases between the two energy levels.

Because $\exp(-\infty) = 0$, $\exp(-1) = 0.368$, and $\exp(0) = 1$, a significant population of the upper level occurs when $T \approx (E_B - E_A)/k$; an excitation potential of 1 eV corresponds to a temperature of 11,600 K. As an example, consider a volume of gas that contains the same number of hydrogen and helium atoms at a temperature such that the number of hydrogen atoms in the first excited state (N_2) equals one-tenth the number in the ground state (N_1), or $N_2/N_1 = 0.1$. On the other hand, the ratio N_2/N_1 for helium will be very low. So at a given temperature, the fraction of atoms in the second level differs widely from element to element, depending on the excitation potential. In this case, the absorption lines arising from transitions from $n = 2$ to $n = 3$ will be strong for hydrogen and very weak for helium. The strength of the line is a function of both the *abundance* of the particular element and the *temperature*. In this example, we have unrealistically ignored the ionization of the atoms; for a complete picture, we must include both ionization and excitation to other levels.

(b) Ionization Equilibrium: Saha's Equation

As the temperature of a gas is increased, more and more energy (either radiative or collisional) becomes available to ionize the atoms. In general, the hot gas consists of neutral atoms, ions, and free electrons. The greater the *electron density* (N_e = number of electrons per unit volume), the greater the probability that an ion will capture an electron and become a neutral atom. These two competing processes — ionization ($\rightarrow$) and recombination ($\leftarrow$) — are written as

$$X \rightleftarrows X^+ + e^-$$

A steady-state condition of *ionization equilibrium* is achieved in the gas when the rate of ionization equals the rate of recombination. A quantitative expression of this ionization equilibrium is given by *Saha's equation* [named after the Indian physicist Meghnad N. Saha (1893–1956)]:

$$N_+ N_0 = [A(kT)^{3/2}/N_e]\exp(-\chi_0/kT) \quad \text{(8-30)}$$

where N_+ is the number density of ions, N_0 is the number density of neutral atoms in the ground state, the constant A includes several atomic constants and incorporates the probability of different states of ionization, T is the absolute temperature, N_e is the electron density, and χ_0 is the *ionization potential* (in electron volts) from the ground state of the neutral atom. Equation 8–30 is very similar to Boltzmann's equation 8–29 except for the dependence upon N_e and the additional factor of $T^{3/2}$, which arises because a continuum of energies above χ_0 will ionize the atom and the liberated electron is more likely to escape the ion as the electron's kinetic energy increases. Note that for pure hydrogen gas, $N_e = N_+$.

The Boltzmann excitation equation (8–29) applies to *any two levels of excitation*, those of an ion as well as those of a neutral atom. Similarly, the Saha ionization equation (8–30) can be generalized to give the ratio N_{i+1}/N_i for any *stage of ionization $i + 1$* and the next lower stage i. The appropriate form of Saha's equation is

$$N_{i+1}/N_i = [A(kT)^{3/2}/N_e]\exp(-\chi_i/kT) \quad \text{(8-31)}$$

where χ_i is the ionization potential of the lower stage (the energy needed to ionize the particle in the i state to the $i + 1$ state. For example, Equation 8–31 applies to the ionization balance between Ca III ($i + 1 = 3$) and Ca II ($i = 2$). The relative population of the upper ionization stage increases rapidly with increasing temperature or smaller values of χ_i.

(c) The Boltzmann and Saha Equations Combined

The Boltzmann equation gives the number of atoms in an excited state relative to the number in the ground state; this applies to both neutral and ionized atoms. The Saha equation tells us the relative populations of two adjacent stages of ionization. We combine these two equations to calculate the number of atoms available to make a certain transition and so produce a given spectral line.

Consider the Balmer absorption lines of neutral hydrogen. Their strength is proportional to the number of atoms in the first excited state (N_2) of the neutral atom relative to the total number of hydrogen atoms in *all* stages of ionization (N). Hydrogen has only two stages of ionization, however: neutral (N_0) and singly ionized (N_+); hence, we know that $N = N_0 + N_+$. The proportion of N_2/N is

$$\begin{aligned} N_2/N &= N_2/(N_0 + N_+) \\ &\approx (N_2/N_1)/[1 + (N_+/N_0)] \quad \text{(8-32)} \end{aligned}$$

where we have used the reasonable approximation $N_0 \approx N_1$ in the last equality. The Boltzmann equation yields N_2/N_1, the ratio of neutral atoms in the first excited state to those in the ground state; the Saha equation gives N_+/N_0, the ratio of ionized to neutral atoms. The approximate form of Equation 8–32 is accurate enough for our present purposes.

A plot of N_+/N_0 as a function of temperature (Figure 8–13) shows that most of the hydrogen is neutral at temperatures below 7000 K, but at higher temperatures, ionization increases to the point where the number of neutral atoms becomes negligible. The exponential increase of N_2/N_1 with increasing temperature is therefore countered by the lack of neutral atoms at high temperatures. As a result, the N_2/N curve has a maximum around 10,000 K. The strength of the Balmer absorption lines of hydrogen is greatest

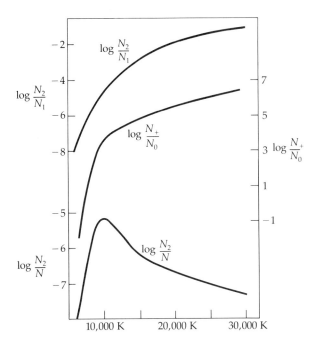

Figure 8-13

Excitation and ionization curves for hydrogen Balmer lines. The relative populations of the energy levels (N_2/N_1) from the Boltzmann equation and the ionization stages (N_+/N_0) from the Saha equation are calculated for equilibrium at the indicated temperatures. The lower curve shows the combination of the upper two with $N = N_0 + N_+$.

near $\approx 10,000$ K, decreasing at both higher and lower temperatures. At 6000 K—the approximate surface temperature of the Sun—the ratio N_2/N_1 is about 10^{-8}, but reasonably strong Balmer absorption lines are seen as a result of the great abundance of hydrogen in the Sun. At 20,000 K—the temperature of very hot stars—the ratio is about 10^{-2}, but the Balmer lines are similar in strength to those of the Sun because most of the hydrogen atoms are now ionized!

The Boltzmann and Saha equations have wide application in astrophysics. Through them, we may interpret stellar absorption (and emission) spectra in order to deduce the surface temperatures and pressures of stars. For example, at temperatures in the range 5000 to 7000 K, calcium should be predominantly in the form Ca II (singly ionized). Stars with strong Ca II lines but weak Ca I lines (our Sun, for example) must then have temperatures of this order. On the other hand, a star that has much lower density than our Sun but produces equally strong Ca II lines

must actually have a lower temperature to compensate for the smaller electron density N_e (Equation 8–31), assuming that the chemical composition is the same.

Now to generalize the previous formulas to obtain $N_{i,s}$, the relative number of atoms in *any* state of excitation s of a stage of ionization i. The ratio of interest is $N_{i,s}/N$, where N is summed over *all stages of ionization*:

$$N = N_0 + N_1 + N_2 + \cdots + N_n = \sum_{i=0}^{n} N_i$$

In general, n is the number of electrons in the neutral atom, but in practice, only two or three stages of ionization need be considered, for the number of ions in other stages will be negligible at a given temperature. To a reasonable first approximation,

$$\begin{aligned} N_{i,s}/N &\approx N_{i,s}/(N_{i-1} + N_i + N_{i+1}) \\ &= (N_{i,s}/N_i)/[(N_{i-1}/N_i) + 1 \\ &\quad + (N_{i+1}/N_i)] \end{aligned} \qquad (8-33)$$

The numerator of the last expression is given by the Boltzmann equation:

$$N_{i,s}/N_i \propto \exp(-\chi_s/kT)$$

and the denominator can be found from the Saha equation:

$$N_{j+1}/N_j \propto [(kT)^{3/2}/N_e] \exp(-\chi_j/kT)$$

Although Equation 8–33 is useful for the strongest spectral lines of a gas, in the general case extensive numerical computations are necessary to accurately reproduce spectral-line strengths. Our approximation holds only when i is the predominant stage of ionization for the prevailing temperature.

8-5 Spectral-Line Broadening

Spectral lines are never perfectly sharp; their profiles always have finite width. The fundamental tenets of quantum mechanics account for the minimum width (natural broadening) of a spectral feature, and various physical processes further broaden the line profile. By interpreting the observed profile of a spectral line in terms of these broadening mechanisms, we can deduce some characteristics of the radiation from a star.

(a) Natural Broadening

Quantum mechanics ascribes a wave nature to all atomic particles; an electron in an atomic energy level is such a particle. The *Heisenberg uncertainty principle* implies that the energy of a given state may not be specified more accurately than

$$\Delta E = (1/2\pi)(h/\Delta t) \qquad (8-34)$$

where h is Planck's constant and Δt is the *lifetime* of the state. So an assemblage of atoms will produce an absorption or emission line with a minimum spread in photon frequencies — the *natural width* — of order $\Delta v = \Delta E/h \approx 1/\Delta t$. Typical excited states live about 10^{-8} s before decaying (in contrast, a ground state may last forever), so that a normal natural width is near 0.5 nm for visible light. Much smaller natural widths occur for the so-called *metastable states*, some of which last more than 1 s ($\Delta t \geq 1$ s).

(b) Thermal Doppler Broadening

Whereas natural broadening depends only upon the intrinsic lifetime of an energy level, *thermal Doppler broadening* depends upon the temperature and composition of a gas. When a gas is at a certain temperature T, gas particles (each of mass m) move about with a Maxwellian distribution of velocities characterized by the mean kinetic energy of Equation 8-28:

$$\overline{mv^2/2} = 3kT/2$$

Atomic motions along our line of sight imply *Doppler shifts* in the radiation absorbed or emitted in atomic transitions. At a given temperature, the spectral lines of heavy elements are narrower than those of light elements since, on the average, the heavy particles move more slowly than the light particles. For example, neutral hydrogen at 6000 K moves at the mean speed $v \approx 12$ km/s, corresponding to a fractional Doppler broadening of $\Delta\lambda/\lambda \approx v/c \approx 4 \times 10^{-5}$; hence, the thermal Doppler width of the Balmer α line (656.3 nm) is approximately 0.025 nm.

(c) Collisional Broadening

The energy levels of an atom are shifted by neighboring particles, especially charged particles, such as ions and electrons (called the *Stark effect*). In a gas, these perturbations are random and result in a broadening of spectral lines. Because the perturba-

tions are larger the nearer the perturbing particle, there is a direct dependence of this *collisional* (or *pressure*) *broadening* upon particle density. The greater the density (and hence pressure) of the gas, the greater the width of the spectral lines.

(d) The Zeeman Effect

When an atom is placed in a *magnetic field*, the atomic energy levels each separate into three or more sublevels — this is the *Zeeman effect*, after the Dutchman Pieter Zeeman (1865–1943). Instead of a single atomic transition and a single spectral feature, we now have three or more closely spaced lines (the spacing is proportional to the magnetic field strength). If the Zeeman components are not resolved, we see only a broadened spectral line. In those cases in which the magnetic field is very strong and is highly ordered and uniform (such as sunspots and magnetic stars), we may resolve the Zeeman splitting and deduce the magnetic field strength and orientation of the source.

(e) Other Broadening Mechanisms

Finally, we mention three macroscopic broadening mechanisms based upon the Doppler effect. Consider a typical star whose image cannot be resolved. Large-scale random motions at the surface of such a star imply Doppler shifts that appear as *turbulence broadening* of spectral lines. If the atmosphere of the star is expanding, we simultaneously see gas moving in all directions; the integrated effect of all the Doppler shifts is *expansion broadening* of the observed spectral lines. A rapidly rotating star (not seen pole-on) will have *rotationally broadened* lines since one limb of the star is approaching us while the other limb is receding. Because all spectral features exhibit the *same* rotational broadening, we may identify the stellar rotation and determine its rate (or period).

8-6 Blackbody Radiation

So far we have dealt with individual atomic transitions and the spectral lines they produce. Where then does a *continuous spectrum* come from? (Recall that spectral absorption lines result when photons are selectively absorbed from such a continuum.) We noted that various emission and absorption continua can originate from individual atoms and that spectral features become more broadened as atoms interact more strongly with one another. When an aggregate of atoms interact so strongly (such as in a solid, a liquid, or an opaque gas) that all detailed spectral features are washed out, a thermal continuum results.

Such a continuous spectrum comes from a *blackbody*, whose spectrum depends only upon the absolute temperature. A blackbody is so named because it absorbs *all* electromagnetic energy incident upon it —it is completely *black.* To be in perfect thermal equilibrium, however, such a body must radiate energy at exactly the same rate that it absorbs energy; otherwise, the body will heat up or cool down (its temperature will change). Ideally, a blackbody is a perfectly insulated enclosure within which radiation has come into thermal equilibrium with the walls of the enclosure. Practically, *blackbody radiation* may be sampled by observing the enclosure through a tiny pinhole in one of the walls. The gases in the interior of a star are opaque (highly absorbent) to all radiation (otherwise, we would see the stellar interior at some wavelength!); hence, the radiation there is blackbody in character. We sample this radiation as it slowly leaks from the surface of the star—to a good approximation, the continuum radiation from stars is blackbody in nature.

(a) Planck's Radiation Law

After Maxwell's theory of electromagnetism appeared in 1864, many attempts were made to understand blackbody radiation theoretically. None succeeded until, in 1900, Max K.E.L. Planck (1858–1947) postulated that electromagnetic energy can propagate only in discrete quanta, or *photons*, each of energy $E = h\nu$. This brilliant German physicist then derived the *spectral intensity relationship*, or *Planck blackbody radiation law:*

$$I_\nu \Delta\nu = (2h\nu^3/c^2)[1/(e^{h\nu/kT} - 1)]\Delta\nu \quad \textbf{(8-35a)}$$

where $I_\nu \Delta\nu$ is the *intensity* (J/m² · s · ster) of radiation from a blackbody at temperature T in the frequency range between ν and $\nu + \Delta\nu$, h is Planck's constant, c is the speed of light, and k is Boltzmann's constant. Note the exponential in the denominator.

Because the frequency ν and wavelength λ of electromagnetic radiation are related by $\lambda\nu = c$, we may also express Planck's formula (Equation 8–35a)

in terms of the intensity emitted per unit wavelength interval:

$$I_\lambda \Delta\lambda = (2hc^2/\lambda^5)[1/e^{hc/\lambda kT} - 1)]\Delta\lambda \quad \textbf{(8-35b)}$$

Equation 8–35b follows because the intensity $I_\nu \Delta\nu$ equals the intensity $I_\lambda \Delta\lambda$ in the corresponding wavelength interval, where

$$\nu = c/\lambda \rightarrow |\Delta\nu| = |c\Delta\lambda/\lambda^2|$$

This expression follows by differentiation or by noting that

$$\lambda\nu = (\lambda + \Delta\lambda)(\nu + \Delta\nu) = c$$

where $\Delta\lambda\Delta\nu$ is negligible.

Equation 8–35b is illustrated in Figure 8–14 for several values of T. Note that both I_λ and I_ν increase as the blackbody temperature increases—the blackbody becomes *brighter*. This effect is easily interpreted in terms of Equation 8–35a, when we note that $I_\nu \Delta\nu$ is directly proportional to the *number of photons* emitted per second near the energy $h\nu$.

To better understand the Planck blackbody formulas (Equations 8–35), let's outline a derivation by Albert Einstein. At a given frequency ν, the radiation will be in thermal equilibrium with the atoms that compose the walls of the blackbody cavity. Because of the strong atomic interactions in the walls, all possible energy states are available there. Consider the upper (U) and lower (L) energy states corresponding to the quantum of energy $E_U - E_L = h\nu$. The Boltzmann equation relates the number of atoms in each state via

$$N_U = (g_U/g_L)N_L e^{-(h\nu/kT)}$$

If the probability per atom for a downward transition (U → L) is a_{UL} and that for an upward transition (L → U) is a_{LU}, then a steady-state equilibrium is attained when

$$N_L a_{LU} = N_U a_{UL}$$

But a_{LU}—"radiative excitation"—depends upon the energy density ρ_ν of radiation at the frequency ν and upon the atomic parameter B_{LU} (the *Einstein absorption coefficient*): $a_{LU} = B_{LU}\rho_\nu$. De-excitation (a_{UL}) can occur in two ways: spontaneously (A_{UL}) or through the stimulating influence of the radiation field ($B_{UL}\rho_\nu$); hence, we have $a_{UL} = A_{UL} + B_{UL}\rho_\nu$. Combining these various results and solving for ρ_ν, we find

$$\rho_\nu = (A_{UL}/B_{UL})/[(B_{LU}g_L/B_{UL}g_U)e^{h\nu/kT} - 1]$$

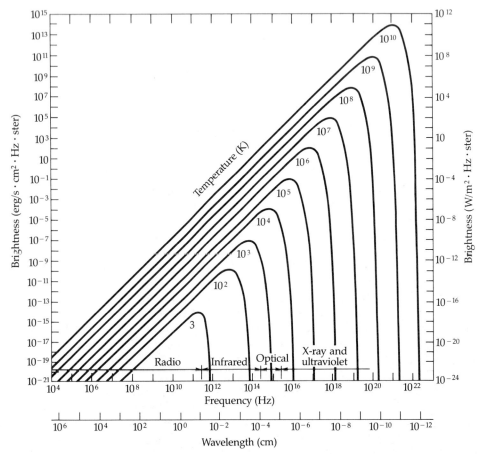

Figure 8–14

Blackbody radiation. A log–log plot of the Planck curves for a wide range of temperatures. Note that the wavelengths run from longer to shorter.

The result is independent of the number of atoms involved. Einstein then substituted suitable expressions for the atomic coefficients A_{UL}, B_{UL}, and B_{LU} to obtain the *Planck blackbody radiation density law*:

$$\rho_\nu \Delta\nu = (8\pi h\nu^3/c^3)[1/(e^{h\nu/kT} - 1)]\Delta\nu \quad \textbf{(8–36)}$$

where $\rho_\nu \Delta\nu$ is an *energy density* (J/m³). In converting to intensity radiated from the blackbody (I_ν), a factor of $c/4\pi$ is introduced (Equation 8–35a).

Before we leave the topic of blackbody radiation, which is so useful in astrophysics, we put forth some useful approximations. We have so far considered formulas for the monochromatic intensities (Equations 8–35). If we evaluate these in the case where the exponentials become very large (much greater than unity), then we find

$$I_\nu(T) = (2h\nu^3/c^2)\exp(-h\nu/kT) \quad \textbf{(8–37a)}$$

$$I_\lambda(T) = (2hc^2/\lambda^5)\exp(-hc/\lambda kT) \quad \textbf{(8–37b)}$$

which is an appropriate approximation when the temperature is low and the wavelengths are short. It

is sometimes called the *Wien distribution*. In the opposite case, when the value of the exponential is very small (much less than unity), we use the power series expansion for the exponential to note that $e^x - 1 \approx x$, and

$$I_\nu(T) = 2\nu^2 kT/c^2 \quad \textbf{(8–38a)}$$

$$I_\lambda(T) = 2ckT/\lambda^4 \quad \textbf{(8–38b)}$$

which is known as the *Rayleigh–Jeans* distribution and works at high temperatures and long wavelengths (low frequencies).

(b) Wien's Law

A blackbody appears to become bluer as its temperature increases. This phenomenon corresponds to the bulk of the radiation being emitted at shorter wavelengths (Figure 8–14) as T becomes larger. Wilhelm Wien (1864–1928) expressed the wavelength λ_{max} at which the maximum intensity of blackbody radiation is emitted—the peak (that wavelength for

TABLE 8–4   The Varieties of Temperatures

Temperature Type	Basic Law or Equation	Necessary Observations
Color	Planck curve	Brightness at two or more wavelengths
Effective or radiation	Stefan–Boltzmann law	Bolometric magnitude* and radius
Excitation	Boltzmann equation	Relative intensities of spectral lines of the same element
Ionization	Saha equation	Relative intensities of spectral lines of adjacent stages of ionization
Kinetic	Thermal Doppler broadening	Widths and profiles of spectral lines

* See Chapter 11 for definition.

which $dI_\lambda/d\lambda = 0$) of the Planck curve (found from taking the first derivative of Planck's law)—by Wien's displacement law:

$$\lambda_{max} = 2.898 \times 10^{-3}/T \qquad (8\text{–}39)$$

where λ_{max} is in meters when T is in kelvins. For example, the continuum spectrum from our Sun is approximately blackbody, peaking at $\lambda_{max} \approx 500$ nm; therefore, the surface temperature must be near 5800 K. Note that because $\lambda_{max}T = $ constant, increasing one proportionally decreases the other.

(c) The Law of Stefan and Boltzmann

The area under the Planck curve (integrating the Planck function) represents the total *energy flux F* (W/m²) emitted by a blackbody when we sum over all wavelengths and solid angles:

$$F = \sigma T^4 \qquad (8\text{–}40)$$

where $\sigma = 5.669 \times 10^{-8}$ W/s · K⁴. The strong temperature dependence of Equation 8–40 was first deduced from thermodynamics in 1879 by Josef Stefan (1835–1893) and was derived from statistical me-

chanics in 1884 by Boltzmann; therefore we call the expression the *Stefan–Boltzmann law*. The brightness of a blackbody increases as the fourth power of its temperature. If we approximate a star by a blackbody, the energy output of the star (in watts) is just $E = 4\pi R^2 \sigma T^4$ since the surface area of a sphere of radius R is $4\pi R^2$.

(d) Temperature

We end with a word of caution concerning *temperature*. If we are dealing with a true blackbody, we may establish its temperature using (1) the shape of the Planck curve or at least two points on the curve, (2) Wien's law, or (3) the Stefan–Boltzmann law. Because no astrophysical object is a perfect blackbody, we will obtain slightly different temperatures from each of these three methods. In addition, temperatures based on such continuum spectra may differ from the temperatures derived from the relative intensities of spectral lines by using the Boltzmann and Saha equations. It is therefore a good practice to stipulate what kind of temperature one is using (see the examples in Table 8–4).

Problems

1. **(a)** Show that a beam of light obliquely incident upon and passing through a plane-parallel piece of glass is simply displaced without changing direction when it emerges from the glass.
   (b) If the glass has thickness d and index of refraction n, what is the linear displacement of the beam as a function of n and θ?

2. What aperture is required to give 1 arc-second resolution for a wavelength of
   (a) 500 nm (visible)
   (b) 21 cm (radio)

Can you make a generalization from your results?

3. **(a)** At what wavelengths will the following spectral lines be observed:
   (i) A line emitted at 500 nm by a star moving toward us at 100 km/s
   (ii) The Ca II line (*undisplaced* wavelength of 397.0 nm) emitted by a galaxy receding at 60,000 km/s
   (b) A cloud of neutral hydrogen (H I) emits the 21-cm radio line (at rest frequency 1420.4 MHz) while

moving away at 200 km/s. At what *frequency* will we observe this line?

4. A simple form of the *binomial theorem* states that

$$(1 + x)^n = 1 + nx + [n(n - 1)x^2/2] + n(n - 1)(n - 2)x^3/6 + \cdots$$

when $x^2 < 1$. Starting with the relativistic expression, use this theorem to derive the classical equation for the Doppler shift in wavelength for the case in which $v \ll c$.

5. (a) What is the energy of one photon of wavelength $\lambda = 300$ nm? Express your answer both in joules and in electron volts.

   (b) An atom in the *second* excited state ($n = 3$) of hydrogen is just barely ionized when a photon strikes the atom. What is the wavelength of the photon if *all* of its energy is transferred to the atom?

6. The emission line of He II at 468.6 nm corresponds to what electronic transition?

7. By applying the Boltzmann equation to the *neutral* hydrogen atom (neglect ionization), derive an expression for the population of the nth energy level relative to that of the ground state, at temperature T. Now, assuming that the multiplicity of each level is unity ($g_n \equiv 1$), construct an appropriate graph showing your results for $T = 6000$ K.

8. To understand the relative importance of the different parameters in the Saha equation, perform the following experiment. Assume that $T = 5000$ K, $N_e = 10^{15}/$cm^3, and $IP = 12$ eV. By what factor does the *ionization ratio* (N_+/N_0) change when we separately

   (a) double the temperature
   (b) double the electron density
   (c) double the ionization potential

Which is more important during the temperature change, the exponential term or the $T^{3/2}$ term?

9. Let N_2 be the number of second-level (first excited state) hydrogen atoms and N_1 be the number in the ground state. Using Figure 8–13, find the excitation ratio (N_2/N_1) and the excited fraction (N_2N) for each of the following stars:

   (a) Sirius, $T = 10,000$ K
   (b) Rigel, $T = 15,000$ K
   (c) the Sun, $T = 5700$ K

Which star will exhibit the strongest Balmer absorption lines? Explain your reasoning in arriving at this answer.

10. (a) What is the *speed* of an electron with just sufficient energy to ionize by collision a sodium atom in the ground state?

    (b) What is the speed of a *proton* ionizing this atom?

    (c) What is the corresponding gas temperature?

    (d) At this temperature, what is the fractional *thermal Doppler broadening* ($\Delta\lambda/\lambda$) of a sodium spectral line?

11. (a) How much more energy is emitted by a star at 20,000 K than one at 5000 K?

    (b) What is the predominant *color* of each star in part a? Use the Wien displacement law, and express your answers in wavelengths.

12. Deduce an *approximate* expression for Planck's radiation law (Equation 8–35a)

    (a) at high frequencies ($h\nu/kT \gg 1$), the Wien distribution.

    (b) at low frequencies ($h\nu/kT \ll 1$), the Rayleigh–Jean approximation.

    You may use the approximation $\exp(h\nu/kT) \approx 1 + h\nu/kT$ when $h\nu/kT \ll 1$. To what *wavelength* does $h\nu/kT = 1$ correspond?

Telescopes and Detectors

We have discussed the nature and properties of light and how to use an understanding of these facts to infer the physical properties of astronomical objects. To underscore that process, we describe briefly in this chapter how astronomers gather and detect light—the basis of observational astronomy. In many respects, astronomy is much more an observational science than an experimental one (as physics and chemistry are).

9–1 Optical Telescopes

As extensions of the human eye, telescopes amplified the power of detection without extending the spectral range of our vision. Today we can sense much more than the visible part of the electromagnetic spectrum, but the discussion in this section restricts itself to optical telescopes—those that manipulate light detectable by the eye. Be warned, however, that the light that reaches the Earth's surface has passed through many filters before it reaches us: intergalactic and interstellar media, the Earth's atmosphere, and always the telescope and detection system.

Optical components, such as lenses and mirrors, are used to control the paths of light rays. For a telescope, such optical components bring light to a *focus*, usually to make an *image*. Light can be concentrated in a focus by either a curved mirror (using reflection) or a lens (by refraction). A lens brings rays from a point source to a point image at the focus of the lens. The lens makes an image of an object of finite size by focusing rays from each point on the object onto a separate point in the image (Figure 9–1). This image is generally smaller than the object and upside down. For objects at large distances, the distance from lens to image is approximately the same for all objects. This distance is termed the *focal length*. A smoothly curved mirror—one whose surface, for instance, follows the curve of a parabola—brings all the light to a

focus. The distance from the mirror's surface to its focus is its focal length.

For either a lens or a mirror, the ratio of focal length to diameter is called the *f ratio*:

$$f \text{ ratio} = f/d$$

where f is the focal length and d the diameter. The f ratio essentially gives the brightness of the image. A small f ratio, such as $f/3$, results in a brighter image than a large f ratio, such as $f/15$. That is, an $f/15$ lens or mirror will spread the image out more than an $f/3$ one will and so decrease the brightness.

Another property related to focal length is the linear size of the image of an extended object. For example, the Moon subtends a $1/2°$ arc in the sky; if its image covers 10 cm at the focal surface of the lens, then the scale is $0.05°/$cm, or 200 cm per degree. For the size s of an image at the focus that corresponds to $1°$ in the sky,

$$s = 0.01745f$$

where f is the focal length; s then comes out in units of f per degree; it is usually called the *plate scale*. An example: Capilla Peak Observatory of the University of New Mexico has an $f/13$ reflecting telescope with a 60-cm-diameter mirror. Its focal length is

$$f/d = 13$$
$$f = 13 \times 60 \text{ cm}$$
$$= 780 \text{ cm}$$

The plate scale is

$$s = 0.01745 \times 780 \text{ cm}$$
$$= 13.6 \text{ cm}/°$$

A telescope is essentially an instrument that gathers light and makes an image at a focus. A lens or mirror, called the *objective*, brings the light to a focus. A lens at the focus, called the *eyepiece*, allows visual examination of the image. There are two basic types of telescopes, distinguished by their objectives: *refracting telescopes* (or *refractors*), which use a lens, and *reflecting telescopes* (or *reflectors*), which use a mirror.

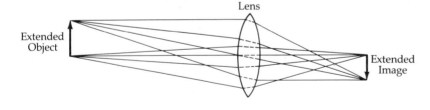

Lens

Extended Object

Extended Image

Figure 9–1

Images. A convex lens will make an image of an extended object that is smaller than the object and upside down.

A telescope is basically a light bucket for collecting photons. That's the main reason astronomers want large telescopes — for their greater light-gathering capacity. A telescope's *light-gathering power* is directly proportional to the square of its diameter. The amount of light a lens or mirror catches depends on its surface area, and its area with diameter d is $(\pi/4)d^2$. Light-gathering power is a relative measure, not an absolute one. It states how two instruments compare with one another, not how much light is gathered. Because the factor $\pi/4$ is a constant, we need use only the diameters of the instruments to arrive at a comparative figure. For example, compared with your eye, which has a diameter of about 0.5 cm, a telescope with a 50-cm objective has a light-gathering power of

$$LGP = (50/0.5)^2 = 100^2 = 10,000$$

Similarly, the 5-m Hale telescope at Mount Palomar outdoes the 0.6-m Capilla Peak telescope by

$$LGP = (5/0.6)^2 = 8.3^2 = 69$$

A second important function of a telescope is to produce an image in which objects that are close together in the sky can be seen as clearly separate. This ability is called *resolving power* and is sometimes expressed as the inverse of the minimum angle there must be between two points in order for them to be easily separated:

$$RP = 1/\theta_{min}$$

The resolving power and the minimum angle depend on the diameter of the objective and also on the wavelength of the light. For the same wavelength, the resolving power depends directly on the objective's diameter. The minimum resolvable angle depends on both the diameter of the telescope's objective and the wavelength of light being observed. In general,

$$\theta_{min} = 206,265\lambda/d$$

where θ_{min} is the *minimum resolvable angle,* or *resolution* in seconds of arc (206,265 is the number of arcseconds in 1 rad), λ is the wavelength, and d is the diameter of the objective in the same length units. As a slight complication, the circular aperture of a telescope produces a diffraction pattern that spreads out the images. To take this diffraction into account, at least at visible wavelengths, requires multiplying θ_{min} by 1.22. For example, a 10-cm (0.1-m) telescope working at a wavelength of 6×10^{-7} m has a minimum resolvable angle of

$$1.22\theta_{min} = (1.22)(206,265)(6 \times 10^{-7}/0.1)$$
$$= 1.5 \text{ arcseconds}$$

This result means that if this telescope is pointed at two stars that are more than 1.5″ of arc apart, you will see two separate stellar images.

The example above gives the *theoretical* resolving power of a 10-cm telescope, but such performance rarely if ever is attained by groundbased telescopes. The resolving power of a large telescope is limited not by its optics but by the Earth's atmosphere. Stars twinkle because turbulence in the air makes the atmosphere act like an imperfect, distorted lens. The motion of blobs of air distorts and blurs images seen through a telescope; this is an effect referred to as *seeing.* Even on the best of nights, the 5-m Hale telescope does not resolve better than a 10-cm telescope; it is a rare night when star images are smaller than 1″ of arc. The limit the Earth's atmosphere sets on the resolving power of big telescopes makes a strong case for placing a large telescope in space. Here a telescope's resolving power would be limited by it optics, not the atmosphere.

The third function of a visual telescope is to magnify the image. *Magnifying power,* the apparent increase in the size of an object compared with unaided visual observation, depends on the ratio of the focal length of the objective to the focal length of the eyepiece:

$$MP = F/f$$

where F is the focal length of the objective and f is the focal length of the eyepiece in the same length units. For example, an eyepiece having a focal length of 5 cm used with an objective having a focal length of 780 cm gives a magnifying power of

$$MP = 780/5 = 156$$

If you insert an eyepiece with *half* the focal length, you *double* the magnifying power. You could choose eyepieces to produce as high a magnifying power as you like, but there is little point in using a magnifying power any greater than that necessary to see clearly the smallest detail in the image. The degree of detail is determined by the resolving power and the seeing. Extremely high magnification of an image produced by a telescope with a low resolving power merely makes the fuzziness worse.

9–2 Invisible Astronomy

The human eye senses only the visible range of the electromagnetic spectrum. To cover the complete range, from radio to gamma rays, requires a variety of detectors sensitive to various wavelengths. This is where current technology can drive new developments in observational astronomy. Invisible astronomy involves techniques that enable us to go beyond the wavelength limits of optical astronomy.

The practice of invisible astronomy also relates to the transparency of the Earth's atmosphere, which effectively absorbs large segments of the electromagnetic spectrum, especially ultraviolet light, X-rays, some infrared wavelengths, and short-wavelength (millimeter) radio waves (Figure 9–2 and Table 8–1). Infrared radiation is primarily absorbed by water vapor, which is found concentrated in the lower portions of the atmosphere, below 20 km. (Carbon dioxide also absorbs a lesser amount.) The ultraviolet and X-ray radiations are primarily absorbed in the ionosphere, at an altitude of 100 km, well above the levels that can be reached by balloons and airplanes. The obvious way to get around atmospheric absorption is to go above the atmosphere. This is space astronomy, which makes use of rockets, balloons, and airplanes as well as satellites and spacecraft. So invisible astronomy has two natural divisions—that which can be done from the ground and that which must be accomplished in space.

(a) Ground-Based Radio Astronomy

Radio astronomy was born in 1930 when Karl Jansky (1905–1950) undertook a study for the Bell Telephone Company to identify sources of static affecting transoceanic radiotelephone communications. Jansky identified one source of noise as a celestial object: the Milky Way in Sagittarius. Jansky's discovery was published in 1932 but had little impact on the astronomers of the day. However, an American engineer, Grote Reber, read Jansky's work and decided to search for cosmic radio static in his spare time. By the 1940s, Reber had made detailed maps of the radio sky. He sensed that a new astronomy was in the making and so took an astrophysics course at the University of Chicago to learn more about astronomy and to discuss his discoveries with astronomers— only a few of whom were impressed. World War II forced technical developments in radio and radar work. John S. Hey in Britain accidentally discovered that the Sun strongly emits radio waves. After the war, he continued his astronomical pursuits at radio wavelengths. So did other groups in Britain, the Netherlands, and Australia.

A common type of radio telescope, the radio dish (Figure 9–3), functions like a reflecting telescope. Essentially, it's a radio-wave bucket with a detector (a radio receiver) at its focus. It reflects and concentrates radio waves in much the same way a mirror does in a reflecting telescope. The radio receiver translates incoming radio waves into a voltage that can be mea-

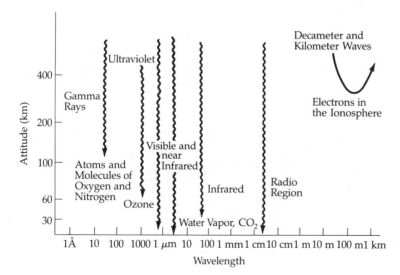

Figure 9–2

Atmospheric transmission. The Earth's atmosphere is transparent to limited regions of the electromagnetic spectrum.

Figure 9–3

A radio dish. A view from the back of the NRAO millimeter-wave radio telescope on Kitt Peak. *(M. Zeilik)*

Radio astronomers use the technique of interferometry to make small radio telescopes function as if they were large ones—and so gain in resolving power. Imagine two radio telescopes placed, say, 10 km apart. If the signals received by the two are synchronized, the separate units can be made to act like a single dish with a diameter of 10 km, but only for a narrow strip across the sky. (That's because they act like two small pieces at the opposite ends of a large dish.) The essence of interferometry is to consider the phase of the wavefronts reaching the two dishes of a simple interferometer. Imagine that a source is directly overhead and that the waves arrive at both telescopes in phase. When the signals are combined (electronically in a mixer), they constructively reinforce to give a strong signal. Now assume that the source has moved somewhat west of overhead so that the path lengths to the two antennas differ by exactly $\lambda/2$. They arrive 180° out of phase and destructively interfere. So the two telescopes "see" the sky as a series of bands of constructive and destructive interference—an interference pattern exactly like that for light passing through two slits. The angular separation of the peaks in the intensity pattern establishes the resolving power of the interferometer.

sured and recorded on magnetic tape. Finally, the measurements are usually made into a contour map (Figure 9–4) or a false-color intensity map.

Radio telescopes have one major drawback: low resolving power. Resolving power depends on both the size of the objective and the wavelength of the gathered radiation (remember, $\theta_{min} \propto \lambda/d$). Radio waves are much longer than the wavelengths of visible light, typically 10^5 times longer. So if an optical and a radio telescope have the same diameter, the radio one will have 10^5 times *less* resolving power. For example, for a radio telescope to have the same resolving power as the 5-m Hale optical telescope, it would require a diameter 10^5 times as great, about 500 km! Obviously, a single dish of this size cannot be built on Earth.

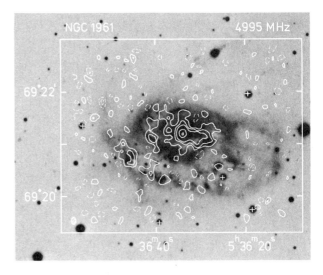

Figure 9–4

Radio contour map. The emission is 21 cm from hydrogen atoms, made with the VLA. The radio contours have been superimposed on a negative optical print. *(E. Hummel, University of New Mexico and National Radio Astronomy Observatory)*

Consider two antennas (Figure 9–5) separated by a distance L, which is an integral multiple of the observing wavelength λ, so that

$$L = n\lambda$$

where n is any integer. Now imagine a radiation source at P. The waves from this source travel along the different path lengths P_1 and P_2, where $P_1 > P_2$ and P' is the extra amount of travel along P_1. Note that

$$P' = L \sin \theta$$

and for constructive interference, P' must be integral multiples of the wavelength, so that

$$L \sin \theta = m\lambda$$

or

$$\sin \theta = m\lambda/L$$

where m is any integer. As the Earth turns, θ varies. The fringes are evenly spaced by the angle for which $m = 1$:

$$\sin \theta_f = \lambda/L = \lambda/n\lambda = 1/n$$

and because θ_f will be small, $\sin \theta_f \approx \theta_f$ and

$$\theta_f \approx 1/n \quad \text{rad}$$

where n is just the number of wavelengths (λ) in the baseline (L). For $L = 21$ km at an observing wavelength of 21 cm,

$$\theta_f \approx (21/21) \times 10^5 \approx 10^{-5} \text{ rad} \approx 2''$$

comparable to an optical telescope. Note that the fringe spacing essentially sets the resolving power of the interferometer.

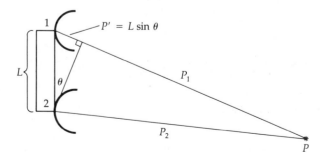

Figure 9–5

Geometry for an interferometer.

A two-antenna interferometer gives only one part of the separation between points in the source (in the same direction as the baseline). To completely map a section of the sky with uniform resolving power requires a little work. The trick: twist the object with respect to the interferometer's pattern, a technique known as *aperture synthesis*. This twisting can be accomplished naturally by observing the source move across the sky; its motion will twist its orientation with respect to the interferometer's baseline. To help this process, the baseline of the interferometer can also be twisted or the antennas can be arranged in a Y pattern [as in the Very Large Array (VLA) in New Mexico, Figure 9–6A]. The Earth's rotation effectively synthesizes an aperture with a resolution comparable to a single antenna with a diameter equal to the largest baseline between the dishes (Figure 9–6B). Eventually, sets of interference maps are combined by a computer to give a composite picture of the source with high resolution. To construct a complete map of a radio source with a two-antenna interferometer is a tedious task that may take months of observing time. The VLA and other interferometer arrays are designed to overcome this difficulty. They consist essentially of many two-antenna arrays at different angles to each other, operating on the same source at the same time. A complete radio picture can be constructed in about 1 day or less, depending on the resolution required and the area of the sky covered.

Even higher resolution can be obtained by the technique known as Very-Long-Baseline Interferometry (VLBI). The VLBI signals received by very distant antennas (even located on different continents) are recorded on magnetic tape and combined later in a computer. The maximum baseline for such VLBI is the diameter of the Earth, $L = 12,000$ km, and so (for the example above) a resolving power of $\theta_f = 3 \times 10^{-3}$ arcsecond is possible at 21 cm. In the future, radio telescopes in space separated by larger baselines will give even better resolution; some proposals even envision one element of an interferometer in Earth orbit or on the Moon—a separation 30 times greater than is possible on the Earth. Resolution could then approach about 100 microarcseconds and would be high enough to see the disks of nearby solar-type stars! As a compromise for this grand scheme, the National Science Foundation has begun construction of a VLBA (Very-Large-Baseline Array) that will operate like the VLA but will have greater

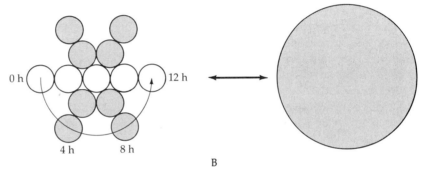

Figure 9–6

Radio interferometry. (A) A view down one arm of the VLA. Each antenna is 25 m in diameter. *(M. Zeilik)* (B) For aperture synthesis, the rotation of the Earth turns one arm of five antennas to cover a full circle in 24 h. *(J. Burns)*

0 h   12 h

4 h   8 h

B

antenna separations (up to a few thousand kilometers). Present plans call for about ten 25-m antennas distributed from Puerto Rico to Hawaii and controlled by telephone links from an operation center. The receivers will be synchronized by atomic clocks, and the data tapes will be shipped to a large computer in New Mexico for processing to form images from the interference patterns.

(b) Ground-Based Infrared Astronomy

Carbon dioxide and water vapor in the Earth's atmosphere absorb much of the incoming infrared radiation. The ground-based infrared astronomer can observe at only a few restricted wavelength ranges: 2 to 25, 30 to 40, and 350 to 450 μm. Such observations are best made from high sites in dry climates, where there is a minimum of water vapor in the atmosphere above the telescope.

An infrared telescope differs from an optical one in the detector at the telescope's focus. Because our eyes and photographic film sense infrared radiation poorly, special infrared detectors are required; sensitive ones suitable for astronomical work have been around for only about ten years. A common infrared detector is a *bolometer*, a tiny chip of germanium (about the size of the head of a very small nail) cooled to about 2 K. When infrared radiation strikes a bolometer, it heats up and its resistance to an electric current changes. Such changes can be measured electronically, and the amount of variation indicates how much infrared energy the bolometer is absorbing.

Infrared observing has at least two distinct advantages over optical observing. First, infrared radiation is hindered less by interstellar dust. Second, cool celestial objects (3000 K and cooler) give off most of their radiation in the infrared (as expected from Wien's law for blackbody radiation). Another practical advantage is that much infrared observing can be done during the day, when telescopes are not used by optical astronomers. This advantage occurs because very little sunlight at infrared wavelengths is scattered by air molecules, leaving the infrared sky dark both day and night (except looking directly at the Sun). However, infrared astronomy has the problem that a telescope emits like a blackbody at 300 K, so that, as expected from Wien's law, it generates an intense infrared flux at 10 μm. The Earth's atmosphere also contributes a strong background flux at this wavelength. Special techniques have been developed to compensate for this strong local flux in order to measure the much smaller infrared flux from celestial objects.

(c) Space Astronomy

Parts of the infrared spectrum, ultraviolet light, and X-rays can be detected only above the Earth's atmosphere from airplanes, balloons, rockets, satellites, or spacecraft. For example, most of the far-infrared radiation (wavelengths longer than about 40 μm) does not penetrate to the ground. At altitudes of 15 to 20 km or so, however, very little of the Earth's atmosphere remains above the observer. Far-infrared observations can be made at these altitudes from airplanes or balloons. The equipment is simply a reflecting telescope equipped with a bolometer.

For ultraviolet astronomy, methods of light gathering and detection are similar to those of optical astronomy. Photographic plates and special television tubes respond well to ultraviolet light, and so detection presents no serious problem. Because glass absorbs ultraviolet light, refracting telescopes cannot be used, but reflectors work perfectly well. Good ultraviolet astronomy must be done from space, for the absorbing layer of the atmosphere (the ionosphere) is higher than balloons or aircraft can reach. Probably the most successful ultraviolet telescope has been the International Ultraviolet Explorer (IUE) launched in 1978 and still (but barely) operational. It has a 0.45-m telescope with detectors that operate from 115 to 320 nm.

For the high-energy realm of X-rays, special gathering and focusing techniques are required. In recent years, X-ray telescopes have used the fact that X-rays can be reflected from certain surfaces if they strike at very small angles, almost parallel to the reflecting surface. Such reflections by a complex series of concentric parabolic and hyperbolic reflecting surfaces produce reasonable images, which are sensed by electronic detectors. Probably the most productive of recent X-ray telescopes was HEAO 2, which is called the Einstein Observatory, and was launched in 1978 and used until 1981. The 58-cm telescope in the satellite could produce high-resolution images of X-ray sources in the wavelength range from 0.3 to 5 nm.

Because of their higher energies, gamma rays present an even more difficult detection and focusing problem. Basically, gamma-ray telescopes do not yet exist because of these problems—no images are made. Crystals that fluoresce when they absorb gamma rays are used. The flash of visible light is then detected by phototubes. Various techniques are used to limit the field of view, which typically covers a few degrees of the sky.

In the near future, the United States will launch (by means of the Space Shuttle) the Hubble Space Telescope. With a 2.4-m mirror, the Space Telescope will orbit the Earth at an altitude of about 500 km and will operate mainly in the ultraviolet and visible regions. Unhampered by the atmosphere, the telescope's resolution will be limited only by its optics. Because it can be serviced by the Space Shuttle, new instruments can be added and old ones updated. Hence, the Space Telescope will serve as a flexible instrument for breakthrough observations in the 1980s and beyond (we hope!).

9–3 Detectors and Image Processing

You may have the romantic image that many people do, of an astronomer glued to the eyepiece of a telescope through a cloudless night working to produce spectacular photographs. Today, most astronomers observe through the telescope on a television screen from a warm room! And most optical observations (which encompass a narrow region of the electromagnetic spectrum) no longer involve direct photography. Here we will give you a taste of modern technology's influence on astronomy.

(a) Photography

This old standby still cannot be beat if the goal is to gather a large amount of storable information in a short time. Special photographic emulsions used by astronomers are usually coated on a glass plate, whose size depends on the characteristics of the telescope employed. (The glass plate does not flex and can be measured accurately.) The emulsion can integrate the light striking it to build up an image of very faint objects over a long period of time (Figure 9–7) —occasionally many hours spread over several nights. The plates are often specially treated to increase their sensitivity to light, and yet they rarely have quantum efficiencies greater than a few percent. (We use *quantum efficiency* to mean the percentage of photons striking a detector that activate it relative to the incoming total.)

Although a photograph's quantum efficiency is relatively low, its large area collects information from a large part of the entire field of view of the telescope. This image can later be converted to a digital form that can be manipulated by computer to enhance

Figure 9–8

Image processing. A computer-processed image of a 1910 photograph (taken at Lowell Observatory) of Halley's Comet. The computer enhances the differences of brightness in the tail, which are not obvious in the original photograph. *(National Optical Astronomy Observatories)*

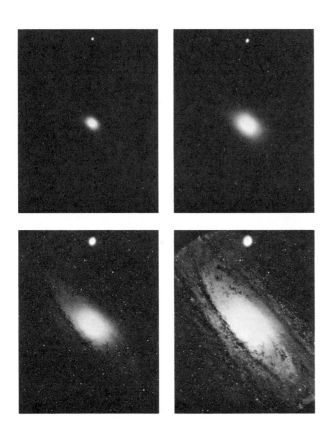

Figure 9–7

Time exposures. Longer times for exposing a photographic plate reveal fainter details. The times here are 1, 5, 30, and 45 min. *(National Optical Astronomy Observatories)*

specific aspects of the original data (Figure 9–8). This process, called *image processing,* plays a central role in all types of astronomy today, especially invisible astronomy, where the computer-processed image quickly conveys significant information to the human brain.

(b) Phototubes

The quantum nature of light shows up dramatically in the *photoelectric effect.* Light striking the surfaces of certain materials can be absorbed and an electron dislodged. The photons must have a certain minimum energy to knock out the electrons. The displaced electrons can flow in a current, or they can be individually counted. A device that does this counting is called a *phototube.* A phototube that can transform *one* electron into many ($\approx 10^5$) for ease of detection is called a *photomultiplier.* Phototubes do not produce images; they simply measure the flux of photons through the focal plane of a telescope, usually through a small aperture that restricts the field of view to, say, a single star or galaxy. Filters can be placed in the light path in front of the phototube to restrict the detection to certain wavelength bands. The active materials in phototubes have quantum efficiencies that sometimes reach 20 percent, though 10 percent is more typical. The response of a photomultiplier is nicely linear — twice the flux gives twice the current. Hence, light intensity can be measured with great accuracy by measuring an electric current or by counting the rate of production of individual photons.

A television camera includes a phototube designed to produce images by a magnetic focusing technique. However, commercial television cameras are not very sensitive, cannot accumulate images over long times, and are not linear in response. (Linearity is important because you want the output to be directly proportional to the number of photons captured.) Astronomers have modified television systems for certain applications — for acquisition and guiding, for example.

(c) Charge-Coupled Devices

Astronomers would love an ideal detector for their demanding low-light applications. It would have a high quantum efficiency (to make good use of those few photons), the ability to integrate, a linear response over a wide range of photon fluxes, and the ability to cover a large angular field of view (as does a photographic plate). Recent advances in solid-state microelectronics have resulted in the development of *charge-coupled devices* (CCDs), which may be the as-

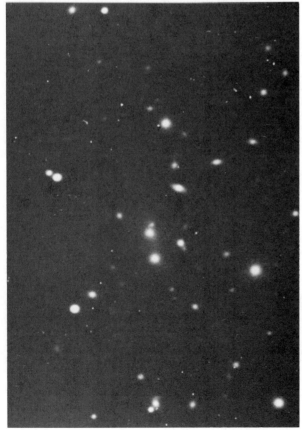

Figure 9–9

A CCD image of a cluster of galaxies in the constellation of Corona Borealis. *(Jay Moody, Capilla Peak Observatory, University of New Mexico)*

tronomers' dream detector. A CCD is a small chip, a few to 10 mm on a side, made of a thin silicon wafer. (It uses the same microelectronic technology employed in integrated circuits.) It consists of a large number of small regions each of which makes up a picture element called a *pixel*. A typical chip may contain 500 by 500 pixels (250,000 total) arranged in rows and columns that can be electronically controlled. Each pixel independently can accumulate a charge as photons are absorbed and dislodge electrons. In a sense, each pixel acts like a very small phototube, but a CCD can integrate the charge in the pixels for a long time and then move the charges out in a regimented way to digitally preserve the spatial pattern of light intensity falling on the chip. Computers then process the image from each exposure of the chip.

CCDs offer the following advantages over other detectors. One, they are highly quantum efficient — close to 100 percent in the red region of the spectrum. This great quantum efficiency means that a small telescope can act like a large one. (That was the main purpose in developing a CCD camera for the Capilla

Peak telescope at the University of New Mexico.) Two, CCDs are very linear, and so they can easily measure light intensity accurately. Three, though small so far, CCDs are area detectors. They cover a good fraction of a telescope's field of view (as if many small phototubes were operating at once, a multichannel detector that makes use of most of the light coming through a telescope at a given time). Fourth, the image emerges in a naturally digital form, ready for computer processing (Figure 9–9).

Along with the inexpensive microcomputers that are a key part in the operation, CCDs are the best current example of technology expanding the capabilities of older, smaller telescopes. The sky subtends a large solid angle, but any one telescope sees only a small piece of that solid angle at any one time. And the number of telescopes is extremely limited; many are in great demand. CCDs enhance the power of small telescopes and may well re-elevate the status of the small observatory. For example, in the ability to detect faint objects, a 1-h exposure on a 61-cm telescope with a CCD equals a 1-h exposure on a 500-cm telescope with a photographic plate.

Problems

1. Compare the resolving power and the light-gathering power of the human eye with those of a
   (a) 10-cm telescope
   (b) 4-m telescope

2. For what astronomical purposes would one use a telescope having a long focal length and a large f ratio?

3. The maximum possible radio telescope separation of the VLA is 40 km. What is its resolving power when operating at 1.5 cm? Why is it that the Earth's atmosphere does not limit the resolving power?

4. The transmission of the Earth's atmosphere at 3 μm is about 10 percent. Astronomers define the optical depth τ at some wavelength λ such that

   $$I_\tau = I_o \exp(-\tau)$$

   where I_o is the original intensity in a light beam and I_τ the intensity after passing through a material with optical depth τ. What is the optical depth through the Earth's atmosphere at 3 μm?

5. Calculate the scale height for water vapor in the Earth's atmosphere. Use this information to state whether telescopes atop Mauna Kea would be better for infrared astronomy than ones at sea level.

6. The shortest wavelength at which VLBI is routinely done is 1.3 cm.
   (a) What is the minimum resolvable angle of an interferometer operating at 1.3 cm if the baseline is about one Earth diameter?
   (b) VLBI experiments are now being attempted at a wavelength of 3 mm. What is the resolving power at this wavelength for an Earth-diameter baseline?

7. One of the disadvantages of an interferometer is that it is insensitive to objects with angular sizes much larger than the minimum resolvable angle of the shortest baseline (essentially because different portions of the source destructively interfere with each other). The flux from a source of angular diameter θ'' measured by an interferometer of baseline L (in kilometers) is approximately

   $$F_{meas} \approx F_{true} e^{-0.3(L\theta/\lambda)^2}$$

   where λ is measured in centimeters. For a wavelength of 6 cm, what is the largest angular size of a source that can be observed with an interferometer of minimum baseline 1000 km if the lowest detectable flux is $0.1 F_{true}$? What would be the properties of sources that one would *not* want to observe using an interferometer?

8. **(a)** What is the resolution of the Space Telescope (objective diameter = 2.4 m) at
   i)　λ = 500 nm (visible)
   ii)　λ = 200 nm (ultraviolet)
   iii)　λ = 2000 nm (infrared)
   (b) Considering your answers to **a**, why would it ever be advantageous to observe in the infrared with the space telescope?

9. What are the observational characteristics of an object that is better studied by the Space Telescope than by a very large ground-based optical telescope? What are the characteristics best suited for a very large ground-based optical telescope?

10. One of the main advantages of photographic plates or CCDs is their ability to integrate, that is, to collect light for a time much longer than the human eye can. The human eye effectively integrates for about 0.2 s (which is why we don't notice the individual frames of movie films). The flux of the faintest object that can be detected is proportional to $1/\sqrt{t}$, where t is the integration time. (The square root arises because the flux of the background sky against which the object must be contrasted also increases with integration time.) How much fainter an object can be detected using a 1-h integration on the Palomar 5-m telescope than can be seen with the naked eye?

11. One of the major difficulties with gamma-ray astronomy is the poor resolving power, with angles less than about 2° being unresolved even for the best instruments. Nevertheless, gamma-ray images of our Galaxy have been produced using satellite data. Discuss the difficulty in interpreting these images in light of the poor resolving power.

The Sun: A Model Star

Our Sun is the nearest star. The fascinating properties and phenomena of the solar surface layers are easily observed and have been studied in great detail. Unfortunately, models for understanding solar phenomena have not kept pace with such detailed data. Because the Sun is a fairly typical star and because it is the only star that is not a mere point of light as seen from the Earth, the discussion here serves as the basis to investigate the other stars.

10–1 The Structure of the Sun

At 1 AU from the Earth, the Sun provides the energy necessary for life. This gigantic sphere of gas, with a radius of 6.96×10^5 km ($\approx 109 R_\oplus$) and a mass of 1.99×10^{30} kg ($333,000 M_\oplus$) has a *luminosity*, or rate of energy radiation, of 3.86×10^{26} W. The average density of the Sun is only 1400 kg/m³, a value consistent with a composition of mostly gaseous hydrogen and helium.

The pressure, and also the temperature, in the solar interior must be extremely high just to support the weight of the Sun. The body of the Sun is gaseous and hot; in fact, the gases (primarily hydrogen and helium) are almost completely ionized (a *plasma*). Temperature, pressure, and density increase from the Sun's surface inward to the center, where energy is liberated by thermonuclear reactions. As hydrogen is transformed to helium in the Sun's *core*, vast quantities of energy are released in the form of photons and thermal motions (Figure 10–1). The photons diffuse outward through the large *radiative zone* until they reach the outer *convective zone*, where most of the transport of energy takes place by boiling motions of the gas. The visible surface of the Sun (the *photosphere*) occurs at the top of the convective layer, where the complex and extended solar atmosphere begins.

The base of the solar atmosphere is the *photosphere* (Figure 10–1), a thin layer of gas that represents the greatest depth to which we may observe and from which emanates the bulk of the visible radiation. Sunspots appear on the photosphere. The next layer outward is the *chromosphere;* from the top of the chromosphere emerge the sharp spicules and the graceful prominences. Beyond this region lies the tenuous extended *corona* with its ghostly light merging into the outward-flowing *solar wind* and the interplanetary medium. Of the entire Sun, only the solar atmosphere is directly observable, but what a rich arena of activity it is!

Although we cannot see directly into the interior of the Sun, there are two classes of observations that allow us to make some educated guesses about the solar depths. One is the observation of the neutrino flux from the Sun. You will see in Section 16–1(d) that the observed flux is only about one-third that predicted by theoretical models. This deficiency has challenged astrophysicists and physicists into proposing a number of explanations, including modifications of the solar models and new theories of the characteristics of the neutrino.

The other probe is the analysis of the motions of gases at the solar surface. It turns out that the Sun pulses! Oscillations are observed as periodic changes in the Doppler shifts of photospheric and chromospheric spectral lines. These oscillations have periods ranging from less than 5 min to 2 h 40 min. The first to be discovered and in many ways the most important are the 5-min oscillations, which are observed as vertical motions of areas of the Sun, much like small boats on ocean waves. At peak amplitude, the velocities are 0.4 km/s. These oscillations are interpreted as resulting from sound waves from the interior, and their characteristics can tell us some of the properties of the interior. Thus it may be said that the Sun rings like a bell or, perhaps more accurately, that its interior can be probed in a manner similar to the way terrestrial seismic waves are used to study the interior of the Earth. This field of research has been given the name *helioseismology*. Acoustic waves within the Sun are visible as oscillations on its surface. Their pattern and period hold clues to the Sun's interior.

From helioseismologic studies have come inferences concerning the details of the convection-layer structure, the opacity of the gases and thus relative atomic abundances, especially helium, and the internal dynamics of the Sun. There is some indication that the rotation of the interior is considerably faster than at the surface. There are still many, many uncertainties, not least of which is the origin of the sound waves.

The remainder of this chapter presents a detailed examination of the more directly observable phenomena in the outer layers of our Sun.

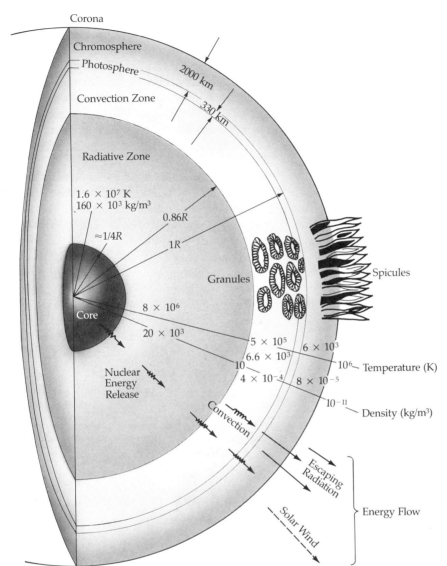

Figure 10-1

The structure of the Sun. This model shows the main regions of the Sun and values of important physical parameters. The granules and spicules are drawn out of proportion.

10-2 The Photosphere

(a) Granulation

We cannot peer deeper into the Sun than the base of the photosphere, which defines the layer of the Sun's atmosphere where the gases become opaque to visible light. Here (Figure 10-2) we see a patchwork pattern of small (average diameter about 700 km), transient (average lifetime from five to tens of min-

utes) *granules:* bright irregular formations surrounded by darker lanes.

This solar granulation is the top layer of the Sun's *convective zone,* a gaseous layer about $0.2R_\odot$ thick located just below the base of the photosphere. In this zone, heat energy is transported by *convection;* hot masses of gas *(convection cells)* rise, appear as bright granules, and dissipate their energy at the photosphere; the cooler gases sink back down. The resulting transfer of energy imparts motions of the order of

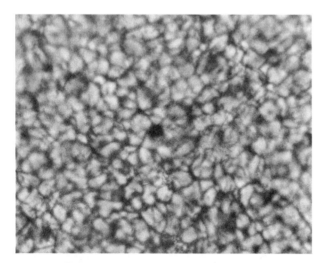

Figure 10-2

Photospheric granulation. The bright areas are about 1500 km across. The dark circle is a small sunspot. *(National Optical Astronomy Observatories)*

tenths of a kilometer per second to the lower layers of the photosphere.

(b) Photospheric Temperatures

The continuum spectrum of the entire solar disk defines a Stefan–Boltzmann effective temperature (Table 8–4) of 5800 K for the photosphere, but how does the temperature vary in the photosphere? A clue to the answer is evident on any white-light photograph of the Sun (Figure 10–3), for we see that the brightness of the solar disk decreases from the center to the limb—this is termed *limb darkening.*

Limb darkening arises because we see deeper, hotter gas layers when we look directly at the center of the disk and higher, cooler layers when we look near the limb (Figure 10–4). Assume that we can see only a fixed distance *d* through the solar atmosphere. The limb appears darkened as the temperature *decreases* from the lower to the upper photosphere because, according to the Stefan–Boltzmann law (Chapter 8), cool gas radiates less energy than does hot gas. The top of the photosphere, or bottom of the chromosphere, is defined as height = 0 km. Outward through the photosphere (Figure 10–5), the temperature drops rapidly then again starts to rise at about 500 km into the chromosphere, reaching very high temperatures in the corona.

At this point, you may have discerned an apparent paradox: how can the solar limb appear darkened when the temperature rises rapidly through the chromosphere? Answering this question requires an understanding of the concepts of *opacity* and *optical depth.* Simply put, the chromosphere is almost optically transparent relative to the photosphere. Hence, the Sun appears to end sharply at its photospheric surface—within the outer 300 km of its 700,000-km radius.

Our line of sight penetrates the solar atmosphere only to the depth from which radiation can escape unhindered (where the optical depth is small). Interior to this point, solar radiation is constantly absorbed and re-emitted (and so scattered) by atoms and ions. To characterize this absorption by the gas at a given wavelength, we speak of the *opacity* k_λ (in m^2/kg) of the gas. The simplest way to understand opacity is to consider what happens when radiation of flux F_λ strikes a slab of gas of thickness dx. Let the mass density ρ (in kg/m^3) describe this gas. Part of this flux is absorbed by the slab, and so

$$dF_\lambda = -k_\lambda \rho F_\lambda dx \qquad (10-1)$$

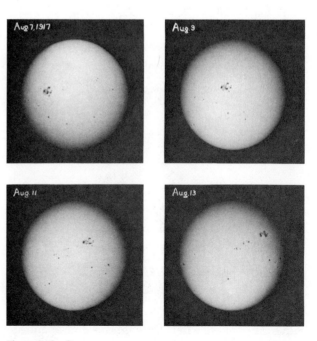

Figure 10-3

The solar photosphere. These photographs show limb darkening and the Sun's rotation over a seven-day period. *(Yerkes Observatory)*

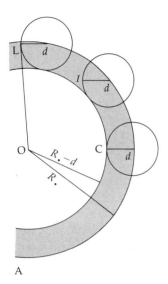

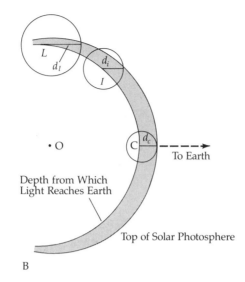

Figure 10-4

Geometry for limb darkening. The center of the disk is denoted by C, the limb by L, and an intermediate position by I. (A) All circles have the same radius d, but the line of sight ends higher in the atmosphere than at the center. (B) High-density gas low in the atmosphere is more opaque than the thinner gas higher up; here each circle corresponds to the same optical depth.

In a uniform medium, Equation 10-1 is integrated to yield

$$F_\lambda(x) = F_\lambda(0) \exp(-k_\lambda \rho x) \qquad (10-2)$$

so that the flux diminishes exponentially with depth of penetration. For convenience, astronomers define another measure of this absorption by the *optical depth*, τ_λ, where

$$d\tau_\lambda \equiv k_\lambda \rho \, dx \qquad (10-3)$$

Note that τ_λ is dimensionless and Equation 10-2 becomes

$$F_\lambda(\tau_\lambda) = F_\lambda(0) \exp(-\tau_\lambda) \qquad (10-4)$$

For $k_\lambda < 1$, the flux is constant, so that the gas is said to be *optically thin* (transparent) at this wavelength. In general, an *optically thick* (opaque) gas is one with optical depth much greater than unity; the base of the photosphere corresponds to this case. Limb darkening arises from photospheric opacity (Figure 10-4B), and k_λ is also comparatively low in the chromosphere.

(c) H⁻ Continuous Absorption

The photosphere emits a blackbody continuum and so must be opaque at visible wavelengths, but the densities here are far less than those needed for the gas to be opaque and generate blackbody continuum radiation. How then does the photospheric continuum radiation arise?

The chief contributor to the continuous opacity is H⁻, *the negative hydrogen ion.* This ion can exist be-

cause the single electron of the neutral hydrogen atom does not completely screen the positive proton. Hence, a second electron may loosely attach itself; the ionization potential is only 0.75 eV (that of normal hydrogen is 13.54 eV). Absorption occurs by the dissociation reaction $H^- \rightarrow H + e^-$, and emission

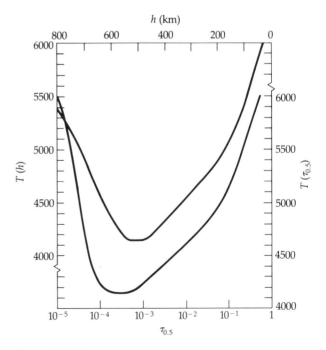

Figure 10-5

Temperature structure of the photosphere and chromosphere.

takes place when an electron attaches itself to a neutral hydrogen atom in the reaction $H + e^- \rightarrow H^-$. The solar continuum in the infrared and optical regions is produced by these reactions, which are enhanced by the relatively high electron and hydrogen densities in the photosphere.

(d) The Fraunhofer Absorption Spectrum

Spectral Lines

In 1814, the German physicist Joseph von Fraunhofer (1787–1826) made the first definitive mapping of the vast array of absorption lines seen on the photospheric continuum—the *Fraunhofer absorption spectrum*. Not knowing the correct identifications of these features, Fraunhofer designated the strongest lines (starting from the red) with capital letters and the weaker lines with lower-case letters. (See Section 8–4 for the definition of line strength.) Today we still refer to the D lines of a resolved doublet of sodium, the b lines of magnesium, and the H and K lines of Ca II. The earliest line identifications included the hydrogen Balmer series and absorption lines of sodium, calcium, and magnesium.

Many more solar lines have come under scrutiny since the advent of rocket and satellite spectroscopy, for now we can study that part of the solar ultraviolet spectrum normally blocked by the Earth's atmosphere. The Fraunhofer absorption line spectrum ex-

tends to wavelengths as short as 165 nm. At wavelengths shorter than the near ultraviolet (<165 nm), the solar spectrum is dominated by *emission lines* that are produced in the chromosphere and corona [Sections 10–3(a) and 10–4(c)].

Because the solar continuum comes from a thin gas of H^- ions, the Fraunhofer absorption lines can and do form at the same solar atmospheric level as does the continuum. The weaker lines originate in the lower photosphere, and the stronger ones form largely in the upper photosphere; in fact, the strongest (the Balmer lines and the H and K lines of Ca II) form primarily in the lower chromosphere.

The opacity is low in the chromosphere for most wavelengths. In strong lines, however, the opacity becomes large even at considerable heights in the solar atmosphere. The reasons include the abundance of the element, the number of atoms in the lower level of the transition producing the line (which in turn depends on the temperature through the Boltzmann equation; Section 8–4), and the transition probability (the intrinsic atomic parameter that determines the probability that an atom will make that particular transition). The depth of formation differs from line to line and within a given line. So the Fraunhofer absorption lines provide a powerful tool for probing different heights of the solar atmosphere (Figure 10–6).

High-dispersion spectra have revealed fine

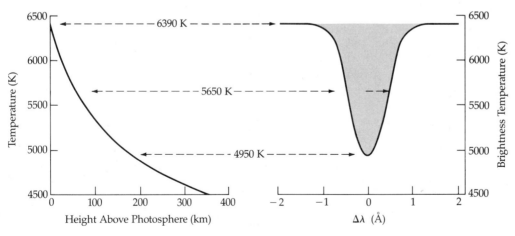

Figure 10–6

Photospheric depth and line profiles. Different parts of an absorption line form at different heights, the line center coming from the coolest, highest region. *(After R. W. Noyes)*

structure wiggles in most Fraunhofer lines. These wavelength oscillations are Doppler shifts arising from vertical motions at about 0.4 km/s of small-scale (1000 km in diameter) structures in the photosphere. They appear to be generated by gas motions induced by the granules at the bottom and continuing up into the chromosphere. In addition, horizontal motions run parallel to the solar surface with about the same velocities. These motions arise in larger structures (30,000 km across) called *supergranules.* Gas flows slowly from the center of a supergranule to the edge. Supergranules are related to higher chromospheric structures, though they actually originate below the granulation. Their flowing gases carry magnetic lines of force to their edges, and so the magnetic fields become more localized and concentrated. As you will see, solar activity is generally associated with strong localized magnetic fields.

Elemental Abundances

Using the ideas outlined in Chapter 8, we can analyze the Fraunhofer lines to infer properties of the photosphere. The characteristic temperatures and pressures are first found, and then the relative line strengths (Section 8–4) tell us the chemical composition (Table 10–1). Hydrogen is by far the most abundant element, helium second; all of the heavier elements account for about 2 percent of the total mass.

Table 10–1   Abundances of the Elements in the Solar Atmosphere Relative to Hydrogen by Number

H	1,000,000
He	63,000
C	420
N	87
O	690
Ne	37
Na	1.7
Mg	40
Al	3
Si	45
P	0.27
S	16
K	0.11
Ca	2.1
Fe	32

10–3 The Chromosphere

The solar *chromosphere* extends about 10,000 km above the photosphere, and its gas density is far less than that of the photosphere. This thin layer has a reddish hue—as a result of the Balmer (H_α) emission of hydrogen—visible during a total solar eclipse.

(a) The Chromospheric Spectrum

The chromospheric spectrum is completely overwhelmed by the photosphere except when the photospheric light is blocked out. Emission lines are visible in the chromosphere, notably those of helium, which requires high temperatures to become excited.

Chromospheric spectra reveal characteristic variations with altitude in the chromosphere. Atomic transitions of low excitation potential, such as some of those of neutral metals, are seen only at the base of the chromosphere. Lines of ionized iron and calcium are evident somewhat higher up. The hydrogen Balmer and neutral helium features are seen for many thousands of kilometers above the photosphere. The chromospheric lines fade with height, but the strongest line of He II (468.6 nm) fades most slowly. This decrease in line strengths has two major causes: (1) the gas density drops sharply with height and (2) the temperature increases rapidly with height above the photosphere (Figure 10–7). The high-excitation-potential lines of helium remain strong as a result of the high temperatures in the outer chromosphere. Helium (after the Greek *helios,* meaning "sun") was found in chromospheric spectra before it was discovered on Earth. Most helium lines in the optical part of the spectrum are too weak (optically thin) to be seen against the solar disk.

The hydrogen Balmer lines are formed mainly in the chromosphere, even though the chromosphere is hotter than the photosphere. The explanation of this effect is that the photospheric continuum radiation (and the temperature there) are not sufficiently energetic to promote many hydrogen atoms from the ground (Lyman) state to the first excited state. (Remember that the Balmer lines must originate from the first excited state.) Only in the hotter chromosphere, in accordance with the Boltzmann equation, does the population of second-level hydrogen atoms become significant, primarily through collisional excitation. Then continuum radiation is absorbed in the chro-

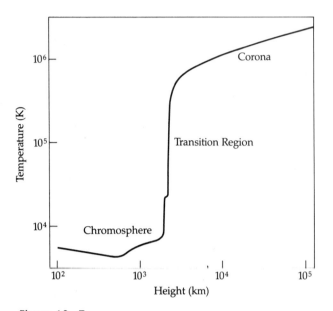

Figure 10–7

Temperature structure in the chromosphere and corona.

mosphere to form the Balmer absorption lines (when viewed in projection against the photosphere) and the Balmer emission lines that color the chromosphere (when seen against dark space at the Sun's limb).

(b) Chromospheric Fine Structure

Certain absorption lines (H_α and the H and K lines of Ca II) in the chromosphere have a large optical depth. Radiation at these wavelengths cannot escape from the photosphere because the chromosphere is essentially opaque here. The chromosphere may be studied at these wavelengths because the absorption lines are not completely black; the center of each line is darker than the adjacent continuum, but some photons are still emitted in our direction from the chromosphere.

Monochromatic photographs of the Sun (Figure 10–8) in H_α and Ca II K reveal large bright and dark patches. These are the *plages* and *filaments* associated with solar activity [Section 10–6(b)]. In addition, distinctive structure appears over the entire solar disk: the bright *network* (Figure 10–9) associated with the magnetic fields at the boundaries of supergranules. A brightening of the Ca II K line correlates with increased magnetic field strengths, which range from

10^{-3} T. What do we mean by a brightening? An H_α brightening (Figure 10–10) corresponds to a shallower profile (more emission at the line center). Beyond the limb, of course, H_α is a true emission line.

At the limb of the Sun, tenuous jets of glowing gas 500 to 1500 km across extend to a distance of 10,000 km upward from the chromosphere (Figure 10–11A). In the *spicules*, which are best observed in H_α radiation, gas is rising at about 20 to 25 km/s. Although spicules occupy less than 1 percent of the Sun's surface area and have lifetimes of 15 min or less, they probably play a significant role in the mass balance of the chromosphere, corona, and solar

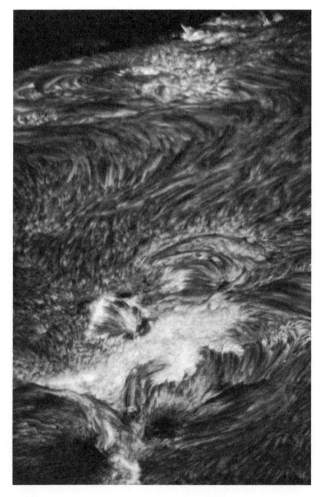

Figure 10–8

The Sun in H_α. This region near the limb shows the structure from strong local magnetic fields. (*R. B. Dunn, National Optical Astronomy Observatories*)

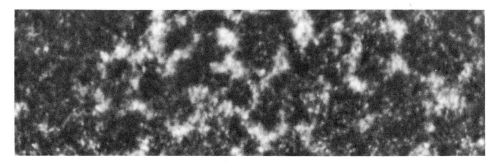

Figure 10–9

Chromospheric network. This photograph in the Ca II K line shows the bright network associated with the magnetic fields at the edges of supergranules. *(N. R. Sheeley and S. Y. Liu, National Optical Astronomy Observatories)*

wind. Spicules are not distributed uniformly over the solar surface (Figure 10–11B) but form a network pattern, a part of the chromospheric network that constitutes the boundaries of the supergranules. So spicules occur only in regions of enhanced magnetic fields.

(c) The Transition Region

Ultraviolet spectral features occur because of the high chromospheric temperatures and are an excellent probe of the upper chromosphere and narrow chromosphere–corona transition region. Temperatures rise very steeply, from about 10,000 K in the chromosphere to 50,000 K within a few hundred kilometers of the transition zone to over 10^6 K in the corona (Figure 10–7). At wavelengths shorter than 150 nm, the photospheric continuum becomes undetectably small (see the Planck curve for 6000 K in Figure 8–14). Because of the weakness of the photospheric continuum in the far ultraviolet, radiative excitation cannot occur. However, at the high temperatures of the Sun's upper atmosphere, atoms and ions become collisionally excited and produce emission lines as they return to their ground states. Moreover, they can be observed in projection on the face of the Sun, for they will not be overwhelmed by the photospheric background.

One of the strongest ultraviolet lines is the Lyman α line. Recall that the strong hydrogen Balmer lines imply an abundance of hydrogen atoms in the first excited state ($n = 2$). Absorption from the continuum will excite these atoms to higher levels ($n = 3$, 4, and so on), producing the Balmer absorption series [Section 10–3(a)]; however, most of the atoms will return immediately to the ground state ($n = 1$), emitting L_α photons. Hydrogen atoms can be excited to $n = 2$ only by collisions or reabsorption of L_α photons (self-absorption) because of the absence of the photospheric continuum at this wavelength. Photographs in L_α show essentially the same chromospheric network seen in Ca II.

Emission lines of C III, peaking at 70,000 K, N III at 100,000 K, and O VI at 300,000 K are used to study the structure through the transition zone. The network apparently continues through the region. It disappears, however, in images made in Mg X at 60 nm, corresponding to 1.6 million K, well into the corona.

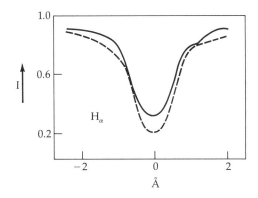

Figure 10–10

Profile of the H_α line. The dashed line shows the absorption profile from the quiet Sun; the solid line shows that from bright regions, such as the network or plages.

A

B

Figure 10–11

Spicules. (A) An H_α photograph of the limb shows the top of the chromosphere and the spicules there. (B) In this H_α photograph, the spicules appear as elongated dark features against the disk. *(R. B. Dunn, National Optical Astronomy Observatories)*

10–4 The Corona

At solar eclipses, the *corona* appears as a pearly white halo extending far from the Sun's limb (Figure 10–12). A brighter inner halo hugs the solar limb, and coronal streamers extend far into space.

(a) The Visible Corona

Coronal continuum radiation at optical wavelengths is composed of two parts. Indeed, the corona itself may be divided into the *K corona* (dominating near the Sun) and the *F corona* (evident farther out to a few solar radii. Part of the low-intensity coronal continuum matches the wavelength dependence of the photosphere. This results from light scattered by electrons, which constitute more than half the particle number density in the K corona. No photospheric absorption lines show in this scattered component, a fact we attribute to Doppler broadening by rapidly moving electrons. Because the absorption lines are completely washed out, these electrons must have extremely high temperatures — average (quiet) coronal temperatures are 1 to 2×10^6 K.

Superimposed on the electron-scattering continuum is the F corona spectrum, which does show the Fraunhofer absorption lines. This component comes from light scattering from dust particles, identical to those grains that pervade interplanetary space. The dust is concentrated in the plane of the ecliptic, for we see the outer extension of the F corona as the *zodiacal light*. We can separate the F and K coronal components because the Fraunhofer lines appear only in the former and because electrons and dust differ radically in the way they polarize the light they scatter.

Solar activity strongly affects the appearance of the K corona (Figure 10–12). At times of *sunspot maximum* [Section 10–6(a)], the corona is very bright and uniform around the solar limb and bright *coronal streamers* (Figure 10–13) and other condensations associated with active regions are much in evidence. At *sunspot minimum*, the corona extends considerably farther at the solar equator than at the poles and the coronal steamers are concentrated at the equator.

Figure 10–12

A quiet-sun corona. Note the polar plumes and the elongation at the equator. *(J. D. Bahng and K. L. Hallam)*

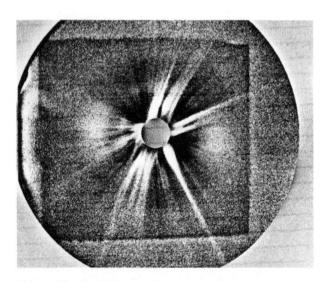

Figure 10–13

Active Sun coronal streamers, showing the magnetic field configuration and solar wind flow far from the photosphere. *(C. Keller, Los Alamos National Laboratory)*

(b) The Radio Corona

Later in this chapter, we will describe the radio bursts associated with solar activity; here we discuss the radio properties of the quiet Sun. In the ionized gases of the solar atmosphere, free electrons provide the emission and absorption of radio radiation. The interactions involved are called *free–free transitions:* an electron sideswipes an ion or atom and absorbs or emits a low-energy photon while the electron's kinetic energy changes slightly. The electron is not captured but is free both before and after the collision. The closer the electron passes the scatterer, the more both the frequency and the strength of the interaction increase. Hence, the character of the free–free photons depends upon the gas density. The denser the gas, the more frequent and energetic the interaction; the resulting photons have higher energies (or shorter wavelengths). This explains why short-wave (1 to 20 cm) radiation characterizes the chromosphere and lower corona and wavelengths longer than 10 cm arise in the outer corona.

At wavelengths longer than about 20 cm, the Sun is observed to be *limb-brightened* (into the corona), confirming that the longer the wavelength observed, the higher in the corona we are looking. (Recall that, at visible wavelengths, the Sun is limb-

darkened, Figure 10–3.) At wavelengths longer than 50 cm, we encounter great radio variability because the emission from the tenuous electron gas fluctuates wildly as the number density of particles varies.

Just as the H^- ion accounts for photospheric opacity, so also does the electron density lead to coronal opacity at radio wavelengths. The wavelength dependence of this electron opacity is the following. The corona is optically thin at short wavelengths (≈ 1 cm), so that such radiation can reach us from the chromosphere; at longer wavelengths, it becomes more and more optically thick, so that the corona is opaque to these wavelengths at greater heights. On the average, we can fit the radio data to the low-energy tail of a Planck blackbody distribution at temperatures of about 10^6 K.

(c) Line Emission

Forbidden Lines

Superimposed on the visible coronal continuum are some emission lines that were unidentified until about 1942, when W. Grotian of Germany and B. Edlén of Sweden interpreted them. (They were long called coronium lines, for they did not fit any known atomic transition.) The two strongest lines are the green line of Fe XIV (530.3 nm) and the red line of Fe X (637.4 nm); both are forbidden lines (Section 8–2).

Two significant obstacles hindered the identification of the coronal emission lines: (1) the responsible transitions are *forbidden* and (2) the temperatures of the corona are unexpectedly high. In quantum mechanics, certain energy levels of an atom are *metastable* because downward transitions from such levels are strongly prohibited by selection rules. While an ordinary permitted transition takes place in about 10^{-8} s, these metastable levels may persist for seconds or even days before a forbidden transition occurs. In most laboratory and astrophysical situations, gas densities are so high that collisional de-excitation empties metastable levels very rapidly — there is just not enough time for a forbidden transition to take place. In the near vacuum of the corona, however, metastable levels populated by either photospheric radiation or collisions can decay and forbidden emission features are formed.

Very energetic collisions are required to ionize iron; say, nine and 13 times, and so the coronal gas must be very hot. To produce Fe X requires a temperature of 1.3×10^6 K, and Fe XIV requires the even higher temperature of 2.3×10^6 K. At times of strong solar activity (such as flares), much higher temperatures occur, for the lines of Ca XV (3.6×10^6 K) are seen. The characteristic range is 10^6 K (quiet corona) to 4×10^6 K (active corona).

Extreme Ultraviolet Lines

Highly ionized atoms, such as those present in the solar corona, have lost many of the electrons that shield the atomic nucleus, and the remaining electrons are strongly attracted and tightly bound to the nucleus. Permitted transitions correspond to very high excitation potentials, and the resulting spectral photons are very energetic—at ultraviolet wavelengths. In fact, the spectral region from 5 to 50 nm (detectable only above the Earth's atmosphere) is dominated by the permitted emission lines from the coronal ions Fe VIII to XVI, Si VII to XII, Mg VIII to X, Ne VIII to IX, and S VIII to XII. This wealth of unambiguous transitions permits us to deduce the relative elemental abundances in the corona. The results are consistent with the photospheric abundances.

Coronal Loops and Holes

Because the coronal gas is so hot, it emits low-energy X-rays and shows up in X-ray photographs of the Sun (Figure 10–14). These pictures show that the coronal gas has an irregular distribution above and around the Sun. The large loop structures indicate where the ionized gas flows along magnetic fields that arch high above the Sun's surface and return to it. The hot gas is trapped in these magnetic loops. Solar physicists now view the corona as consisting primarily of such loops.

Note also that some regions of the corona appear dark (Figure 10–14), especially at the top pole and down the middle part of the Sun. Here the coronal gas must be much less dense and less hot than usual; these regions are called *coronal holes.* The coronal holes at the poles do not appear to change very much, but those above other regions seem somehow related to solar activity. X-ray bright points, lasting only a matter of hours, are scattered within these coronal holes. These are themselves apparently short-lived active regions and are also loops, albeit very small ones.

What makes a coronal hole? Solar astronomers believe that coronal holes mark areas where magnetic fields from the Sun continue outward into space

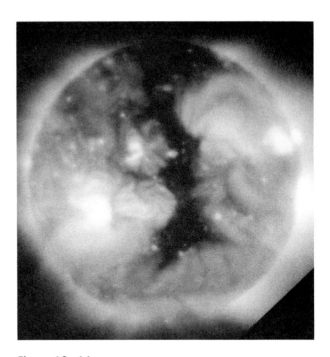

Figure 10–14

The X-ray sun. The hot plasma in the corona appears bright; note the loops, shaped by magnetic fields. The dark region down the middle is a coronal hole. *(G. Viana, Harvard College Observatory)*

rather than flow back to the Sun in loops. So the coronal gas, not tied down in these regions, can flow away from the sun out of the coronal holes; this flow makes the solar wind (Section 10–5).

The coronal gas does *not* follow the differential rotation of the photosphere. Rather it rotates at the same angular speed at all latitudes (as the Earth does). This fact implies that the bottoms of the magnetic loops are anchored deep below the photosphere, perhaps at the very bottom of the convective zone, where such fields might be generated by a solar dynamo.

Now back to why the corona (and chromosphere) are so hot. The magnetic loops (Figure 10–15) rising from the photosphere up to 400,000 km into the corona are believed to play the key role. The magnetic field in a loop is twisted by photospheric motions at its base. If the twisting happens slowly, it generates electric fields that then heat the coronal gas. It does not take much energy to do so because the coronal gas is so thin that it has a small heat capacity.

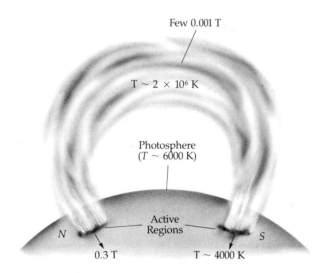

Few 0.001 T

$T \sim 2 \times 10^6$ K

Photosphere
($T \sim 6000$ K)

Active
Regions

N

S

0.3 T $T \sim 4000$ K

Figure 10–15

Coronal loops. A general model for the loop structure in the lower corona.

10–5 The Solar Wind

The high coronal temperatures tend to blow the corona away from the Sun. The Sun's gravitational attraction on this gas is insufficient to retain it, and so a continuously flowing *solar wind* leaves the Sun. This flowing gas, composed of approximately equal numbers of electrons and protons, is a *plasma* (it is electrically neutral on a large scale). The thermal conductivity of the plasma is very high, so that high temperatures prevail over great distances from the Sun. Hence, the wind accelerates as it expands (large-scale speeds near 300 km/s at $30R_\odot$ and 400 km/s at 1 AU) and the particle density decreases to an average of a few million electrons and protons per cubic meter at 1 AU. These characteristics of the solar wind have been measured directly by interplanetary space probes.

A plasma couples tightly to lines of magnetic force; in fact, the magnetic field is essentially frozen into the gas. Therefore, the solar wind drags the extensions of solar magnetic fields into interplanetary space. The large-scale solar magnetic fields are directly related to the interplanetary fields by a sectored structure (Figure 10–16). The latitude band within 30° of the solar equator may frequently be divided into extended longitude regions of one magnetic po-

larity or the other. The radially outflowing solar wind conveys these fields away from the Sun in sectors; the sector boundaries are distorted into spirals because the Sun rotates away from the receding gas and magnetic field.

The solar wind shows considerable complexity and variability. For instance, at the Earth's orbit, the proton density varies from 0.4 to 80×10^6 m^{-3} and the speed ranges from 300 to over 700 km/s. This variation is closely tied to coronal holes, where both the density and the temperature are less than in the normal corona; the density may be only 7 percent of the usual value. Some of the energy that normally goes to heating the coronal material trapped within the magnetic loops is used to accelerate the gas outward through the open field lines. So bursts of solar wind may reach speeds in excess of 700 km/s.

Solar activity, especially flares, changes the magnetic field structure, sometimes dramatically. Whereas the normal solar wind is composed of low-energy protons and electrons (10^3 eV), solar flares eject clouds of high-energy protons (10^7 to 10^{10} eV). These clouds rush through the solar wind, altering its speed and density locally and distorting the magnetic field structure. These clouds are dangerous to unshielded astronauts, and they cause magnetic distur-

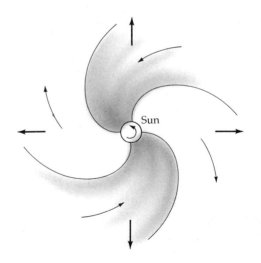

Sun

Figure 10–16

Magnetic fields and the solar wind. The interplanetary magnetic field extends from the Sun. The thin arrows show the direction of magnetic fields, and the thick ones show the direction of the solar wind. The shading shows two sectors of the structure.

bances in the Earth's magnetosphere a few days after they leave the Sun.

10–6 Solar Activity

(a) The Solar Cycle

The Sun is near enough that we can easily observe transient phenomena in its atmosphere. Such phenomena are manifestations of *solar activity,* and they are linked through the solar rotation and magnetic field in the *solar cycle.* The remainder of this chapter discusses the *active Sun,* using the generic term *active region* to specify an area with sunspots, prominences, plages, and flares (see Table 10–2 for a summary of solar activity).

Sunspots

Prior to 1610, Europeans believed the Sun to be a spotlessly perfect radiant sphere; in that year Galileo Galilei published his *Siderius Nuncius (Sidereal Messenger),* wherein he reported the existence of dark spots—*sunspots*—on the Sun's surface. (Oriental astronomers had reported sunspots centuries before.) Sunspots are photospheric phenomena (Figure 10–3) that appear darker than the surrounding photosphere (at about 5800 K) because they are cooler

(sunspot continuum temperatures are about 3800 K, and sunspot excitation temperatures are about 3900 K). The darkest, central part (with the temperatures just mentioned) is termed the *umbra* (Figure 10–17); the umbra is usually surrounded by the lighter *penumbra* with its radial filamentary structure. Small sunspots develop from *pores,* larger-than-usual dark areas between bright granules. Although most pores and small spots soon disintegrate, some grow into true sunspots of huge proportions. The largest have umbral diameters of 30,000 km and penumbral diameters more than twice as large.

The most important characteristic of a sunspot is its magnetic field. Typical field strengths are near 0.1 T, but fields as strong as 0.4 T have been measured. (The field strengths are deduced from the observed Zeeman splitting of spectral lines.) These fields may inhibit the convective transport of energy to the photosphere, so that the sunspot is cooler than its surroundings. Related to the magnetic field is a horizontal flow of gas in the sunspot penumbra: gas moves out along the lower filaments and inward along the higher filaments (at speeds up to 6 km/s).

A given sunspot has an associated *magnetic polarity.* Lines of magnetic force diverge from a north magnetic pole and converge at a south pole; you are familiar with this characteristic of bar magnets and our Earth. A magnetic pole cannot exist in isolation, however, because magnetic lines of force must be

Table 10–2 Summary of Solar Activity

Photosphere	
Sunspots	Strong magnetic fields, lower temperature than photosphere
Faculae*	Denser, hotter, brighter than photosphere (in white light)
Bipolar magnetic regions	Medium to weak magnetic fields
Chromosphere	
Plages*	Brighter than chromosphere in H_α and Ca II lines
	Denser, hotter than chromosphere
Prominences (filaments)	Chromospheric material in corona
	Exhibit motions associated with magnetic fields
Flares†	Brief brightenings in plages (in H_α and Ca II lines)
Corona	
Condensations*	White-light features due to increased electron density, increased emission in forbidden and UV lines, associated with slowly varying radio emission
Radio bursts due to fast electrons trapped in high corona†	
Particle emission (solar cosmic rays) and enhanced solar wind†	

* Phenomena related to plages at different heights in the solar atmosphere.
† Flare-related phenomena.

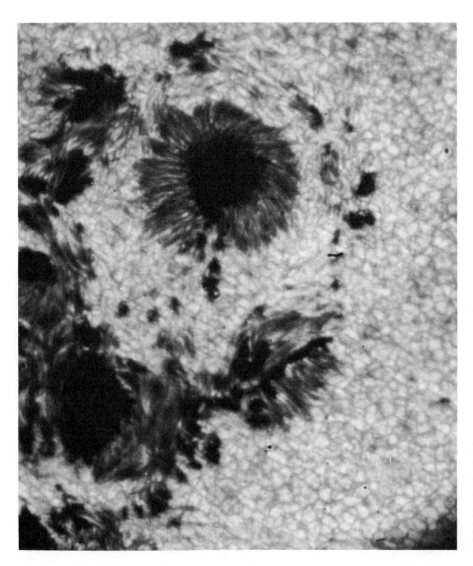

Figure 10–17

Sunspot group. A classic sunspot has an almost circular central umbra and a radial penumbra, determined by magnetic fields. Note the distorted spots in the lower left. Small spots without penumbras are pores. (*Project Stratoscope, Princeton University*)

complete (according to Maxwell's equations). So, two sunspots of complementary polarity are generally found together in a *bipolar* spot group (see the following paragraphs). Exceptions to this rule occur. Sometimes the second magnetic region is so diffuse that only one lone sunspot is seen; at other times, large, complicated groups of many sunspots appear — such a group may become the nucleus of a large active region [Section 10–6(b)] on the solar disk.

Sunspot Numbers

Counts of the number of sunspots visible at any given time have been recorded since Galileo's time (Figure 10–18). The sunspot number changes with time. A cyclic phenomenon is taking place, for successive sunspot maxima (or minima) occur every 11 years on the average (there may be a variation of as much as two or three years from cycle to cycle). A new cycle begins when the number is a minimum.

Recent investigations, especially by solar physicist John Eddy, indicate that some historical evidence exists to show an absence of the 11-year cycle in sunspot activity in the period before 1700 — a lull in activity called the *Maunder minimum*. Hardly any sunspots were seen in the 60-year period from 1645 to 1705 (Figure 10–18). The relative consistency of the cycle in modern times may be a phase in changes that take place over longer times.

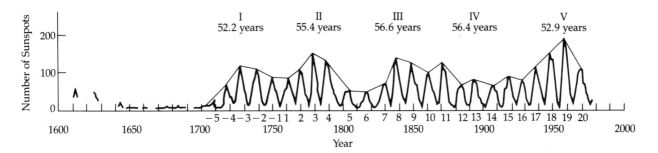

Figure 10-18

Sunspot cycles. Plotted here are the number of sunspots observed annually from 1600 to the 1970s. Note how few sunspots were seen prior to 1700; that was the time of the Maunder minimum. *[H. Yoshimura, Astrophysical Journal 230:905 (1979)]*

Latitude Variation

The distribution of sunspots in solar latitude varies in a characteristic way during the 11-year sunspot-number cycle. Sunspots tend to reside at high latitudes ($\pm 35°$) at the start of a cycle (Figure 10–19), most spots are near $\pm 15°$ at maximum, and the few spots at the end of the cycle cluster near $\pm 8°$. Very few sunspots are ever found at latitudes greater than $\pm 40°$. The lifetime of a sunspot ranges from a few days (small spots) to months (large spots). In fact, a sunspot dies at the same latitude where it was born (a characteristic that permits us to determine the solar rotation; see below). What takes place is this: as the cycle progresses, new spots appear at ever lower latitudes. Note that the first high-latitude spots of a cycle appear even before the last low-latitude spots of the previous cycle have vanished.

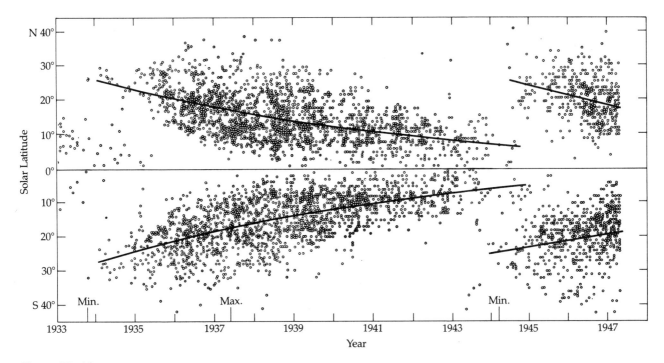

Figure 10-19

A diagram showing the change in latitude of sunspots as a function of phase in the solar activity cycle. *(G. Abetti, The Sun, Faber and Faber, 1969)*

Sunspot Polarity

Because most sunspot groups tend to be magnetically bipolar, it is useful to refer to *preceding spots* and *following spots* (in the sense of solar rotation). The Sun rotates eastward (as does the Earth), so that a preceding spot lies west of the following spot *as seen from the Earth*. After George E. Hale (1868–1938) discovered the magnetic character of sunspots in 1908, it was realized that all bipolar groups in one solar hemisphere have the same polarity and those in the other hemisphere have the opposite polarity. For example, in one solar cycle, preceding spots in the northern solar hemisphere have negative (south) polarity and preceding spots in the southern hemisphere have positive (north) polarity. Moreover, the sense of polarities reverses with each cycle, so that northern hemisphere leading spots will be positive in the next cycle. So a 22-year solar cycle is the true length of sunspot magnetic activity.

Solar Rotation

By observing sunspots, Galileo determined that the Sun's surface rotates eastward (synodically) in about one month. Today, the same method is used in the sunspot zone (other methods, such as Doppler shifts, are necessary above latitude $\pm 40°$), and we know that the Sun rotates *differentially*. That is, the rotation period is shorter at the solar equator (about 25^d) than at higher latitudes (about 27^d at $40°$ and 30^d at $70°$). The average sidereal period adopted for the sunspot zone is 25.4 days, with a corresponding synodic period (appropriate in discussions of solar–terrestrial events) of 27.3 days.

To sum up, sunspots reveal on a small scale the complexity and variability of solar magnetic phenomena. The parts of a sunspot are all transient magnetic structures — basically a sheaf of magnetic flux tubes filling the umbra and penumbra and fanning out above them. The secret to sunspots is the evolution of the fine details of their structure — a spatial resolution of 0.1″ means a linear resolution of 70 km, a resolving power to be achieved by the Hubble Space Telescope.

(b) Active Regions

As the sunspot number increases, so also does solar activity. With each sunspot group is associated a large *active region*, several hundred thousands of kilometers across (Table 10–2). Magnetic activity on the Sun concentrates in these active regions.

Bipolar Magnetic Regions and Plages

The most significant property of active regions is their magnetic fields: 0.1 T in sunspots and 0.01 T overall. Even when the sunspot group is a tangle of different polarities, the enveloping region is usually bipolar in character, and so we refer to *bipolar magnetic regions (BMRs)*.

A *magnetograph* maps out the magnetic field structure of BMRs (Figure 10–20). In a magnetic field, the Zeeman effect splits a spectral feature into several components, each with a characteristic optical polarization. When the splitting and intensity of these components are compared with a magnetograph based on the different polarizations, a map of the magnetic field strength and direction is produced.

In white light, photospheric *faculae* are brightenings that mark active regions. The enhanced brightness is from greater temperatures and densities than those found in nearby regions of the photosphere, similar to the chromospheric plages to which the faculae are related. Limb darkening renders faculae visible near the solar limb, though they are invisible near the center of the disk.

Above active regions in the photosphere, the bright *plages* float in the chromosphere. Plages are regions in which the density and temperature are higher than in the surrounding chromosphere and are caused by the magnetic fields of the active regions. These features appear on spectroheliograms taken in the light of the H_α line and the spectral lines of Ca II. In many respects, they look like concentrations and intensifications of the chromospheric network (Figure 10–20).

In the corona, active regions manifest themselves again in the higher densities and temperatures of *coronal streamers* and condensations of the white-light corona. Coronal line emission is stronger over plage regions than elsewhere, and enhanced radio emission arises from increased electron densities. This radio emission characterizes the long-lived active regions.

Prominences and Other Displays

Spectacular markers of active regions are the *prominences*, which appear as long, dark *filaments* when seen projected on the solar disk. Though visible

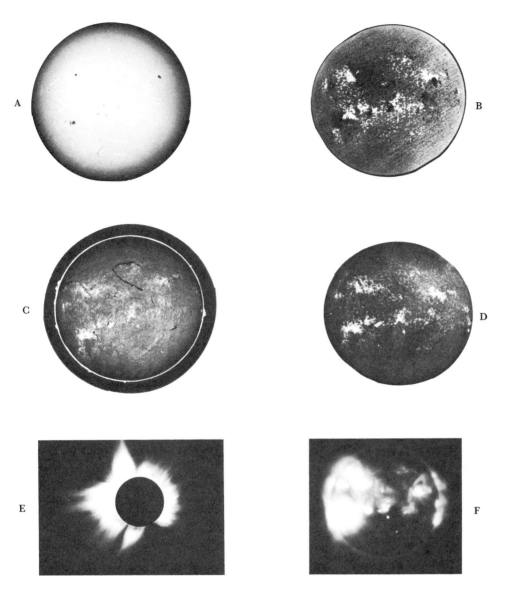

Figure 10–20

The Sun at various wavelengths in March 1970. (A) White-light photograph. *(Culgoora Solar Observatory)* (B) Magnetogram of bipolar magnetic regions. *(W. Livingston, National Optical Astronomy Observatories)* (C) A photograph in H$_\alpha$. *(P. S. McIntosh, National Oceanic and Atmospheric Laboratories)* (D) A Ca II spectrogram. *(National Optical Astronomy Observatories)* (E) The corona during an eclipse. *(G. Newkirk, High Altitude Observatory)* (F) X-rays from the corona. *(American Science and Engineering)*

in white light at a total solar eclipse, these displays are best recorded in H$_\alpha$ or Ca II lines.

Prominences are streams of chromospheric gas occupying coronal regions tens of thousands of kilometers above the chromosphere. Two characteristic types of prominences are *quiescent* and *active*.

Quiescent prominences (Figure 10–21) last for weeks and look like curtains with gas slowly descending from the corona into the chromosphere; they tend to lie along the neutral line separating the two poles of a BMR. Most active prominences survive only a few hours. Among the most active are the *loop*

Figure 10–21

A quiescent prominence. Note the fire-like structure. This time-lapse photograph shows that the material of the prominence streams downward *(National Optical Astronomy Observatories)*

prominences (Figure 10–22), which are closely associated with solar flares and which survive only an hour or so, during which time gas streams down the magnetic field lines joining the BMR poles. Flares will also upon occasion disrupt quiescent prominences, causing them to erupt and be ejected through the corona at high velocities.

(c) Solar Flares

Among the most puzzling, spectacular, and energetic phenomena associated with active regions are *solar flares*. Although these transient outbursts liberate tremendous quantities of energy, we still do not truly know how they originate. Flares radiate at many frequencies, from the X-rays and gamma rays to long-wavelength radio waves; in addition, they emit high-energy particles called *solar cosmic rays* (protons, electrons, and atomic nuclei; see Solar Cosmic

Rays in this section). Flare X-rays and ultraviolet radiation disrupt terrestrial radio communication by disturbing the Earth's ionosphere. The high-energy particle clouds, which are lethal to unprotected astronauts, reach the Earth in 30 min; clouds of low-energy particles and disturbances in the solar wind require from six to 24 h to transit from Sun to Earth.

Optical Manifestations

Flares usually appear in plages as brightenings in H_α (Figure 10–23). In fact, the H_α line becomes an emission feature, reaching maximum brightness within 5 min and decaying in about 20 min (3 h for the largest flares). Flares vary in size from 10,000 km to over 300,000 km; in general, the larger the flare, the more energetic it is and the longer-lived. At the peak of the solar cycle, the average occurrence of small flares is hourly and that of large flares is

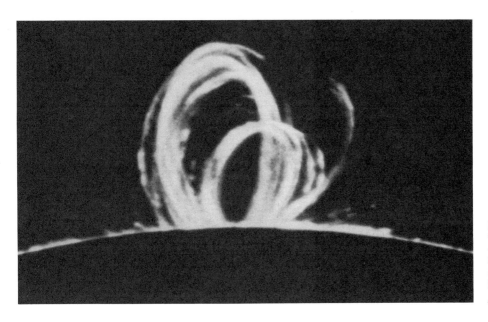

Figure 10-22

A loop prominence. Here the magnetic loops in the corona control the flow of the ionized gas. (*National Optical Astronomy Observatories*)

Figure 10-23

A flare in H_α. This strong flare (upper left) appears to shine brightly in H_α. Note the association with two sunspots in a BMR. Dark filamentary prominences are also visible above the photosphere. (*National Optical Astronomy Observatories*)

monthly. They are virtually nonexistent at solar minimum. Even during maximum, attempts to predict major flares are still rough.

X-Ray and Radio Bursts

In some flares, X-rays and centimeter radio waves occur together; they probably originate in the upper chromosphere or corona (Figure 10–24). Both consist of two components: (a) a *slow* component lasting about 30 min and (b) a *sudden* component (or *burst*) lasting a few minutes. The burst phenomenon is partly *nonthermal* (probably synchrotron), with X-rays more energetic than 20 keV and radio emission corresponding to 8×10^7 K, and partly thermal, causing extreme ionization. The less-energetic slow components result from a heating of the corona to 4×10^6 K; enhanced densities also stimulate emission in ultraviolet and forbidden lines.

Higher in the corona, a flare produces synchrotron radio bursts at meter wavelengths. These disturbances travel through the corona at speeds up to $0.3c$. It appears that energetic electrons are trapped by coronal magnetic fields having a strength of about 10^{-4} T, leading to the radio emission.

Solar Cosmic Rays

Solar flares accelerate atomic particles leaving the Sun. A blast wave propagates through the solar wind at 1500 km/s, disturbing the solar wind flow at the Earth (Section 10–5). Protons, electrons, and atomic nuclei are accelerated to high energies in flares—we call these *solar cosmic rays*. Most of the particles observed are protons, for the electrons lose much of their energy in exciting radio bursts in the corona and the solar abundance of other nuclei is low. *Alpha particles* (helium nuclei) are the second most abundant nuclei after the protons. Solar-particle energies range from keV (10^3 eV) to about 20 GeV (1 GeV = 10^9 eV), with the bulk of the particles in the MeV (10^6 eV) range. The highest-energy particles arrive at the Earth within 30 min of the H_α flare maximum, followed by the peak number of particles 1 h later, with the low-energy cosmic rays bringing up the rear hours later. About half the flare energy (some 10^{25} J for the largest flares!) is in the H_α emission, half in the shock wave, and only 1 percent in solar cosmic rays.

A Flare Model

Present thinking suggests the following picture for the development of a flare (Figure 10–25). Magnetic loops through the corona connect the two parts of a bipolar active region; field lines from the outer parts of the region extend out indefinitely. A prominence may define a neutral line in the BMR. A stress on the magnetic field (in the convection zone?) causes an instability. Energy is released instantaneously at the top of the loops as magnetic field lines reconnect —a process similar to that in the Earth's magneto-

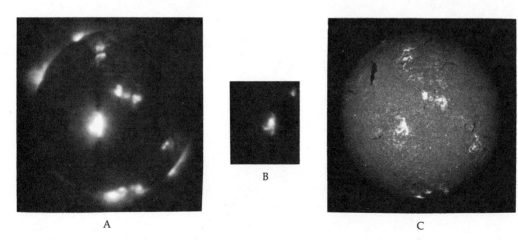

A

B

C

Figure 10–24

Active regions in X-rays. (A) Active regions in this X-ray photograph include a solar flare near the center of the disk. (B) A closeup of the flare at shorter X-ray wavelengths. *(American Science and Engineering)* (C) An H_α photograph for comparison; note that the bright active regions show up in X-rays. *(ESSA)*

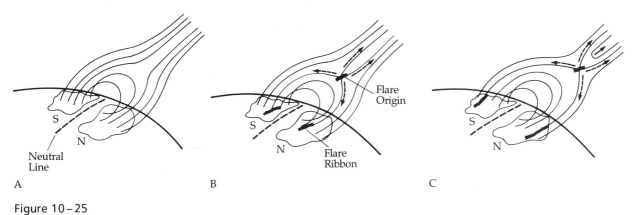

Figure 10–25

Model of a flare. (A) Preflare stage of a BMR. Note the unconnected magnetic field lines above the coronal loop. (B) Start of the flare as the field lines reconnect and a current flows down to the feet of the BMR. (C) Site of the flare moves up into the corona. *(After R. Noyes)*

sphere (Section 4–6). According to Lenz's law, electric currents that oppose this change in the magnetic field are established. Ohmic dissipation of the currents heats the gas. The resulting heating of that region of the corona causes the emission of X-rays, extreme ultraviolet emission, and centimeter radio bursts while electrons and protons are accelerated downward along the loops into the chromosphere, heating it to produce the H_α brightening (two-ribbon flares are quite common, corresponding to the two sides of the loop), more lower-energy X-rays, and transition-zone extreme ultraviolet radiation. At the same time, other particles are accelerated outward, with the electrons producing meter-wavelength radio bursts and the protons and other nuclei becoming solar cosmic rays. Only major flares produce all these manifestations; the most energetic may even emit gamma rays. Most flares, however, are much more modest.

(d) A Model of the Solar Cycle

Since 1960, we have been struggling with a detailed magnetic model of the solar cycle that satisfies at least some of the observations. One model (Figure 10–26) produces a 22-year solar cycle by coupling the Sun's magnetic fields to its differential rotation.

Prior to the start of solar activity, the Sun has a weak dipolar magnetic field, with the lines of force running along meridians about $0.1R_\odot$ below the surface. Each line completes itself by emerging near the poles. This configuration is called a *poloidal field*. The Sun's differential rotation draws each line out along the equator, wrapping it around the Sun many times. A *toroidal field* develops. As the density of lines increases, so also does the associated magnetic pressure; inside the *flux tubes* that contain the lines, the gas pressure must decrease (to have pressure equilibrium), but then the tubes are less dense than their surroundings—they experience *magnetic buoyancy*. The field strength further increases as gas motions (convection) twist the lines.

The amplification of the magnetic field is maximum at higher latitudes (about $\pm35°$), so that a critical field strength is first attained there. At this point, a *flux loop* rises to the solar surface and appears as a BMR with the strongest fields in the sunspots. Supergranules probably facilitate the emergence of a flux tube since spots generally originate at the boundaries of supergranules. The opposite poles are a consequence of the continuity of the field lines from north to south: because the following spot is nearer to the Sun's poles both in latitude and along the flux tube, its polarity is opposite that of the nearest pole.

At lower latitudes, the critical field strength is attained later. A sunspot maximum corresponds to the time of greatest field strength over the largest latitude band; eventually, lines become mixed and recombine near the equator, so that fewer spots occur later in the cycle. Supergranulation pushes the flux tubes about, dissipating the magnetic fields. The differential rotation causes the following magnetic regions to move toward the poles while the preceding regions move toward the equator. At the equator,

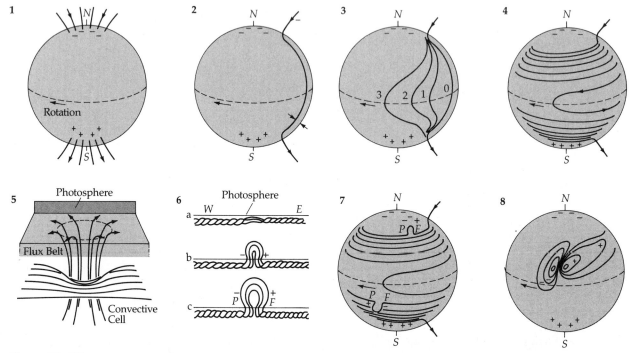

Figure 10–26

Model of the solar cycle. At the start (1), the Sun has a simple dipolar field with lines of force emerging at latitudes of 60° and lying at the base of the convective zone (2). Differential rotation (3) stretches the field lines out along the equator until they are tightly wound (4). Rising convective cells (5) twist the field lines into ropes (6), where kinks develop in which the magnetic pressure increases, causing the kinks to rise to the surface and form bipolar regions (7) with preceding (P) and following (F) fields. The following fields migrate toward the poles (8), first neutralizing and then reversing the polar fields.

regions of opposite polarity from different hemispheres meet and annihilate. The following regions converge on the poles, where their opposite polarity first neutralizes and then reverses the polar fields. After 11 years, the polarity of the Sun's overall dipolar field is reversed and the stage is set for the second half of the 22-year cycle.

In summary, the analysis of the solar cycle rests on the development of solar magnetic dynamo models. To date, such models have focused on the interplay of poloidal and toroidal field configurations. The shear of differential rotation produces a toroidal field from a poloidal one. In turn, cyclonic motions in the convective zone may regenerate a poloidal field. If so, then all stars with an outer convective zone and differential rotation should have magnetic activity cycles. We now have good observations that such is the case (Section 18–3). Hence, we can conclude that we have pinned down the basic physics even though we do not know all the details.

Problems

1. Show that the Sun is not at rest in the center of the Solar System by calculating the distance from the Sun's center to the center of mass of the Sun–Jupiter system.

2. Find the thermal Doppler width of a spectral line at 500 nm formed in the Sun's photosphere ($T \approx 5400$ K).

3. **(a)** Given that the photosphere is at a temperature of

6000 K, would you expect collisional or radiative excitations to be more important in exciting hydrogen atoms to the second ($n = 2$) level?

(b) Would you expect the Lyman α line to appear in emission or absorption?

4. Consider the following two lines of similar excitation potential: Fe I at 414.4 nm and Fe II at 417.3 nm. Explain in general terms (with reference to the Boltzmann and/or Saha equation) why the 414.4-nm line is the stronger of the two in the photospheric spectrum and the weaker of the two in the chromospheric spectrum.

5. From our description of chromospheric and coronal radio opacities, describe how you would determine the motion of a solar radio burst through the solar atmosphere.

6. How can one determine the temperature structure of a sunspot using only its continuous spectrum?

7. How would you unambiguously assign a sunspot near the sunspot minimum to the old or the new cycle?

8. Some prominences are said to have speeds greater than the escape speed from the Sun at the chromosphere. What is the critical speed?

9. Using the data on the solar wind given in this chapter, compute the average rate of mass loss of our Sun ($M_\odot$/year) from **(a)** the solar wind and **(b)** energy generation.

10. Use the equation of hydrostatic equilibrium to estimate the pressure in the Sun's center.

11. Calculate and compare the scale heights for hydrogen in the Sun's photosphere, chromosphere, and corona.

Stars:
Distances
and
Magnitudes

Having discussed the Sun, the nearest star, we now jump far beyond the Solar System to the stars. This chapter presents some of the methods by which the distances to the stars are determined and quantifies stellar brightness in astronomical terms. The key point is this: by comparing other stars with the Sun, we can infer their physical properties.

11–1 The Distances to Stars

Within the Solar System, we can determine absolute distances by using Newtonian celestial mechanics (space-probe trajectories and Kepler's laws) and radar. However, even the nearest stars are so distant, in terms of familiar measures such as the astronomical unit, that other methods must be employed to determine their distances. As you will see, some of these methods are very indirect.

(a) Trigonometric (Heliocentric) Parallax

As the Earth orbits the Sun, the nearest stars appear to move relative to the more distant stars. The lack of any observation of this heliocentric *parallax* effect led astronomers such as Tycho Brahe to be skeptical of the Copernican heliocentric model of the Solar System. It was not until 1838, when F. Bessel detected the parallax of the star 61 Cygni and F. Struve detected that of Vega, that Copernicus' model was finally vindicated. (Note that Bradley's discovery of stellar aberration in 1729 also proved that the Earth is in motion.)

The parallactic displacement of a star on the sky as a result of the Earth's orbital motion permits us to determine the distance from the Sun to the star by the method of *trigonometric parallax* (Figure 11–1). We define the trigonometric parallax of the star as the angle π subtended as seen from the star by the Earth's orbit of radius 1 AU. If the star is at rest with respect to the Sun, the parallax is half the maximum apparent annual angular displacement of the star as seen from the Earth. Letting a denote the Sun–Earth distance and d the Sun–star distance, we have

$$\pi \text{ (rad)} = a/d \qquad (11-1)$$

Recall that there are 2π rad in a circle (360°), so that 1 rad equals 57°17′44.81″ (206,264.81″). Hence, if we agree to measure all angles in arcseconds and all dis-

tances in *parsecs* (abbreviation "pc"; 1 pc = 206,265 AU), Equation 11–1 becomes

$$\pi'' = 1/d \qquad (11-2)$$

Note that 1 pc = 3.086×10^{16} m = 3.26 lightyears, where 1 *lightyear* is the distance light travels in one year.

The measurement and interpretation of stellar parallaxes are a branch of *astrometry*, and the work is exacting and time-consuming. Consider that the nearest star, Alpha Centauri, at a distance of 1.3 pc, has a parallax of only 0.76″ (Table A4–1); all other stars have smaller parallaxes. Today, stellar parallaxes can be determined with a probable error of order ±0.0004″, which means that a parallax 0.100″ ± 0.004″ has a 50 percent probability of actually being between 0.096″ and 0.104″. So present trigonometric parallaxes are believable only to distances of about 100 pc ($\pi'' = 0.01''$), which is miniscule relative to the 8.5-kpc (1 kpc = 10^3 pc) distance to the center of our Galaxy. Technological advances including the Hubble Space Telescope will improve parallax accuracy to 0.001″ within a few years. Fewer than 10,000 stellar parallaxes have been measured, but there are about 10^{12} stars in our Galaxy! Space observations planned by the European Space Agency will determine the parallaxes of about 100,000 stars. The method of trigonometric parallax is important because it is our only direct distance technique for stars.

The trigonometric parallax of a star is determined by photographing a given star field from a number (about 20) of selected points in the Earth's orbit. The comparison stars selected are distant background stars of nearly the same apparent brightness as the star whose parallax is being measured. Corrections

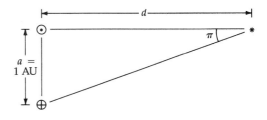

Figure 11–1

Trigonometric parallax. The angle subtended by the Earth's orbit of radius 1 AU is π arc-seconds at a star–Sun distance of d (in parsecs).

are made for atmospheric refraction and dispersion and for detectable motions of the background stars; any motion of the star relative to the Sun is then extracted. What remains is the smaller annual parallactic motion; it is readily recognized because it cycles in 1 year. Because a seeing resolution of 0.25" is considered exceptional (more typical is 1"), it may seem strange that a stellar position can be determined to $\pm 0.01"$ in one measurement; this accuracy is possible because we are determining the center of the fuzzy stellar image.

(b) Other Geometric Methods

To reach distances greater than 100 pc using geometric methods, we can use (1) the Sun's motion through the nearby stars and (2) the motions of *star clusters* that are not too far away. Both methods depend upon stellar motions; because this topic is extensively discussed in Chapter 15, we only summarize the techniques here.

The motion of the Sun among the nearby stars — the *solar motion* — is 20 km/s (4.1 AU/year) toward the constellation Hercules. This baseline grows year by year, so that over an interval of ten years, we could measure stellar distances to about 2000 pc if the nearby stars were stationary in space. All stars do move, however, just as the Sun does, and so only average parallaxes of groups of stars are possible. By assuming that the peculiar motions (Chapter 15) of the stars in a large sample (preferably the same spectral type; Chapter 13) average to zero, we can deduce the *mean parallax* of that sample. Moreover, because the solar motion affects only that component of a star's proper motion parallel to the solar motion, we may use the other (perpendicular) component to find *statistical parallaxes* (Chapter 15). Note that we have been forced to sacrifice accuracy in order to attain greater distances and that the distances obtained relate to a group of stars.

The *moving cluster* method leads to precise individual stellar distances greater than 100 pc; unfortunately, there are very few clusters (the Hyades cluster in the constellation Taurus is the most famous) where the method is really applicable. A star cluster consists of many stars moving as a group through space because the stars are bound together gravitationally. If the cluster subtends an appreciable angle on the sky, the individual stellar proper motions appear to converge to (or diverge from) a single point on the sky. By

measuring the average radial velocity of the cluster and using trigonometric calculations, we can determine the distance to each star in the cluster (Chapter 15).

(c) Luminosity Distances

Finally, we preview distance determinations based upon stellar luminosity. We must rely upon these rather indirect methods to truly probe our Galaxy (and other galaxies). Each method yields only *relative* distances until we calibrate the stars involved. The calibration is accomplished when we find a nearby representative star (or cluster) for which the distance and stellar luminosity can be determined by means of trigonometric parallax or the moving-cluster method. Because radiant flux decreases as the square of the distance from the source [Section 8–1(c)], the luminosity follows from the observed flux once the distance is known. Conversely, if we can estimate the luminosity and measure the flux, we can infer the distance from the inverse-square law of light. You will see in detail how these methods work in the next two chapters.

The subject of stellar distances underlies all of astronomy and astrophysics. Now to discuss stellar brightnesses quantitatively and link them to stellar distances.

11–2 The Stellar Magnitude Scale

The first stellar brightness scale — the *magnitude scale* — was defined by Hipparchus and refined by Ptolemy almost 2000 years ago. In this qualitative scheme, naked-eye stars fall into six categories: the *brightest* (Table A4–2) are of *first* magnitude, and the *faintest* of *sixth* magnitude. *Note that the brighter the star, the smaller the value of the magnitude.* In 1856, N. R. Pogson verified William Herschel's finding that a first-magnitude star is 100 times brighter than a sixth-magnitude star and the scale was quantified. Because an interval of five magnitudes corresponds to a factor of 100 in brightness, a one-magnitude difference corresponds to a factor of $100^{1/5} = 2.512$. (This definition reflects the operation of human vision, which converts equal *ratios* of actual intensity to equal *intervals* of perceived intensity. In other words, the eye is a logarithmic detector.) The magnitude scale has been extended to positive magnitudes larger than $+6.0$ to include faint stars (the 5-m telescope on Mount Palomar can reach to magnitude $+23.5$) and to negative magnitudes for very bright objects (the star Sirius is magnitude -1.4). The limiting magnitude of the Hubble Space Telescope is expected to be about $+25$.

Astronomers find it convenient to work with logarithms to base 10 (Mathematical Appendix) rather than with exponents in making the conversion from brightness ratio to magnitude and vice versa. Consider two stars of magnitude m and n with respective apparent brightnesses l_m and l_n. The ratio of their brightnesses l_n/l_m corresponds to the magnitude difference $m - n$. Because a one-magnitude difference means a brightness ratio of $100^{1/5}$, $m - n$ magnitudes refer to a ratio of $(100^{1/5})^{m-n} = 100^{(m-n)/5}$, or

$$l_n/l_m = 100^{(m-n)/5} \qquad (11-3)$$

Taking the logarithm to base 10 of both sides of Equation 11–3 yields, because $\log x^a = a \log x$ and $\log 10^a = a \log 10 = a$,

$$\log(l_n/l_m) = [(m - n)/5]\log 100 = 0.4(m - n) \qquad (11-4)$$

or

$$m - n = 2.5 \log(l_n/l_m) \qquad (11-5)$$

Equation 11–5 defines apparent magnitude; note that $m > n$ when $l_n > l_m$, that is, brighter objects have numerically smaller magnitudes. Also note that when the brightnesses are those observed at the Earth, physically they are fluxes. *Apparent magnitude* is the astronomically peculiar way of talking about fluxes.

To help you use this key equation (11–5), here are a few worked examples:

(a) The apparent magnitude of the variable star RR Lyrae ranges from 7.1 to 7.8—a magnitude amplitude of 0.7. To find the relative increase in brightness from minimum to maximum, we use

$$\log(l_{\max}/l_{\min}) = 0.4 \times 0.7 = 0.28$$

so that

$$l_{\max}/l_{\min} = 10^{0.28} = 1.93$$

This star is almost twice as bright at maximum light than at minimum.

(b) A binary system consists of two stars a and b, with a brightness ratio of 2; however, we see them unresolved as a point of magnitude $+5.0$. We would like to find the magnitude of *each* star. The magnitude *difference* is

$$m_b - m_a = 2.5 \log(l_a/l_b) = 2.5 \log 2 = 0.75$$

Since we are dealing with brightness ratios, it is not right to put $m_a + m_b = +5.0$. The sum of the luminosities $(l_a + l_b)$ corresponds to a fifth-magnitude star. Let us compare this to a 100-fold brighter star, of magnitude 0.0 and luminosity l_0:

$$m_{a+b} - m_0 = 2.5 \log[(l_0/(l_a + l_b)]$$

or

$$5.0 - 0.0 = 2.5 \log 100 = 5$$

but $l_a = 2l_b$, so that $l_b = (l_a + l_b)/3$; therefore

$$m_b - m_0 = 2.5 \log(l_0/l_b)$$
$$= 2.5 \log 300 = 2.5 \times 2.477 = 6.18$$

The magnitude of the fainter star is 6.18, and from our earlier result on the magnitude difference, that of the brighter star is 5.43.

11–3 Absolute Magnitude and Distance Modulus

So far we have dealt with stars as we see them, that is, their fluxes, or apparent magnitudes, but we frequently want to know the luminosity of a star. A very luminous star will appear dim if it is far enough away, and a low-luminosity star may look bright if it is close enough. Our Sun is a case in point: if it were at the distance of the closest star (Alpha Centauri), the Sun would appear slightly fainter to us than that star does. Hence, *distance* is the link between fluxes and luminosities.

The luminosity of a star relates to its *absolute magnitude*, which is the magnitude that would be observed if the star were placed at a distance of 10 pc from the Sun. By convention, absolute magnitude is capitalized (M) and apparent magnitude is written lower-case (m). The inverse-square law of radiative flux links the flux l of a star at a distance d to the luminosity L it would have it if were at a distance $D = 10$ pc:

$$L/l = (d/D)^2 = (d/10)^2$$

If M corresponds to L and m corresponds to l, then

Equation 11–5 becomes

$$m - M = 2.5 \log(L/l)$$
$$= 2.5 \log(d/10)^2 = 5 \log(d/10)$$

Expanding this expression, we have the useful alternative forms

$$m - M = 5 \log d - 5 \qquad \textbf{(11–6)}$$

$$M = m + 5 - 5 \log d \qquad \textbf{(11–7)}$$

$$M = m + 5 + 5 \log \pi'' \qquad \textbf{(11–8)}$$

Here d is in parsecs and π'' is the parallax angle in arcseconds. The quantity $m - M$ is called the *distance modulus*, for it is directly related to the star's distance in Equation 11–6. In many applications, we refer only to the distance moduli of different objects rather than converting back to distances in parsecs or light-years. Note that absolute magnitude is the astronomically peculiar way of talking about luminosity.

11–4 Magnitudes at Different Wavelengths

(a) Magnitude Systems

Detectors of electromagnetic radiation (such as the photographic plate, the photoelectric photometer, and the human eye) are sensitive only over given wavelength bands. So a given measurement samples but part of the radiation arriving from a star. Because the flux of starlight varies with wavelength, the magnitude of a star depends upon the wavelength at which we observe. Originally, photographic plates were sensitive only to blue light, and the term *photographic magnitude* (m_{pg}) still refers to magnitudes centered around 420 nm (in the blue region of the spectrum). Similarly, because the human eye is most sensitive to green and yellow, *visual magnitude* (m_v) or the photographic equivalent *photovisual magnitude* (m_{pv}) pertains to the wavelength region around 540 nm.

Today we can measure magnitudes in the infrared, as well as in the ultraviolet and extreme ultraviolet (via satellite and rocket), by using filters in conjunction with the wide spectral sensitivity of photoelectric photometers. So systems of many different magnitudes (color combinations) are possible. A widely used magnitude system is the *UBV system:* a combination of ultraviolet (U), blue (B), and visual

(V) magnitudes. These three bands are centered, respectively, at 365, 440, and 550 nm; each wavelength band is roughly 100 nm wide (Figure 11–2). In this system, apparent magnitudes are denoted by B or V and the corresponding absolute magnitudes are subscripted: M_B or M_V. Note that to be useful in measuring fluxes, the *UBV* system must be calibrated in energy units for each of its band passes.

Finally, we note that the *UBV* system has been extended into the red and infrared (in part because of the development of new detectors, such as CCDs, sensitive to this region of the spectrum). The extensions are not as well standardized as that for the *UBV* system developed by Harold Johnson, but they tend to include R and I in the far red and J, H, K, L, and M in the infrared (which have band centers at roughly 1.2, 1.6, 2.2, 5, and 10 μm). Other photometric systems exist for different purposes; most observations, at least of stars, have been done in the *UBV* system.

(b) Color Index

A quantitative measure of the *color of a star* is given by its *color index* (CI), which is defined as the difference between magnitudes at two different effective wavelengths. For example,

$$CI = m_{pg} - m_v = M_{pg} - M_v \qquad \textbf{(11–9)}$$

where the last equality follows from Equation 11–7

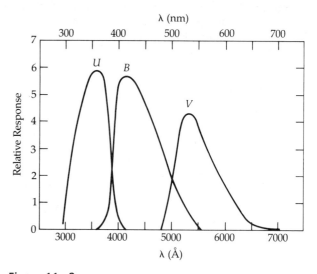

Figure 11–2

UBV filters. The profile of the standard filters in the *UBV* system for a source with equal flux at all wavelengths.

because we are talking about one star. Note that the long-wavelength magnitude is always subtracted from the short-wavelength magnitude and that a star's color index does not depend upon distance. Similarly, the quantities $B - V$ and $U - B$ are also color indices. Also, note that because the color index is a magnitude difference, it is equivalent to a flux *ratio* at the specific wavelengths involved.

Because stars have different temperatures, their spectral energy (Planck) curves peak at different wavelengths (Chapter 8 and Figure 11–3); therefore, hot stars are bluish and cool stars are reddish. Using the color index $B - V$, we see that a bluish star (20,000 K) has a negative color index because it is brighter in the blue (smaller B magnitude) than at longer wavelengths (larger V magnitude). A reddish star (3000 K) has a positive color index because it is brighter in V than in B.

Magnitude systems, and the color indices derived therefrom, are observer-dependent because the percentage of the incident stellar energy measured depends upon both the wavelength (Figure 11–3) and the particular instruments used. So that all observations may be compared on the same basis, the systems are set so that the color index of stars (such as Vega) with surface temperatures about 10,000 K are equal to zero. Hotter stars have a negative color index, and cooler ones have a positive index.

We have assumed that the observed color indices are intrinsic to the stars. However, interstellar space is pervaded by dust grains that absorb and scatter star-

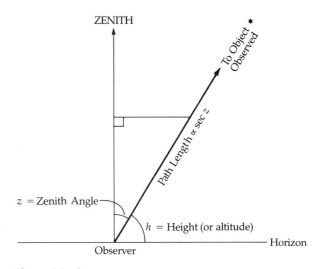

Figure 11–4

Geometry for atmospheric extinction.

light, and so the observed light appears redder than when it was emitted (Chapter 19). This *interstellar reddening* is wavelength-dependent, and it therefore affects color indices. We term the difference between the observed and the intrinsic color indices the *color excess* (Chapter 19). We mention this here in order to warn you that a star's temperature is not uniquely determined by a *single* color index measurement between two wavelength bands—three-color (or more) photometry is necessary to separate out the effects of interstellar reddening.

It is fairly straightforward to correct for *atmospheric extinction,* that is, scattering and absorption in the Earth's atmosphere. Current practice is to express magnitudes after such correction, as though they were observed above the Earth's atmosphere.

Assume that the Earth's atmosphere is plane-parallel (that is, ignore the Earth's curvature). Define h as the angular height of a star above the horizon and z as the angle between the direction to the zenith and the direction to the star ($z = 90° - h$); z is the *zenith angle.* The path length traversed by light through the atmosphere is proportional to sec z (Figure 11–4). So at the zenith, a star suffers minimal extinction, which increases rapidly close to ($\approx 30°$) the horizon. If possible, you want to avoid making measurements at such low elevations, where the plane-parallel approximation no longer holds true.

On a given night, selected standard stars and unknown sources suffer the same extinction. (You

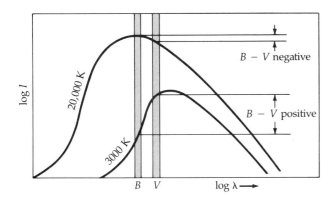

Figure 11–3

Color index. Blackbody curves for 20,000 K and 3000 K, along with their intensities at B and V wavelengths. Note that $B - V$ is negative for the hotter star, positive for the cooler one.

hope that the extinction does *not* vary over short time periods). To find the *extinction coefficient* for the night, you measure the flux of a standard star over a large range of zenith angles (from close to the horizon up to the zenith) and plot measured values against sec z. If things have been done correctly, you get a straight line that tells you the amount of extinction (in magnitudes) per unit sec z (which has the name of *air mass*), that is, magnitudes per air mass. You then correct your other measurements with this extinction coefficient, which depends on the wavelength of observation.

(c) Bolometric Magnitudes and Stellar Luminosities

The complete spectral energy distribution of a star is only sampled by the color indices we have been discussing. Of greater significance for the overall structure of the star, however, is its total rate of energy output (in watts) at all wavelengths. If we place ourselves outside the Earth's obscuring atmosphere, the radiative flux from the star per unit wavelength l_λ (W/m$^2 \cdot$ Å) permits us to define the total *bolometric flux* (W/m^2; referred to as brightness earlier in the chapter):

$$l_{bol} = \int_0^\infty l_\lambda \, d\lambda \qquad (11-10)$$

The *apparent bolometric magnitude* of the star m_{bol} follows from Equation 11–5 as

$$m_{bol} = -2.5 \log l_{bol} + \text{constant} \quad (11-11)$$

where the constant is an arbitrary zero point. The *absolute bolometric magnitude* of the star M_{bol} is the bolometric magnitude if the star were at the standard distance of 10 pc. In the notation of Equation 11–10, the *visual flux* l_v, for example, is just

$$l_v = \int_0^\infty l_\lambda S_\lambda \, d\lambda \qquad (11-12)$$

where S_λ expresses the spectral sensitivity of the visual photometric system (the V pass band of Figure 11–2). In analogy to Equation 11–11, we can then define the *visual magnitude* as

$$m_v = -2.5 \log l_v + \text{constant} \quad (11-13)$$

Our Sun is the only star for which l_λ has been accurately observed. Indeed, l_{bol} is related to the *solar constant*: the total solar radiative flux received at the

Earth's orbit outside our atmosphere (1370 W/m^2). The *solar luminosity* $L_\odot$ (3.86 × 10^{26} W) follows from the solar constant in the following manner. Using the inverse-square law, we find the radiative flux at the Sun's surface $R_\odot$. Then $L_\odot$ is just $4\pi R_\odot^2$ times this flux. The solar energy distribution curve may be approximated by a Planck blackbody curve (Chapter 8) at the *effective temperature* T_{eff}, defined as the temperature of a blackbody that would emit the same total energy as an emitting body, such as the Sun or a star. Then the Stefan–Boltzmann law (Chapter 8) implies

$$L_\odot = 4\pi R_\odot^2 \sigma T^4_{eff} \quad \text{J/s} \qquad (11-14)$$

where σ is the Stefan–Boltzmann constant.

If we know the absolute bolometric magnitude of a star, we can use Equation 11–5 to find that star's luminosity:

$$M_{bol}(\odot) - M_{bol}(*) = 2.5 \log(L_*/L_\odot) \quad (11-15a)$$

Since $M_{bol}(\odot) = +4.7$, this becomes

$$\log(L_*/L_\odot) = 1.89 - 0.4M_{bol}(*) \quad (11-15b)$$

Usually $M_{bol}(*)$ is *not* directly observed (although this is now becoming possible with space probes and satellites), but L_* may be deduced by studying the star's spectrum (Chapter 13); then the absolute bolometric magnitude follows from Equation 11–15a.

In practice, we use the *bolometric correction BC*, which is the difference between the bolometric and visual magnitudes, to determine a star's bolometric magnitude. For example,

$$BC = m_{bol} - m_v = M_{bol} - M_v$$
$$BC = 2.5 \log(l_v/l_{bol}) \qquad (11-16)$$

Bolometric corrections are inferred from ground-based observations by using theoretical stellar models; these corrections have been checked and improved with the ultraviolet data from orbiting satellites. In the UBV magnitude system, the bolometric correction is a minimum for stars with $T_{eff} = 6500$ K; $BC = -0.07$ for our Sun. (Table A4–3 gives, among other data, the bolometric corrections for stars of different temperatures.) For stars with surface temperatures of 6700 K, the spectral energy peaks in the V wavelength band, so that the greatest percentage of the star's energy is detected. For other stellar temperatures, a smaller percentage of the total radiative energy is measured in the V band; hence, their bolometric corrections are larger (in absolute value) than that for 6700-K stars.

Finally, a word of caution. The light we receive and measure from stars has been filtered many times before we record: by the matter in interstellar space, by the Earth's atmosphere, by the optics of our telescope, and by the detector and filters used. The goal of the observer is to correct for this filtration so that the light has the same characteristics as that just emitted by the star at its photosphere.

Problems

1. Astronomers living on Jupiter would define their astronomical unit in terms of the orbit of Jupiter. If they defined parsec in the same manner as we do, how many Jovian astronomical units would such a parsec contain? How many Earth astronomical units would equal a Jovian parsec? How many Earth parsecs are there in a Jovian parsec?

2. A variable star changes in brightness by a factor of 4. What is the change in magnitude?

3. What is the combined apparent magnitude of a binary system consisting of two stars of apparent magnitudes 3.0 and 4.0?

4. If a star has an apparent magnitude of −0.4 and a parallax of 0.3″, what is
   (a) the distance modulus?
   (b) the absolute magnitude?

5. What is the distance (in parsecs) of a star whose absolute magnitude is +6.0 and whose apparent magnitude is +16.0?

6. What are the absolute magnitudes of the following stars:
   (a) $m = 5.0$, distance $d = 100$ pc
   (b) $m = 10.0$, $d = 1$ pc (is there such a star?)
   (c) $m = 6.5$, $d = 250$ pc
   (d) $m = -3.0$, $d = 5$ pc
   (e) $m = -1.0$, $d = 500$ pc
   (f) $m = 6.5$, parallax $\pi = 0.004''$

7. What would the expression for absolute magnitude be, in terms of apparent magnitude and distance, if absolute magnitude were defined as the magnitude a star would have at 100 pc?

8. The Sun has an apparent magnitude of −26.5.
   (a) Calculate its absolute magnitude.
   (b) Calculate its magnitude at the distance of Alpha Centauri (1.3 pc).
   (c) The Palomar Sky Survey is complete to magnitudes as faint as +19. How far away (in parsecs) would a star identical to the Sun have to be in order to just barely be bright enough to be visible on Sky Survey photographs?

9. Using the data from this chapter, determine how much brighter than Sirius the Sun is, as seen from the Earth. How much more luminous is Sirius than the Sun?

10. A certain globular cluster has a total of 10^4 stars; 100 of them have $M_v = 0.0$, and the rest have $M_v = +5.0$. What is the integrated visual magnitude of the cluster?

11. The V magnitudes of two stars are both observed to be 7.5, but their blue magnitudes are $B_1 = 7.2$ and $B_2 = 8.7$.
    (a) What is the color index of each star?
    (b) Which star is the bluer and by what factor (in brightness) is it bluer than the red star?

12. What is the color index of a star at a distance of 150 pc with $m_v = 7.55$ and $M_B = 2.00$?

13. What is the absolute bolometric magnitude of a star of luminosity 10^{33} W?

14. Given the expressions for the luminosity of a star (Equation 11–14) and its bolometric magnitude in terms of that of the Sun (Equation 11–15a), find an expression for the bolometric magnitude of the star as a function of its temperature and radius. The effective temperature of the Sun is 5780 K.

15. The bolometric correction for a star is −0.4, and its apparent visual magnitude is +3.5. Find its apparent bolometric magnitude.

Stars:
Binary
Systems

Most stars clearly visible in optical telescopes turn out to be binary or multiple star systems. The stars in a multiple system are physically related; they orbit one another under the influence of their mutual gravitational attraction. That Newton's law of universal gravitation is applicable beyond our Solar System became evident when William Herschel detected the orbital motion of the Castor binary system in 1804. As you will see, known physical laws can be coupled with suitable observations of binary systems to tell us about many important stellar characteristics: masses, radii, densities, surface temperatures and luminosities, and rotation rates. That *most* stars in the neighborhood of our Sun (well over 50 percent, perhaps close to 80 percent) belong to multiple systems luckily allows us to infer the physical properties of many of them. This fact is especially important for the estimate of stellar masses, which can be done directly *only* in binary systems.

12–1 Classification of Binary Systems

For physical and observational reasons, we classify stellar binary systems into several different types:

Apparent binary   Two stars that are not physically associated but appear close together on the sky because they lie along the same line of sight. Their uncorrelated space motions soon reveal that they are not members of a physical binary system. (Sometimes these are called *optical* binaries.)

Visual binary   A bound system that can be resolved into two stars at the telescope. The mutual orbital motions of these stars are observed to have periods ranging from about 1 year to thousands of years.

Astrometric binary   Only one star is seen telescopically, but its oscillatory motion on the sky reveals that it is accompanied by an unseen companion. Both bodies are orbiting about their mutual center of mass.

Spectroscopic binary   An unresolved system whose duplicity is revealed by periodic oscillations of the lines in its spectrum. In some cases, two sets of spectral features are seen (one for each star) oscillating with opposite phases; in other cases, one of the stars is too dim to be seen, so that only one set of oscillating spectral lines is recorded. Typical orbital periods here range from hours to a few months.

Spectrum binary   An unresolved system in which spectral features do not reveal orbital motion but two clearly different spectra are superimposed. We infer that two members of a binary system are producing the observed composite spectrum.

Eclipsing binary   A binary system whose two stars periodically eclipse one another, leading to periodic changes in the apparent brightness of the system. Such systems may also be visual, astrometric, or spectroscopic binaries.

Finally, we point out that the fact that the Sun is a single star is an oddity. Among solar-type stars, the observed ratio of single : double : triple : quadruple systems is 45 : 46 : 8 : 1. For binary systems, orbital separations range uniformly from 3×10^9 to 3×10^{15} m (orbital periods from one day to 3×10^6 years). Roughly 10 percent of all stars are binaries with orbital periods of from one to ten days, another 10 percent with periods of from ten to 100 days, and so on.

In what follows, keep these ideas in mind. The *apparent orbit* is that traced out on the sky; most likely, it is tilted with respect to the line of sight. The *true orbit* is corrected for this tilt. When one of the stars is considered fixed while the other moves around it, the orbit is then the *relative orbit*. The *absolute orbit* is that traced out by both stars around the center of mass of the system.

12–2 Visual Binaries

As a result of the Earth's turbulent atmosphere, the seeing image of a star is seldom less than 1″ in diameter. The two stars of a binary system are easily resolved telescopically as a *visual binary* if their centers are separated by more than 1″. The members of a visual binary must be well separated in angle at some point in their orbital motion; otherwise, the duplicity will not be resolved. So the observed orbital periods are necessarily long (years to hundreds or thousands of years), as expected from Kepler's third law.

(a) The Determination of Stellar Masses

A single observation of a visual binary (Figure 12–1) is specified by giving the apparent angular *separation* (in arcseconds on the sky) and the *position angle* (angle measured eastward from north, in degrees) of the fainter star (the *companion*) relative to the brighter star (the *primary*). As time passes, these points trace out the *apparent relative orbit* of the binary system on the celestial sphere.

Two gravitating bodies orbit one another, as well as their center of mass, in accordance with Kepler's laws. So the orbit is an ellipse, and the orbital motion satisfies the law of equal areas and the third law. We do not generally see the true orbit, however, for the orbital plane of a binary system may be inclined at any angle to the plane of the sky. (The *inclination* is 0° when the two planes coincide and 90° when the orbit is seen edge-on.) Fortunately, the law of equal areas holds (but with a different constant of proportionality) for the apparent (projected on the sky) orbit and the elliptical true orbit always projects into an elliptical apparent orbit. The foci of the apparent orbit do not correspond to the true foci (in particular, the primary does not lie at one focus of the apparent ellipse). By measuring the displacement of the primary from the apparent focus, we can determine the inclination of the orbit to the local tangent plane on the celestial sphere; the true eccentricity and the true semimajor

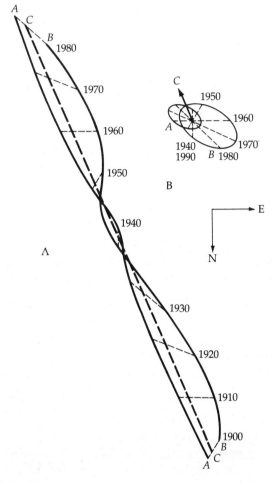

Figure 12–2

Motions of Sirius A and B. (A) The apparent motions relative to background stars of Sirius *(A)*, its companion *(B)*, and the center of mass of the system *(C)*. (B) Orbital motions of Sirius A and B relative to the system's center of mass.

axis *a″* (in arcseconds) can then be determined as well.

Having determined the true orbit of the visual binary, we may now apply Kepler's third law to deduce the masses of the member stars. The general form of the third law is

$$(M_1 + M_2)P^2 = a^3 \qquad \textbf{(12–1)}$$

where mass M is measured in solar masses $(M_\odot)$, the orbital period P is measured in years, and the true orbital semimajor axis a is measured in astronomical units. Although we may observe P directly, a follows from $a″$ only when we know the distance (or parallax

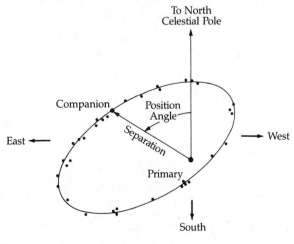

Figure 12–1

Apparent relative orbit of a visual binary. The orbit is apparent because it is projected on the sky. It is relative because one star (the primary) is assumed to be the center of the motion of the other (the companion).

π'') to the visual binary. Geometrically, we have (Figure 12–1)

$$a = a''/\pi'' \tag{12–2}$$

so that Equation 12–1 may be written in terms of observables as

$$(M_1 + M_2)P^2 = (a''/\pi'')^3 \tag{12–3}$$

An accurate value for the sum of the stellar masses follows from Equation 12–3. To determine the individual masses, we must find the relative distance of each star from the center of mass of the system because

$$M_1 a_1 = M_2 a_2 \quad \text{(where } a_1 + a_2 = a) \tag{12–4}$$

On the sky, the center of mass travels in a straight path with respect to the background stars, and the binary components weave periodically about this path (Figure 12–2). By eliminating the center-of-mass motion and correcting for the orbital inclination, we obtain a''_1 and a''_2 and therefore a''_1/a''_2, which equals a_1/a_2.

Let's illustrate how binary star masses are found by means of a hypothetical example. The two stars of a visual binary are observed to have a maximum separation of 3.0″ and a trigonometric parallax of 0.10″; the apparent orbit is completed in 30 years. Both stars orbit a common center of mass (Figure 12–2). Neither star orbits at the focus of the true orbits in space, but each can lie at the focus of the relative orbits. Here the primary coincides with the focus of that orbit. Then we are seeing the true orbit, and the sum of the stellar masses is $30M_\odot$ (from Equation 12–3).

$$M_1 + M_2 = (3.0/0.1)^3/30^2 = 30$$

The companion is observed to be five times farther from the center of mass than the primary, so that $a_1/a_2 = 1/5$ and thence (Equation 12–4)

$$M_1 \text{ (primary)} = 25M_\odot$$

$$M_2 \text{ (companion)} = 5M_\odot$$

(b) The Mass–Luminosity Relationship

Just as the determination of the period and size of the Earth's orbit (by Kepler's third law) leads to the Sun's mass, so also have we deduced binary stellar masses. Because it is necessary to know the distance

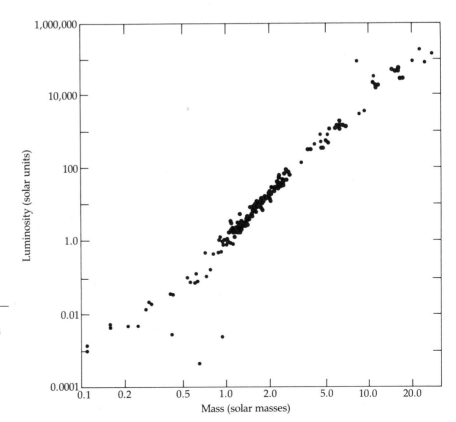

Figure 12–3

Mass–luminosity relationship. Masses and luminosities are shown here for stars in binary systems for which good values can be inferred. The stars off to the right of the trend line (at low luminosities) are white dwarfs made of degenerate rather than ordinary gases.

to the binary system in order to establish these masses, we need only observe the radiant flux of each star to find its luminosity.

When the observed masses and luminosities for stars in binary systems are graphed, we obtain the correlation (Figure 12–3) called the *mass–luminosity relationship* (or *M–L law*). In 1924, Arthur S. Eddington conjectured that the mass and luminosity of normal, main-sequence stars (Chapters 13 and 16) should be related by the equation

$$L/L_\odot = (M/M_\odot)^\alpha \qquad \textbf{(12–5)}$$

His first crude theoretical models indicated that $\alpha \approx 3$. On a log–log plot (Figure 12–3), Equation 12–5 graphs as a straight line with a slope of α. So main-sequence stars do seem to conform to Equation 12–5, although the exponent varies from $\alpha \approx 3$ for luminous and massive stars through $\alpha \approx 4$ for solar-type stars to $\alpha \approx 2$ for dim red stars of low mass. Today, astrophysical theories of stellar structure explain these results in terms of the different internal structures of stars of different mass and the opacities of stellar atmospheres at different temperatures. Note that the M–L law does not apply to highly evolved stars, such as red giants (with extended atmospheres) and white dwarfs (with degenerate matter; Figure 12–3). While most stellar masses lie in the narrow range from $0.05M_\odot$ to $50M_\odot$, stellar luminosities cover the vast span $10^{-4} \le L/L_\odot \le 10^6$!

12–3 Spectroscopic Binaries

If a binary system cannot be optically resolved at the telescope, its duplicity may show in its spectrum. Although orbital motion may not be detectable, we know that we are dealing with a *spectrum binary* when two different sets of line features are seen superimposed in the spectrum. A more useful, and interesting, case is the *spectroscopic binary*: here two stars orbit their center of mass closely (≤ 1 AU) and rapidly ($P \approx$ hours to a few months) and the orbital inclination is not $0°$.

The spectrum of a spectroscopic binary (Figure 12–4) exhibits lines that oscillate periodically in wavelength. If the companion is so dim that its spectral features are not detected, we have a *single-line spectroscopic binary* (Figure 12–4A); two stars of more nearly equal luminosity produce two sets of

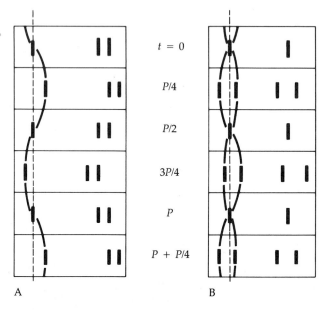

Figure 12–4

Spectra of spectroscopic binaries. (A) In a single-line system, only one set of lines shows an oscillation in wavelength from the Doppler shift. (B) In a double-line system, two sets of lines oscillate out of phase as the two stars revolve around their center of mass.

spectral features that oscillate in opposite senses (in wavelengths)—we call this system a *double-line spectroscopic binary* (Figure 12–4B). About a few thousand spectroscopic binaries are known, and good orbits have been determined for some 400.

(a) The Velocity Curve

To obtain useful information from the spectrum of a spectroscopic binary, we must interpret the behavior of the spectral lines. Because the two stars orbit in a plane inclined (at angle i) to the celestial sphere, that component of their velocity in the line of sight *Doppler-shifts* their spectral features. (Note that no Doppler shift can occur as a result of orbital motion alone when $i = 0°$; the system may then appear as a spectrum binary.) In addition, the center of mass of the system moves with respect to the Sun, so that the entire spectrum may be Doppler-shifted by some constant amount.

From Equation 8–13, the Doppler-shift formula is

$$\Delta\lambda/\lambda_0 \equiv (\lambda - \lambda_0)/\lambda_0 = v_r/c \qquad \textbf{(12–6)}$$

where λ_0 is the *emitted* wavelength (laboratory wavelength) of a spectral feature, λ is the *observed* wavelength, v_r is the radial speed (positive for recession, negative for approach) of the star, and $c = 3 \times 10^5$ km/s is the speed of light. Because of the finite width of spectral lines, we are limited at visible wavelengths to a shift resolution of $\Delta\lambda \geq 0.001$ nm; hence, the radial speed must be $v_r \geq 1$ km/s to be detectable. So the periods of observable spectroscopic binaries are necessarily short.

When we convert (using Equation 12–6) the Doppler shifts to radial velocities and plot the results as a function of time, we obtain the *velocity curve*. The simplest case is circular stellar orbits at the inclination $i = 90°$ (edge-on); the two curves (one for each star) are sinusoidal and oscillate with exactly opposite phases about the center-of-mass velocity in a period P (Figure 12–5). In this case, we find the distances to the center of mass by noting that in one period, the primary traverses the circumference $2\pi r_1$ at constant speed V. Hence, $VP = 2\pi r_1$ and

$$r_1 = VP/2\pi \quad \text{and} \quad r_2 = vP/2\pi \quad (12\text{–}7)$$

The ratio of stellar masses is

$$M/m = r_2/r_1 = v/V$$

the relative semimajor axis a is $r_1 + r_2$, and, from Equation 12–1, the sum of the stellar masses is

$$M + m = a^3/P^2$$

The individual stellar masses follow from their sum and ratio, and the dynamical characteristics of this spectroscopic binary are completely determined.

In general, this simple picture does not occur. If the case shown in Figure 12–5 were a single-line spectroscopic binary (only the primary spectrum is seen), we could determine only r_1 and the "mass function" $m^3/(M + m)^2$ (see the next section for details); a reasonable value for M might be obtained from the primary spectral type; then the system could be approximately deciphered. A greater difficulty is that, unless the system is also an *eclipsing* binary (Section 12–4), we have no idea what the orbital inclination is. If the velocity curve is purely sinusoidal, we know only that we are dealing with a circular orbit whose plane is tilted at *some* angle i to the celestial sphere. The amplitudes of the velocity curves give (by trigonometry) the *observed* (denoted by primes) circular speeds:

$$V' = V \sin i$$
$$v' = v \sin i$$

Hence, we may determine the mass ratio exactly since

$$M/m = r_2/r_1 = v/V = v'/V'$$

but only the *lower limit*, $a \sin i$, to the relative semimajor axis is accessible.

If the orbit is not circular but has an eccentricity e, the velocity curves are distorted from pure sinusoids (Figure 12–6). The double-line curves are mirror images of one another but have different amplitudes —an orbital inclination i merely reduces all radial

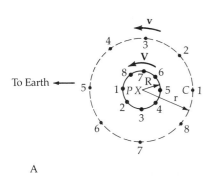

A

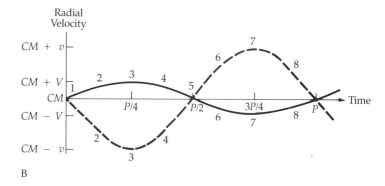

B

Figure 12–5

A binary velocity curve. (A) The primary *(P)* and companion *(C)* stars orbit the center of mass *(X)* in circular orbits of inclination 90°. (B) The center of mass is receding at a constant speed *(CM)* relative to the Sun, and the stars move at speeds V (primary) and v (companion) relative to the center of mass in one orbital period P.

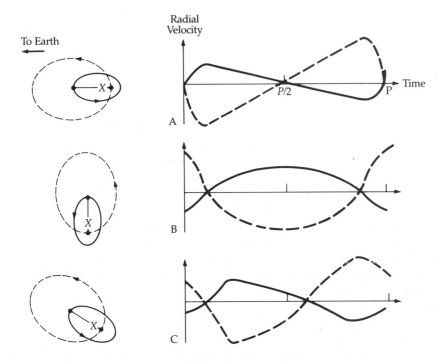

Figure 12–6

Velocity curves for elliptical orbits. The primary (mass M) and companion (mass m) have elliptical orbits of the same eccentricity but with semimajor axes in the ratio of M/m. The inclination is 90°, and the orbital period is P. (A) Major axes along the line of sight. (B) Major axes at 90° to the line of sight. (C) Major axes at 45° to the line of sight.

velocities by the same factor $\sin i$. The periodicity and characteristic shapes of these curves allow us to find P, e, and Ω (the orientation of the major axis with respect to the line of sight) immediately. When $i = 90°$, the relative semimajor axis and both stellar masses may be obtained.

(b) The Mass Function

The strong tidal interactions between the component stars in short-period spectroscopic binaries ($P \leq 10$ days) circularize their orbits (in about 10^8 years). Consider a circular (or small eccentricity, $e \ll 1$) relative orbit at inclination i. What can we say about the masses of the stars? Because we can obtain P, r_1', and r_2' (thence $a' = r_1' + r_2'$) from the velocity curve of a double-line binary, Kepler's harmonic law gives

$$(M + m) \sin^3 i = (a')^3/P^2$$

Recall that M/m was also found in this case. If we see only the primary in a single-line binary, we can find only the *mass function* $f(M, m)$ by

$$(M + m)P^2 = a^3 = (r_1 + r_2)^3 = r_1^3(1 + r_2/r_1)^3$$
$$= r_1^3(1 + M/m)^3 = (r_1')^3(M + m)^3/m^3 \sin^3 i$$

or

$$f(M, m) \equiv m^3 \sin^3 i/(M + m)^2 = (r_1')^3/P^2 \quad \textbf{(12–8)}$$

where r_1' and P are observed and we have used the relation $Mr_1 = mr_2$.

If the orbital inclination is unknown, of what use is the mass function? We cannot evaluate individual stellar masses, but by combining many data, *statistical masses* may be obtained. If the orbital planes are randomly distributed in i, then the mean value of $\sin^3 i$ is 0.59; however, we are more likely to detect spectroscopic binaries with $i \approx 90°$ (almost edge-on), and so we correct for this observational selection effect by assigning a somewhat larger value to the mean of $\sin^3 i$, say $\approx 2/3 \approx 0.67$. In addition, we have spectral information on the visible components, which can suggest appropriate masses.

12–4 Eclipsing Binaries

When the inclination of a binary orbit is close to 90°, each of the stars can eclipse the other periodically—we call these *eclipsing binaries*. A few thousand such systems are known; most are also spectroscopic binaries, and very few are visual binaries. For a relative orbit of radius ρ, tilted an angle ϕ

Figure 12–7

Eclipse geometry. (A) Front view from Earth; note that the companion must pass in front of the primary for an eclipse to occur. (B) Side view showing the permissible range of the companion for eclipses.

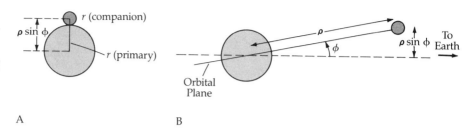

A   B

to the line of sight ($\phi = 90° - i$), an eclipse can occur (Figure 12–7) only when $\rho \sin \phi < R$ (primary) $+ R$ (companion), where R is the stellar radius. So small orbits are favored; because these small orbits have short periods and high orbital velocities, they imply spectroscopic binaries.

(a) Interpreting the Light Curve

Eclipsing binaries are most readily detectable by their periodically varying brightnesses. If we plot the magnitude or flux of such a binary as a function of time, we obtain the *light curve*, which generally exhibits two brightness minima of different depths (corresponding to the two possible eclipses per orbital period).

When the stars are not in eclipse, the observed brightness is essentially constant in time. The deeper minimum—*primary eclipse*—occurs when the hotter star passes behind the cooler star; the other eclipse—the *secondary*—is shallower. Several types of eclipses are possible: (1) when $i = 90°$, both the *total* eclipse (smaller star behind larger star) and the *annular* eclipse (smaller star in front) are termed *central*; (2) when $\rho \cos i < [R$ (primary) $- R$ (compan-

ion)], we still have total and annular eclipses; and (3) when $[R$ (primary) $- R$ (companion)$] < \rho \cos i < [R$ (primary) $+ R$ (companion)], only *partial* eclipses take place. Note that in every case exactly the same stellar area is covered at both primary minimum and secondary minimum, if the orbits are circular or $i = 90°$.

Consider the light curve associated with central eclipses and a circular relative stellar orbit for the situation where the larger star has a lower surface temperature than the smaller one (Figure 12–8). Four points (in the time during one eclipse) exist where the limbs of the two stars are tangent; we speak of *first contact* (t_1) when the eclipse begins, *second contact* (t_2) when brightness minimum is reached, *third contact* (t_3) when the smaller star begins to leave the disk of the larger star, and *fourth contact* (t_4) when the eclipse ends. Both primary and secondary minima are flat, and they occur exactly half an orbital period apart. If we denote the stellar radii by R_1 (larger star) and R_s (smaller star) and the relative orbital speed of the smaller star by v, the geometry implies

$$2R_s = v(t_2 - t_1) = v(t_4 - t_3) \quad \textbf{(12–9a)}$$

Figure 12–8

Central eclipses for circular orbits. The smaller star is assumed the hotter of the two. The four numbered contact points define the duration of the eclipse. These central eclipses have flat bottoms. (A) During secondary eclipse, the smaller star passes in front of the larger one. (B) During primary eclipse, one-half an orbital period later, the smaller star passes behind the larger one.

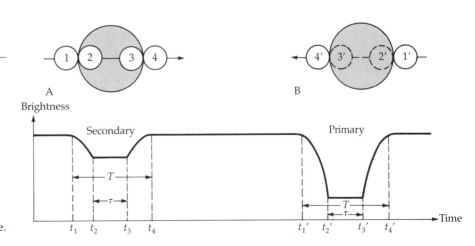

$$2(R_s + R_1) = v(t_4 - t_1) \qquad (12-9b)$$

However, the radius a of the relative circular orbit is

$$a = vP/2\pi \qquad (12-10)$$

where P is the orbital period. By combining Equations 12–9 and 12–10, we can determine the stellar radii relative to the orbital radius only:

$$R_s/a = \pi(t_2 - t_1)/P$$
$$R_1/a = \pi(t_4 - t_2)P \qquad (12-11)$$

Without consulting the stellar spectra, we may also determine the ratio of the effective surface temperatures of the two stars. Let these effective (blackbody) surface temperatures be denoted by T_1 and T_s. Surface brightness is equal to σT^4_{eff} by the Stefan–Boltzmann law (Chapter 8); since the same stellar area (πR_s^2) is covered at each eclipse minimum, the relative depths of the two eclipse minima give us $(T_1/T_s)^4$ directly — the hotter star is eclipsed at primary minimum.

When the eclipses are *partial* for a circular orbit (Figure 12–9), both eclipses are still of equal duration (though briefer than for central eclipse) and the brightness minima are not flat. Because the two eclipses still occur exactly half an orbital period apart, we know that the orbit is circular. In this case, it is possible to determine (1) the orbital inclination i, (2) the relative stellar radii R_s/a and R_1/a, and (3) the relative effective surface temperatures T_1/T_s.

For elliptical orbits, different times elapse from primary to secondary eclipse and from secondary to primary eclipse. Also, in general, the eclipse durations are not equal. These features permit us to determine the eccentricity e, orientation Ω, and inclination i of the orbit.

(b) Eclipsing-Spectroscopic Binaries

We have seen that light curves yield only relative results. It is indeed fortunate that most eclipsing binaries are also spectroscopic binaries, for we may determine speeds in kilometers per second from their velocity curves. For Equations 12–9 and 12–10, we find absolute values (in kilometers) for a, R_s, and R_1. Because the orbital inclination follows from the light curve, we can evaluate $\sin i$ and determine the stellar masses. By plotting these results, we obtain the mass–radius correlation for these stars (Figure 12–10). Mean stellar densities $\bar{\rho}$ may then be computed by

$$\bar{\rho}_1 = 3M/4\pi R_1^3$$

Knowing the stellar radii, we may find the ratio of stellar luminosities (from the effective temperature ratio) and the total luminosity of the system; the flux of the system then tells us the distance to the binary. Finally, we can infer the masses and luminosities of each star (Figure 12–11; see also Figure 12–3). Table 12–1 summarizes the various data we can obtain from binary stars.

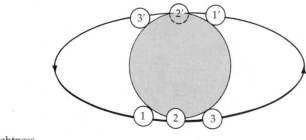

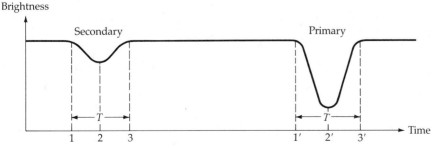

Figure 12–9

Partial eclipses for an inclined circular orbit. The smaller star is assumed the hotter, and the eclipses occur one-half a period apart, but note that the light curve is not flat during the eclipses.

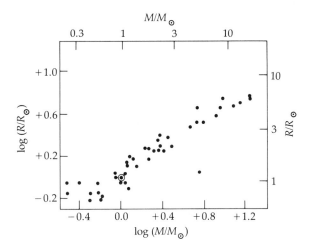

Figure 12-10

Main-sequence mass–radius relationship. Note that the more massive stars are larger.

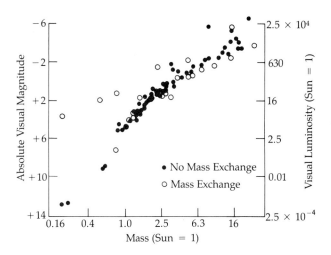

Figure 12-11

Mass–luminosity relationship for eclipsing binaries. The open circles represent systems that have mass flowing between the two stars. *[C. Lacy, Astrophysical Journal 213:458 (1977)]*

(c) Other Useful Information

Simple light curves do not tell the whole story of eclipsing binaries. Much useful information, including knowledge about stellar interiors, may be gleaned from the unusual light curves of more complicated systems (Figure 12-12).

From our experience with the Sun (Chapter 10),

we expect that stellar disks will exhibit *limb darkening* (Figure 12-12A). This feature rounds off the edges of the eclipses. In tight binary orbits, the hotter star will heat that part of the cooler star's atmosphere that is nearest (on the line between the stellar centers). This hotter gas is more luminous and leads to the *reflection*

Table 12-1   Stellar Data from Binary Systems

Type of Binary	Observations Performed or Needed	Parameters Determined
Visual	(a) Apparent magnitudes and π''	Stellar luminosities
	(b) P, a'', and π''	$\begin{cases} \text{Semimajor axis } (a) \\ \text{Mass sum } (M + m) \end{cases}$
	(c) Motion relative to CM	M and m
Spectroscopic	(a) Single-line velocity curve	Mass function $f(M, m)$
	(b) Double-line velocity curve	$\begin{cases} \text{Mass ratio } (M/m) \\ (M + m) \sin^3 i \\ a \sin i \end{cases}$
Eclipsing	(a) Shape of light curve eclipses	$\begin{cases} \text{Orbital inclination } (i) \\ \text{Relative stellar radii } (r_{l,s}/a) \end{cases}$
	(b) Relative times between eclipses	Orbital eccentricity (e)
	(c) Light loss at eclipse minima	Surface temperature ratio (T_l/T_s)
Eclipsing-Spectroscopic	(a) Light and velocity curves	$\begin{cases} \text{Absolute dimensions } (a, r_s, r_l) \\ e \text{ and } i \\ M \text{ and } m \text{ (also densities)} \end{cases}$
	(b) Spectroscopic parallax + apparent magnitude	$\begin{cases} \text{Distance to binary} \\ \text{Stellar luminosities} \\ \text{Surface temperatures } (T_l, T_s) \end{cases}$

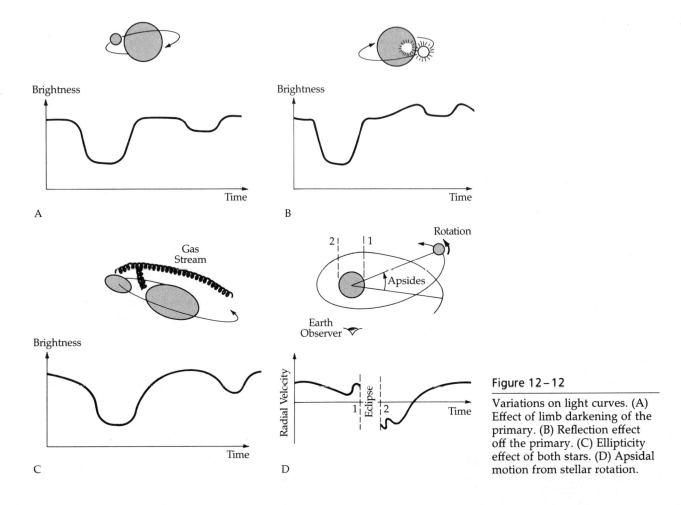

Figure 12-12

Variations on light curves. (A) Effect of limb darkening of the primary. (B) Reflection effect off the primary. (C) Ellipticity effect of both stars. (D) Apsidal motion from stellar rotation.

effect: just before and after secondary minimum, the system appears brighter than we would otherwise expect (Figure 12–12B).

In close binaries, the fluid (gaseous) bodies of the stars are distorted into prolate spheroids (like footballs) oriented along their lines of center so that they are also in synchronous rotation as a result of the strong tidal forces. This gives rise to the *ellipticity effect* (Figure 12–12C); the observed brightness varies continuously and not just during eclipses. Such systems are usually in circular orbit, also because of tidal effects. If the distortion is great enough, gas streams may be drawn from the stellar atmospheres into the space between and around the stars —the system Beta Lyrae is one famous example. This gas is most readily noticeable in spectra, for bright emission lines can be seen.

Stellar rotation shows up in two ways. The velocity curve exhibits anomalous bumps (Figure 12–12D) just before and just after each eclipse—in one

case we see the receding limb of the rotating star (because the rest is already eclipsed) and in the other we see the approaching limb. In the light curve, stellar rotation appears in the rotation of the direction of the semimajor axis of elliptical orbits (Figure 12–12D). Because a rotating star is oblate (flattened), its gravitational attraction is not that of a point mass; two such stars in a binary interact to change the orientation of their semimajor axes with time.

Finally, we expand the idea of close binaries to that of *contact binaries.* Eclipsing systems with extremely short periods—only a few hours—are in physical contact. Their light curves show this interaction because their maxima are rounded and their minima have almost the same depths. In such systems, the two stars share a common envelope of material and both are severely distorted by tidal effects.

We picture the interactions of these systems by considering the effective gravity at many points locally. The effective gravity results from the combina-

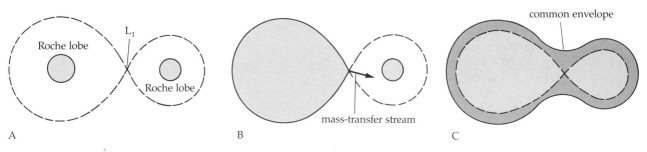

Figure 12–13

Contact binaries. (A) A detached system, in which both stars are smaller than their Roche lobes. (B) In a semidetached system, one star fills its Roche lobe and mass flows to the companion. (C) Both stars fill their Roche lobes in a contact system, and they are surrounded by a common envelope of uniform temperature.

tion of the real gravitational attractions and the centripetal force from orbital motions. If you explore the space around the stars, you will find a certain region, shaped like a figure 8, where the effective gravities of the two stars touch (L_1 in Figure 12–13A). Here the effective gravity is zero. Each half of the figure 8 indicates the regions controlled by the effective gravity of each star; these are called *Roche lobes.*

We can now classify close binaries on the basis of how large each star is relative to its Roche lobe. If both stars are smaller than their Roche lobes, the system is *detached* (Figure 12–13A). If one fills its Roche lobe, the system is *semidetached;* matter can flow through the contact point L_1 to the other star (Figure 12–13B). One star eats up the mass of the other. Finally, if both stars fill their Roche lobes, they are in *contact* and a common envelope of material enshrouds them both (Figure 12–13C).

12–5 Interferometric Stellar Diameters

Finally, we briefly mention two methods for determining stellar diameters that do *not* require a binary system. All such methods are basically *interferometric;* they depend upon the constructive and destructive interference of the light waves from a star (Section 8–1).

To see why such indirect procedures are necessary (except for our Sun), consider the following. At a distance of 1 pc, a star with a diameter of 1 AU (radius = $109R_\odot$) subtends an angle of 1.0″ on the sky, but this angle is just the same as the size of the seeing image (as a result of the Earth's turbulent at-

mosphere) seen at a telescope, so that the star's size is unresolvable. Besides that, no star is as close as 1 pc.

Within a band 10° wide centered on the ecliptic, stars may be occulted by our Moon. In such a *lunar occultation,* the star does not disappear instantaneously; instead it fades away in a few seconds (Figure 12–14). Electromagnetic wavefronts from the star are progressively screened out as time goes on,

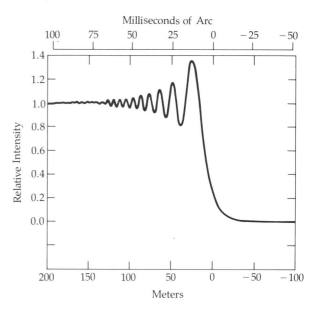

Figure 12–14

Occultation of a star by the Moon. As the limb of the Moon cuts in front of the star, a diffraction pattern appears before the light is completely cut out. The top scale is an angular one, the bottom a linear one.
[S. Ridgeway et al., Astronomical Journal 84:247 (1979)]

but the unobstructed portions interfere (actually, diffract) to produce a characteristic intensity-versus-time pattern at the Earth. This pattern depends directly upon the angular size of the star; if we know the stellar distance, we can deduce the stellar diameter.

In the 1920s, A. E. Michelson invented and used a *stellar interferometer* to measure the angular diameters of large nearby stars. In this device, widely separated (many meters) mirrors deflect the starlight to an ordinary focusing telescope, where the wavefronts from different parts of the star produce a characteristic interference pattern (Figure 12–15). This pattern depends upon the angle between the wavefronts from opposite limbs of the star; this intersection angle increases as the stellar angular diameter increases. So to find a star's angular diameter, we use the simple small-angle approximation:

$$\alpha = D/d$$

where α is in radians, D is the star's physical diameter, and d is its distance (D and d are in the same units).

One star measured this way is Aldebaran (Alpha Tauri). The results average about 21 milliarcseconds for the angular diameter. The distance d of Aldebaran is 20.8 pc. Then

$$D = \alpha d$$

with 21×10^{-3} arcsecond $= 1.02 \times 10^{-7}$ rad, so that

$$D = (1.02 \times 10^{-7})(67.8 \text{ lightyears})$$
$$= 6.92 \times 10^{-6} \text{ lightyears}$$
$$= 6.54 \times 10^{7} \text{ km}$$

So Aldebaran's radius is 3.27×10^{7} km, about 50 times larger than the Sun's radius.

Another type of interferometry, called *speckle interferometry*, exploits the unavoidable seeing effects of the Earth's atmosphere. The image of a star observed through a telescope and the atmosphere has a diffraction pattern that splits up into many bright

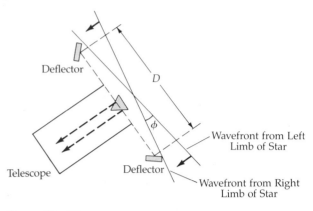

Figure 12–15

Schematic of a Michelson interferometer. The widely spaced mirrors send starlight into the telescope to produce the interference patterns. Light waves from opposite ends of a star arrive simultaneously, intersecting one another at angle ϕ, which is the angular diameter of the star.

granules, or speckles, because of the random interference of light traveling through different atmospheric cells. The result is different phases of the incoming waves. Because the speckle effect is a function of wavelength, observations must be made at a narrow band of wavelengths. In practice, many short exposures of the speckle pattern are made rapidly. These are then averaged and the speckle pattern analyzed by Fourier techniques.

Speckle interferometry not only provides resolved images of single stars; it also offers a way to resolve the components in very close multiple star systems. This technique is especially useful for spectroscopic binaries, where measurements of the radial velocity curve combined with the angular separation most directly determine the stellar masses. The method is fast and accurate. On a large telescope, a few hundred stars can be observed in one night and errors are typically a few milliarcseconds.

Problems

1. From the information given in Section 12–2(a), show that Kepler's harmonic law holds for the apparent orbit of a visual binary.

2. Demonstrate the correctness of Equation 12–2. Use a diagram if necessary.

3. What is the sum of the stellar masses in a visual binary of period 40 years, maximum separation 5.0", and parallax 0.3"? Assume an orbital inclination of zero and a circular apparent orbit.

4. Find the distance in parsecs to a visual binary that

consists of main-sequence stars of absolute bolometric magnitudes of $+5.0$ and $+2.0$. The mean angular separation is 0.05″, and the observed orbital period is ten years. What assumptions have you made to arrive at your answer?

5. Show that binary systems with small orbits have high orbital speeds.

6. The velocity curves of a double-line spectroscopic binary are observed to be sinusoidal, with amplitudes of 20 km/s and 60 km/s and a period of 1.5 years.
    (a) What is the orbital eccentricity?
    (b) Which star is the more massive, and what is the ratio of stellar masses?
    (c) If the orbital inclination is 90°, find the relative semimajor axis (in astronomical units) and the individual stellar masses (in solar masses).

7. An eclipsing binary has an orbital period of 2^d22^h, the duration of each eclipse is 18^h, and totality lasts 4^h.
    (a) Find the stellar radii in terms of the circular orbital radius a.
    (b) If spectroscopic data indicate a relative orbital speed of 200 km/s, what are the actual stellar radii (in kilometers and solar radii)?

8. The surface temperature of one component of an eclipsing binary is 15,000 K, and that of the other is 5000 K. The cooler star is a giant with a radius four times that of the hotter star.
    (a) What is the ratio of the stellar luminosities?
    (b) Which star is eclipsed at primary minimum?
    (c) Is primary minimum a total or an annular eclipse?
    (d) Primary minimum is how many times deeper than secondary minimum (in energy units).

9. The star Sirius A has a surface temperature of 10,000 K, a radius of $1.8R_\odot$, and $M_{bol} = 1.4$; the radius of its white dwarf companion, Sirius B, is $0.01R_\odot$ and $M_{bol} = 11.5$.
    (a) What is the ratio of their luminosities?
    (b) What is the ratio of their effective temperatures?
    (c) If they orbit at $i = 90°$, which star is eclipsed at primary minimum?
    (d) If your photometer can measure magnitudes to an accuracy of ≈ 0.001, would you be able to detect the hypothetical primary eclipse? (*Hint:* Use $\log_{10}(1 + x) = x/2.3$ for $x \ll 1$.)

10. Derive an expression giving the stellar angular diameter in milliarcseconds when the actual stellar diameter (in solar radii) and distance from us (in parsecs) are known.

Stars:
The
Hertzsprung – Russell
Diagram

We infer the properties of stars from their light —in bulk for fluxes and spread out for spectra. This chapter deals with the wealth of information that can be discerned by studying *stellar spectra.* First we consider stellar atmospheres, for here the stellar spectra originate. Then we tell the story of spectral observations—how they have been made, correlated, and interpreted. Finally, we present that famous and crucial synthesis—the Hertzsprung–Russell diagram —and some of its implications. This discussion will lead to an understanding of the stars themselves (Chapter 16).

13–1 Stellar Atmospheres

The spectral energy distribution of starlight is determined in a star's *atmosphere,* the region from which radiation can freely escape. To understand stellar spectra, we first discuss a model stellar atmosphere and investigate the characteristics that determine the spectral features.

(a) Physical Characteristics

The stellar photosphere, a thin, gaseous layer, shields the stellar interior from view. The photosphere is thin relative to the stellar radius, and so we regard it as a *uniform* shell of gas. The physical properties of this shell may be approximately specified by the average values of its *pressure P, temperature T,* and *composition μ* (chemical abundances).

We make the reasonable assumption that the number density n (number/m³) of gas particles (molecules, atoms, ions, and electrons) is high enough so that *thermodynamic equilibrium* holds. This means that particles collide so frequently that both the Boltzmann and the Saha equation apply. We also assume that the gas obeys the *perfect-gas law:*

$$P = nkT \qquad (13-1)$$

where k is Boltzmann's constant. The particle number density is related to both the *mass density* ρ (kg/m³) and the composition (or *mean molecular weight*) μ by the following definition of μ:

$$1/\mu = m_H n/\rho \qquad (13-2)$$

where $m_H = 1.67 \times 10^{-27}$ kg is the mass of a hydrogen atom. For a star of pure atomic hydrogen, $\mu = 1$. If the hydrogen is ionized, $\mu = 1/2$ because electrons

and protons (hydrogen nuclei) are equal in number and electrons are far less massive than protons. In general, stellar gases are ionized and $1/\mu \approx 2X + (3/4)Y + (1/2)Z \approx 2$, where X is the *mass fraction* of hydrogen, Y is that of helium, and Z is that of all heavier elements. The mass fraction is the percentage by mass of one species relative to the total.

We also assume a steady-state atmosphere: although the individual gas particles move about rapidly, nothing changes with time on the macroscopic scale (there are no mass motions). This implies *hydrostatic equilibrium,* wherein a typical volume of gas experiences no net force. We have already derived this relationship in Chapter 4 and so simply repeat the equation here:

$$dP/dr = -(GM/R^2)\rho = -g\rho \qquad (13-3)$$

where g is the gravitational acceleration (m/s²), or gravity, at the photosphere. Note that the pressure decreases continuously outward through the star.

By parameterizing the pressure in terms of optical depth τ (Chapter 10) instead of radius r, where $d\tau = -\kappa\rho\, dr$ (κ is the atmospheric opacity in m²/kg), Equation 13–3 takes the useful form

$$dP/d\tau = g/\kappa \qquad (13-4)$$

In our uniform approximation, then, we integrate Equation 13–4 to

$$P = (g/\kappa)\tau \qquad (13-5)$$

so that the atmospheric gas pressure depends upon g and κ. We take $\tau = 1$ as the atmospheric level where spectral features are formed because this opacity is the minimum needed for line formation.

(b) Color Temperatures

From the pressure and composition of the stellar atmosphere, we now turn to the temperature. We have already noted (Chapter 8) that the continuous spectrum, or continuum, from a star may usually be well approximated by the Planck blackbody spectral-energy distribution. For a given star, the continuum defines an effective, or color, temperature by the appropriate Planck curve. We use the word *color* because of Wien's displacement law:

$$\lambda_{max} T = 2.898 \times 10^{-3} \text{ m} \cdot \text{K} \qquad (13-6)$$

which states that the peak intensity of the Planck curve occurs at a wavelength λ_{max} that varies inversely with the Planck temperature T. So most of the radiation intensity occurs at wavelengths near λ_{max}, and the star has the *color* appropriate to λ_{max}. This implies that hot stars (smaller λ_{max}) are *bluer* than cool stars (larger $\lambda_{max} \rightarrow$ redder). Also note here that the hotter a star is, the greater will be its *surface brightness* (or luminous flux in W/m²), in accordance with the Stefan–Boltzmann law:

$$F = \sigma T^4 \qquad (13-7)$$

where $\sigma = 5.67 \times 10^{-8}$ W/m² · K⁴.

A word of caution: the effective Planck temperature of a star is usually not identical to its spectral-line temperature because spectral-line formation depletes radiation from the continuum. This effect is called *line blanketing* (Figure 13–1) and becomes important when the numbers and strengths of spectral lines are large. When spectral features are not numerous, we can detect the continuum between them and obtain a reasonably accurate value for the star's effective surface temperature.

(c) Spectral-Line Formation

Chapter 8 describes how spectral absorption features are formed when the molecules, atoms, and ions of a gas absorb continuum photons and re-emit

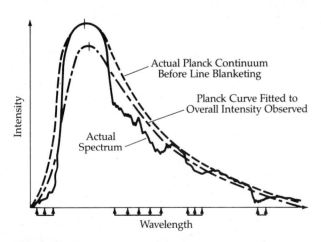

Figure 13–1

Line blanketing. The color temperature of a star should correspond to its Planck continuum (dashed line), but absorption lines (arrows) blanket the continuum to produce the observed spectrum (solid line). A Planck curve fitted to the observed intensity (dotted line) results in too low a color temperature.

fewer of these photons toward the observer. The composition of the gas determines which species are available to absorb the photons, and the temperature and pressure determine which spectral features are formed. For example, molecular spectral features can originate only in a cool gas, for molecules are easily dissociated by collisions with other particles; neutral atoms and their spectral lines predominate at intermediate temperatures; at high temperatures, all species are ionized so that only the spectral features arising from ions are seen.

Let's put these statements on a more quantitative basis. Take a cubic meter of gas in which the number of particles of each elemental species is specified by the composition μ. Now consider the particles of a particular element (such as hydrogen). Continuum photons at discrete wavelengths are absorbed when the particles (neutral atoms or ions) are excited from one atomic energy level to another—the strength of each absorption feature is essentially proportional to the number of particles populating a given energy level (*equivalent width*, Chapter 8). The relative number of atoms in energy levels B (N_B) and A (N_A), where B > A, is determined by the *Boltzmann excitation-equilibrium equation* (Section 8–4):

$$N_B/N_A \propto \exp[(E_A - E_B)/kT] \qquad (13-8)$$

where E is the energy of the level and T is the gas temperature. At a given temperature, the higher levels are *less* populated than the lower levels, but the relative number of *ions* in a given ionization stage i (i = the number of electrons lost by an atom) is determined by both the temperature and the electron number density by the *Saha ionization-equilibrium equation* (Section 8–4):

$$N_{i+1}/N_i \propto [(kT)^{3/2}/N_e] \exp(-\chi_i/kT) \qquad (13-9)$$

where χ_i is the ionization potential from stage i to stage $i + 1$.

It is useful to express Equations 13–8 and 13–9 in logarithmic form; that is, we take the base-10 logarithm of both sides of each equation. Noting that $\log_{10} e = 0.4343$ and expressing T in kelvins and all energies in electron volts, we have, from Equation 13–8,

$$\log(N_B/N_A) = (-5040/T)|E_A - E_B| \\ + \text{constant} \qquad (13-10)$$

and from Equation 13–9,

$$\log(N_{i+1}/N_i) = (3/2) \log T - (5040/T)\chi_i$$
$$- \log N_e + \text{constant}' \qquad \textbf{(13–11)}$$

In this way, the dependence upon each parameter is clearly manifested. Since the electrons also constitute a perfect gas, we may write Equation 13–11 in terms of the *electron pressure* P_e (where $P_e \propto N_e T$) as

$$\log(N_{i+1}/N_i) = (5/2) \log T - (5040/T)\chi_i$$
$$- \log P_e + \text{constant}'' \qquad \textbf{(13–12)}$$

The Boltzmann and Saha equations must be combined to deduce the population of every energy level and thence the spectral-line strengths. Figure 8–13 shows that the temperature behavior of the combination is the following. At low temperatures, the atoms are all neutral and in their ground states. As the temperature rises, the higher energy levels of the neutral atoms become more populated until $T \approx \chi_0/k$, when single ionization produces a significant number of ions (depleting the number of neutrals). At still higher temperatures, these ions predominate, but when $T \approx \chi_1/k$, double ionization (two electrons lost) becomes important. At the top of the ionization stages, the temperature is so high that only bare nuclei and free electrons remain—the atoms are fully ionized—and no more spectral-line absorption can take place. You should remember that this sequence is strongly dependent upon the exact energy-level structure of each atomic species, so that the spectral features produced at any temperature uniquely characterize the species.

The following verbal flow chart expresses the spectral-line behavior to be expected of a given stellar atmosphere. Denoting the strength of a spectral feature by its equivalent width w, we have the functional dependence

$$w = w(n, T)$$

since the rate of absorption is determined by the number density n of atoms and the intensity of the continuum, which relates to the temperature T. However, the equation of state of the gas (Equation 13–1) and the composition of the gas (Equation 13–2) imply that

$$n = n(P, T, \mu)$$

Hydrostatic equilibrium (Equations 13–3 and 13–4) determines the pressure:

$$P = P(M, R, \kappa)$$

The opacity κ is clearly a function of number density, gas composition, and ionization–excitation state determined by T:

$$\kappa = \kappa(n, T, \mu)$$

while the star's luminosity L depends upon its temperature and radius through the relationship

$$L = 4\pi R^2 \sigma T^4 \qquad \textbf{(13–13)}$$

If we combine the dependences of these five equations, then in general

$$w = w(L, T, M, \mu) \qquad \textbf{(13–14)}$$

For stars of a given composition, Equation 13–14 reduces to

$$w = w(L, T, M) \qquad \textbf{(13–15)}$$

and if we consider, for example, only stars with a unique mass–luminosity relation, we have

$$w = w(L, T) \qquad \textbf{(13–16)}$$

Equation 13–16 is the theoretical justification for seeking temperature sequences and two-dimensional Hertzsprung–Russell diagrams in the remainder of this chapter: Equation 13–14 covers those cases in which stellar luminosity is not related to stellar mass and those in which compositional differences are important. We note, however, that Hertzsprung–Russell diagrams were first arrived at observationally; the theoretical understanding came later.

13–2 Classifying Stellar Spectra

(a) Observations

A single stellar spectrum is produced when starlight is focused by a telescope onto a *spectrometer* or *spectrograph,* where it is dispersed (spread out) in wavelength and recorded photographically or electronically. If the star is bright, we may obtain a *high-dispersion* spectrum, that is, few nanometers per millimeter on the spectrogram, because there is enough radiation to be spread broadly and thinly (the solar spectrum of Figure 13–2 is a good example). At high dispersion, a wealth of detail appears in the spectrum, but the method is slow (only one stellar spectrum at a time) and limited to fairly bright stars. Dis-

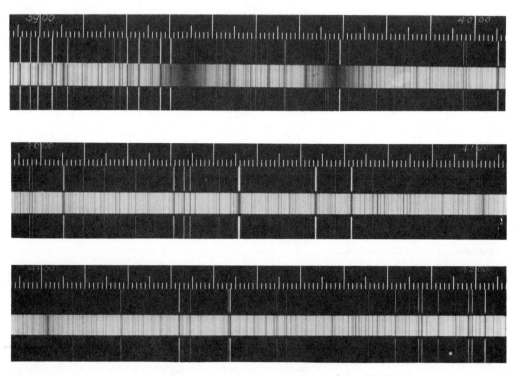

Figure 13–2

A solar spectrum. This high-dispersion spectrum (0.5 Å/mm) shows a wealth of detail. The bright lines above and below the absorption-line spectrum are used for wavelength calibration. *(Mount Wilson Observatory, Carnegie Institute of Technology)*

persion is the key to unlocking the information in starlight.

(b) The Spectral-Line Sequence

At first glance, the spectra of different stars seem to bear no relationship to one another. In 1863, however, Angelo Secchi found that he could crudely order the spectra and define different *spectral types.* Alternative ordering schemes appeared in the ensuing years, but the system developed at the Harvard Observatory by Annie J. Cannon and her colleagues was internationally adopted in 1910. This sequence, the *Harvard spectral classification system,* is still used today. (About 400,000 stars were classified by Cannon and published in various volumes of the *Henry Draper Catalogue,* 1910–1924, and its Extension, 1949. The catalog is now being reobserved for updated classifications by astronomers at the University of Michigan, headed by Nancy Houk.)

At first, the Harvard scheme was based upon the strengths of the hydrogen Balmer absorption lines in stellar spectra, and the spectral ordering was alphabetical (A through P). The A stars had the strongest Balmer lines, and the P stars had the weakest. Some letters were eventually dropped, and the ordering was rearranged to correspond to a sequence of decreasing temperatures (recall the effects of the Boltzmann and Saha equations): OBAFGKM. Stars nearer the beginning of the spectral sequence (closer to O) are sometimes called *early-type* stars, and those closer to the M end are referred to as *late-type.* Each spectral type is divided into ten parts from 0 (early) to 9 (late); for example, . . . F8 F9 G0 G1 G2 . . . G9 K0. . . . In this scheme, our Sun is spectral type G2.

Many mnemonics have been devised to help students retain the spectral sequence. We urge you to devise your own. A variation of the traditional one is "Oh, Be a Fine Guy, Kiss Me." Our students have come up with others, including such masterful phrases as "Only Bold Astronomers Forge Great Knowledgeable Minds" and "Optical Binary Affairs Fundamentally Generate Keplerian Marriages."

Figure 13–3A shows exemplary stellar spectra

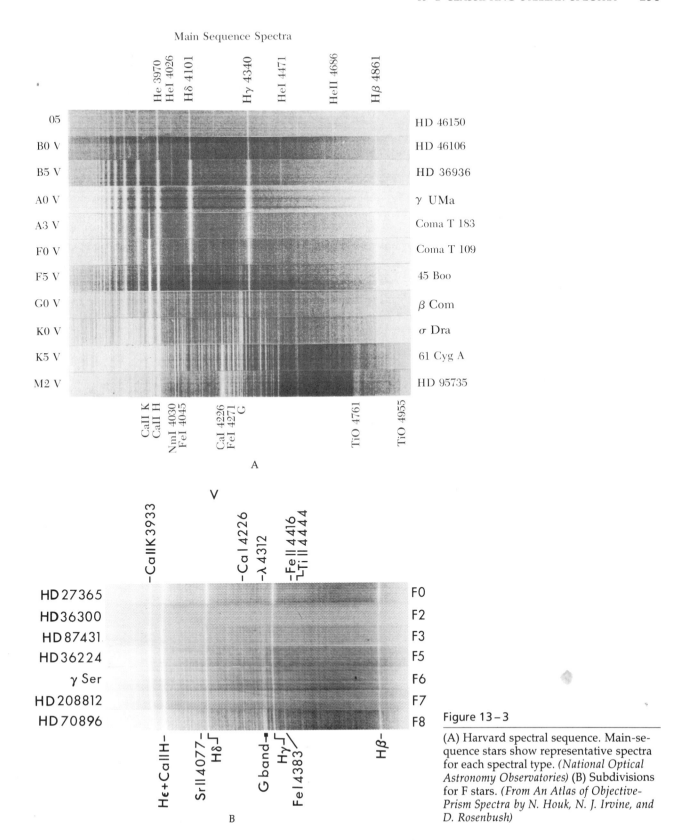

Figure 13-3

(A) Harvard spectral sequence. Main-sequence stars show representative spectra for each spectral type. *(National Optical Astronomy Observatories)* (B) Subdivisions for F stars. *(From An Atlas of Objective-Prism Spectra by N. Houk, N. J. Irvine, and D. Rosenbush)*

arranged in order; note how the conspicuous spectral features strengthen and diminish in a characteristic way through the spectral types. Table 13–1 summarizes those spectral characteristics that define each spectral type. Study this table and Figure 13–3B shows the subdivisions for the F spectral type.

(c) The Temperature Sequence

The spectral sequence is by definition a temperature sequence, but we must carefully qualify this statement. There are many different kinds of temperatures and many ways to determine them. In Figure 13–4, the strengths of various spectral features are plotted against excitation–ionization (or Boltzmann–Saha) temperature; the spectral sequence does correlate with this temperature.

Theoretically, the Planck *color* temperature should correlate with spectral type. From the spectra of intermediate-type stars (A to K), we find that the (continuum) color temperature does so, but difficulties occur at both ends of the sequence. For O and B stars, the continuum peaks in the far ultraviolet, where it is undetectable by ground-based observations. Through satellite observations in the far ultraviolet, we are beginning to understand the ultraviolet spectra of O and B stars. For the cool M stars, not only does the Planck curve peak in the infrared, but numerous molecular bands also blanket the spectra of these dim stars.

Table 13–1   The Harvard Spectral Sequence

Spectral Type	Principal Characteristics	Spectral Criteria
O	Hottest bluish stars; relatively few lines; He II dominates	Strong He II lines in absorption, sometimes emission; He I lines weak but increasing in strength from O5 to O9; hydrogen Balmer lines prominent but weak relative to later types; lines of Si IV, O III, N III, and C III
B	Hot bluish stars; more lines; He I dominates	He I lines dominate, with maximum strength at B2; He II lines virtually absent; hydrogen lines strengthening from B0 to B9; Mg II and Si II lines
A	Bluish stars; ionized metal lines; hydrogen dominates	Hydrogen lines reach maximum strength at A0; lines of ionized metals (Fe II, Si II, Mg II) at maximum strength near A5; Ca II lines strengthening; lines of neutral metals appearing weakly
F	White stars; hydrogen lines declining; neutral metal lines increasing	Hydrogen lines weakening rapidly while H and K lines of Ca II strengthen; neutral metal (Fe I and Cr I) lines gaining on ionized metal lines by late F
G	Yellowish stars; many metal lines; Ca II lines dominate	Hydrogen lines very weak; Ca II H and K lines reach maximum strength near G2; neutral metal (Fe I, Mn I, Ca I) lines strengthening while ionized metal lines diminish; molecular G band of CH becomes strong
K	Reddish stars; molecular bands appear; neutral metal lines dominate	Hydrogen lines almost gone; Ca lines strong; neutral metal lines very prominent; molecular bands of TiO begin to appear by late K
M	Coolest red stars; neutral metal lines strong; molecular bands dominate	Neutral metal lines very strong; molecular bands prominent, with TiO bands dominating by M5; vanadium oxide bands appear

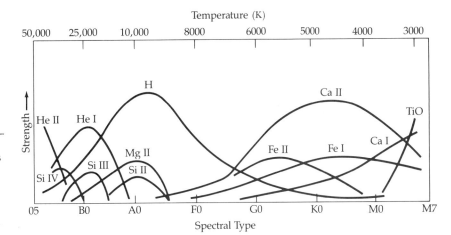

Figure 13–4

Absorption lines and temperature. The strengths of the absorption lines for various ionic species are shown as a function of stellar temperature. These changes result in ionization–excitation equilibria as described by the Boltzmann–Saha equation.

In practice, we measure a star's *color index, CI = B − V*, to determine the effective stellar temperature (Chapter 11). If the stellar continuum is Planckian and contains no spectral lines, this procedure clearly gives a unique temperature (Figure 13–5), but observational uncertainties and physical effects do lead to problems: (a) for the very hot O and B stars, *CI* is small and negative and small uncertainties in its value lead to very large uncertainties in *T*; (b) for the very cool M stars, *CI* is large and positive, but these faint stars have not been adequately observed and so *CI* is not well determined for them; (c) any instrumental deficiencies, calibration errors, or unknown blanketing in the *B* or *V* bands affect the value of *CI*—and thus the deduced *T*.

13–3 Hertzsprung–Russell Diagrams

Examine Figure 13–6. In 1911, Ejnar Hertzsprung plotted the first such two-dimensional diagram (absolute magnitude versus spectral type) for observed stars, followed (independently) in 1913 by Henry Norris Russell; today, this plot is called a *Hertzsprung–Russell (H–R) diagram*. As will soon become clear to you, this simple diagram represents one of the great observational syntheses in astrophysics.

(a) Magnitude Versus Spectral Type

The first H–R diagrams considered stars in the solar neighborhood and plotted *absolute magnitude M* versus *spectral type Sp*, which is equivalent to lumi-

nosity versus spectral type or luminosity versus temperature.

Figure 13–7 shows this type of plot for the stars with well-determined distances within about 5 pc of the Sun. Note (a) the well-defined main sequence (class V) with ever-increasing numbers of stars toward later spectral types and an absence of spectral classes earlier than A1 (Sirius), (b) the absence of

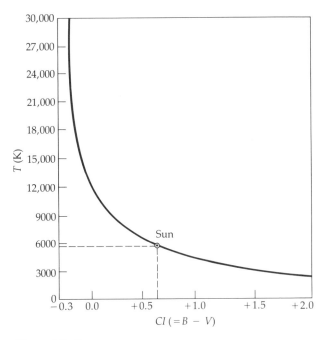

Figure 13–5

Blackbody temperature and color index. The *B* minus *V* color index relates to the blackbody temperature of a star.

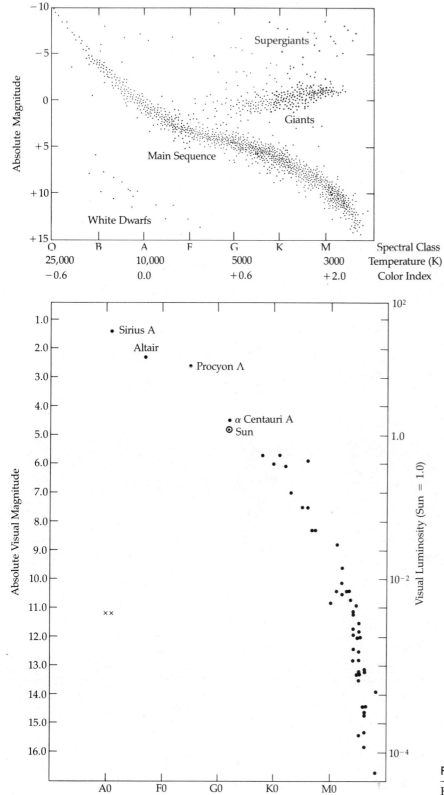

Figure 13–6

Hertzsprung–Russell diagram for bright stars. This information is biased because the brighter stars are easier to observe, and so it does not represent the stars in the Milky Way overall. *(Adapted from G. O. Abell, Exploration of the Universe, 4th edition, Saunders, 1982)*

Figure 13–7

Hertzsprung–Russell diagram for stars within 5 pc of the Sun.

giants and supergiants (classes III and I), and (c) the few white dwarfs at the lower left. In contrast, the H–R diagram for the brightest stars includes a significant number of giants and supergiants as well as several early-type main-sequence stars (Figure 13–8). Here we have made a selection that emphasizes very luminous stars at distances far from the Sun. Note that the H–R diagram of the nearest stars is most representative of those throughout the Galaxy: the most common stars are low-luminosity spectral type M.

Finally, Figure 13–9 presents an H–R diagram (as well as actual numbers at each spectral type) for 36,000 stars. Note that the low-luminosity stars now appear. (White dwarfs are missing from these last two figures because of their very low luminosities.)

(b) Magnitude Versus Color

Because stellar colors and spectral types are roughly correlated, we may construct a plot of absolute magnitude versus color—called a *color–magnitude H–R diagram*. The relative ease and convenience with which color indices (such as $B - V$) may

be determined for vast numbers of stars dictates the popularity of color–magnitude plots. The resulting diagrams are very similar to the magnitude–spectral type H–R diagrams considered above. Let's see what information we can glean from them.

(c) Stellar Populations

There are two extreme stellar populations: the young, metal-rich Population I and the old, metal-poor Population II. How has this been determined? Consider a stellar cluster that, either by its appearance or by the common motion through space of its member stars, is known to be a self-gravitating group of stars formed at about the same time. In addition, the distance to every cluster member is about the same, so that a plot of apparent magnitude versus color is an H–R diagram. Figure 13–10 shows such a color–magnitude diagram for the Pleiades, a young *open* (or *galactic*) *cluster* in the constellation Taurus. Notice (a) the well-defined main sequence, (b) the absence of giants (luminosity classes I to IV), (c) the curving up of the early end of the main sequence, and (d) the few subdwarfs (erroneously included because

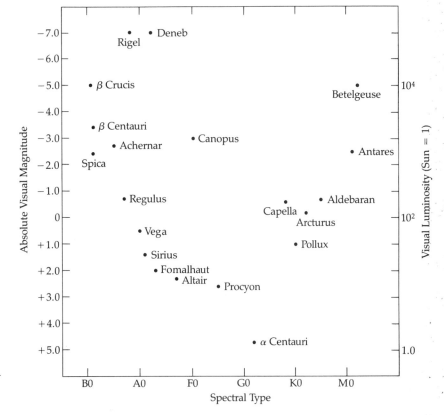

Figure 13-8

Hertzsprung–Russell diagram for the brightest stars in the sky.

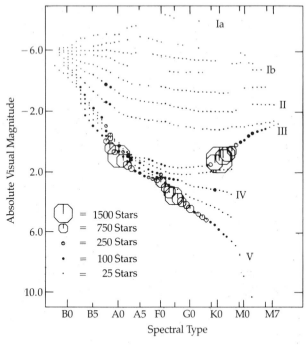

Figure 13–9

Hertzsprung–Russell diagram from the Michigan Spectral Catalog. The size of the symbol represents the number of stars of that spectral type. *(N. Houk)*

of the problems inherent in determining which stars *actually belong* to the cluster). The stellar spectra exhibit high metal abundance ($Z \approx 0.01$)—they are Population I. Figure 13–11 shows the color–magnitude (again, apparent magnitude) diagram of an older (about 100 million years) open cluster called the Hyades. Again we see the main sequence, but with few stars below $B - V \approx 0.2$; in addition, there are several stars in the giant region in this Population I cluster.

Our Galaxy contains many *globular clusters,* which are extremely compact and spherically symmetric balls of old (12 to 15 billion years) stars (up to 500,000 stars in some). These clusters are found at great distances from the central plane of our Galaxy. Figure 13–12 is the color–magnitude diagram of the globular cluster M 3 (the third object listed in Charles Messier's catalog of the year 1781). The stellar spectra reveal a very low metal abundance ($Z \leq 0.001$), and so we assign the stars to Population II. Notice the main sequence running from $B - V = 0.8$ up to the turnoff at $B - V \approx 0.4$, the heavily populated giant

branch, and the high-luminosity branch running toward the left.

So we have two well-defined stellar populations exemplified by these clusters. Population I stars (open clusters) have a higher metal abundance than do Population II stars (globular clusters). As discussed in Chapter 16, the color–magnitude diagram can be understood in terms of stellar evolution in the solar neighborhood: the main sequence represents young (Population I) stars; the giants are older, more-evolved stars (Populations I and II); and the white dwarfs are stars at the end point of stellar evolution. The subdwarfs are Population II interlopers passing through the solar neighborhood. Note that Population II is associated with old stars.

(d) Luminosity Classifications

So far, we have been discussing *one-dimensional* temperature sequences. As shown in Section 13–1, this description can represent stars only if they all have the same mass, radius, and composition. Chapter 11 shows, however, that stars at a given temperature clearly differ in their luminosities, and so a *two-dimensional* (L, T) representation is absolutely necessary (Equation 13–16). Over 90 percent of the stars in the solar neighborhood do define a single band—the *main sequence*—on such an (L, T) plot, but many stars do not reside on the main sequence.

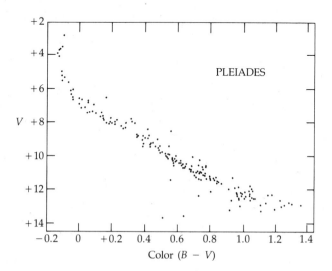

Figure 13–10

Color–magnitude diagram for the Pleiades. *(H. L. Johnson and R. I. Mitchell)*

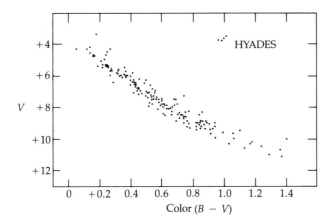

Figure 13-11

Color–magnitude diagram for the Hyades. Because the Hyades cluster is older than the Pleiades, some red giants appear in this H–R diagram. *(H. L. Johnson)*

As early as 1897, Antonia Maury at Harvard recognized distinctly different spectra for stars of a given color temperature; certain spectral features were successively sharper (that is, narrower) than for main-sequence stars. In 1905–1907, Hertzsprung confirmed that the narrow-line stars are much more luminous than the corresponding main-sequence stars. Between 1914 and 1935, the Mount Wilson luminosity classification appeared. This scheme, originated by W. S. Adams and A. Kohlschutter, orders stellar spectra (of the same temperature!) according to the strengths or weaknesses of certain spectral features. After 1937, W. W. Morgan and P. C. Keenan at Yerkes Observatory introduced the currently used *M–K luminosity classification* scheme, which defines six stellar luminosity classes and their subclasses directly in terms of observational criteria. Let's describe the M–K scheme in detail.

Before we do so, a word about the philosophy behind the M–K system and its limitations. Morgan and Keenan tried to devise an empirical system based only on the observable features in stellar spectra. They applied their classification to a homogeneous group of stars and did so through a group of standards to which other stellar spectra could be carefully compared. These standards seem to fall into natural groupings. In general, the M–K scheme works best for Population I stars in the solar neighborhood. For other kinds of stars (and for wavelengths outside the visible region of the spectrum), new classification

systems will have to be developed. That can be done by using the same process underlying the M–K system but applying them to different stars as standards, different wavelength regimes, and different spectral resolutions.

Six M–K *luminosity classes* are differentiated, with finer subdivisions indicated by an appended "a" or "b," as illustrated in Table 13–2 (note the spectral criteria) and Figure 13–13A. If we make a two-dimensional plot of absolute visual magnitude M_v versus spectral type, these classes appear as line segments (Figure 13–13B). In this scheme, our Sun is a G2 V star (that is, yellowish main-sequence), and its radius is much smaller than those of the giants (I to IV) of spectral type G2. Typical designations are B1 III (Beta Centauri), A3 V (Fomalhaut), F0 Ib (Canopus), K1 IVa (37 Librae), and M5 V (Barnard's Star). Note that the "a" and "b" subdivision notations are rarely used.

Figure 13–13B is an absolute magnitude–spectral type Hertzsprung–Russell diagram. Let's concentrate on the physical interpretation of the different luminosity classes. For a given spectral type, the equivalent terminology *luminosity effect, surface gravity effect* or *pressure effect* distinguishes the luminosity classes. Because spectral type corresponds to temperature, Equation 13–13 tells us that classes I to IV represent stars with radii much larger than those

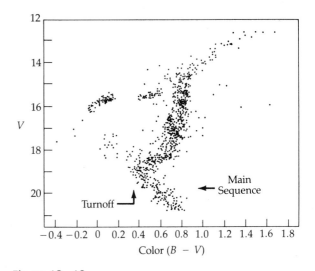

Figure 13-12

Color–magnitude diagram for a globular cluster. Note the many red giants. *(A. R. Sandage and H. L. Johnson)*

Table 13–2   The Morgan–Keenan Luminosity Classes

Class	I	II	III	IV	V	VI
Subclasses*	Ia, Iab, Ib	IIa, IIab, IIb	IIIa, IIIab, IIIb	IVa, IVab, IVb	Va, Vab, Vb	VI
Name	Supergiant	Bright Giant	Giant	Subgiant	Dwarf†	Subdwarf
Spectral Type			Spectral Criteria‡			

Spectral Type	Spectral Criteria‡
O	No criteria earlier than O9; near O9, ratio of equivalent widths of lines of He I, He II, C III, O III, and Si IV
B	Ratios of lines of He I, N II, and Si III—especially near B2; after B3, absorption line strengths of hydrogen Balmer, especially H_δ and H_γ
A	Until A3, same Balmer strengths; after A2, ratios of lines of Fe II, Mg II, and Ti II. For late-type A, three O I lines in the infrared ($\lambda = 777.1$ to 777.5 nm)
F	Balmer hydrogen strengths ineffective after F5; ratios of lines of hydrogen Balmer, Fe I, and Ca I around F5; in general, line-strength ratios of hydrogen Balmer to Sr II lines
G and K	Strength of molecular G band of CH; enhancement of H_δ and H_γ; relative line intensities of Sr II and Fe I lines; strong blue molecular band of CN and its other absorption bands; line ratios of Fe I, Fe II, Mn I, Ca I
M	Line ratios of Fe I, Cr I, H_δ, Sr II, and Y II; also Fe II, Ni I, Ti I, K I, and Ca I; infrared CN bands

* In a given spectral type, luminosity *decreases* along the sequence a, ab, b.

† These are the *main-sequence* stars.

‡ The luminosity classes are discerned by studying these spectral features, which depend upon temperature and hence upon spectral type.

of main-sequence stars. For example, a G2 supergiant is about 12.5 magnitudes brighter than our Sun; this then implies a luminosity ratio of 10^5, or a supergiant radius of about $300R_\odot$! Because stellar masses don't exceed $100M_\odot$, this supergiant must be about 10^6 times less dense than our Sun, on the average. From Equation 13–3, we see that the *surface gravity* of the supergiant is about 10^{-4} $g_\odot$ so that (by Equation 13–5) the photospheric gas pressure and electron number densities are also about 10,000 times lower than in the Sun's photosphere. Hence, the spectral features of the supergiant must be different from those of the Sun—in accordance with the Saha equation—even though both stars exhibit essentially the *same* color temperature. The pressure effect is somewhat less important than the temperature effect for it appears in the equation only linearly, whereas the temperature enters exponentially. A giant will exhibit almost the same spectrum as does a main-sequence star of the same spectral type, as long as the giant's surface temperature is lowered slightly to compensate for its lower electron density (by the Saha equation, the ratio N_{i+1}/N_i will remain the same if both N_e and T decrease appropriately). Even in this case, however, the spectral lines of the giant will be sharper than those of the main-sequence star since

the giant's features suffer much less *pressure broadening* (Section 8–5).

To give you some idea of the observed characteristics of stars of different luminosity classes, we present the following data in Table A4–3: absolute visual magnitudes, color indices, effective surface temperatures, bolometric corrections, stellar radii, and stellar masses. Study the values and trends in this table in conjunction with Figures 13–7 and 13–9. Note that the most common stars are G and K giants and A and F main-sequence stars.

(e) Elemental Abundance Effects

You may have noticed that we have not yet considered the effects of stellar chemical composition upon spectral classification, nor have we discussed the subdwarfs. The reason for this is simple. The vast majority of the stars in the solar neighborhood have the same composition: $X \approx 0.70$, $Y \approx 0.28$, $Z \approx 0.02$ (high metal abundance by mass). They belong to the so-called Population I [Section 13–3(c)]. The subdwarfs (Population II, low metal abundance, $Z \approx 0.001$) reside below the main sequence because of their metal deficiency. Fewer heavy elements implies less line blanketing, so that these stars appear to be

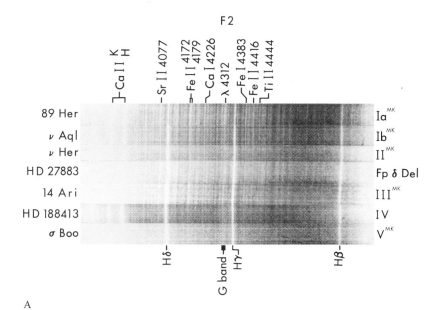

A

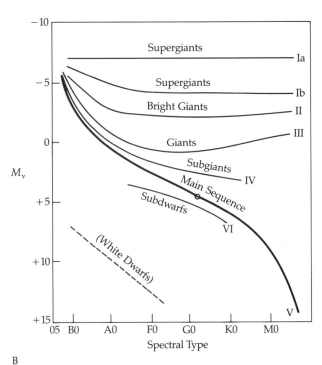

B

Figure 13-13

(A) Luminosity classes as indicated by differences in spectral line intensities for spectral type F2. *(From A Second Atlas of Objective Prism Spectra by N. Houk and M. V. Newberry)* (B) The Morgan-Keenan luminosity classes.

hotter and bluer (A to F) than they actually are (probably G to K). Let us now consider those rare cases of Population I where elemental abundance effects do appear in the spectra.

Wolf-Rayet Stars In 1867, C. Wolf and G. Rayet discovered three O stars with anomalously strong and wide *emission* lines. Today, only about 200 of these hot (up to 10^6 K), luminous (absolute magnitudes from -4.5 to -6.5) stars are known—the so-called *Wolf-Rayet stars*. As you will see in Chapter 17, the exceedingly wide (a few nanometers) emission features of ionized He, C, N, and O seen in the spectra of these stars arise in an atmospheric envelope expanding from the star at about 2000 km/s. Two abundance branches are distinguished: (a) the *WC stars*, with an apparent overabundance of carbon (spectral features of carbon and oxygen, up to C IV and O VI, are prominent), and (b) the *WN stars*, with an apparent excess of nitrogen (N III to N V lines dominate).

Hot Emission-Line Stars   In spectral types O, B, and A, we find the *Of, Be*, and *Ae stars*, with bright emission lines of hydrogen. Similar to the Wolf–Rayet stars, these stars are thought to be slowly losing mass in the form of expanding atmospheric envelopes (where the emission lines arise).

Peculiar A Stars   In the spectra of *peculiar A*, or *Ap, stars*, the lines of ionized Si, Cr, Sr, and Eu are selectively enhanced. In many cases, this enhancement is time-varying (the so-called *spectrum variables*), and it appears to be associated with strong *magnetic fields* (≈ 1 T) at the stellar surface.

Carbon Stars   Mixed in with ordinary G, K, and M stars (a temperature range from 4600 to 3100 K), we find the rare *carbon*, or *C, stars*, giants that appear to be overabundant in carbon relative to oxygen. In the early Harvard classification, these were divided into (a) the hotter R stars, distinguished by the bands of C_2 and the bands of cyanogen (CN) and (b) the cooler N stars, exhibiting C_2, CN, and CH bands with little TiO evident.

Heavy-Metal-Oxide Stars   Finally, among the M stars we find a significant number of *S stars*, also known to be giants. These stars are distinguished spectroscopically by their enhanced CN absorption bands but more importantly by the presence of the molecular bands of the heavy-metal oxides ZrO, LaO, and YO instead of TiO.

(f) Distance Determinations

We end by discussing two methods by which reasonably accurate stellar distances may be determined using H–R diagrams. Both methods depend upon an accurate calibration of the *absolute-magnitude* H–R diagram. [The *moving-cluster method* (Chapter 15) gives the best calibration.]

For individual stars (or clusters in which only one star is observed thoroughly), we employ the *method of spectroscopic parallaxes* (Figure 13–14A). From a

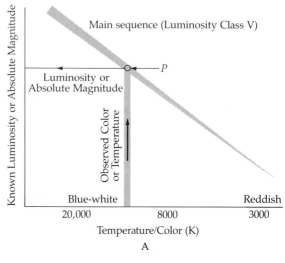

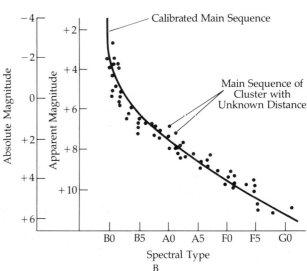

Figure 13–14

Distances from spectral types. (A) Estimating luminosity from a calibrated H–R diagram. (B) Matching the main sequence for a cluster whose distance is not known to that of a calibrated H–R diagram.

star's spectrum, we determine both its spectral type and its luminosity class. These data fix a position in the H–R diagram, from which we can read off the star's absolute magnitude. From the observed apparent magnitude, we compute the distance modulus (Equation 11–6) and then the stellar distance. Observational errors and the scatter of the H–R diagram imply a magnitude uncertainty of about ±1.0, so that the distance is known to within 50 percent. Spectroscopic parallaxes permit us to probe the vast distances of our Galaxy as well as nearby galaxies.

As an example, consider an MO III star with a measured apparent visual magnitude of $+10$. What is its distance? From Figure 13–9, note that the absolute visual magnitude of such a star is roughly 0. Then

$$m - M = 5 \log d - 5$$
$$10 - 0 = 5 \log d - 5$$
$$\log d = 15/5 = 3$$
$$d = 10^3 \text{ pc}$$

Greater accuracy in distance determination is possible by using the *main-sequence fitting technique* for an entire *cluster* of stars. Here we plot the color–apparent magnitude diagram of the test cluster and shift this plot up and down (in magnitude) on a calibrated H–R diagram until the two main sequences overlap at the same spectral types (Figure 13–14B). The difference between the test-cluster apparent magnitudes and the calibrated absolute magnitudes $(m - M)$ is the same for every star in the cluster; this number is the distance modulus of the test cluster. In Figure 13–14B, $m - M$ is 5.5, and so

$$m - M = 5 \log d - 5$$
$$5.5 = 5 \log d - 5$$
$$2.1 = \log d$$
$$d = 126 \text{ pc} = 410 \text{ lightyears}$$

By using many stars, we can cancel out random errors and achieve good accuracy (± 0.2) in determining the distance modulus.

(g) X-Ray Emission

In Chapter 10, you saw that X-ray emission from the Sun comes from hot plasma trapped in coronal loops of magnetic flux. The solar dynamo that generates the magnetic field is thought to arise from a complex combination of convection and differential rotation. The interaction of surface magnetic fields and the local plasma governs solar coronal activity. We expect the same for other stars; X-ray observations confounded this expectation. Almost all stars emit X-rays, and most far more energetically than the Sun.

Specifically, the X-ray data imply that all main-sequence stars of spectral types F through M emit X-rays with powers ranging from 10^{20} to 10^{25} W. And all stars of spectral type earlier than B5 also emit X-rays, with energy outputs that range from 10^{23} to 10^{28} W. In fact, for main-sequence stars, only the narrow range from B8 to A5 shows no evidence for X-ray emission. Finally, most giant and supergiant stars emit X-rays. Those giants that do not, range from spectral type A to G; the supergiants that do not are cooler than type G.

These results suggest that almost all stars have hot coronae controlled by large and numerous magnetic flux tubes. For cool stars that means convective zones and differential rotation. For hot stars, the physics is not yet clear; it may relate to vigorous stellar winds.

Problems

1. The absorption spectra of four stars exhibit the following characteristics. What are the appropriate spectral types?
   (a) The strongest features are titanium oxide bands.
   (b) The strongest lines are those of ionized helium.
   (c) The hydrogen Balmer lines are very strong, and some lines of ionized metals are present.
   (d) There are moderately strong hydrogen lines, and lines of neutral and ionized metals are seen, but the Ca II H and K lines are the strongest in the spectrum.

2. To *approximately* which spectral types may the following stars be assigned if their continuous spectra are of maximum intensity at

   (a) 50 nm
   (b) 300 nm
   (c) 600 nm
   (d) 900 nm
   (e) 1.2 μm
   (f) 1.5 μm
   (*Hint:* Plot λ_{max} from Wien's law as a function of spectral class for the main sequence.)

3. If the parallax of a main-sequence star is in error by 25 percent, how far and in what direction will this star be displaced from the main sequence in an H–R diagram?

4. Which parameter in the Saha ionization–equilibrium

equation is *most* important in explaining the spectral differences between

(a) giants and dwarfs of spectral type G

(b) B and A dwarfs

5. (a) What is the ratio of the surface gravities of a K0 V star and a K0 I star?

(b) If both stars had the same atmospheric temperatures and opacities, what would be the approximate ratio of strengths of the Ca II K lines in their spectra? (The ionization potential of calcium is 6.1 eV.)

(c) What assumptions did you make to answer part (b)?

(d) What is the ratio of the mean densities of these two stars?

(e) If the atmospheric electron densities of these stars were directly proportional to their mean densities, would your answer to (b) remain unchanged? Why?

6. (a) Describe two characteristics that distinguish sub-dwarfs from ordinary Population I main-sequence stars.

(b) Why is the H–R diagram of stars in the solar neighborhood (within 500 pc) *not* an unambiguous two-dimensional plot?

7. Why is an absolute magnitude–spectral type H–R diagram quite different (in principle) from an apparent magnitude–$(B - V)$ H–R diagram?

8. In determining distances via the main-sequence fitting technique, why must we refrain from comparing the observed H–R diagram of a galactic cluster with the "calibrated" H–R diagram of M 3?

9. (a) It is sometimes said that the spectral type of a star depends only upon luminosity and surface temperature; under what conditions is this statement approximately true?

(b) Give three generic examples that violate the statement that stellar masses are uniquely determined by their colors.

10. In Section 13–1(a), we wrote the mean molecular weight μ for a fully ionized gas as $1/\mu = 2X + (3/4)Y + (1/2)Z$, where, for example,

$X \equiv$ hydrogen mass fraction

$$= \frac{\text{mass density of hydrogen}}{\text{mass density of } all \text{ constituents}}$$

Derive this relationship for μ, indicating the assumptions and approximations used at each step.

11. The number 5040 appears in Equations 13–10 and 13–11.

(a) Show where this number comes from (how it is derived).

(b) What are the units of this number in these equations?

12. Use hydrostatic equilibrium to compare the central pressure of the Sun and (a) a B0 V star, (b) a G2 III star, (c) a G2 I star.

13. Using Figures 13–7 and 13–9, estimate the distance to an M Ib star of apparent magnitude +1.0.

14. Estimate the distances to the following clusters from Figures 13–7, 13–8, and 13–9 (use Table A4–3 to convert $B - V$ to spectral type):

(a) the Pleides (Figure 13–10)

(b) the Hyades (Figure 13–11)

(c) M 3 (Figure 13–12)

PART THREE

The Milky Way Galaxy

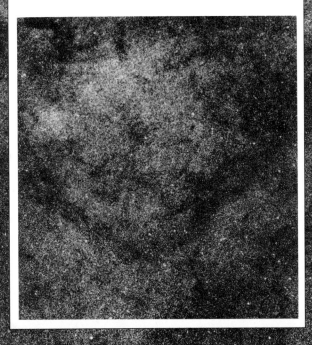

Our Galaxy: A Preview

We have touched upon some of the basic characteristics of the stars. Now we expand our horizons one step farther (not the last step, by any means; see Part 4) and consider that magnificent whirl of stars — *our Galaxy*. In this chapter, we introduce you to the Galaxy, presenting a broad picture so that Chapters 15 to 18 may be considered in the proper context. In Chapter 20, this context will be reiterated and the material summarized.

14–1 The Shape of the Galaxy

(a) Observational Evidence

When you get away from brightly lighted urban areas, you can rediscover an irregular band of diffuse light about 10° broad that encircles the celestial sphere approximately along a great circle. We call this band *the Milky Way*. The wide-angle vision field of

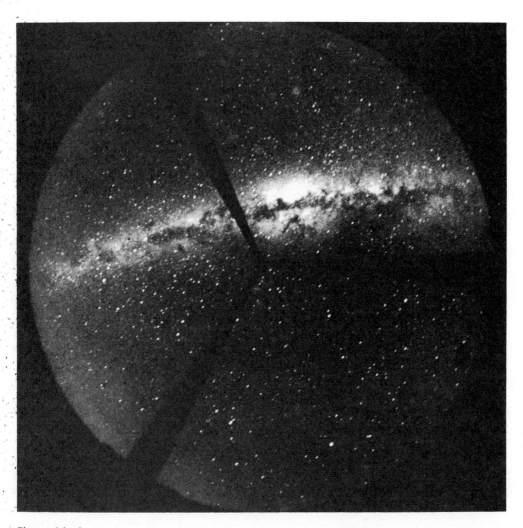

Figure 14–1

Wide-angle view of the Milky Way. The center of the Galaxy lies in the direction of the bright bulge in the upper center. Note the dark lanes through the plane. The three skinny, dark triangles are parts of the camera support. (*A. D. Code and T. E. Houck*)

the naked eye has been mimicked and improved upon by the use of specially designed wide-angle cameras (Figure 14–1). The brightest part of the Milky Way lies in the constellation Sagittarius (visible to northern hemisphere observers near the southern horizon during the summer), and dark lanes of obscuration are evident along the midline of the Milky Way. Directly opposite on the sky, in the constellations Auriga, Perseus, and Orion, the Milky Way appears unspectacular—in fact, the anticenter lies in Auriga.

How are we to determine the structure of the Milky Way? Early in this century, gigantic telescopes revealed that the luminous clouds of the Milky Way are *star clouds* (Figure 14–2); that is, these vague blobs of light were resolved into collections of millions of stars. (Galileo made this discovery in the seventeenth century.) Section 14–2 will show how the stars of our Galaxy (from the Greek *galaktikos,* meaning "milky white") are distributed in space so that we perceive the Milky Way. In 1781, the French astronomer Charles Messier published a catalog listing 107 fuzzy (nonstellar-appearing) objects he observed while searching for comets; some of these were later found to be glowing clouds of gas in our Galaxy (Chapter 19), open clusters, and distant globular clusters, but some others—the so-called *nebulae* (from the Latin singular *nebula,* meaning "cloud") —remained an enigma until well into the twentieth century. Today, we understand that some nebulae are *galaxies,* separate extragalactic entities similar to our own Galaxy (Chapter 21). One of the nearest, the Andromeda galaxy (Messier 31), turns out to be similar to our Galaxy (Figure 14–3). Our Galaxy is appar-

Figure 14–2

Star clouds in Sagittarius. Note again the dark lanes. (*Hale Observatory, California Institute of Technology*)

Figure 14–3

M 31, the Andromeda galaxy. This is a spiral galaxy similar to our own. *(Lick Observatory)*

ently a highly flattened (like a pancake) stellar system that is being viewed (from the Sun) edge-on from a point far from its center. Note that the galactic center lies in the direction of the *central nuclear bulge,* that luminous swelling evident in Figures 14–1 and 14–3. Within this nuclear bulge lies the *nucleus* of the Galaxy.

(b) The Galactic Coordinate System

To map the Galaxy, we disregard the geocentric coordinates based on the Earth's celestial equator and celestial poles (Appendix 10), as well as the heliocentric ecliptic system, and define a *galactic coordinate system* (Figure 14–4). Here the center line of the Milky Way (more accurately, the mass centroid of the galactic plane) defines a great circle on the sky called the *galactic equator,* along which *galactic longitude l* is measured in degrees (0 to 360) eastward from the direction to the center of the Galaxy in Sagittarius.

Galactic latitude b is the angular distance on the celestial sphere (in degrees from 0 to ±90) either north or south from the galactic equator. Hence, the galactic anticenter is at $l = 180°$, $b = 0°$, the north galactic pole (NGP) at $b = +90°$, and the south galactic pole (SGP) at $b = -90°$. In terms of right ascension α and declination δ—in the geocentric celestial equatorial system—the (epoch 1950) coordinates of the NGP are 12^h49^m, $+27.4°$; hence, the galactic equator is inclined about 63° to the celestial equator and is almost perpendicular to the ecliptic. Figure 14–4 shows how these three important angular-coordinate systems are related: (a) the celestial equatorial (α, δ) system, (b) the ecliptic (λ, β) system, and (c) the galactic (l, b) system. Note in particular the location of the galactic center $(l = 0°, b = 0°)$ at $\alpha = 17^h42.4^m$, $\delta = -28°55'$. Observations show the position of the center of the Galaxy, and the galactic coordinate system is defined with the presumed position of the center as the zero point.

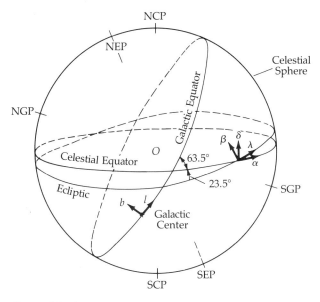

Figure 14–4

Coordinate systems. Three coordinate systems are centered on the observer *(O)*: (1) celestial equatorial with the north celestial pole (NCP), south celestial pole (SCP), and equator; (2) ecliptic, inclined 23.5° to the celestial equator, with north ecliptic pole (NEP) and south ecliptic pole (SEP); (3) galactic, with the galactic equator inclined 63.5° to the celestial equator, and a north galactic pole (NGP) and a south galactic pole (SGP).

14–2 The Distribution of Stars

The most direct way to determine the size and shape of our Galaxy is to investigate the *spatial distribution* of stars. Here you will learn how to "count" the stars, see how misconceptions can arise and have arisen in interpreting the data, and uncover the stellar indications of the Galaxy's size and its spiral-arm features.

(a) The Star-Count Method

Gauging the Stars

How to find out the extent and layout of the Galaxy? One technique involves counting stars in different directions in the sky. Assuming a uniform distribution in space, the directions in which more stars are seen are the directions in which the Galaxy extends greater distances. Consider the solid-angular area ω (in units of square degrees) on the celestial sphere. At a distance r from the observer, the solid angle subtends an area A,

$$A = \omega r^2$$

The volume contained between r and a distance dr farther out is

$$dV = \omega r^2 dr \qquad \textbf{(14–1)}$$

If $D(r)$ is the number density (number of stars per unit volume) at the distance r, then the number of stars in this volume is

$$n(r) = D(r)\,dV = \omega D(r)r^2\,dr \qquad \textbf{(14–2)}$$

$D(r)$ may be treated as constant if one uses the number density in the vicinity of the Sun. Integrating over all volume elements from the observer up to the distance r, we get the total number of stars contained in the solid angle ω within that distance:

$$N(r) = \int_0^r n(r)dr = \int_0^r \omega D(r)r^2 dr$$

$$N(r) = (1/3)D(r)r^3\omega \qquad \textbf{(14–3)}$$

To study a large number of stars, however, it is much easier to consider their *apparent magnitudes* rather than their distances. Assume that we consider only stars of the absolute magnitude M (selected by spectral type, for example); then apparent magnitude and distance are related by (Equation 11–6)

$$\log r = (m - M + 5)/5$$

$$= 0.2m + \text{constant}$$

or

$$r = 10^{0.2m + \text{constant}} \qquad \textbf{(14–4)}$$

We now use this in Equation 14–3 to derive an expression for the number of stars of a given absolute magnitude within a certain area of the sky brighter than apparent magnitude m:

$$\log N(m) = 0.6m + C \qquad \textbf{(14–5)}$$

where the constant C incorporates the dependence on M, ω, and D. This equation tells us that, assuming a uniform density, there should be $10^{0.6} = 3.98$ times as many stars of a given absolute magnitude at apparent magnitude $m + 1$ as at m. A spread in absolute magnitude can be allowed for by an appropriate adjustment in C.

The assumption of uniform star density is the most questionable and has been found to be invalid by comparisons of observed and predicted star counts. Two factors cause this discrepancy: the non-uniform distribution of stars and interstellar absorption.

Interstellar Absorption

We now know that we are not located at the center of the Galaxy, but early star-count data indicated that our Sun was at the center. The reason that people believed the Sun was in the center of the Galaxy (wrongly!) was found in 1930 by the work of Robert J. Trumpler, who, while studying star clusters, noted that many (those near the galactic plane) appeared strangely faint for their observed angular sizes. By Equation 11–6, these dim clusters would be very distant, but then their linear sizes (in parsecs) would be very large. All these problems disappear when we assume, as Trumpler did, that interstellar space is not a perfect vacuum but instead is filled with obscuring material (dust) that is concentrated toward the galactic plane. Note the dark lanes of obscuration defining the galactic plane in Figure 14–1.

Starlight is scattered and absorbed by this dust, and so we refer to the phenomenon as *interstellar absorption* (Chapter 19). Because the stars observed at high galactic latitudes (perpendicular to the plane of the Galaxy, at the Sun's position) farther than 500 pc from the plane do not strongly exhibit the characteristic effects of interstellar absorption, we can believe that the disk of the Galaxy is thin, about 1 kpc at the Sun. At low galactic latitudes (near the galactic plane), however, the obscuration becomes severe. We characterize this absorption by a medium of uniform density and opacity (Section 10–2); then Equation 10–2 implies an exponential diminution of starlight with distance traveled through the medium. Because stellar magnitude is proportional to the logarithm of the observed flux (Equation 11–5), the increase in apparent magnitude from interstellar absorption is proportional to the distance to the star. Near the galactic equator, the visual absorption is about 1.0 mag/kpc; in very dense interstellar clouds, it is much higher. So a star 1 kpc distant in the galactic plane has an apparent distance modulus of 11.0 mag even though its actual distance modulus is 10.0 mag; this star appears to be about 1.6 kpc distant. A star at 5 kpc suffers 5.0 mag of absorption, so that not only is it dim, but its distance is overestimated by a factor of 10 if we neglect the effect of interstellar absorption. (Note that 5 mag is a factor of 100 brightness, and $100^{1/2} = 10$, by the inverse-square law for light.)

You can now understand why the Galaxy appeared small and heliocentric to astronomers early in this century; the interstellar obscuration in the galactic plane was the culprit. By including absorption in the analysis of the star-count data, we find that we are midway to the edge of a very large (40 to 50 kpc diameter) stellar system, and the number density of stars increases rapidly toward the galactic center.

Luminosity Function

In addition to star counts, we can characterize the stars in our region of the Galaxy by their *luminosity function*: the number of stars per unit spatial volume with a given absolute magnitude M (or luminosity L). We can infer a star's absolute magnitude by studying its spectrum and placing it appropriately on a calibrated H–R diagram; that is, we estimate its luminosity from its spectral type.

As a crude, and incorrect, first attempt, we can consider the 50 brightest visible stars, plotting a histogram of the numbers occurring in the absolute magnitude intervals M to $M + 1/2$ (Figure 14–5). We expect that those stars that appear the brightest are also intrinsically bright (note the preponderance at negative M), but Figure 14–5 is biased since very luminous stars can be seen to great distances. A truer picture of the actual luminosity function is given by

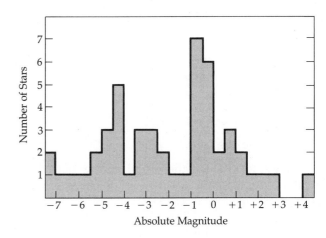

Figure 14–5

Luminosity function for the 50 brightest stars.

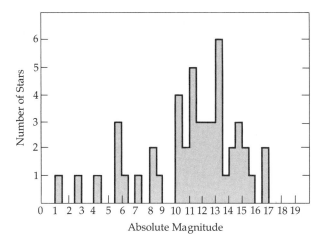

Figure 14-6

Luminosity function for the 50 nearest stars. Note that most of the stars have low luminosity.

the histogram of the 50 nearest stars (Figure 14–6); note the overwhelming predominance of intrinsically faint stars ($10.0 \leq M \leq 15.0$) and the dearth of very luminous stars. A stellar sample complete to a limiting magnitude of $M \approx 19.0$ at V band (Figure 14–7) shows that most stars are very dim ($M > 10.0$), with few being brighter than our Sun, which is by no means a very luminous star ($M = 5$). Faint stars are in the overwhelming majority (the distribution has a broad peak near $M \approx 13$), with only a sparse sprinkling of luminous stars. These results apply to the solar neighborhood and to similar regions in the Galaxy. The luminosity function differs in locations, for instance, far above and below the midplane and in the central regions of the nucleus.

(b) Luminous Stars and Stellar Clusters

Luminous stars and stellar clusters are discernible to very great distances. For example, if we survey to a limiting magnitude of $m = 15.0$, B stars of absolute magnitude $M = -5.0$ can be seen to a distance of about 100 kpc; interstellar absorption in the galactic plane will, however, limit us to an actual distance of 5 kpc. Stellar type O and B stars are very young, and they tend to occur in small clusters called *associations*; moreover, these hot, luminous stars can ionize the hydrogen gas in which they are imbedded, producing a bright region of ionized hydrogen (Chapter 19).

Such associations and ionized hydrogen regions give an indication of the *spiral-arm* structure in the galactic disk. In Chapter 19, you will see that interstellar molecular hydrogen gas best delineates these spiral features.

Recall that a *globular cluster* is a dense spherical cluster of about 10^5 to 10^6 stars; some 200 such globular clusters are known to be associated with our Galaxy, and they are very luminous ($M \approx -4$ to -10). Because the nearest globular cluster is 3 kpc distant, we have a good probe of galactic distances. The distance to a given globular cluster may be found (1) by the main-sequence fitting technique, (2) by the apparent magnitudes of certain well-known types of stars in the cluster, and (3) by the apparent angular diameter of the cluster. When the distance and direction of globular clusters are plotted, we find that they define a spherical system of radius ≈ 25 kpc, with its center 8 to 10 kpc away from the Solar System in Sagittarius. Harlow Shapley discovered this effect in 1917, and he correctly concluded that the Sun is far from the galactic center (the center of the globular cluster system). This interpretation receives strong support from the observation of a globular cluster system, of similar size and shape, centered on the middle of the Andromeda galaxy (Figure 14–3). The current conventional standard puts the Sun at about 8.5 kpc from the center, which gives the galactic disk a diameter of at least 50 kpc.

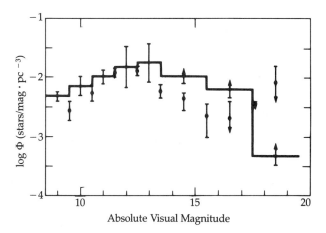

Figure 14-7

Local luminosity function. The sample here is complete to an absolute magnitude of 19 (solid circles). The solid line is a step function that fits the data fairly well. (*G. Gilmore, Royal Observatory, Edinburgh*)

14–3 Stellar Populations

Chapter 13 stated that stars could be assigned to different *stellar populations,* characterized by their observed metal abundances, which are the abundances of all elements *except* hydrogen and helium (Z values). In our Galaxy, we find that the stars in the globular clusters are extremely metal-deficient (Z ≤ 0.001); we call these stars *extreme Population II* stars; most of the individual stars seen far from the galactic plane (in what is called the *galactic halo*) are of this population. Within about 500 pc of the galactic plane, the spatial density of stars has increased so markedly that we speak of the pancake-shaped *galactic disk* with its *bulge* at the galactic center. Here the metal abundance is much greater than in Population II stars, and it *increases* as we approach the galactic plane. We call these the *disk Population* stars. Both Population II and disk Population stars are very old (billions of years), and their brightest representatives are *red giants.* However, in the galactic plane (especially in the spiral arms) we find the young, luminous blue stars of *Population I* (greatest metal abundance; Z ≥ 0.01). Population I stars are seen in *open (galactic) clusters,* in O and B associations, and near concentrations of interstellar dust and gas.

The abundance of heavy elements increases continuously from the halo to the spiral arms buried within the disk. In Chapter 16, you will see that hydrogen and helium are processed into heavier elements in the centers of stars and returned to the interstellar medium upon the demise of these stars. So our Galaxy likely began as a spherical cloud of H and He, the metal-poor Population II stars formed, the cloud was enriched with metals as it collapsed toward the galactic plane, and only in the denser regions of the spiral arms is star formation now producing the metal-rich stars of Population I.

In summary, we can roughly divide the stars in the Galaxy into two general groups, Population I and Population II, on the basis of their metal abundances. These can be further subdivided by location into disk, bulge, and halo subgroups. The point of this scheme is to try to group together stars of similar age, composition, and (as you will see in detail later) kinematics. The key to this choice of collective characteristics is to study the history of the Galaxy as recorded in its stars. Metal abundance and orbital characteristics are fossil properties; they are unchanged by galactic evolution.

14–4 Galactic Dynamics: Spiral Features

We have described the stellar content of our Galaxy, but what about the *dynamics* of the system? (Stellar motions are considered in detail in Chapter 15.) Because theoretical models based upon Newtonian gravitation account well for the observed properties of globular clusters, which contain almost 1 million stars, we assume that our Galaxy is held together by the mutual gravitational attraction of its stars. The spherical part of the system (the halo with its host of globular clusters) is then analogous to a spherical stellar cluster. The flattened disk, however, clearly implies a rapidly rotating entity that formed *after* the halo of globular clusters.

Every star associated with our Galaxy moves within the gravitational attraction produced by all of the other stars. Assume circular stellar orbits about the galactic center in the galactic plane. In particular, our Sun is in such an orbit, with a radius of 8.5 kpc. In Chapters 15 and 20, we show that, near the galactic center, the orbital angular speed ω (radians per second) is approximately constant (*rigid-body rotation*), whereas near the Sun, the circumgalactic speed slowly rises with distance from the center of the Galaxy and the Sun's speed is about 220 km/s toward $l = 90°$. At the Sun's position in the Galaxy, we may approximate the solar motion as a circular Keplerian orbit about a massive central body of mass M_G. Since the centripetal acceleration maintaining this circular orbit is produced by the gravitational attraction between the core (M_G) and the Sun ($M_\odot$), we have

$$v_\odot^2/R_\odot = GM_G/R_\odot^2 \qquad (14-6)$$

where $v_\odot$ is the Sun's circular speed (220 km/s), $R_\odot$ is the distance to the galactic center (8.5 kpc), and G is the constant of universal gravitation. Equation 14–6 may be evaluated to give the mass of our Galaxy within the solar orbit:

$$
\begin{aligned}
M_G &= v_\odot^2 R_\odot/G \\
&= (2.20 \times 10^5 \text{ m/s})^2 (2.6 \times 10^{20} \text{ m})/ \\
&\quad (6.7 \times 10^{-11} \text{ m}^3/\text{kg} \cdot \text{s}^2) \\
M_G &= 1.9 \times 10^{41} \text{ kg} \approx 10^{11} M_\odot \qquad (14-7)
\end{aligned}
$$

Hence, if an average star has about solar mass, our Galaxy consists of approximately 100 billion stars; the most massive spiral galaxies known have masses near $10^{13} M_\odot$.

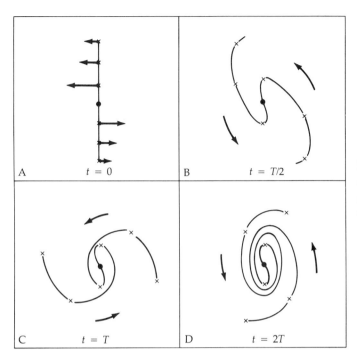

A $t = 0$

B $t = T/2$

C $t = T$

D $t = 2T$

Figure 14–8

The wind-up problem. Here are two hypothetical "spiral arms" made of six stars following Keplerian orbits. After two cycles, the arm configuration has disappeared into a tightly wound pattern.

Today we believe that the spiral arms of our Galaxy are also produced by dynamical effects. If this were not the case, these features would wind up and disappear (Figure 14–8) in only a few revolutions of the Galaxy. The Galaxy is about 15×10^9 years old, and at the Sun's location, it rotates once every 240 million years ($\approx 2\pi R_\odot/v_\odot$); it is unlikely that we could observe a spiral structure that was not a steady dynamical feature of the galactic disk. A spiral arm is thought to be a manifestation of a rotating *density wave* in the galactic disk (Chapter 20). The disk is initially unstable to density perturbations (essentially sound waves), which can grow and gravitationally attract material along spiral paths, but these waves rotate only half as fast as the disk, so that material passes through the density pattern in the direction of galactic rotation. Star formation starts as the dust and gas permeating the disk are compressed at the spiral-arm feature; that is why hot, young Population I stars outline spiral structure.

14–5 A Model of the Galaxy

We close this chapter by describing a schematic model of our Galaxy that ties together the observations and considerations presented above (Figure 14–9). The entire Galaxy is contained within its spherical *halo*, with a diameter of roughly 50 kpc. Here reside the few old Population II halo stars and the globular clusters (which contain Population II stars). These objects define the spherical envelope while they move along highly eccentric and inclined orbits about the galactic center with orbital periods near 1 to 3×10^8 years—a close analogy to the spherical comet cloud centered upon our Sun (Chapter 7).

The outer halo of the Galaxy has no definite boundary. Observations of globular clusters (Figure 14–10) indicate that the halo extends to at least 100 kpc and then includes the Magellanic Clouds and dwarf galaxies that are companions to the Milky Way (Figure 14–10). The halo also includes gas in the shape of an expanded disk with a thickness of no more than 10 kpc. This gas is clumpy, contains both neutral and ionized areas, and corotates with the rest of the Galaxy. Because we can easily see out of the Galaxy, the amount of dust in the halo must be small.

Bisecting the spherical halo is the circular *galactic plane*. Inward to the galactic plane, the spatial density of stars increases and their metal abundance rises. We have now reached the oblate spheroidal *galactic disk*, with a perpendicular thickness of about 1 kpc and the large *central bulge* at its center. About half the mass of

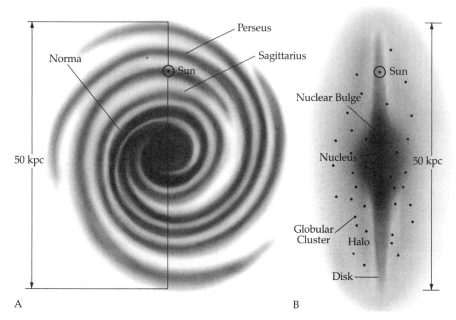

A

B

Figure 14 – 9

Model of the galaxy. (A) Top view, showing the spiral arm structure and the names of a few nearby arms. (B) Side view, showing the nuclear bulge and halo.

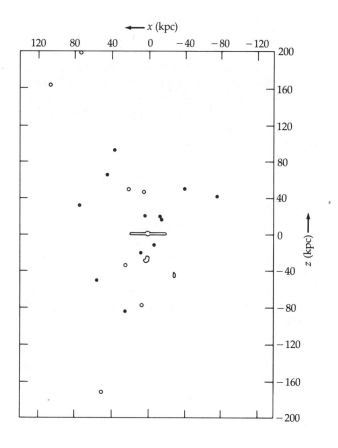

Figure 14 – 10

A cross section (in the x, z plane) of the distribution of material in the outer halo of the Galaxy: globular clusters (dots), dwarf galaxies (circles), and the Magellanic Clouds (kidney-shaped). The scale is in kiloparsecs with distances relative to the Galaxy's center. *[Adapted from a diagram by B. Carney, Publications of the Astronomical Society of the Pacific 96:841 – 855 (1984)]*

the Galaxy ($\approx 10^{11}M_\odot$) resides here, in disk Population stars that move in low-eccentricity orbits about the galactic center.

At the very center of the Galaxy is the small (<1 pc diameter) massive (about $10^6 M_\odot$) *nucleus.* In other spiral galaxies, the nucleus has a star-like appearance, but in our Galaxy it is hidden from our view by the interstellar obscuration in the galactic plane. It makes its presence known by emitting thermal and nonthermal radio continuum emission, infrared radiation, X-rays, and even gamma rays. In the infrared at wavelengths near $2\,\mu$m, the galactic nucleus is 30" in diameter (1.5 pc in linear dimensions!) with several nearby sources in a sphere of radius 10 pc. The galactic nucleus may be an extremely compact stellar cluster wherein events of great violence are and have been taking place, or it may be a single very massive object, possibly a black hole.

At the midline of the disk is the *galactic plane,* with its veneer of *spiral arms* in a region only about 500 pc thick (consider how thin this really is; the ratio of diameter to thickness is about 100 : 1). Newborn

stars of Population I, as a result of their high luminosities and short lifetimes, signal the presence of the spiral density waves propagating around the galactic plane. Myriad clouds of hydrogen gas and dust occupy the galactic plane, with some tendency to collect near the spiral features. The *interarm* regions contain some disk Population and old Population I stars, with stellar number densities about 10 percent that in an arm.

Our Sun, a Population I star about 5 billion years old, is located 8.5 kpc from the galactic center, on the inner edge of one spiral arm. Although it is not exactly in the galactic plane, it is only a small distance above it. Carrying the Solar System with it, the Sun orbits the galactic center once every 240 million years in a path of very low eccentricity. As we gaze out from the Earth upon the celestial flywheel that is our Galaxy, we see the neighboring stars participating with us in the galactic rotation toward $l = 90°$ and the breathtaking band of the Milky Way encircling the sky from its center in Sagittarius.

Problems

1. Using a simple diagram, explain why our Galaxy appears as the Milky Way in the night sky.

2. Using Figure 14–4, describe the various aspects of the Milky Way on evenings of different seasons for an observer in New Mexico. Use the altitude–azimuth horizon system of coordinates, and mention the tilt of the Milky Way to the horizon and the observability of Sagittarius.

3. The integral count ratio $N(m+1)/N(m) = 3.98$ with the assumption that stars are uniformly distributed in space. Make a schematic diagram plotting $\log N(m)$ versus m and show what effect each of the following would produce:
   (a) a uniformly distributed interstellar obscuration of 1 mag/kpc
   (b) a strongly absorbing interstellar dust cloud localized at the apparent magnitude m_{cloud}

4. What stellar population would you expect to find (and why)
   (a) in the nucleus of our Galaxy
   (b) in the spiral arms of the Andromeda galaxy
   (c) in the Pleiades star cluster
   (d) in intergalactic space (beyond the halo)
   (e) in the galactic bulge

5. (a) A globular cluster is in elliptical orbit ($e = 0.9$) about the center of our Galaxy, reaching *apogalacticon* (farthest distance from the center) at the distance 40 kpc. What is the *perigalacticon* (nearest) distance, and how long will this cluster require to complete one orbit?
   (b) What is the approximate speed of escape from the Galaxy in the solar neighborhood if the Sun's circular orbital speed about the galactic center is 220 km/s?

6. Consider the center of our Galaxy to be a spherical star cluster of radius R and uniform mass density ρ.
   (a) What is the total mass M of this cluster?
   (b) What is the mass contained within the sphere of radius $r < R$?
   (c) In terms of M, what is the angular speed ω of a star in circular orbit at a distance r ($< R$) from the cluster's center? Note that $\omega = v/r$, where v is the circular orbital speed of the star in kilometers per second.
   (d) Using an analogy, explain why we refer to the result of part (c) as *rigid-body rotation.*

7. Assume that the galactic disk may be approximated by a plane-parallel slab 500 pc thick and having the ga-

lactic plane at its midline. Take the Sun to be located in the galactic plane.

(a) How many magnitudes of absorption are there at $b = 90°$?

(b) How many magnitudes of absorption are there at the general galactic latitude b?

(c) Explain why the region $b \leq 10°$ is called the "zone of avoidance" (essentially total obscuration in terms of apparent magnitudes).

8. In the direction perpendicular to the galactic plane, approximately how thick (in kiloparsecs) is the galactic bulge?

9. This chapter suggests that stars are formed when disk materials (gas and dust) catch up with a density wave. Assume that the newborn stars continue about the galactic center with a circular speed appropriate to their distance.

(a) How far will an O star move from the spiral arm of its birth in 1 million years?

(b) Are you surprised to find our Sun in a spiral arm (or near one, at least)?

10. If interstellar absorption averages 1 mag/kpc, calculate the optical depth for this absorption in the galactic plane.

11. For each of the bins in Figure 14–6, convert absolute magnitude to luminosity (in solar units). Multiply the luminosity by the number of stars for each bin. Make a table of the product number of stars × luminosity versus absolute magnitude. If these stars are representative of the Galaxy, at what absolute magnitude (counting upward) is 95 percent of the Galaxy's luminosity accounted for? Roughly what range of spectral types contribute the bulk of the Galaxy's luminosity?

12. Repeat the previous exercise except convert luminosity to mass by the rough formula $M \approx L^{1/3}$, where M and L are in solar units. Is most of the mass contained in high-, low-, or intermediate-mass stars? Roughly what range of spectral types contribute most of the mass of the Galaxy?

Galactic Rotation: Stellar Motions

Our Solar System resides in a rotating spiral galaxy known as the Milky Way Galaxy, the Sun one of its many stars. This chapter explains the observable motions of the stars in our Galaxy (especially those in the solar neighborhood), verifying the galactic rotation and evaluating its characteristics quantitatively. These results imply a total mass and mass distribution in the Galaxy. We begin with the observational properties of stellar motions. Then we investigate the characteristic rotational velocity of our Sun about the center of the Galaxy and the differential galactic rotation.

15–1 Components of Stellar Motions

Before we plunge in, we need to clarify coordinates and reference systems (Appendix 10). The most familiar are *geocentric* (Earth-centered) coordinate systems, such as the system of right ascension and declination, and the *galactic* coordinate system. However, dynamical investigations are best referred to a *heliocentric* (Sun-centered) coordinate system. In practice, astronomers distinguish between apparent celestial positions (the geocentric positions of objects determined from Earth-bound observations) and true celestial positions (the heliocentric positions relative to our Sun). True stellar positions (such as heliocentric right ascension and declination) differ from apparent positions because the latter exhibit periodic displacements, such as geocentric parallax orbits, as a result of the Earth's orbital motion around the Sun. We can correct terrestrial observations to a heliocentric basis because the Earth's orbit is well known. So we discuss only heliocentric celestial positions. The second section of this chapter introduces yet another dynamical reference system, called the *local standard of rest*, which is based upon the rotational dynamics of our Galaxy.

Stellar motions arise from the velocities of stars through space. These velocities are vectors, which may be resolved into two perpendicular components: the radial velocity along the line of sight and the tangential velocity in the plane of the sky.

(a) Radial Speed

The *radial speed* of a star is its speed of approach or recession; it is easily obtained from the Doppler shift of stellar spectral lines. To determine this Doppler shift, a laboratory comparison spectrum is photographed adjacent to the stellar spectrum and the relative positions of lines in the two spectra are measured (Figure 15–1). The measured Doppler shift, $\Delta\lambda = \lambda - \lambda_0$, permits us to deduce the radial speed of the star from the Doppler formula (Equation 8–13):

$$v_r = (\Delta\lambda/\lambda_0)c \qquad (15-1)$$

where c is the speed of light. When $\lambda > \lambda_0$, the spectrum is *redshifted* and v_r is the radial speed of *recession;* when $\lambda < \lambda_0$, the spectrum is *blueshifted* and the star is *approaching* us. To refer these motions to the Sun, we must correct for the component of the Earth's orbital velocity (speed ≈ 30 km/s) along the line of sight to the star.

The radial speed may be determined for any star for which a spectrum may be obtained. The distance to the star is irrelevant, for it is only the flux of the star that determines whether it is bright enough to produce a spectrum when observed through a telescope–spectrograph combination. From Equation 15–1, we see that we cannot measure radial speeds smaller than about 1 km/s since $\Delta\lambda$ usually cannot be discerned to accuracies better than ±0.001 nm at visible wavelengths.

(b) Proper Motion

The motion of a star in the plane of the celestial sphere is called the *proper motion μ*, and it is usually expressed in seconds of arc per year. For a given

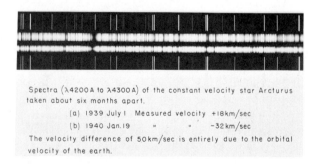

Spectra ($\lambda4200\text{Å}$ to $\lambda4300\text{Å}$) of the constant velocity star Arcturus taken about six months apart.
(a) 1939 July 1   Measured velocity +18 km/sec
(b) 1940 Jan.19   "   "   −32 km/sec
The velocity difference of 50 km/sec is entirely due to the orbital velocity of the earth.

Figure 15–1

Doppler shifts of Arcturus. The two spectra were taken six months apart. The difference in the shifts arises from the Earth's orbital motion. The bright lines above and below are the reference spectra. *(Palomar Observatory, California Institute of Technology)*

speed perpendicular to the line of sight, the proper motion will be greater the closer the star is to us, an obvious analogy with trigonometric parallax (Chapter 11). Very distant stars exhibit no measurable proper motion, and they may therefore be used as *reference*, or *background*, *stars*. Most proper motions are very small; the largest known is that of Barnard's Star, an exceptional 10″/year (Figure 15–2). Compared with the cyclic nature of parallax orbits, proper motions have the distinct advantage of being cumulative; measurements may be made many years apart, so that small annual angular displacements can accumulate to an easily measured amount. An accuracy of ±0.003″/year is attainable when observations are spaced over decades, but we must exercise great care in the selection of standards of reference. As a result of galactic rotation (Section 15–4), the ideal reference system is based upon distant quasars (Chapter 23). Because observations of proper motions are made many years apart, precautions similar to those taken for parallax observations are necessary. In particular, corrections must be made for stellar parallax and for the aberration of starlight. In addition, fundamental reference systems are required over the entire sky, and these are not easily established. Yet proper-motion measurements are well worth the trouble, for they are a key to our knowledge of the structure of the Galaxy.

(c) Tangential Speed

The proper motion of a star is due to its *tangential speed* v_t, which is its linear speed in the direction perpendicular to the line of sight. To convert the angular measure of proper motion to a linear speed (in kilometers per second) transverse to the line of sight, we must know the distance d to the star, for

$$v_t = d \sin \mu \approx \mu d \qquad \text{(15–2)}$$

The last part of Equation 15–2 follows because μ is very small (less than 5×10^{-5} rad/year). In this equation, you must be careful to remember the units, for d is normally given in parsecs and μ in seconds per year. Equation 15–2 gives v_t in parsecs per year when μ is in radians per year, but v_t is found in astronomical units per year when μ is in arcseconds per year; we use d in parsecs in either case. By applying the appropriate conversion factors (astronomical units to kilometers and years to seconds), we find that

$$v_t = 4.74\mu''d = 4.74(\mu''/\pi'') \text{ km/s} \qquad \text{(15–3)}$$

where the distance d is given in parsecs, the parallax π'' in seconds of arc, and the proper motion μ'' in arcseconds per year. For example, a star whose distance is 100 pc and whose proper motion is 0.1″/year has a tangential speed of 47.4 km/s.

(d) Space Motion

The *space velocity* **V** of a star with respect to the Sun has been decomposed into two perpendicular components (Figure 15–3): (1) the radial velocity,

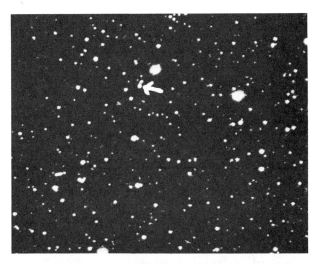

Figure 15–2

Proper motion of Barnard's Star. Indicated by an arrow, Barnard's Star shows considerable proper motion in the time interval from August 24, 1894 (top) to May 30, 1916 (bottom). *(Yerkes Observatory)*

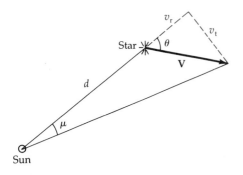

Figure 15–3

Components of space velocity.

whose magnitude is v_r (the radial speed), and (2) the tangential velocity, with magnitude v_t (tangential speed). From the Pythagorean theorem (and the law of vector addition), we find

$$V^2 = v_r^2 + v_t^2 \qquad (15–4)$$

Once again we caution you that all speeds in Equation 15–4 must be in the same units; this need implies (from Equation 15–3) that we must know the distance to the star. The angle that the space velocity makes with the line of sight is θ, and it is found from

$$\tan \theta = v_t / v_r$$

Hence, these two components can be added (Equation 15–4) to find the total speed and direction of a star in space (relative to the Sun). For stars in the Sun's neighborhood, the magnitude of their space velocities average 25 km/s.

15–2 The Local Standard of Rest

Our Galaxy has two physically preferred frames of reference. The first is the *galactocentric* system: centered at the nucleus of the Galaxy, its reference plane is the galactic plane and its reference axis is the galactic rotation axis. We are primarily interested in the second system here—the *local standard of rest (LSR).*

One way to define the LSR is by the *dynamical LSR*: the reference frame, instantaneously centered upon the Sun, which moves in a circular orbit about the galactic center at the circular speed appropriate to its position in the Galaxy. So all stars in the solar neighborhood that are in circular galactic orbits are

essentially at rest in the dynamical LSR. Any deviations from circular motion in the solar neighborhood will appear as stellar peculiar motions with respect to the dynamical LSR.

The Sun's galactic orbit is not perfectly circular. So, with respect to the LSR, a *solar motion* of 19.5 km/s occurs toward the constellation Hercules ($l = 56°$, $b = 23°$). On the celestial sphere, the Sun is moving toward the *solar apex* and away from the *solar antapex*. The nature and extent of the solar motion were first demonstrated by William Herschel in 1783 using statistical methods; let's paraphrase his analysis in modern terms.

The stars in the solar neighborhood exhibit *peculiar motions* with respect to the LSR; that is, they swarm about like sluggish bees. If the Sun were at rest in the LSR, the average of these peculiar velocities would also be zero with respect to the Sun. If the Sun moves with respect to the LSR, however, each star will (in addition to its peculiar motion) reflect this solar motion to an extent dependent upon its position on the celestial sphere (Figure 15–4). Stars on the great circle 90° from both the apex and the antapex will, on the average, exhibit the largest proper motions toward the antapex; hence, proper-motion averages (out to about 50 pc) reveal the locations of

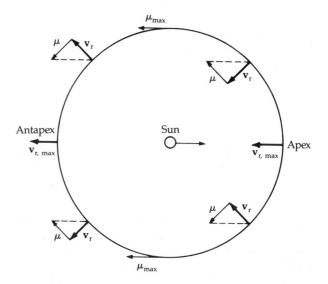

Figure 15–4

Reflected solar motion. The solar motion with respect to the LSR affects stars around the celestial sphere. Hence the stars reflect the Sun's motion toward the apex and away from the antapex.

the apex and antapex. We determine the speed of the solar motion by averaging the radial speeds of stars near the apex and stars near the antapex; the average radial speed of approach is greatest at the apex (another way to locate the apex) and that of recession is greatest at the antapex (another way to locate the antapex). Finally, note that the average proper motion vanishes at both the apex and the antapex. Because the stars sampled for radial speed can be much more distant than those sampled for proper motion, a different apex location will result from each sample; in practice, the two locations are almost identical.

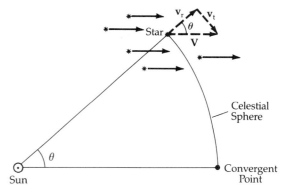

Figure 15-5

Moving-cluster geometry. A cluster of stars (*) moves with space velocity V, and the proper motions appear directed to the convergent point.

15-3 Moving Clusters

We briefly mentioned *moving stellar clusters* in Chapter 11. Recall that such a cluster is a gravitationally bound group of stars traveling through the Galaxy. Hence, all the cluster members exhibit the same peculiar motion relative to the LSR because the stellar motions are completely correlated, not random. We can identify the members of a given cluster by this property. Random observational errors permit a small number of stars that are not part of the cluster to slip into our chosen sample and limit us—by Equation 15-3—to those clusters nearer than about 500 pc.

Consider a small, or distant, cluster which subtends a small solid angle on the sky. Its radial speed v_r and angular proper motion μ''—with respect to the LSR—may be obtained by studying a single star (and correcting for the reflected solar motion). To avoid errors from the possible inclusion of stars not part of the cluster, we usually take averages over several stars. The direction of the cluster's space velocity is unknown, however (it is too distant for parallax measurements), so that we cannot find the cluster distance from Equation 15-3. Some other distance criterion must be used in this case (see the following).

On the other hand, the situation improves when we have a cluster (usually nearby) that subtends a large solid angle on the sky. Then (Figures 15-5 and 15-6), because all cluster members are moving in the same spatial direction, their proper motions appear to converge toward (or diverge from) a single point on the celestial sphere. We call this the *convergent point*. This phenomenon occurs from perspective effects that cause parallel motions of stars to appear to radi-

ate from a single point on the sky. Again, we know the radial speed of the moving cluster once we have determined it for a single member star, but now we also know the direction of the cluster's space motion. When we extend the proper motions of the member stars on the sky, they intersect at, for example, the convergent point of the cluster (Figure 15-6). The angular distance θ from the convergent point to a member star is the same as the angle between the line of sight to that star and the star's space velocity vector **V** (Figure 15-5):

$$V = v_r/\cos\theta \qquad (15-5)$$

but we know that $v_t = V\sin\theta$, and so using Equations 15-3 and 15-5 gives us the *cluster distance* from its derived parallax:

$$\pi'' = 4.74\mu''/v_r\tan\theta \qquad (15-6)$$

These *moving-cluster parallaxes* apply to individual stars; that is the extraordinary power of this technique.

Recently, radial velocities for about 40 stars in the Hyades have been reobserved to obtain a new cluster radial velocity of 39.1 km/s and a new convergent-point position of right ascension = 95.3° and declination = 7.2°. The distance modulus based on four recent works then comes out to $m - M = 3.23$ for a distance of 44.3 pc. This result coincides almost exactly with the distance found from trigonometric parallaxes: $m - M = 3.25$. Hence, we know the distance to the Hyades with an error of a little less than 1 pc, or an accuracy of about 2 percent.

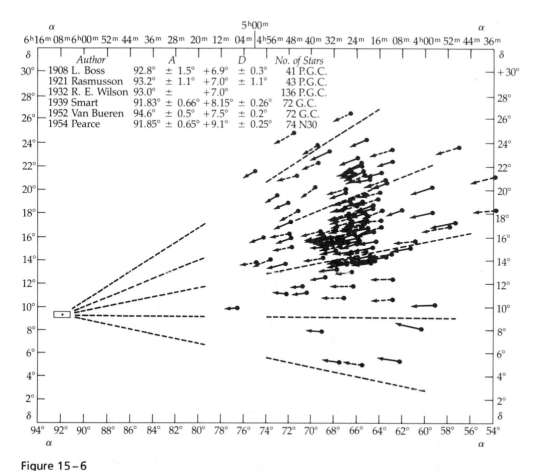

Figure 15–6

Hyades motions. This nearby cluster has a well-defined group proper motion. *(From O. Struve, B. Lynds, and H. Pillans, Elementary Astronomy, Oxford University Press, 1959)*

Once we have obtained an accurate cluster distance, we can make an absolute calibration of the H–R diagram of that cluster (Chapters 11 and 13); the Hyades distance of 44.3 pc (obtained by the moving-cluster method) now provides the fundamental distance scale upon which all distances greater than about 100 pc (even to the limits of our Universe!) are based. Distances less than 50 pc are based on trigonometric parallaxes. By using the Hyades H–R diagram and the main-sequence fitting technique (Chapters 11 and 13), astronomers have extended the scale to the Praesepe (Beehive) Star Cluster in Cancer at 159 pc and to the Double Cluster (h and Chi Persei) in Perseus at 2330 pc—with an accuracy of about 10 percent in the distances. Chapter 21 shows how this distance scale may be further extended to extragalactic objects and the Universe.

15–4 Galactic Rotation

(a) Differential Galactic Rotation

Stars in the solar neighborhood orbiting the galactic center in perfectly circular orbits would be at rest in the LSR. This statement implies *rigid-body rotation* ($\omega = constant$ angular speed about the galactic center) of our region of the Galaxy, however. The particles a rigid body is made up of remain at fixed distances from one another, and every particle moves around the center of the body in the same period. Now $\omega = v/r$, where v is the circular orbital speed and r is the orbital radius; hence, for a rigid body, $v \propto r$ if the stars orbit in such a way that Equation 14–6 holds. Now, the Sun orbits in a roughly Keplerian fashion, but then $v \propto r^{-1/2}$ and $\omega \propto r^{-3/2} \neq$

constant. We say that there is *differential galactic rotation* when the orbital angular speed is a function of distance from the galactic center, $\omega = \omega(r)$.

Let's explicitly show the difference between solid-body and Keplerian motion. Consider most of the Galaxy's mass concentrated at its center. Then Kepler's third law,

$$P^2 = (4\pi^2/GM)R^3$$

where R is the distance to the center at which a mass orbits with circular velocity V and period P such that

$$P = 2\pi R/V$$

gives us

$$4\pi^2 R^2/V^2 = (4\pi^2/GM)R^3$$

and

$$V = (GM/R)^{1/2}$$

$$\propto R^{-1/2}$$

for Keplerian motion.

In contrast, picture a spherical mass with a uniform density ρ throughout. Then the centripetal acceleration a at a distance R from the center is

$$a = V^2/R = GM(R)/R^2$$

where $M(R)$ is the mass within R, or

$$M(R) = 4\pi R^3 \rho/3$$

so that

$$V^2/R = G(4\pi R^3 \rho/3)/R^2$$

and

$$V = (4\pi G\rho/3)^{1/2}R$$

$$\propto R$$

for solid-body rotation ($\omega = $ constant).

Let's now follow Jan Oort's brilliant work of 1927 and derive the effects of such differential galactic rotation (particularly in the solar neighborhood). At the end, you will see that these effects are indeed observable.

First, assume for simplicity *circular* galactic orbits in the galactic plane (Figure 15–7). Here we have defined R as, for example, the star's distance from the galactic center, R_0 as the Sun's distance from the center, d as the Sun–star distance, Θ as the star's circular orbital speed, Θ_0 as the orbital speed of the LSR, l as the galactic longitude of the star, α as the

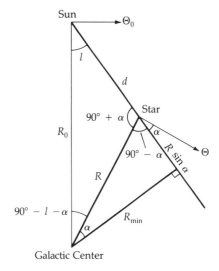

Figure 15–7

Geometry for galactic rotation. The Sun, galactic center, and a star define the galactic plane; all motions are circular.

angle between the line of sight to the star and its orbital velocity, ω as the star's galactic *angular* speed, and ω_0 as that of the LSR.

We begin with v_r, the star's radial speed with respect to the LSR. From Figure 15–7, we have

$$v_r = \Theta \cos \alpha - \Theta_0 \sin l \qquad (15\text{–}7)$$

and the law of sines (Mathematical Appendix) gives

$$\sin(l/R) = \sin[(90° + \alpha)/R_0] \equiv \cos(\alpha/R_0) \qquad (15\text{–}8)$$

but $\omega = \Theta/R$ (and $\omega_0 = \omega_0/R_0$), so that using Equations 15–7 and 15–8,

$$v_r = R_0(\omega - \omega_0) \sin l \qquad (15\text{–}9)$$

For a rigid rotation, $\omega \equiv \omega_0$, so that $v_r \equiv 0$; differential galactic rotation implies a finite radial speed for the star.

How about v_t, the star's tangential speed with respect to the LSR? Again, from Figure 15–7:

$$v_t = \Theta \sin \alpha - \Theta_0 \cos l \qquad (15\text{–}10)$$

and now the law of sines gives

$$\sin(l/R) = \sin(90° - l - \alpha)/d \equiv [\cos(l + \alpha)]/d$$

$$\equiv (\cos \alpha \cos l - \sin \alpha \sin l)/d \qquad (15\text{–}11)$$

where the last equality follows from the identity

$\cos(x + y) = \cos x \cos y - \sin x \sin y$. We solve Equation 15–11 for $\sin \alpha$ and use Equation 15–8 for $\cos \alpha$ to put Equation 15–10 in the form

$$v_t = R_0(\omega - \omega_0) \cos l - d\omega \quad (15–12)$$

Now rigid rotation ($\omega = \omega_0$) implies that $v_t = -d\omega_0$, which means that the LSR coordinate system is *inertial*—its axes always point in the same directions relative to all other galaxies and the Universe.

The *Oort formulas*, Equations 15–9 and 15–12, are completely general (for concentric, circular, co-planar orbits); you will see them extensively in Section 15–4(c) and Chapter 20. Here, however, we specialize to the solar neighborhood, where $d \ll R_0$. To do so makes use of an approximation valid for regular functions.

If $f(x)$ is a smoothly varying curve (function of x), then we can find the value of this curve near some point $x = x_0$ by using the Taylor expansion:

$$f(x) = f(x_0) + (df/dx)_{x_0}(x - x_0) \\ + (1/2)(d^2f/dx^2)_{x_0}(x - x_0)^2 + \cdots$$

Now $(df/dx)_{x_0}$ is the curve's *slope* at x_0, and $(d^2f/dx^2)_{x_0}$ is its curvature, or the rate of change of the slope with x; recall that $x - x_0$ is small. The first two terms in this expansion, with $(df/dx)_{x_0} = $ constant, excellently approximate $f(x)$ when the curvature is negligible. Near the Sun, ω is approximately equal to ω_0, and so, to a first approximation,

$$\omega - \omega_0 \approx (d\omega/dR)_{R_0}(R - R_0) \quad (15–13)$$

where $(d\omega/dR)_{R_0}$ is the rate of change of the orbital angular speed with respect to distance evaluated at the distance $R = R_0$. Historically, we use the *Oort constant A*, defined as

$$A \equiv -(R_0/2)(d\omega/dR)_{R_0} \quad (15–14)$$

to write the radial speed (Equation 15–9) as

$$v_r = -2A(R - R_0) \sin l \quad (15–15)$$

but you can see that, for small d, Figure 15–7 implies that $R_0 - R \approx d \cos l$, so that Equation 15–15 takes the final form

$$v_r = Ad \sin 2l \quad (15–16)$$

where the identity $\sin 2l = 2 \sin l \cos l$ has been used. The tangential speed equation (15–12), if the same approximation (Equation 15–13) and the relation $R_0 - R \approx d \cos l$ are assumed and terms like d^2 or smaller are neglected, becomes

$$v_t = d(A \cos 2l + B) \quad (15–17)$$

where the *Oort constant B* is given by

$$B \equiv A - \omega_0 \quad (15–18)$$

and where the identity $\cos^2 l = (1/2)(1 + \cos 2l)$ has been used.

If the radial and tangential speeds, given by Equations 15–16 and 15–17, are plotted versus galactic longitude, the resulting curves are termed double-sinusoids (sinusoidal with a period of 180°, not 360°). Figure 15–8 shows observed stellar motions plotted in this way—it is clear that our Galaxy is *rotating differentially*. Figure 15–9 illustrates the physical cause for the results shown in Figure 15–8. Because A is positive (why?), the angular speed $\omega(R)$ decreases with increasing R in the solar neighborhood. In fact, it is decreasing rapidly enough that the following effects are observable: (a) stars with $R < R_0$ are moving about the galactic center more rapidly than the LSR and (b) stars with $R > R_0$ are orbiting at lower angular speeds than the LSR. So both sets of stars appear to be moving toward larger values of l; those closer in are passing us by, and those farther out are being left behind by the LSR. The resulting apparent stellar motions with respect to the LSR are shown in Figure 15–9, decomposed into radial and tangential parts; compare this with Figure 15–8.

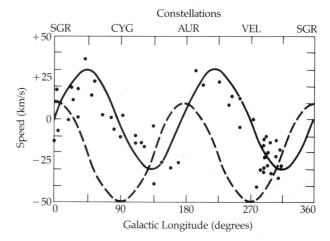

Figure 15–8

Observed galactic rotation. The radial velocities of nearby Cepheids are plotted as a function of galactic longitude. These are motions with respect to the LSR. The solid curve is the expected motions in the Oort model.

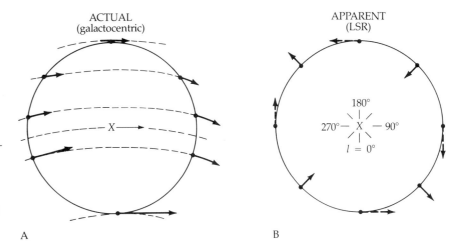

ACTUAL
(galactocentric)

APPARENT
(LSR)

Figure 15–9

Differential galactic rotation. (A) Stars and the LSR (X) orbit the galactic center with smaller orbital speeds at greater distances. (B) With respect to the LSR, we expect a double-sinusoidal pattern in stellar radial velocities.

A

B

(b) Characterizing the LSR

The LSR is specified by its distance R_0 from the center of the Galaxy and its circular orbital speed Θ_0 around the center of the Galaxy. How to determine these parameters? The stellar effects of differential galactic rotation (Figure 15–8), for example, may be evaluated using Equations 15–16 and 15–17 to find Oort's constants A and B. Since radial velocity data permit us to reach great distances, the reasonably accurate value of $A = 15$ km/s · kpc has been found from B stars and Cepheid variables (Chapter 18). Proper-motion data give us the more uncertain value of $B = -10$ km/s · kpc. Using Equation 15–18, we have $\omega_0 = \Theta_0/R_0 = A - B = 25$ km/s · kpc. Clearly, an independent determination of either R_0 or Θ_0 is needed before the other unknown can be found. Recent observations suggest that $A = 14$ and $B = -12$, so that $A - B = 26$ km/s · kpc. A value of $A = -B = 13$ is well within the errors of the observations.

In the next section (and in Chapter 20), you will see that a good value for the combination AR_0 is obtainable from radio astronomical observations of the neutral atomic hydrogen (H I) and carbon monoxide (CO) in our Galaxy. Using the value for A, we find $R_0 \approx 8.5$ kpc and $\Theta_0 \approx 220$ km/s. If we could somehow determine $\omega(R)$ for our Galaxy and thence $(d\omega/dR)_{R_0}$, Equation 15–14 would give us R_0; unfortunately, we must already know R_0 if we are to obtain $\omega(R)$. The values we use here are, at the time of writing, those recommended to the International Astronomical Union in 1985 to be adopted for R_0 and Θ_0. The old values were $R_0 = 10$ kpc and $\Theta_0 = 250$ km/s.

An orbital speed of $\Theta_0 \approx 300$ km/s has been determined by studying the radial velocities of nearby galaxies (those close enough that the Hubble expansion is not important; Chapter 21) and assuming that the LSR is moving only with respect to the average frame of rest of these galaxies. That is, the galactic center is assumed to be the local standard of rest of these galaxies.

To better understand differential galactic rotation, let's derive the Oort constants for the LSR in Keplerian orbit about the mass $M_G = (1.5 \times 10^{11})M_\odot$. Then

$$\omega^2 R = GM_G/R^2$$

or

$$\omega(R) = (GM_G/R^3)^{1/2} \qquad (15\text{--}19)$$

Differentiating Equation 15–19 with respect to R gives

$$d\omega/dR = -(3/2)(GM_G)^{1/2}R^{-5/2} = -3\omega/2R$$

so that Oort's first constant is (by Equation 15–14)

$$A = (3/4)\omega_0 = 19 \text{ km/s · kpc} \quad (15\text{--}20a)$$

Finally, from Equation (15–18),

$$B = A - \omega_0 = -(1/4)\omega_0 = -6.4 \text{ km/s · kpc} \qquad (15\text{--}20b)$$

Note that $R_0 = 10$ kpc was used in Equation 15–19 to obtain Equations 15–20. The calculated values of A and B assuming Keplerian orbits do *not* agree with the observed values. Why not?

(c) The Rotation Curve of Our Galaxy

As a result of differential galactic rotation, $\omega = \omega(R)$, so that $\Theta = \Theta(R)$—the latter relationship is called the *rotation curve* of our Galaxy. The rotation curves for other (nearby) galaxies are found by measuring the radial velocities of H I features in these galaxies, but for our own Galaxy the problem is more intricate.

Return to Figure 15–7 and notice that the maximum radial speed $v_{r,max}$ observed at a given galactic longitude occurs when the line of sight passes closest (R_{min}) to the galactic center. Here the line of sight is tangent to the orbit, and from Figure 15–7 we have

$$R_{min} = R_0 \sin l \qquad \textbf{(15-21)}$$

Now Equation 15–9 is inverted to the general forms

$$\Theta(R_{min}) = v_{r,max} + \Theta_0 \sin l \qquad \textbf{(15-22a)}$$

and

$$\omega(R_{min}) = \omega_0 + (v_{r,max}/R_0 \sin l) \qquad \textbf{(15-22b)}$$

where $v_{r,max}$ is observed and R_{min} may be found from Equation 15–21. Note that in Equations 15–22 at least two of the following must be known before we can determine the rotation curve: R_0, Θ_0, and ω_0. If we confine our attention to the radial velocities of nearby objects ($d \ll R_0$), then Equations 15–15 and 15–21 lead to

$$v_{r,max} = 2AR_0(\sin l)(1 - \sin l) \qquad \textbf{(15-23)}$$

and we obtain the combination AR_0 by observing at galactic longitudes near (but smaller than) 90° and near (but larger than) 270°. (These provisos follow because the line of sight can be tangent only to orbits interior to R_0.)

Equation 15–23 is useful for distances of a few kiloparsecs, but the interstellar obscuration (Chapter 19) severely impedes stellar studies at greater distances. Fortunately, H I and CO can be seen by radio techniques to vastly greater distances, and then we must use Equations 15–22. Because H I and CO clouds tend to delineate spiral features, only a few places occur where the line of sight is tangent to a spiral arm, and we may evaluate $v_{r,max}$ from these observations, as discussed in Chapter 20. Also, emission from the CO molecule can be used together with data from the associated bright stars to determine the rotation curve beyond the solar circle ($R > R_0$). By combining optical and radio data, we can deduce a rotation curve for our Galaxy (Figure 15–10). From these data, $A = 17.7$ and $B = -8.1$.

Note that the actual galactic rotation curve does *not* follow that expected from simple Keplerian motions. From close to the galactic center out to 300 pc, the curve rises steeply, then drops, bottoming out at about 3 kpc. It then rises slowly out to near the position of the Sun. Carbon monoxide observations show

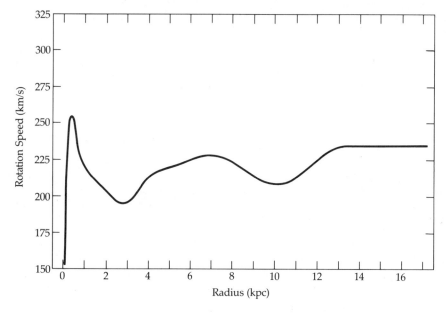

Figure 15–10

Rotation curve of the Galaxy. Based primarily on CO observations, this scale is based on a distance of 8.5 kpc from the galactic center. [D. P. Clemens, *Astrophysical Journal* 295:422 (1985)]

the rotation curve in the outer parts of the Galaxy. The curve rises more beyond the Sun, reaching almost 300 km/s at 18 kpc. What does this curve tell us? Because even the outer parts of the Galaxy do not revolve in a Keplerian fashion, much of the Galaxy's material must lie out beyond the Sun's orbit. From the rotation curve out to 18 kpc, the Galaxy's mass is 3.4×10^{11} solar masses (Equation 14-6). So at least as much mass lies exterior to the solar circle as interior to it. Much of this matter is invisible; the Galaxy seems to have a massive halo of nonluminous matter.

Note that, roughly, the rotation curve is flat outside the central region of the Galaxy. A flat rotation curve requires that $A = -B$, and a value of 13 for each falls within the present error range. However, it is not yet clear that a flat rotation curves applies to the local area in which the rotation constants are derived.

(d) High-Velocity Stars

Now to return to the solar neighborhood to complete our study of stellar motions. Nearby stars are observed to be moving in every possible direction with respect to the LSR (consider the solar motion, discussed in the preceding sections); these stars are in eccentric orbits about the galactic center, and so they cannot be at rest in the LSR. Decompose a star's peculiar motion with respect to the galactic center into three mutually perpendicular components: (1) Π, the speed radially outward (toward $l = 180°$), (2) Θ, the speed toward $l = 90°$ as before, and (3) Z, the speed perpendicular to the galactic plane (positive toward $b = 90°$). Now the stellar-motion components with respect to the LSR are $(\Pi, \Theta - \Theta_0, Z)$; for example, in this notation the solar motion is $(-10.4, 14.8, 7.3)$ km/s. So the Sun is moving inward toward the galactic center and forward toward Cygnus and is rising out of the galactic plane relative to the LSR.

In Figure 15-11, we schematically show the velocity distributions of several types of stars in the galactic plane $(\Pi, \Theta - \Theta_0)$. Here we can clearly see that young stars are almost at rest in the LSR, that older stars are moving faster with respect to the LSR, and that very old Population II stars are in rapid motion relative to the LSR. The Z components of the stellar velocities behave similarly, so that the older a star is, the more rapidly and the farther it moves out of the galactic plane. This phenomenon is associated with the origins of the stars: young stars are born in spiral arms and move in nearly circular orbits in the

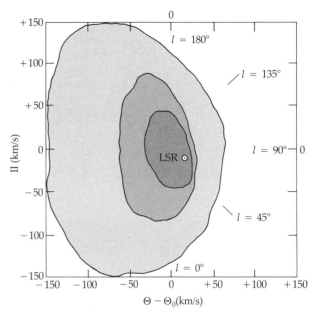

Figure 15–11

Stellar motions in the galactic plane. The observed velocities with respect to the LSR are plotted for stars in the solar neighborhood. The darkest shading are A stars; the intermediate shading, older K giants; the lightly shaded area, very old disk and Population II stars.

galactic plane, whereas the older disk and Population II stars were born far from the galactic plane (and even in the galactic halo).

We can compute velocity curves for stellar orbits of different eccentricities and semimajor axes (Figure 15–11). This procedure shows that young stars and our Sun are in nearly circular galactic orbits at $R \approx R_0$ and that very old stars (including subdwarfs) and the globular clusters are in highly eccentric orbits with $R_0/2 \leq R \leq R_0$ (here R represents the semimajor axis).

Note that the stellar velocities seem to avoid the direction of galactic rotation ($l = 90°$), and no star moves faster than about 65 km/s in this direction. This effect is most noticeable for the so-called *high-velocity stars*, those with in-plane speeds greater than 65 km/s relative to the LSR. Two separate phenomena are involved here. Apparently, a star whose total speed is greater than 315 km/s relative to the galactic center will escape from the Galaxy; hence, the deficiency of stars with $\Theta - \Theta_0 \geq 65$ km/s. The high-velocity stars arise not because they are moving ex-

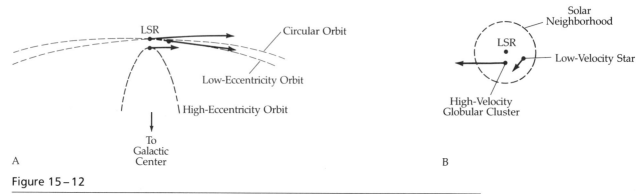

Figure 15–12

High-velocity objects. (A) Three galactic orbits are shown: the circular LSR, a low-eccentricity orbit of a nearby A star, and a high-eccentricity globular cluster near apogalacticon. (B) Relative to the LSR, these same objects in the solar neighborhood appear as a low-velocity A star and a high-velocity globular cluster.

tremely rapidly through the Galaxy but because they are moving much more slowly than the LSR. A star moving slowly near *apogalacticon* (farthest distance from the Galaxy's center; at *perigalacticon*, the star is closest to the center) in a highly eccentric orbit (Figure 15–12) appears to be moving rapidly with respect to the LSR. For example, the LSR (moving at $\Theta_0 = 220$ km/s) seems to be receding at 100 km/s from a star at apogalacticon (in the solar neighborhood) with $\Theta = 150$ km/s. Such high-velocity objects also include the RR Lyrae stars and the high-latitude (large *b*), high-velocity H I clouds.

Problems

1. The Fe I emission lines (at 441.5 and 444.2 nm) in a comparison spectrum are located 15.00 and 15.43 mm, respectively, from an arbitrary reference point. If a stellar Ca I line (of rest wavelength 442.5 nm) is measured to be at 15.27 mm,
    (a) what is the *observed wavelength* of the Ca I line?
    (b) what is the *radial velocity* of this star?

2. By making the appropriate conversions on units, show that Equation 15–3 follows from Equation 15–2.

3. A star located 90° from the solar antapex on the celestial sphere is at rest in the LSR 10 pc from the Sun. As seen from the Sun,
    (a) by what angle (in arcseconds) will this star appear to move on the celestial sphere in ten years?
    (b) in what direction will the star appear to move?

4. The star Delta Tauri is a member of the Taurus moving group. It is observed to have a proper motion of 0.115″/year and a radial velocity of 38.6 km/s and to lie 29.1° from the convergent point of the group.
    (a) What is this star's parallax?
    (b) What is its distance in parsecs?
    (c) Another star belonging to the same group lies only 20° from the convergent point. What are *its* proper motion and radial velocity?

5. Refer to the data given in Problem 15–4. Assume that a probable error of ±0.005″/year is associated with the proper motion of Delta Tauri. If we independently measure the trigonometric parallax of this star (with a probable error of ±0.005″), what are the uncertainties (in parsecs) in the distances determined from these two separate parallaxes?

6. Assume that the mass of our Galaxy is 1.5×10^{11} solar masses and that it is *all* concentrated in a point at the galactic center.
    (a) Plot a *rotation curve* (Θ versus R), with appropriate units and exemplary values along each axis, for this Keplerian case.
    (b) Indicate the rotation *period* at $R = 5$, 8.5, and 20 kpc.
    (c) What is the speed of escape from $R = 8.5$ kpc?

7. A star in a Keplerian galactic orbit of eccentricity 0.8 and semimajor axis 7 kpc moves through the solar neighborhood on its outward journey in the galactic

plane. What is the velocity of this star with respect to the LSR?

8. Use Figure 15–10 to calculate the Galaxy's interior mass to the distance out to which the curve extends. (*Hint:* The motion is *not* Keplerian!)

9. The star BS 1828 has a proper motion of 0.24″/year along position angle 48° (east of north) and a parallax of 0.012″. The H_β line ($\lambda_0 = 486.1$ nm) appears at $\lambda = 485.9$ nm. What is the magnitude of the star's space velocity and what angle does the velocity make to the line of sight (which points away from the Sun)?

10. Determine the proper motion (relative to the LSR) in arcseconds per year of a star in circular motion about the galactic center 4 kpc from the Sun and at a galactic longitude of 60°. Use the rotation curve given in Figure 15–10. (Expect your answers to be small!)

11. The distance from the convergent point to the center of the Hyades cluster is 29.9°, and the cluster's radial velocity is 39.1 km/s. Use this information and that in the chapter to calculate a moving-cluster distance to the Hyades. If the error in the radial velocity determination is 0.2 km/s, what is the error in the distance?

The Evolution of Stars

Stars play a central role in cosmic evolution. The observational characteristics of stars were our primary concern in Chapters 10 through 13; here we discuss the physical laws that govern the structure and evolution of stars. By suitably combining these laws into theoretical stellar models, we can understand the equilibrium configurations of stars, their evolution in time, and the astrophysical basis for cluster H–R diagrams.

We will concentrate upon normal stars, which are in the vast majority, and return later (Chapter 18) to rare peculiar stars with their fascinating variability. These are not two distinctly different classes of objects, for a normal star generally exhibits several brief phases of peculiarity during its lifetime. If we lived a billion years, we would see this, but at this instant of cosmic time we happen to catch most stars in their sedate states—only a few are in active or violent phases now.

By its very nature, a normal star is so hot that it must be entirely gaseous. Let's use the physical precepts applicable to stellar-sized masses to see how this comes about.

16–1 The Physical Laws of Stellar Structure

(a) Hydrostatic Equilibrium

A star is a massive object held together by its self-gravitation and supported against collapse by its internal pressures. The simplest stellar model is a static, spherically symmetric ball of matter; any parameter of physical interest depends not upon time or angle but only upon the radial distance r from the star's center. *Hydrostatic equilibrium* ensues when the inward gravitational attraction exactly balances the outward pressure forces at every point (r) within the star (Figure 16–1). As we approach the center of the star, the pressure must steadily increase to counterbalance the increasing weight of the material that lies above. This physics is expressed in the *equation of hydrostatic equilibrium*:

$$dP/dr = -GM(r)\rho(r)/r^2 \qquad (16\text{–}1)$$

The pressure steadily *decreases* at greater values of r.

Equation 16–1 appears to have three independent variables: $P(r)$, $\rho(r)$, and $M(r)$, but $M(r)$ increases by the amount (Figure 16–1) $dM(r) = 4\pi r^2 \rho(r)\, dr$

when we add the indicated spherical shell (from r to $r + dr$). So $M(r)$ is determined from $\rho(r)$ by

$$dM/dr = 4\pi r^2 \rho(r) \qquad (16\text{–}2)$$

our second fundamental equation. If the star's outer radius is R, then its total mass is

$$M = \int_{r=0}^{r=R} dM(r) = 4\pi \int_0^R \rho(r) r^2 \, dr \qquad (16\text{–}3)$$

We have added together the masses of all the onion-like shells in the model star.

Hence we need know only $\rho(r)$ to determine first $M(r)$ by Equation 16–2 and then $P(r)$—the pressure profile within the star—by Equation 16–1. Given the star's radius, we also find its total mass using Equation 16–3.

The Sun's Central Pressure

To exercise our physical intuition, let's roughly calculate the pressure at the center of our Sun, using the equation of hydrostatic equilibrium. We know that $G = 6.67 \times 10^{-11}$ N · m²/kg², $M_\odot = 1.989 \times 10^{30}$ kg, and $R_\odot = 6.96 \times 10^8$ m; therefore the *mean* solar density is $\bar{\rho}_\odot = 3M_\odot/4\pi R_\odot^3 = 1410$ kg/m³. Taking the surface pressure as zero, letting $r = dr = R_\odot$, and using $M(r) = M_\odot$ in Equation 16–1 yields

$$P_c \approx GM_\odot \rho_\odot/R_\odot = 2.7 \times 10^{14}\ \text{N/m}^2$$

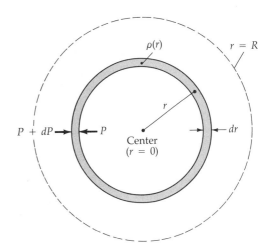

Figure 16–1

Hydrostatic equilibrium in a star. A spherical star is divided into a series of shells of thickness dr, each of which is in equilibrium.

The atmospheric pressure at the Earth's surface is 1 atm $= 1.01 \times 10^5$ N/m², however, and so the gas pressure $P_c \approx 2.7 \times 10^9$ atm at the Sun's center supports the weight of its mass.

Our computed mean solar density is slightly greater than that of water (1000 kg/m³), but since the Sun is centrally condensed, the actual central density is more like 1.6×10^5 kg/m³. The tremendous pressures supporting stars imply very high interior temperatures — high enough to maintain the gaseous state.

(b) Equations of State

To solve our stellar model requires the density profile $\rho(r)$. The composition and local state of the stellar materials must be examined in detail — are we dealing with a gas, a liquid, or a solid? In Chapter 17, you will see that quantum liquids and solid lattices of atomic nuclei become important when we consider stellar corpses (such as white dwarfs and neutron stars); in the vast majority of cases, however, stars are gaseous.

For normal stars, we assume that the material is a *perfect gas*, which obeys the *perfect-gas law*:

$$P(r) = n(r)kT(r) \qquad (16-4)$$

Here the pressure $P(r)$ is related directly to the gas particle number density $n(r)$ (particles/m³), Boltzmann's constant $k = 1.381 \times 10^{-23}$ J/K, and the gas temperature $T(r)$. Now $n(r)$ can be expressed in terms of $\rho(r)$ and the *gas composition* $\mu(r)$:

$$n(r) = \rho(r)/\mu(r)m_H \qquad (16-5)$$

where $m_H = 1.67 \times 10^{-27}$ kg is the mass of a hydrogen atom. Recall that, in terms of the *mass fractions* of hydrogen (X), helium (Y), and all heavier elements —"metals"—(Z), the composition (or *mean molecular weight*) is

$$\mu = [2X + (3/4)Y + (1/2)Z]^{-1} \approx 1/2 \quad (16-6)$$

So the perfect-gas equation of state becomes

$$P(r) = \rho(r)kT(r)/\mu(r)m_H \qquad (16-7)$$

As you will see later, $\mu(r)$ is usually specified for a stellar model, so that only $T(r)$ remains to be determined. In massive stars, the gas pressure is significantly augmented by *radiation pressure*: $P_{rad}(r) = (a/3)T^4(r)$, where $a = 7.564 \times 10^{-14}$ J/m³ · K⁴ is the radiation constant.

The Sun's Central Temperature

By inverting the perfect-gas law of Equation 16–7, using the P_c and $\rho_\odot$ values found in Section 16–1(a), and making the same crude but simplifying approximations, we estimate the required central temperature T_c of our Sun:

$$T_c \approx P_c\mu m_H/\rho_\odot k \qquad (16-8)$$

Since μ is near $1/2$ in our Sun (mostly hydrogen), we have

$$T_c \approx 12 \times 10^6 \text{ K} \qquad (16-9)$$

which is close to the computer-determined value of about 16 million! At these temperatures, the gas is dissociated into ions and electrons — a neutral mixture termed a *plasma*. Our earlier statement that stars are gaseous spheres is consistent and true. The gaseous state persists throughout the star since $\rho(r)$ decreases approximately as rapidly as $T(r)$ outward through the star.

(c) Modes of Energy Transport

To determine $T(r)$, consider how energy is transported from the stellar interior to the surface, where it is radiated away into space. The star's self-gravitation forces the stellar center to be at a much higher temperature (16 million K for our Sun) than the stellar surface (about 6000 K for the Sun); the heat energy must flow from the higher-temperature regions to the lower-temperature regions (the second law of thermodynamics).

Three processes transport energy: conduction, convection, and radiation. *Conduction* occurs when energetic atoms communicate their agitation to nearby cooler atoms by collisions; this mode works extremely well in solids (especially metals) but poorly in gases because of the low thermal conductivities. (The atoms are too far apart.) *Convection* transports heat energy by means of mass motions in fluids. When $T(r)$ varies rapidly enough with distance (that is, there is a steep temperature gradient dT/dr), the fluid becomes unstable and boils. This process takes place in limited regions of most stars, with hot fluid masses rising, releasing their heat energy, and sinking again to pick up more energy. The top of such a *convective zone* is observed at the base of the Sun's photosphere (Chapter 10). Unfortunately, no completely adequate mathematical theory of convective transport has yet been devised. However, a useful and reasonable formulation of convection has been applied to stars.

The third mode, *radiative transport*, is an important means of energy flow in sections of most stars. Here high-energy photons flow outward from the star's interior, losing energy by scattering and absorption in the hot plasma of the *radiative zone*. At the extreme temperatures of stellar interiors, the most important sources of such *opacity* (Chapters 10 and 13) are (1) *electron scattering*—the scattering of radiation (photons) by free electrons and (2) *photoionization*—the use of radiant energy to detach electrons from ions (Chapter 8).

Let's now crudely derive the *equation of radiative transport*. At the base of a thin shell, the spherical surface is essentially a blackbody emitter at temperature $T(r)$, so that by Equation 8–40 we have the outward radiative flux (J/m² · s), $F(r) = \sigma T^4(r)$, where $\sigma = 5.67 \times 10^{-8}$ J/m² · s · K⁴ is the Stefan–Boltzmann constant. At $r + dr$, the temperature is $T + dT$, however, and the outward flux is $F + dF = \sigma(T + dT)^4 \approx \sigma(T^4 + 4T^3\, dT)$. Now dT is negative since the exterior of the shell must be cooler than its interior, so that the flux absorbed within the shell is

$$dF = 4\sigma T^3(r)\, dT \qquad (16\text{–}10a)$$

This absorption is from the *opacity* $\kappa(r)$ of the shell's material and from Equation 10–1:

$$dF = \kappa(r)\rho(r)F(r)\, dr \qquad (16\text{–}10b)$$

Combining Equations 16–10 and defining the *luminosity* (J/s) by $L(r) = 4\pi r^2 F(r)$, we find that the total energy flow per second through a thin spherical shell is

$$L(r) = [-16\pi\sigma r^2 T^3(r)/\kappa(r)\rho(r)](dT/dr) \quad (16\text{–}11)$$

The complete machinery of the theory of radiative transfer introduces an additional factor of $\frac{4}{3}$ into Equation 16–11, so that the correct equation of radiative transport is

$$L(r) = [-64\pi\sigma r^2 T^3(r)]/[3\kappa(r)\rho(r)](dT/dr) \quad (16\text{–}12)$$

The stellar model is now complete with Equation 16–12 since $\kappa(r)$ depends only upon $\mu(r)$, $T(r)$, and $\rho(r)$; at the star's surface, $L(r)$ equals the observed (bolometric) luminosity of the star.

Solar Luminosity from Radiative Transfer

For our Sun, a rough estimate of dT/dr is $-T_c/R_\odot$, which equals a gradient, -2×10^{-2} K/m.

Use Equation 16–12 to estimate the solar luminosity. Setting $r = R_\odot$, $T(r) = T_c$, and $\rho(r) = \rho_\odot$, we have

$$\begin{aligned}L_\odot &\approx (-64\pi\sigma R_\odot^2 T_c^3/3\kappa\rho_\odot)(-T_c/R_\odot) \\ &= (9.5 \times 10^{29}/\kappa)\ \text{J/s} \qquad (16\text{–}13)\end{aligned}$$

where we have yet to determine a reasonable opacity (m²/kg). From its dimensions, κ is the interaction area per gas particle multiplied by the number of particles per kilogram of the stellar material; a mass of 1 kg of completely ionized hydrogen contains 6×10^{26} protons and the same number of electrons. For electron scattering, the interaction area per electron is approximately 10^{-30} m²; for hydrogen photoionization, this area per atom is near 10^{-20} m². In the solar interior, the latter opacity source dominates, so that (very approximately) $10^{-3} \ll \kappa \leq 10^7$. Accurate opacities are extremely difficult to determine. Our prediction for the solar luminosity falls in the broad range $10^{22} \leq L_\odot \ll 10^{32}$ J/s; the mean value of 10^{27} J/s is very close to the measured value of 3.90×10^{26} J/s, implying an opacity value of about 2.4×10^3.

(d) Energy Sources

Because stellar luminosity represents energy loss, no star is perfectly static; a stellar model is, however, an excellent approximation for times that are short relative to stellar evolution time. In fact, stars *must* evolve because they lose energy to space. How long will the star remain in essentially steady state, and what energy source maintains this stability? Geological and paleontological evidence indicates that our Sun has been radiating energy at a fairly steady rate for a few billion years; such energy generation takes place in stellar interiors.

The rate of energy production per unit mass of stellar material (J/s · kg) is denoted by $\epsilon(r)$. (In fact, the energy production rate also depends on temperature and density; we write $\epsilon(r)$ as a shorthand for the temperature and density at r.) Now, $\epsilon = 0$ except in stellar cores and in certain localized spherical shells. For our Sun, we estimate the average value of ϵ needed to maintain the solar luminosity to be

$$\epsilon_\odot \approx L_\odot/M_\odot = 2.0 \times 10^{-4}\ \text{J/s} \cdot \text{kg}$$

We can find out how such energy generation within our thin spherical shell augments the stellar luminosity (Figure 16–1). The luminosity $L(r)$ enters the bottom of the shell while the greater luminosity $L + dL$ leaves the top—from the energy produced in the shell's mass $4\pi r^2 \rho(r)\,dr$. The additional luminosity is

$$dL = 4\pi r^2 \rho(r)\epsilon(r)\,dr \qquad \textbf{(16-14)}$$

Equation 16–14 expresses the balance between net energy lost from the shell dL and net energy generated within the shell, an energy, or *thermal, equilibrium*.

In a quasi-static gaseous star, energy may be generated only by gravitational contraction and/or thermonuclear fusion reactions. Each process is important at some stage in a star's evolution. Let's consider these energy sources in detail.

Gravitational Contraction

Gravitational potential energy can be transformed to kinetic energy of motion (as when a rock is dropped near the Earth's surface); the bulk form of kinetic energy is *heat*. Consider a very slowly contracting star. The heat energy of its interior provides the pressure—from the random motions of the gas particles—that supports the star against its self-gravitation. When the star contracts to a smaller radius, the self-gravitation increases so that the internal pressures (and hence temperatures and heat energy) must also increase to maintain approximate hydrostatic equilibrium. The gravitational potential energy decreases about twice as fast as the heat energy increases, however, and so to conserve the total energy of the system, approximately half of the potential energy change must be radiated into space—the star's luminosity.

This energy-conversion process may be illustrated by a simple analogy. A small satellite of mass m moves in a circular orbit of radius r at speed v about a great mass M. From Equation 1–32, the satellite's kinetic energy is $mv^2/2$, and its gravitational potential energy is $-GMm/r$. Since the centripetal acceleration v^2/r maintaining the orbit is provided by the mutual gravitational attractive acceleration $GM/r^2 = v^2/r$, the kinetic energy $mv^2/2 = GMm/2r$, or *half* the magnitude of the potential energy. If we now move the satellite to a smaller (stable) orbit at $r - dr$, the increase in kinetic energy is certainly only half the decrease in potential energy (which becomes more negative). To conserve total energy (potential plus kinetic), the other half of the potential energy change must be transmitted to the agent that alters the satellite's orbit—in the case of a star, this energy is radiated away. This result applies generally and is

called the *virial theorem*. It implies that the gravitational contraction of a mass results in the conversion of the gravitational potential energy half to thermal energy and half to radiative energy, that is, $U = KE + PE$.

Let's apply these concepts to our Sun. For each kilogram of solar material, the average gravitational potential energy available for radiation is

$$GM_\odot/2R_\odot = 9.54 \times 10^{10}\ \text{J/kg} \qquad \textbf{(16-15)}$$

Comparing this with $\epsilon_\odot$, we see that *gravitational contraction* can sustain the Sun at its present luminosity for only 15 million years; some other energy source must be sought if we are to account for billions of years of sunshine. In Section 16–3, you will see when gravitational contraction is important in stellar evolution.

Thermonuclear Reactions

Only after about 1938—largely because of the work of Hans Bethe—did astronomers realize that the long-term energy source for stars must be *thermonuclear fusion reactions*. In this process, light atomic nuclei collide with such violence and frequency in the high-temperature, high-density stellar interior that they fuse into heavier nuclei and release tremendous quantities of energy (such as in a hydrogen bomb). We say that the lighter elements "burn" to form heavier elements in this process of *nucleosynthesis*.

In atomic nuclei (Section 8–2), the strong nuclear force overcomes the electrostatic repulsion of the positively charged protons and binds from one to 260 nucleons (protons and neutrons) in a region about 10^{-15} m in diameter. Two nuclei will fuse to form one larger nucleus if they approach within 10^{-15} m of one another, but their mutual electrostatic repulsion—all nuclei have a positive charge—amounts to a 1-MeV potential barrier. In contrast, at 10^7 K the average thermal energy of a proton is only 1 KeV. Classically, protons cannot fuse because of the strong *coulombic barrier*. Fusion *does* happen, however, because quantum physics allows the protons to tunnel through the barrier rather than go over it. The easiest fusion reaction involves two protons (hydrogen nuclei); such reactions become significant at temperatures around 10 million K.

The great abundance of hydrogen makes it the key constituent in stellar nuclear reactions. The next stable nucleus is helium, ${}_2^4\text{He}$, with atomic weight 4.

Since the hydrogen nucleus (one proton) only has atomic weight 1, four protons are required to make one helium nucleus. The atomic weights do not exactly match because the more exact atomic weight of a proton is 1.0078, and four of them add to 4.0312, while the weight of $^4_2$He is 4.0026, leaving a *mass defect* of 0.0286. This mass is converted to an amount of energy given by Einstein's equation for the equivalence of mass and energy,

$$E = mc^2 \qquad \textbf{(16-16)}$$

where c is the speed of light. Because a unit atomic weight is 1.66×10^{-27} kg, the energy released by the conversion of four $^1_1$H nuclei to one $^4_2$He nucleus is

$$E = 0.0286(1.66 \times 10^{-27})(9 \times 10^{16})$$
$$= 4.3 \times 10^{-12} \text{ J}$$

We can also use Equation 16–16 to determine the total energy store of the Sun if we assume that it originally consisted of pure hydrogen, all of which will eventually be converted to helium. The mass liberated in the form of energy in this thermonuclear conversion is the fraction $0.0286/4.0312 = 0.0071$ of the available mass of original hydrogen. Because it is only in the core that the temperature and pressure are high enough to permit nuclear reactions, only about 10 percent of the mass of the Sun is available for energy conversion. So the total thermonuclear energy available in the Sun is

$$E_{\text{total}} = m(4\,^1_1\text{H} - \,^4_2\text{He})(c^2)(0.1 M_\odot)/m(4\,^1_1\text{H})$$
$$= 0.0071(9 \times 10^{16})(0.2 \times 10^{30})$$
$$= 1.28 \times 10^{44} \text{ J}$$

which, with the present solar luminosity of 3.90×10^{26} J/s, will last about 10 billion years. The best estimates of the age of the Solar System yield figures around 5 billion years, so there is ample energy from this reaction to sustain our Sun for another 5 billion years.

Recall that four hydrogen nuclei are required to produce one $^4_2$He nucleus. A simultaneous collision fusing four independent particles is extremely improbable, even at the densities prevalent in the centers of stars, so we must seek a reaction chain. Experimental data pertaining to the types and rates of nuclear reactions probable in stellar interiors are obtained in the laboratory by bombarding target nuclei with projectile particles in high-energy accelerators.

Two different processes lead to the conversion of hydrogen to helium: the proton–proton (PP) chain and the carbon (CNO) cycle. The PP chain dominates at temperatures lower than 2×10^7 K, and the CNO cycle is prominent at higher temperatures (Figure 16–2). In the Sun, for instance, both processes take place but the PP chain is the more important. The CNO cycle plays a negligible role in stars lower on the main sequence than the Sun but predominates in stars hotter than F stars.

The *proton–proton chain* (PP I) consists of the following reactions:

$$^1_1\text{H} + \,^1_1\text{H} \longrightarrow \,^2_1\text{H} + e^+ + \nu \qquad 1.44 \text{ MeV}$$

$$^2_1\text{H} + \,^1_1\text{H} \longrightarrow \,^3_2\text{He} + \gamma \qquad 5.49 \text{ MeV}$$

$$^3_2\text{He} + \,^3_2\text{He} \longrightarrow \,^4_2\text{He} + \,^1_1\text{H} + \,^1_1\text{H} \qquad 12.9 \text{ MeV}$$

where $^2_1$H is heavy hydrogen (deuterium), whose nucleus contains one proton and one neutron, e^+ is a positron, ν a neutrino, and γ a photon. A *positron* has the same mass as an electron, but its charge is positive. *Neutrinos* have no (small?) mass or charge, only energy and spin, and are therefore difficult to detect. Conservation of charge is maintained in the first reaction by the emission of the positron. Note that the first two steps must occur twice before the last can take place and that a total of six protons are involved even though two are again released in the final step. Other reactions may occur instead of the last step of this chain — for example,

$$^3_2\text{He} + \,^4_2\text{He} \longrightarrow \,^7_4\text{Be} + \gamma$$

Then there are two possible branches from $^7_4$Be, both

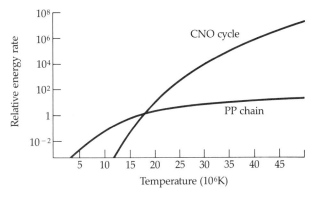

Figure 16–2

Energy-generation rates. The rates for the PP chain and CNO cycle are compared as a function of temperature for Population I stars. Note the crossover at about 18 million K.

resulting in $^4_2$He. All three chains operate simultaneously in a star, but the PP I is the most important; it occurs 91 percent of the time in the Sun, according to theoretical models. On the average, the neutrinos carry off 0.26 MeV.

The other PP reactions occur less frequently and so contribute a minor amount to the Sun's luminosity. The PP II chain is

$$^1H + {}^1H \longrightarrow {}^2H + e^+ + \nu \qquad 1.44 \text{ MeV}$$

$$^1H + {}^2H \longrightarrow {}^3He + \gamma \qquad 5.49 \text{ MeV}$$

$$^3He + {}^4He \longrightarrow {}^7Be + \gamma \qquad 1.59 \text{ MeV}$$

$$^7Be + e^- \longrightarrow {}^7Li + \nu \qquad 0.861 \text{ MeV}$$

$$^7Li + {}^1H \longrightarrow 2\ {}^4He \qquad 17.3 \text{ MeV}$$

with the neutrinos taking away as much as 0.86 MeV. The PP III reaction involves the same first three steps as PP II and then goes on as

$$^1H + {}^7Be \longrightarrow {}^8B + \gamma \qquad 0.14 \text{ MeV}$$

$$^8B \longrightarrow {}^8Be + e^+ + \nu$$

$$^8Be \longrightarrow 2\ {}^4He \qquad 18.1 \text{ MeV}$$

with the neutrinos escaping with 7.2 MeV.

The neutrinos produced by the PP chain should escape the Sun without interacting with any other solar material. Since 1965, Robert Davis has attempted to detect high-energy solar neutrinos (from PP III) by means of an elaborately shielded tank (located far underground in the Homestake mine in South Dakota) filled with 100,000 gallons of tetra-chloroethylene (C_2Cl_4). Although neutrinos react only very weakly with matter, they should transmute some of the ^{37}Cl in the tank to ^{37}A. These radioactive argon atoms can then be collected and counted. The measured neutrino flux is only one-third that predicted by current theoretical solar models. This discrepancy remains one of the great unsolved mysteries of astrophysics; it is called the solar neutrino problem.

The *CNO cycle* also converts hydrogen to helium, but it requires a carbon nucleus as a catalyst:

$$^{12}_6C + {}^1_1H \longrightarrow {}^{13}_7N + \gamma$$

$$^{13}_7N \longrightarrow {}^{13}_6C + e^+ + \nu$$

$$^{13}_6C + {}^1_1H \longrightarrow {}^{14}_7N + \gamma$$

$$^{14}_7N + {}^1_1H \longrightarrow {}^{15}_8O + \gamma$$

$$^{15}_8O \longrightarrow {}^{15}_7N + e^+ + \nu$$

$$^{15}_7N + {}^1_1H \longrightarrow {}^{12}_6C + {}^4_2He \qquad (16\text{–}17)$$

Each step in the chain need occur only once to convert four protons to an *alpha particle* (helium nucleus). The second and fifth steps take place because $^{13}_7N$ and $^{15}_8O$ are unstable isotopes of their respective elements, with halflives of only a few minutes. (Recall that *halflife* is the time in which half of the original quantity of an isotope has disintegrated into its more stable nuclear form.) The cycle starts with the reaction between carbon and hydrogen but ends with the release of an identical carbon nucleus; the $^{12}_6C$ acts as a catalyst. Although the temperature may be sufficiently high, the CNO cycle cannot operate in a star unless carbon is available. Thus the evolution of Population II stars, which formed with essentially no carbon, must differ from Population I stars, which began with a significant amount of carbon.

Higher temperatures are required for the carbon cycle because the coulombic barriers of the carbon and nitrogen nuclei are higher than those of protons and helium nuclei. As a result, temperature dependence goes roughly as T^4 for the PP reaction and as T^{15} for the carbon cycle (Figure 16–2).

At very high temperatures, around 10^8 K, other reactions begin transmuting helium to heavier elements, namely, carbon, oxygen, and neon. Three alpha particles ($^4_2$He has historically been called an alpha particle) will form carbon:

$$^4_2He + {}^4_2He \rightleftharpoons {}^8_4Be + \gamma$$

$$^8_4Be + {}^4_2He \longrightarrow {}^{12}_6C + \gamma \qquad (16\text{–}18)$$

This reaction is known as the *triple-alpha process*, the first stage of helium burning. The intermediate beryllium-8 is not very stable, and the back reaction occurs readily. Nevertheless, an equilibrium is established where some $^8_4$Be takes part in the second step. Other reactions follow with the addition of further alpha particles, leading to the formation of $^{16}_8O$, $^{20}_{10}Ne$, $^{24}_{12}Mg$, and even heavier elements, up to iron. Light elements other than hydrogen, helium, and carbon are rare deep inside stars because such elements (deuterium, lithium, beryllium, and boron) quickly combine with protons at temperatures of only a few million degrees to form one or two helium nuclei—for example,

$$^7_3Li + {}^1_1H \longrightarrow 2\ {}^4_2He$$

The triple-alpha and other helium-burning reactions play a major role in the evolution of stars. These reactions make it possible to start with stars of pure hydrogen and eventually produce most heavy elements up to iron.

16–2 Theoretical Stellar Models

(a) Recapitulating the Physics

The physical principles basic to stellar structure are hydrostatic equilibrium, the perfect-gas equation of state, the various modes of energy transport, and the gravitational and thermonuclear sources of stellar energy. These are the tools used by astrophysicists in computing *stellar models* — theoretical stars in which physical parameters and their rates of change throughout the star are described. These mutually dependent parameters include temperature $T(r)$, mass $M(r)$, density $\rho(r)$, pressure $P(r)$, luminosity $L(r)$, rate of energy production $\epsilon(r)$, and chemical composition in terms of the mean molecular mass $\mu(r)$. Their interdependence is given by the basic equations of stellar structure:

Hydrostatic equilibrium
$$dP/dr = -GM(r)\rho(r)/r^2 \quad (16\text{–}1)$$

Mass equation
$$dM/dr = 4\pi r^2 \rho(r) \quad (16\text{–}2)$$

Energy transport
$$dT/dr = [-3\kappa(r)\rho(r)/64\pi\sigma r^2 T^3(r)]L(r) \quad (16\text{–}12)$$

Energy generation (thermal equilibrium)
$$dL/dr = 4\pi r^2 \rho(r)\epsilon(r) \quad (16\text{–}14)$$

Equation of state
$$P(r) = k\rho(r)T(r)/\mu(r)m_H \quad (16\text{–}7)$$

These equations describe how the parameters vary through the star *only* if we know their values at some particular points (or shells) in the star, such as at the center and the surface. These values constitute the *boundary conditions*. At the center, for example, where $r = 0$, the boundary conditions for the mass and luminosity must be $M(r) = 0$ and $L(r) = 0$. So that the theoretical models bear some relationship to real stars, we use observed stellar characteristics for the boundary conditions at the surface. So at $r = R$, the radius of the star, $M(R) = M$, $L(R) = L$, $T(R) = T_{eff}$, the effective surface temperature (or photospheric temperature), and both $\rho(r)$ and $P(r)$ approach zero.

The properties of the stellar material ultimately depend on the *chemical composition* of the star, which determines the type of energy generation and the opacity.

Chemical composition plays a central role in stellar structure, as may be seen from the basic equations. The equation of state shows that $P(r)$ depends on μ; since the hydrostatic equation indicates that $\rho(r)$ is strongly dependent on $P(r)$, it follows that $\rho(r)$ depends on μ. Note that $\rho(r)$ appears in all the other equations. The difficulty lies in knowing how the composition varies through the star. At best, this must be estimated, and the parameters resulting from the calculations must be compared with observed quantities.

Many theoretical models have been computed for our Sun, for it is the best-observed star and serves as the prototype for all others. These models differ in the relative abundances of hydrogen, helium, and the heavy elements assumed for the newly formed Sun and in the degree of mixing of the elements and their participation in thermonuclear reactions in the solar interior. Compositional mixing becomes more thorough as the extent of the *convection zone* increases. Current solar models that best fit various observations indicate that a large fraction of the hydrogen at the center of the Sun has already been converted to helium (the composition is about 50 percent H, 50 percent He), so that there is a *central helium-enriched core*. Its energy generation, hydrogen burning by the PP chain, still occurs primarily in the core. Energy transport is *radiative* for most of the interior, but beyond about $0.8R_\odot$ the temperature gradient becomes sufficiently steep to maintain *convection*. The solar granulation is direct evidence for such a convection zone.

Models for other stars differ from that for the Sun in the extent of the convection zone and so also in the degree of mixing. In stars more massive than the Sun (those higher on the main sequence), convection occurs at the core and the envelope is entirely radiative.

(b) The Russell–Vogt Theorem

In 1926, H. N. Russell and H. Vogt independently showed that the structure of a star [and hence $M(r)$] are uniquely determined by chemical composition and mass. This *Russell–Vogt theorem* shows that the values of the other parameters — temperature, pressure, density, luminosity, and distance from the center — are completely determined for any point in the star specified by $M(r)$. Moreover, only *one* solution is possible, given the total mass and composition. Actually, because of the interdependence of the several parameters, the Russell–Vogt theorem may be generalized since any one of the three parameters M, R, or L will determine the structure if the chemical composition is known.

The importance of the theorem arises in interpreting the H–R diagram, for it tells us that stars of the same composition but different masses should lie along a smooth line — the greater the mass, the higher the luminosity and the surface temperature. This line, as we shall see, is the zero-age main sequence. Stars that lie elsewhere on the H–R diagram, such as the red giants, have chemical compositions different from the main-sequence stars as a result of the gradual accumulation of products of nuclear reactions. Such compositional changes occur naturally in stellar evolution.

(c) The Physical Basis of the M–L Relationship

The empirical mass–luminosity relationship must have a physical basis in the equations of stellar structure if they are correct. Start with hydrostatic equilibrium (Equation 16–1) and let $dP \rightarrow \Delta P$ and $dr \rightarrow \Delta r$; then

$$\Delta P = P_s - P_c = 0 - P_c$$

where P_s = pressure at the surface, P_c = pressure at the center, and $\Delta r = R$, so that

$$P_c \propto M\rho/R$$

For a perfect gas,

$$P \propto \rho T$$

and so

$$\rho T_c \propto M\rho/R$$

and $T_c \propto M/R$. We follow the same approximation with the radiation transport equation (16–12) so that

$$L \propto R^2(T_c/\kappa\rho)(T_c/R)$$
$$\propto RT_c^4/\kappa\rho$$

Now $\rho \propto M/R^3$, and so

$$L \propto R^4 T_c^4/\kappa M$$

into which we substitute T_c from hydrostatic equilibrium:

$$L \propto R^4(M/R)^4/\kappa M$$
$$\propto M^3/\kappa$$

which is close to the observed relation, $L \propto M^{3.3}$. The difference lies in the dependence of the opacity on temperature and density.

16–3 Stellar Evolution

The study of the physical changes that take place in stars as they alter their composition because of thermonuclear reactions is the subject of *stellar evolution*. All stars follow the same general sequence in their evolution: protostar, pre-main sequence, main sequence, and post-main sequence. Basically, a star's evolution is determined primarily by its mass. Chemical composition plays a secondary role, so that Population I and Population II stars of the same mass follow somewhat different histories. Stellar evolution aims to understand how the luminosity and surface temperature (two observables) change with time. A plot of the points representing different evolutionary stages on an H–R diagram is called a star's *evolutionary track*. This section examines theoretical evolutionary tracks calculated from the basic equations of stellar structure.

(a) The Birth of Stars: Protostars and PMS Stars

Stars are born from the gravitational contraction of interstellar clouds of gas and dust. Chapter 19 provides observational evidence for this process; here we concentrate on the theoretical understanding. Be warned that we do not yet understand the complete process of star formation, but the basic theme is clear: as an interstellar cloud contracts, gravitational potential energy is converted in part (50 percent) to thermal energy and in part (50 percent) to radiative energy. Eventually, the core heats up to the ignition temperature of fusion reactions, and a star is truly born. Prior to that event, the star goes through protostar and pre-main-sequence stages. The contracting cloud is a *protostar* before it establishes hydrostatic equilibrium. Between that stage and the ignition of fusion reactions, it is called a *pre-main-sequence (PMS) star.* The track traced on the

H–R diagram before the star hits the main sequence is called its *PMS evolutionary track.*

The evolutionary tracks for protostars of different masses differ. For this reason, the discussion splits into two parts: solar-mass protostars and massive protostars (ten solar masses or more). Despite differences that come with different masses, the theoretical calculations have the following common features: (1) the collapse starts out in free fall; that is, it is controlled only by gravity (with negligible pressure; "free fall" means that the particles in the cloud do *not* collide as the cloud collapses, so that the internal pressure is zero); (2) it proceeds very unevenly, for the central regions collapse more rapidly than the outer parts and a small condensation in hydrostatic equilibrium forms at the center; (3) once the core forms, it accretes material from the infalling envelope; (4) the star becomes visible to us either by accreting all the surrounding material onto itself or by somehow dissipating it.

Solar-Mass Protostellar Collapse

Now let's take a look at one model for the formation of a sun-like star. Imagine a huge interstellar cloud of dust and gas, mostly in the form of molecular hydrogen (H_2), with sufficient mass to contract gravitationally. Observational data indicate that interstellar dust grains such as those that are part of this cloud are composed of graphite, silicates, and ices (Chapter 19). The initial density of the cloud is 10^{10} particles/m^3, and the initial diameter is 5×10^6 solar radii. During the collapse (assumed pressureless, so particles don't collide), material at the cloud's center increases in density faster than at the edge. Because of the density increase, the collapse time at the center is decreased; it collapses faster, grows denser, and so collapses still faster. The rest of the cloud's mass is left behind in a more slowly contracting envelope. This part of the collapse takes place in free fall.

Let's work out free-fall collapse explicitly. Consider a test particle of mass m at the edge of a cloud of mass M, radius R, and initial density ρ_0. Imagine that the particle falls straight into the center, so that it follows an elliptical orbit of semimajor axis $a = (1/2)r$ and $e = 1$. Then

$$M = (4/3)\pi R^2 \rho_0$$
$$= (4/3)\pi(2a)^3 \rho_0$$
$$= (32/3)\pi a^3 \rho_0$$

so that Kepler's third law,

$$P^2/a^3 = 4\pi^2/GM$$

becomes

$$P = (3\pi/8G\rho_0)^{1/2}$$

The free-fall time t_{ff} equals half this time, P, so that

$$t_{ff} = (3\pi/32G\rho_0)^{1/2}$$
$$t_{ff} = (6.64 \times 10^4)/\rho_0^{1/2} \text{ s} \qquad \textbf{(16–19)}$$

for ρ_0 in kilograms per cubic meter. For the cloud being considered here, $\rho_0 = (10^{10})(3.3 \times 10^{-27} \text{ kg}) = 3.3 \times 10^{-17} \text{ kg/m}^3$, so that

$$t_{ff} = (6.64 \times 10^4)/(3.3 \times 10^{-17})^{1/2}$$
$$\approx 10^5 \text{ years}$$

With the rapid fall of material in the core, the hydrogen molecules gain kinetic energy. They hit each other and also strike dust grains and so transfer kinetic energy to the grains. Heated by such collisions, the dust grains radiate at infrared wavelengths. As long as this heat radiation can escape into space, the kinetic energy is dissipated, the cloud stays cool, the pressure stays low, and the collapse continues in free fall. At some time, however, the density of the core reaches a critical value, at which point the cloud becomes opaque (optical depth ≥ 1) and traps infrared radiation. Then the core's collapse slows down dramatically as hydrostatic equilibrium is established, some 4×10^5 years after the collapse began.

Meanwhile, the envelope continues to fall inward, showering mass onto the core. As the matter piles up, it increases the core's mass and temperature. At about 2000 K, the hydrogen molecules break up into atomic hydrogen and soak up the energy used to dissociate them. The pressure doesn't rise fast enough to keep the star in equilibrium, and gravity takes command again. The protostar again goes into free fall. When all the hydrogen molecules have broken up, the pressure can rise again and the collapse stops. The star slowly contacts as a PMS star. The total evolutionary time from the start of collapse to this stage is on the order of 1 million years.

We can trace the evolutionary track of a solar-mass protostar on an H–R diagram (Figure 16–3), beginning when the cloud has reached a temperature of 2000 K (point A). As it contracts more, its luminosity and surface temperature both increase (points A to C) until they reach a maximum (point D). Note that

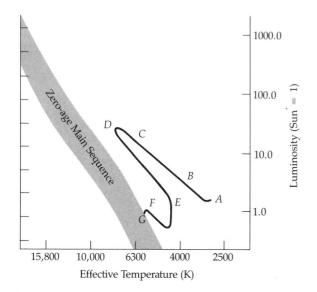

Figure 16–3

Pre-main-sequence evolution of a solar-mass star. *(Based on theoretical calculations by R. Larson)*

the protostar is very luminous yet generally cooler than its main-sequence temperature, and so it must be larger in radius (recall that $L = 4\pi R^2 \sigma T^4$).

Eventually, the interior heats up enough to establish hydrostatic equilibrium (point *D*); it is then a PMS star. Its temperature is so low that its opacity is high and convection transports the energy outward — that's the physical reason for its large size, as large as that of a red giant. As the star shrinks more (points *D* to *E*), the interior reaches a high enough temperature so that the opacity becomes low enough for radiation rather than convection to serve as the energy transport mode. This change causes the evolution path to bend sharply to the left toward the main sequence (points *E* to *F*). Finally, core temperatures hit the minimum required for the PP reaction (point *G*). The heat from fusion reactions sustains hydrostatic equilibrium, and the star stops contracting.

Massive Protostellar Collapse

The collapse of a massive protostar follows the same general scheme as that for a solar-mass one. The two important differences are (1) fusion reactions begin *before* the accretion of the envelope stops and (2) massive protostars do not accrete all of their original cloud's material; some is blown away.

Start with a dense cloud having a molecular hy-drogen density of 10^{12} atoms/m³ and containing 20 solar masses of both gas and dust. Silicate and graph-ite grains (which evaporate at 1500 to 1700 K) are coated with ice mantles (which evaporate at 150 K) at the start. After 4.7×10^5 years, a starlike core 10^{10} m in radius (15 times the size of the Sun) has formed. A shock layer, from accretion of material, surrounds the core (Figure 16–4). Outside the shock is a region (about 10^{12} m in radius, 1500 $R_\odot$) that is dust-free because the heating has evaporated all the dust. In the next region, about ten times bigger (10^{13} m in radius) and somewhat cooler, graphite and silicate cores exist without ice mantles. These grains absorb the photons generated by the accretion and emit ra-diation in the infrared at a temperature of roughly 1000 K.

Eventually fusion reactions begin in the core, which becomes a massive main-sequence star (in this case, roughly a B0 star of mass $12M_\odot$). Its surface temperature (28,000 K) is hot enough to emit photons shortward of 91.2 nm to ionize the hydrogen. A small H II region forms 6.4×10^5 years after the start of the collapse. The H II region expands and helps to blow away some of the accreting material, and the star appears to our optical view. (An H II region is a zone of singly ionized hydrogen surrounding a hot star; see Section 19–2.)

Collapse with Rotation

The models described so far have all assumed that the clouds are *not* spinning — that their angular

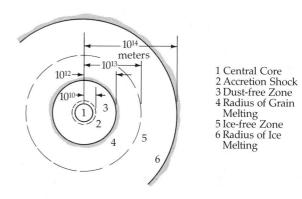

Figure 16–4

Region around a protostar of mass $20M_\odot$. For a nonrotat-ing massive star, a shock layer forms around the core and a region outside of it is free of dust grains. *(Based on theoretical calculations by H. Yorke)*

momentum is zero — but interstellar clouds probably rotate at least a bit. An isolated rotating cloud must conserve angular momentum. As the cloud collapses, each particle moves closer to the axis of rotation and the cloud must rotate faster. At some point, the rotational velocity may become so high that the centripetal acceleration v^2/r balances the gravitational force per unit mass, and the collapse halts. Note that it is the distance from the *axis* of rotation, not from the center of the cloud, that determines the angular momentum. A particle that starts out near the rotation axis can fall a large distance toward the center without changing its distance from the axis much at all, and it will be able to move much closer to the center before its rotational velocity is high enough to stop it. Those parts of the cloud originally near the rotation axis will collapse more than those near the equator, and the cloud will flatten into a disk.

The addition of spin to theoretical models of protostar collapse makes the calculations much harder and the results less conclusive. One key point emerges: in some cases, a disk-like core of rings of material results. These rings turn out to be unstable in some instances, and they break up into several blobs (Figure 16–5). Sometimes these blobs coalesce into fewer ones. If each blob eventually becomes a star, we then have a natural explanation for the occurrence of multiple star systems.

(b) Evolution on and off the Main Sequence

When thermonuclear reactions become the main source of energy generation, the internal stellar pressures quickly attain the level necessary for hydrostatic equilibrium. At this time, contraction stops altogether and the star is stable; it is now truly a star. We refer to this moment of stellar development as age zero. Zero-age stars have surface temperatures and luminosities corresponding to main-sequence stars; in other words, they lie on the main sequence of the H–R diagram. As hydrogen is converted to helium, the composition and therefore the mean molecular weight change, altering the structure of the star. These changes are gradual at first, and the star does not deviate greatly from its initial position on the main sequence. To specify the position in the H–R diagram for stars just starting their hydrogen burning (hydrogen to helium), we refer to the *zero-age main sequence (ZAMS)*.

The development of stars on and from the ZAMS

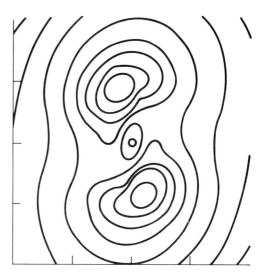

Figure 16–5

Collapse of a cloud with rotation. These are density contours in the equatorial plane of a rotating cloud about 25,000 years after the start of its collapse. Note how two blobs are forming. *[A. Boss and P. Bodenheimer, Astrophysical Journal 234:289 (1979)]*

depends primarily on two characteristics: mass and initial composition. For protostars, the rate of stellar development depends strongly on the total mass. The more massive stars lie higher on the main sequence. Massive stars cannot remain on the main sequence as long as smaller-mass stars can; this is readily seen by comparing rates of energy output (luminosities). Hot stars (O and B stars) expend their energies at far higher rates in proportion to their masses than does the Sun. In contrast, a main-sequence M star expends its energy sedately.

Let's look at a star's lifetime quantitatively. From the mass–luminosity relationship for main-sequence stars (Section 12–2b)

$$L_*/L_\odot = (M_*/M_\odot)^{3.3}$$

A star's lifetime t depends on its store of energy (mass) and on the rate at which it spends that energy (luminosity). So relative to the Sun,

$$
\begin{aligned}
t_*/t_\odot &= (M_*/M_\odot)/(L_*/L_\odot) \\
&= (M_*/M_\odot)/(M_*/M_\odot)^{3.3} \\
&= (M_*/M_\odot)^{-2.3}
\end{aligned}
$$

So the more massive the star, the shorter its lifetime.

Recall that the structure of a star depends on the mean molecular weight, or composition. Moreover, if there is not enough $^{12}_{6}C$ available to act as a catalyst for the CNO cycle, that chain of reactions will not occur, however high the temperature is. Stars now forming out of the interstellar medium will accrete a significant fraction of their mass (on the order of 2 percent) in the form of elements heavier than hydrogen and helium. If they are slightly more massive than the Sun, such stars will then generate energy predominantly by the CNO cycle. These are the young Population I stars in the spiral arms, where the dust and gas are found.

A Population I Star of Mass 5M⊙

Let's follow the evolution of a Population I star with a mass five times that of the Sun. The details of the track will change as new nuclear-reaction data and model calculations become available, but this general description fits most of the observational data. We have chosen a $5M_{\odot}$ star as an example because of its more rapid evolution compared to a $1M_{\odot}$ star. Table 16–1 summarizes the important characteristics and phases of stellar evolution.

The start of interval A at O in Figure 16–6 represents the main-sequence position of the star in the H–R diagram at age zero, when hydrogen burning started at the center of the star. Initially, only temperatures at the very center are sufficiently high for the CNO cycle to operate, and fresh hydrogen is supplied

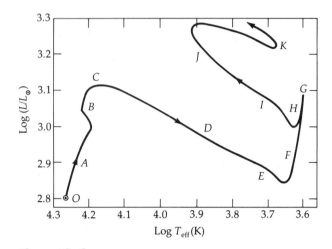

Figure 16–6

Evolutionary track for a star of mass $5M_{\odot}$. *(Based on theoretical calculations by I. Iben Jr.)*

by convective mixing throughout a core. Conditions in the core change in a rather complex fashion, both as a result of the depletion of hydrogen and because of other nuclear reactions involving oxygen and nitrogen. When only a small fraction of all the material in the core is hydrogen, first the core and then the whole star contracts (stage B).

Eventually the core hydrogen is exhausted. As a result of the gravitational contraction of the core, ma-

Table 16–1   Stages of Stellar Evolution (Figure 16–6)

H–R Position	Stage	Physical Processes
$<O$	Protostar	Dust and gas cloud collapses rapidly, accompanied by heating of interior and ionization of atoms
	PMS	Semihydrostatic equilibrium; contraction and heating continue
O	ZAMS	Hydrogen burning commences
A, B	Initial evolution on main sequence	Hydrogen consumed in core; some contraction occurs
C	Evolution off main sequence	Hydrogen depleted in core, isothermal helium core and hydrogen-burning shell established
D, E	Evolution to the right in H–R diagram	Core rapidly contracts, envelope expands, hydrogen-burning shell narrows
F, G, H	Red giant	Energy output increases, convective envelope forms, helium burning begins
I, J	Cepheid	Convective shell contracs, core helium burning becomes major energy source
K, etc.	Supergiant	Helium-burning shell forms

terial just beyond the core is pulled into higher-temperature regions and hydrogen burning starts in a shell surrounding the (primarily) helium core. This shell is fairly thick at first (stage C), but as a larger fraction of the total mass of the star is concentrated into the core, it narrows (stage D). Through stage C, the core is so dense that, to a large extent, it supports the weight of material above it. When this equilibrium ends, the core contracts much more rapidly and heats up. When this happens, energy generation in the shell is accelerated and the outer envelope expands. Such an expansion must be accompanied by cooling at the surface, and the position of the star in the H–R diagram moves to the right (stages C to E).

During the last part of the expansion phase, convection develops in the envelope and changes the direction of the evolutionary track from decreasing to increasing luminosity (region E) by carrying a greater part of the energy outward to the star's surface. This begins the *red-giant* phase of the star's life (stages F to H). Temperatures in the interior continue to rise owing to core contraction; finally a point is reached (about 10^8 K) where the triple-alpha process can begin at the star's center. Helium burning at this stage is short-lived; the star's surface cools, and it drops from position G to position H. The sharp peak at G is called the *helium flash;* it signals the rapid ignition of helium burning in the core.

Why a helium flash rather than a slow ignition? During the core's contraction, the matter there has been compressed to densities greater than 10^8 kg/m³ and no longer acts as an ideal gas; the matter becomes *degenerate* (Chapter 17). One property of degenerate matter is that it has a very high thermal conductivity, so that heat flows through it very quickly. Hence, when the helium ignition temperature is reached in one part of the core, the turnon spreads throughout the core in a flash (by astrophysical standards) — perhaps in just a few minutes!

Once again gravitational contraction takes over until temperatures are sufficiently high to reignite helium. The balance between helium burning in the core and hydrogen burning in the shell gradually shifts in favor of helium burning, and the surface of the star becomes both hotter and brighter (I, J).

When the star reaches this region of the H–R diagram, the outer layers become quite unstable and alternately expand and contract. The pulsations cause the star to change its luminosity periodically; it is a variable star. The phase just before point I corresponds to a Cepheid variable (Section 18–2). When the helium is exhausted and the core consists of carbon, contraction again occurs and a helium-burning shell is formed; by now the star may be a supergiant (K). We are most uncertain about these last phases of stellar evolution.

A Population I Star of Mass 1$M_\odot$

Now we turn to a star having the same mass as our Sun to see what post-main-sequence evolutionary track the Sun might follow. The star reaches the ZAMS when the PP chain dominates energy production. After about 10 billion years, the main-sequence phase ends when almost all of the hydrogen in the core has been converted to helium. During this time, the temperature in the core increases gradually and the star contracts slightly. This results in a greater production of energy, and the star's luminosity increases (A to B in Figure 16–7; see Table 16–2).

When the hydrogen in the core is used up, the thermonuclear reactions cease there. However, they keep going in a shell around the core, where fresh hydrogen still exists. With the end of fusion reactions in the core, the core contracts. This heats up the layer of burning hydrogen, and the reactions produce more energy, but the layer of burning hydrogen heats up the surrounding envelope and causes it to expand. So the radius of the star increases, and its surface temperature decreases. The temperature drop increases the opacity, and convection carries the energy outward in the star's envelope. Note that convective energy transport becomes effective at high opacities and that the convective outflow causes the star's structure to change.

The star is now a red giant; its position on the H–R diagram moves to the region of lower surface temperatures and higher luminosities (C to D). In an additional 500 million years, a solar-mass star ends up with a luminosity of about 1000$L_\odot$, a surface temperature of about 3000 K, and a radius of about 100$R_\odot$. A red giant has a strikingly different structure than a main-sequence star. Most of its mass is concentrated in a dense core only a few Earth radii in size, at temperatures of some 50 million K. The red-giant core is so dense that it no longer behaves like an ordinary gas. The electrons in the core become a *degenerate gas.* In this state, they produce a *degenerate gas pressure,* which depends only on density, not temperature, and this enables the core to attain a pressure

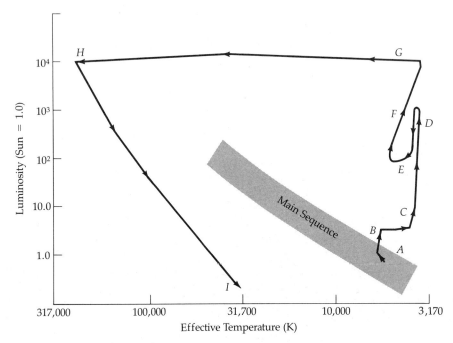

Figure 16 – 7

Evolutionary track for a star of mass $1M_\odot$.

sufficient to balance its gravitational force even though no fusion reactions are going on in it.

As the bloated star attains its red-giant status (point D), the core temperature—which has been steadily increasing as the core contracts—hits the minimum necessary to start helium burning by the triple-alpha process. This helium core is degenerate. Once part of it ignites in the triple-alpha reaction, the heat generated by the fusion spreads rapidly throughout the core by conduction. The rest of the core quickly ignites. If the core were an ordinary gas, this explosive ignition would expand it as a result of the rapid increase in temperature and pressure. The core is degenerate, however, and increased temperature does not increase the pressure in a degenerate gas. So the core does not expand. Instead, the in-

Table 16 – 2   Evolutionary Phases of a Solar Mass Star (Figure 16 – 7)

H – R Position	Stage	Physical Processes
A	ZAMS	Core hydrogen burning begins
B	Evolution on main sequence	Core hydrogen burning ceases; shell hydrogen burning begins
C	Evolution off main sequence	Shell hydrogen burning continues; convection dominates energy transport
D	Red giant	Helium flash occurs; core helium burning begins
E	Subgiant	Core helium burning continues along with shell hydrogen burning
F	Red giant again	Thermonuclear reactions in core end; shell helium and hydrogen burning continues
G	Variable star	Expansion and contraction throw off outer layers
H	Planetary nebula	Star enters planetary nebula stage
I	White dwarf	All thermonuclear reactions stop; slow cooling

creased temperature increases the rate of the triple-alpha process, generating more energy, further increasing the temperature, and so on. This out-of-control process in the core is called the *helium flash*. When the core temperature finally reaches about 350 million K, the electrons become nondegenerate. Then the core expands and cools.

After the helium flash, the star's radius and luminosity decrease a little and its point on the H–R diagram moves slightly downward and to the left. The star quietly burns helium in the core and hydrogen in a layer around the core (point *E*). This phase is the core-helium-burning analog of the star's main-sequence phase (core hydrogen burning).

Eventually the triple-alpha process converts the core to carbon. The reaction stops in the core but continues in a layer around it. This situation—core shut down but thermonuclear reactions going on in a layer—resembles that when the star first evolved off the main sequence. The physical processes force the same evolution; the burning layer makes the star expand. The star again becomes a red giant (point *F*). The electrons in the core—this time carbon-rich—become degenerate again.

Because the rate of the triple-alpha reaction is very sensitive to changes in temperature, the helium-burning shell causes the star to become unstable. Here's how: Suppose the star contracts a little. The temperature and energy production in the layer increase; the pressure also increases. However, the increase in pressure more than compensates for gravity, and so the outer parts of the star expand. The expansion leads to decreases in temperature, pressure, and—most dramatically—energy generation rate. The star contracts, the energy generation increases, the star expands, and the cycle repeats. The star pulsates slowly—once every tens of thousands of years. The pulsations gradually grow larger and more unstable (because of the thinness of the helium-burning shell). A final violent one ejects the cool outer layers of the star (point *G*). A hot core is left behind. The ejected envelope forms a planetary nebula (point *H*; see Sections 18–4 and 19–2).

For a star of roughly 1 solar mass or less, the core never reaches the ignition temperature of carbon burning because it has become degenerate and cannot contract and heat up to ignite carbon burning. In about 75,000 years, such a star becomes a white dwarf, composed mostly of carbon (*H* to *I*). Without

energy sources, the white dwarf cools to a black dwarf in a few billion years.

Extremely Massive Stars

Let's look at recent theoretical work on the evolution of really massive stars—50 to 100 solar masses. This work points out how an effect that is usually ignored dramatically changes the evolution of certain stars. That effect is *mass loss*. The Sun loses mass at a rate of about 10^{-14} solar mass per year, by the solar wind. Other stars are known to lose mass at much greater rates in outflows that are called *stellar winds*. Red giants, for instance, blow off their envelopes at rates of 10^{-7} to 10^{-6} solar mass per year. Massive O stars also have stellar winds at about 10^{-7} to 10^{-6} solar mass per year for the strongest winds. Note that an O star will lose a few solar masses of material during its main-sequence lifetime of a few million years.

The mass loss changes the evolutionary tracks of stars with masses of $50M_{\odot}$ to $100M_{\odot}$. Such stars lose 50 to 60 percent of their initial mass by the end of their main-sequence lifetime. The outer layers of the stars are stripped off—so much, in fact, that the core is revealed and the products of the CNO cycle (such as nitrogen) lie exposed at the surface. Such stripped cores may never become red giants because the layers above the core, where shell burning would take place, have been removed. (The peculiar objects known as *Wolf–Rayet stars*, which are hot stars with strong emission lines in their spectra, may be examples of these stars. They have abnormally large abundances of nitrogen and carbon.)

The evolutionary tracks that result from these calculations (Figure 16–8) start with the stars on the initial main sequence. Take the star of mass $100M_{\odot}$ as the extreme case. From the initial main sequence (point *A*), the star evolves to slightly lower luminosities and lower effective temperatures (*A* to *B*). After 2.8 million years, the star has a mass of only $43M_{\odot}$, a luminosity of $(8.6 \times 10^5)L_{\odot}$, a surface temperature of 4.3×10^4 K, and a radius of $16.7R_{\odot}$. The evolutionary track then swings blueward; 60,000 years later (point *C*) the star has crossed the initial main sequence! At this point, it has a mass of $42M_{\odot}$, a luminosity of $(9.7 \times 10^5)L_{\odot}$, a surface temperature of 5.6×10^4 K, and a radius of $10.5R_{\odot}$. The evolutionary track then swings redward until helium ignition

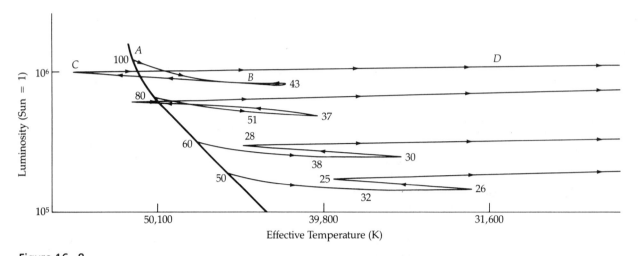

Figure 16–8

Evolutionary tracks for very massive stars. The initial masses (50, 60, 80, and 100$M_\odot$) are indicated along the main sequence. The decrease in mass from stellar winds is indicated along the tracks. (*Based on theoretical calculations by C. de Loore, J.-P. De Greve, and D. Vanbeveren*)

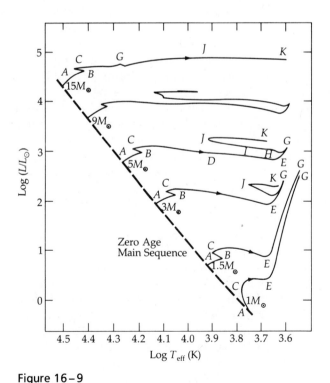

Figure 16–9

Evolutionary tracks for Population I stars of different masses. (*Based on theoretical calculations by I. Iben Jr.*)

occurs (point *D*), 3 million years from the initial main sequence.

Differences Due to Mass

Stars of reasonable masses follow the same general development as that described for the 5$M_\odot$ star. The details and especially the time scales differ (Figure 16–9). Less massive stars, like the Sun, have more extensive hydrogen-burning regions but no convective cores. Less massive stars suffer fewer of the structural changes and subsequent contractions that more massive stars undergo; hence their tracks are considerably simpler and smoother. In solar-type stars, hydrogen burning extends over a relatively large fraction of the star, both in the core-burning phase (points *A* and *B* in Figure 16–9) and during the shell-burning phase (up to *E*). As in the massive stars, the red-giant phase (*F* to *H*) corresponds to the establishment of a deep convective envelope.

In massive stars (15$M_\odot$), the temperatures in the core become high enough to start a triple-alpha process shortly after the star leaves the main sequence. Helium burning in the core starts well before the star has become a red giant (note the location of stage *G* for the 15$M_\odot$ star). The star moves horizontally rather than upward across the H–R diagram. Stars may move in and out of the red-giant region as well as the unstable zone (as Cepheid variables) several times;

massive stars probably do this more often than less massive ones.

Stars less massive than the Sun enter the red-giant region twice. After the first red-giant phase, they evolve through the instability zone, which in this case corresponds to the RR Lyrae variables. Of the low-mass stars, only old Population II stars, like those in the globular clusters, have had time to develop to this stage.

Chemical Composition and Evolution

Population II stars contain only 0.01 percent heavy elements. Do their evolutionary tracks differ much from those of Population I stars with the same mass? The overall sequence is the same, but these stars exhibit a significant difference in position on the H–R diagram during core helium burning. What happens to a star with a heavy-element abundance of 0.01 percent that leaves the main sequence with mass of $0.7M_\odot$ (Figure 16–10)? Almost 16 billion years after the start of core hydrogen burning, PP reactions have consumed the core's hydrogen fuel. Shell burning takes over the energy production, and the star rises to the red-giant region, at first slowly and then swiftly. While a red giant, the star blows off some mass by a strong stellar wind. After the helium flash, the star settles down to core helium burning; it then has a mass of about $0.625M_\odot$ because of the mass lost

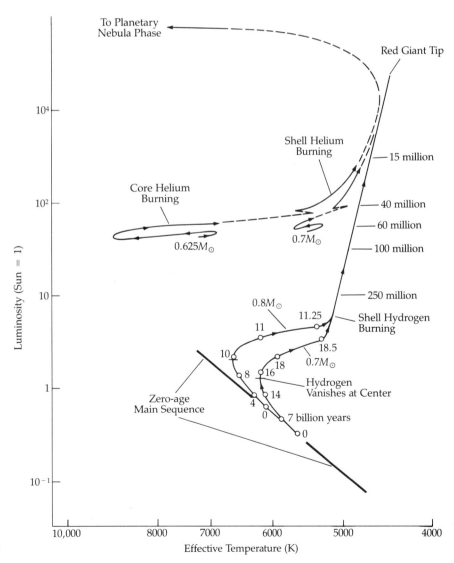

Figure 16–10

Evolutionary tracks for stars with low metal abundances. The two tracks are for two stars, one of mass $0.8M_\odot$ and one of mass $0.7M_\odot$, leaving the main sequence. Both stars lose mass through the red-giant phase. *(Based on theoretical calculations by I. Iben Jr.)*

by a stellar wind. The energy production goes on in a hydrogen-burning shell and a helium-burning core, where the density is roughly 10^7 kg/m³ and the temperature about 100 million K. As this star evolves, its luminosity stays roughly constant and its surface temperature changes, first to higher temperatures and then to cooler ones before the core burning stops. The star's evolutionary track makes a zigzag on the H–R diagram.

Stars with fewer heavy elements (only 0.001 percent) and a range of masses also form a core-helium-burning horizontal branch, but relative to stars of the same mass but higher heavy-element composition

(0.01 percent as above), these stars are shifted to the left (higher temperatures) on the H–R diagram. In clusters, the reason for the range of masses along the horizontal branch is the different amounts lost while the stars are red giants. All the horizontal-branch stars in such a group start out with about the same mass and end up with helium cores of the same size, but some lose more mass than others as red giants and end up with very thin hydrogen envelopes around their helium cores. They are very much like main-sequence stars composed of pure helium and lie far to the left on the H–R diagram. Those that lose very little mass as red giants end up with thicker

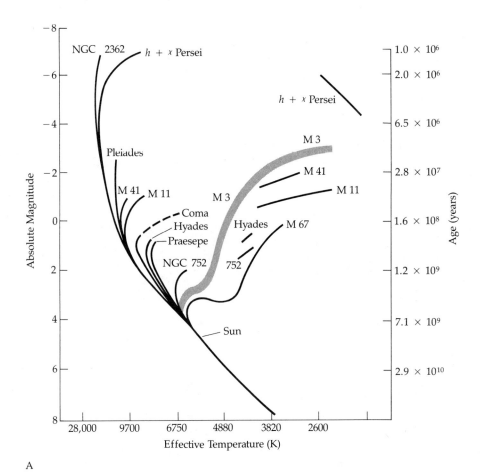

A

Figure 16–11

(A) H–R diagrams for clusters. The observed turnoff points give the approximate ages for a variety of clusters. *(Adapted from a diagram by A. Sandage)* (B) H–R diagram for the Pleiades. The solid lines are the theoretical calculations for the ZAMS and for the best fit for an evolved cluster. (C) H–R diagram for M 67. Same comparison as for the Pleiades. The color excess and apparent distance modulus are given for each cluster. *[Adapted from diagrams by D. A. VandenBerg, Astrophysical Journal Supplement Series 58:711–769 (1985)]*

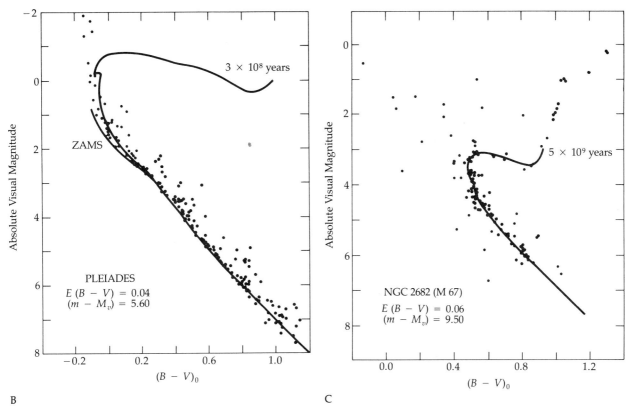

B

C

Figure 16-11 *(continued)*

hydrogen envelopes. They are similar to ordinary red giants, ending up farther to the right on the H–R diagram.

(c) Interpreting the H–R Diagrams of Clusters

If a cluster of stars is formed such that all members contract out of the gas cloud more or less simultaneously, the locations of the stars in the H–R diagram will depend upon the *time* elapsed since the initial formation. For example, 10^8 years after the formation of the cluster, stars with masses greater than about $8M_\odot$ will have evolved beyond the end of the track shown in Figure 16–9. The $5M_\odot$ stars will have reached stage *K*, and less massive stars will still lie on the main sequence. The rapidity of the post-main-sequence phase relative to evolution on and near the main sequence allows us to date a cluster on the basis of its H–R diagram. In the above example of 10^8 years, the cluster has stars on the main sequence

up to a luminosity corresponding to $3M_\odot$ or slightly more, say $L/L_\odot \approx 100$ or absolute magnitude ≈ 0. Somewhat more massive stars lie a little to the right of the main sequence, and some stars have reached the giant branch but have not yet gone beyond it. Such a cluster closely resembles M 11 (Figure 16–11).

The *turnoff* from the main sequence reflects the time elapsed since the stars first arrived on the ZAMS. Ages of clusters may be determined by comparing the turnoff points in the theoretical H–R diagram (Figure 16–11) with the scale on the right of the diagram. The observed H–R diagrams of various clusters represent the loci of the ends of the evolutionary tracks for member stars to that particular time since formation. These loci are called *isochrones,* constant-time lines. The fact that computed isochrones closely resemble observed H–R diagrams is one of the major triumphs of modern astrophysics—it demonstrates that the basic physics of stellar models is correct.

The time taken up by the protostar contraction is so short that it can usually be neglected, particularly

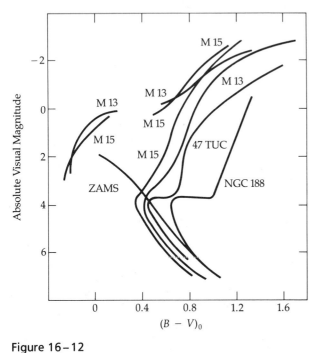

Figure 16–12

H–R diagram for globular clusters. The old galactic cluster NGC 188 has been included for comparison. *(Adapted from a diagram by A. Sandage)*

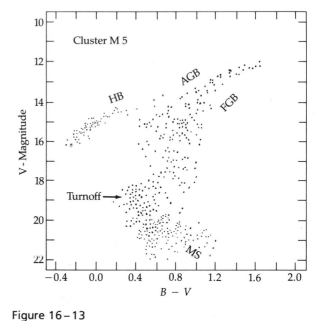

Figure 16–13

Schematic H–R diagram for globular clusters. *(Adapted from a diagram by V. Castellani)*

for well-developed clusters. In many clusters, however, the contraction time for massive stars is so much faster than that for low-mass stars that the massive stars will have already started to evolve off the main sequence by the time the low-mass stars reach it. Such a young cluster is the Pleiades, which still has massive luminous stars on the main sequence. The turnoff at the very upper end of the H–R diagram of this cluster (Figure 16–11A, B, and C) and the appearance of the supergiant branch have usually been attributed to the rapid evolution of these massive stars. The Pleiades have an age of 3×10^8 years. At the other extreme, we have an old open cluster (Population I) whose stars are highly evolved, M 67. It has stars of about $1.25 M_\odot$ evolved onto the red-giant branch and stars of $1 M_\odot$ just about to leave the main sequence. One estimate gives the age of M 67 as 5×10^9 years (Figure 16–11C).

The H–R diagrams of Population II globular clusters (Figure 16–12) differ from the H–R diagrams typical of both intermediate Population I clusters (like M 11) and old Population I clusters (like M

67). The globular clusters are older than M 67, but because they started with almost pure hydrogen and helium and virtually no heavy elements, their evolutionary tracks took different forms than Population I clusters did. The present H–R diagrams for globular clusters represent the end points of the evolutionary tracks (Figure 16–13). Differences from one cluster to another are attributed to differences in the initial chemical composition. Globular clusters clearly show the horizontal branch (HB in Figure 16–13) that results from different mass losses and chemical compositions. Note that a globular cluster's H–R diagram shows stars at all evolutionary phases: main sequence (MS in Figure 16–13), lasting 10^{10} years with core hydrogen burning; first red giant (FGB) for 10^8 years with shell hydrogen burning; horizontal branch (HB) for 10^8 years with core helium burning and shell hydrogen burning; and the asymptotic giant branch (AGB) for 10^7 years with shell double burning. (Compare these stages with those in Figure 16–10.)

Problems

1. Verify that about 6×10^{11} kg of hydrogen is converted to helium in our Sun every second.

2. **(a)** If a star is characterized by $M = 2 \times 10^{32}$ kg and $L = 4 \times 10^{32}$ W, how long can it shine at that luminosity if it is 100 percent hydrogen and converts all of the H to He?
   (b) Do a similar calculation for a star of mass 10^{30} kg and luminosity 4×10^{25} W.

3. Briefly describe the evolution of the following stars from a cloud of gas and dust to their demise:
   (a) $M = 10M_\odot$
   (b) $M = 0.1M_\odot$
   Clearly indicate which stages of the evolution are highly uncertain.

4. Using Figures 16–7 and 16–9 and the data in Table 16–2, sketch the H–R diagrams for star clusters of ages 10^7, 10^8, and 10^9 years (these are *constant-time* lines!) Clearly label the axes, and comment upon the significance of your results (turnoff points).

5. Although detailed models of stellar structure require the use of complex computer codes, simple scalings can be obtained by making rough approximations. For a variable x, we can substitute $\Delta x / \Delta r$ for dx/dr in order to obtain a crude result. (This method is a rough version of a numerical technique called "finite differences.")
   (a) Use the equation of hydrostatic equilibrium (16–1) to show that the central pressure scales as $P_c \propto M^2/R^4$. Substitute $\Delta P / \Delta r$ for dP/dr and take this difference between $r = 0$ and $r = R/2$, that is, $\Delta P / \Delta r \approx [P(r = R/2) - P_c]/(R/2 - 0)$. You may assume that $P(r = R/2)$ is negligible compared with P_c. Also, substitute the mean density of the star $\bar{\rho}$ for $\rho(r)$.
   (b) Now use the radiative transport equation and the same method as in part (a) to approximate dT/dr and obtain the theoretical mass–luminosity relationship, $L \propto M^3$. Assume that $\kappa(r)$ is constant and that $T(r) \propto T_c$.
   (c) Use the same method to obtain a rough relationship between the temperature and mass of a main-sequence star. (No other variables should appear in the proportionality.)
   (d) Combine your answers to parts (b) and (c) to obtain a relationship between T and L. Use the H–R diagrams (Figures 13–7 and 13–9) and Table A4–3 to compare the temperature of a star of $L = 10L_\odot$ with that of the Sun. How does this compare with your theoretical T–L relationship (be quantitative)?

6. Assume that the amount of hydrogen mass available for nuclear reactions in the core of a star is $M_c \approx 0.5M$. Further assume for simplicity that the only energy-generating nuclear reaction is 4 ^{1}H $\rightarrow$ ^{4}He + energy (ignore the fact that some of the energy is in the form of positrons and neutrinos). Obtain an expression for the hydrogen-burning lifetime of a star in years as a function of mass in solar units. (Assume $L \approx$ constant during the hydrogen-burning phase and use the mass–luminosity relationship $L \approx M^3$ in solar units.)

7. Use Figures 16–10, 16–11, and 16–12 to obtain approximate ages for the clusters M 13, M 15, 47 Tuc, NGC 188, and M 5.

Star
Deaths

We have brought you to the brink of the deaths of stars. We have alluded to the processes of post-main-sequence evolution that result in stellar demises and final states—stellar corpses. This chapter examines the violent deaths of stars in detail and the three types of bizarre corpses that can be left behind: white dwarfs, neutron stars, and black holes. These final states are forever; very little can happen to change them.

17–1 White Dwarfs

Several roads result in a white dwarf, depending on the mass of the star. For a star like our Sun [Section 16–3(b)], contraction of the carbon core may not produce temperatures sufficiently high to bring about burning of carbon. The core will nevertheless contract to a highly compressed state, and the increasing temperatures will accelerate the shell helium-burning rates. The star expands and contracts until it ejects its outer layers (Figure 17–1A). The outer envelope expands and cools. Now, however, the expansion is so great that the envelope becomes separated from the core as a thin shell; this is a *planetary nebula* (Figure 17–1B and Chapter 18). The core, having lost its envelope, now stands revealed as a hot, very dense star—a *white dwarf.*

In less massive stars, helium burning may not become appreciable because of low core temperatures. Such stars continue to contract gravitationally to become white dwarfs even without ejecting an envelope. More massive stars end as white dwarfs by some kind of mass loss, perhaps by exchange of mass between members of a binary system [Section 12–4(c)]. Observational evidence indicates that main-sequence stars with $M \leq 7M_\odot$ finish their lives as white dwarfs. Many white dwarfs are in fact faint companions to larger, more massive stars; Sirius B and 40 Eridani B are two examples.

(a) Physical Properties

Because white dwarfs are very dense, the stellar material no longer behaves like an ordinary gas. It becomes so tightly packed that the electrons cannot move completely at random. Hence, their motions are subject to limitations imposed by the proximity of other electrons. Some electrons may still move at very

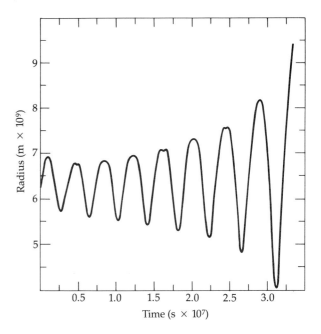

A

B

Figure 17–1

Planetary nebula. (A) Model of a variable star of mass $1.4M_\odot$ pulsating more and more violently until it ejects its outer layers. *(Based on theoretical calculations by D. King)* (B) M 57, the Ring Nebula in Lyra, was formed by the expulsion of the atmosphere of a red giant. The star visible in the center will cool to a white dwarf. *(Lick Observatory)*

high velocities, but they cannot change their velocities by collisions as in an ordinary gas; the electrons may change velocity only by exchanging orbits with other electrons. The laws of quantum mechanics still hold, and the *Pauli exclusion principle* remains valid: only two electrons with opposite spins can have a given energy in a given volume at one time. Because of the close packing, less space is available and the number of possible velocities or energies permissible for an electron becomes smaller. Such a material is called a *degenerate electron gas.* In such a degenerate gas, the electrons are distributed more or less uniformly throughout the medium, surrounding the nuclei. The nuclei themselves are regularly spaced and become more tightly constrained as the pressures increase, until they are so fixed with respect to each other that they resemble a crystalline lattice. Under such conditions, the material more closely resembles a solid than a gas.

The cause of these high densities lies in the fact that all available nuclear energy has been expended and the star contracts gravitationally until stopped by the pressure of the degenerate electron gas. Only stars of mass smaller than about $1.4M_\odot$ (called the *Chandrasekhar limit*) can be stable white dwarfs because of limitations imposed by the stellar structure, which in turn depends upon both hydrostatic equilibrium and the nature of the degenerate electron gas. All the peculiar properties of white dwarfs can be traced to the fact that they are made of degenerate material. A main point to keep in mind is that, the more massive a white dwarf, the smaller its size. In contrast, for main-sequence stars, the more massive ones are larger. (Can you reason why?) Let's examine the mass–radius relationship for white dwarfs, keeping in mind that a degenerate electron gas cannot support a mass greater than $1.4M_\odot$.

The exact relationship between pressure and density in completely degenerate, *nonrelativistic* matter is

$$P = K\rho^{5/3} \qquad (17\text{–}1)$$

where K is a constant. (For a *relativistic* gas, $P \propto \rho^{4/3}$.) This is the equation of state of such material. Contrast this to the equation of state of an ideal gas,

$$P = nkT$$

or $P \propto \rho T$.

Now from hydrostatic equilibrium,

$$P \propto M^2/R^4$$

This result applies to any star. Then using the above equation of state in the density equation,

$$\rho \propto M/R^3$$

we get

$$P \propto \rho^{5/3} \propto M^{5/3}/R^5$$

Now use

$$P \propto M^2/R^4$$

so that

$$M^2/R^4 \propto M^{5/3}/R^5$$

$$R \propto 1/M^{1/3}$$

So as M gets larger, R gets smaller. This result hints at the idea that white dwarfs may have a maximum mass limit. (They do—the Chandrasekhar limit.)

(b) Observations

In 1862, the American optician Alvin Clark observed Sirius B (Figure 17–2), the faint companion to Sirius A. Later, this star was found to be a white dwarf. Because Sirius B is part of a binary system, its mass (in terms of the Sun's mass) can be calculated

Figure 17–2

A white dwarf. Sirius B (arrow), the companion to Sirius A. *(Lick Observatory)*

using Kepler's third law (Section 12–2). This calculation gives a value of about $1M_\odot$ for the mass of Sirius B. This star has a low luminosity, about $(3 \times 10^{-3})L_\odot$, and a high surface temperature, about 29,500 K and so, from $L = 4\pi R^2 \sigma T^4$, it has a radius of around $(7 \times 10^{-3})R_\odot$. Sirius B has an average density of about 3 billion kg/m^3. By coincidence, the brightest star in Canis Minor (the Little Dog, near Canis Major), Procyon, also has a white-dwarf companion. The existence of this companion was predicted in 1862 from the motion of Procyon and was observed in 1882. Called Procyon B, it has a mass of about $0.65M_\odot$.

White dwarfs tend to fall into two general categories: those with spectra showing strong hydrogen lines and those with spectra showing strong helium lines. White dwarfs with the strongest hydrogen lines are put in class DA—D for dwarf and A to indicate that the spectra resemble those of A stars (strongest hydrogen Balmer lines).

An H–R diagram of DA white dwarfs for which reliable observational data are available (Figure 17–3) shows that (1) the DA white dwarfs actually fall along a temperature sequence from 6000 to 31,000 K and (2) they lie parallel to lines of constant radius drawn on the H–R diagram. (Recall that $L = 4\pi R^2 \sigma T^4$. If R is kept constant, L will be high where T is high and low where T is low.) The average radius of these stars is $0.013R_\odot$. Typical values for the physical properties of white dwarfs are mass $0.7M_\odot$, radius $0.01R_\odot$ (7×10^6 m), and density 10^8 kg/m^3.

(c) White Dwarfs and Relativity

The extremely high densities of white dwarfs provide a test for the general theory of relativity, for their surface gravities are high enough to produce a detectable *gravitational redshift* in their spectra. A gravitational redshift occurs whenever light moves from a stronger to a weaker gravitational field. As it does so, it must do work since a photon has an equivalent mass ($E = mc^2$) and a gravitational field can affect it. In such a situation, an ordinary particle loses kinetic energy (as it gains gravitational potential energy) and slows down. Photons cannot slow down, however; they travel only at the speed of light. Instead of slowing down, a photon's loss of energy shows up as a decrease in its frequency (an increase in its wavelength), that is, as a redshift, because $E = h\nu$.

The gravitational redshift produced by a star depends on its mass-to-radius ratio. The larger this ratio, the larger the gravitational redshift. To calculate the shift, consider a photon leaving a mass and traveling to infinity. Its total energy is

$$TE = PE + KE = \text{constant}$$

but $PE < 0$ initially and $PE = 0$ at infinity, so that

$$KE_f = KE_i + PE_i < KE_i$$

If we use Newtonian gravitation, then

$$\Delta KE = \Delta(h\nu) = -GmM/R$$

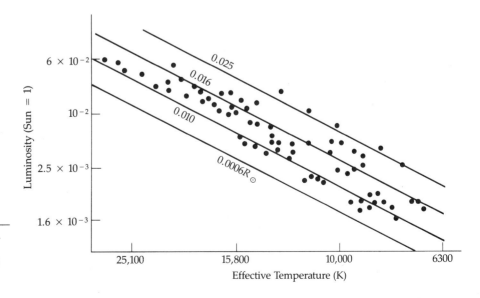

Figure 17–3

H–R diagram for white dwarfs. The lines indicate stars of constant radius. *(Adapted from a diagram by V. Weidmann)*

For a photon,

$$m = E_i/c^2$$
$$= h\nu_i/c^2$$

so that

$$h \Delta\nu = -G(h\nu_i/c^2)M/R$$

and

$$\Delta\nu/\nu_i = -GM/c^2R$$

For strong gravitational fields, we must use general relativity rather than Newton's theory. This application adds a factor of 2:

$$\Delta\nu/\nu_i = -2GM/Rc^2 \qquad (17-2)$$

or

$$\nu_f/\nu_i = 1 - 2GM/Rc^2 \qquad (17-3)$$

where G is Newton's gravitational constant, M is the object's mass, R its radius, and c the speed of light.

The observation is complicated by the motion of the star relative to the Earth because any radial velocity produces a Doppler shift (to the red if the star is receding). Therefore we see both shifts (gravitational and Doppler) together. The two can be separated only if the star's velocity through space can be determined, which is possible for binary systems because their velocities in space can be determined from the spectrum of the primary star. With the velocity known, the Doppler redshift is subtracted from the total redshift to leave any gravitational redshift. For a typical white dwarf with $M = 0.7M_\odot$ and $R = 0.01 R_\odot$, the gravitational redshift amounts to about

$$\Delta\nu/\nu_i = \frac{2(6.67 \times 10^{-11})(0.7 \times 2 \times 10^{30})}{(3 \times 10^8)^2(0.01 \times 7 \times 10^8)} \approx 10^{-4}$$

The measured redshift for Sirius B is 3.0×10^{-4}. The predicted theoretical value is 2.8×10^{-4}. Within experimental error, the redshift observation confirms general relativity.

(d) Magnetic White Dwarfs

Some white dwarfs have intense magnetic fields — 10^2 to 10^4 T at their surfaces. (Recall that the global magnetic field of the Sun is about 10^{-4} T.) These strong fields are probably relics from the time before the star became a white dwarf. The basic physical concept that supports this idea is called *conservation of magnetic flux*. Consider a star with a magnetic field.

The *magnetic flux* is essentially the number of field lines (field strength) times the surface area through which they thread. Imagine compressing the star to a smaller size. The number of field lines remains the same but the surface area decreases, and so the field lines draw closer together. The magnetic field strength increases since the separation of the field lines indicates the intensity of the field.

Because a star's surface area depends on the square of its radius, its magnetic field strength must depend (if flux is conserved) on the inverse square of its radius. An example: Start with a star like the Sun, magnetic field about 10^{-4} T, radius = 7×10^5 km. Imagine collapsing it to the size of a white dwarf, with a radius of 7×10^3 km. What does the conservation of magnetic flux predict for the white dwarf's field strength? We have

$$B_{wd}/B_\odot = (R_\odot/R_{wd})^2$$

where B_{wd} is the white dwarf's field strength, $B_\odot$ the Sun's field strength, R_{wd} the white dwarf's radius, and $R_\odot$ the Sun's radius. Then

$$B_{wd} = B_\odot(R_\odot/R_{wd})^2$$
$$= (10^{-4})[(7 \times 10^5)/(7 \times 10^3)]^2$$
$$= 1 \text{ T}$$

Although lower than the strongest observed fields, this result shows that the simple collapse idea is plausible.

In summary, a degenerate electron gas supports a white dwarf against gravity. The maximum mass of a white dwarf is $1.4 \ M_\odot$. No thermonuclear reactions go on in it; it has come to the end of the line of energy production. Very slowly (in billions of years), its stored internal heat radiates into space. Eventually, it becomes a *black dwarf*.

(e) Brown Dwarfs

There is another category of nonnuclear-energy-generating "stars" that must be distinguished from black dwarfs, white dwarfs, and even red dwarfs; these are the *brown dwarfs*. Rather than representing a true star death, these objects more closely resemble gigantic planets than stars. They result from the gravitational collapse and contraction of protostellar nebulae but have insufficient mass to trigger nuclear reactions in their cores. Somewhat arbitrarily, an object is considered a planet if its mass is less than $0.002M_\odot$ and a brown dwarf if its mass lies in the range $0.002M_\odot$ to $0.08M_\odot$. Objects more massive than $0.08M_\odot$ can develop sufficiently high central temperatures to sustain nuclear fusion. The only energy source for a brown dwarf, therefore, is gravitational contraction. Brown dwarfs are cool and have very low luminosities and so are difficult to observe. One possible candidate is an object detected using infrared speckle interferometry, the companion to the nearby star Van Biesbroeck 8. Its infrared properties imply an object with $T_{eff} = 1360$ K, $R = 0.09R_\odot$, and $L = (3 \times 10^{-5})L_\odot$, values consistent with a large Jovian-type extrasolar planet.

17–2 Neutron Stars

For contracting stellar corpses with masses greater than $1.4M_\odot$, the degenerate electron gas pressure cannot hold off gravity. Matter is crushed to such high densities that inverse-beta decay occurs:

$$p^+ + e^- \rightarrow n + \nu$$

Literally, protons and electrons are squeezed into neutrons; a neutron gas forms. At about 10^{17} kg/m³, the neutrons become subject to quantum laws and become a degenerate gas. In analogy to the behavior of a degenerate electron gas, a degenerate neutron gas provides internal pressure to form a stable entity: a *neutron star*. Because the equation of state for a degenerate neutron gas is almost the same as for an electron one, neutron stars of greater mass will have smaller radii—and an upper mass limit (thought to be about 5 solar masses).

(a) Physical Properties

Neutron stars have diameters of 10 to 20 km, depending on their masses (Figure 17–4). In a typical neutron star with a radius of about 15 km, the inner 12 km consists of a neutron gas at such high densities that it is a fluid. The outer 3 km is a mixture of a neutron superfluid and neutron-rich nuclei arranged in a solid lattice. The structure is that of a crystalline solid, similar to the interior structure of a white dwarf. In the outer few meters, where the density falls off quickly, the neutron star has an atmosphere of atoms, electrons, and protons. The atoms here are mostly iron.

Because a neutron star is so dense, it has an enormous surface gravity. For example, a solar-mass neutron star with a radius of 12 km has a surface gravity 10^{11} times greater than that at the Earth's surface. This intense gravitational field results in a huge escape velocity, as much as $0.8c$. Also, objects falling onto a neutron star from a great distance have at least the escape velocity when they hit. That means that even a small mass carries a huge amount of kinetic energy.

The gravitational redshift from a neutron star is

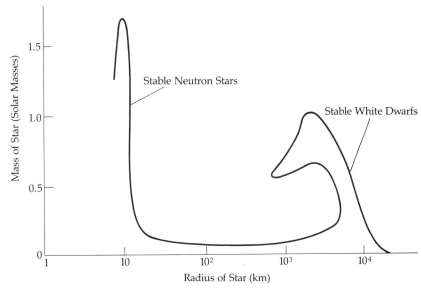

Figure 17–4

Mass–radius diagram for cold matter. Given is the radius for a mass. Note the regions for stable white dwarfs and neutron stars. In these regimes, the internal degenerate pressure supports the star. *(Based on theoretical calculations by K. Thorne and colleagues)*

substantial. For a solar-mass neutron star about 7 km in radius,

$$\Delta\lambda/\lambda_0 = 2\,GM/c^2R$$
$$= \frac{(6.67 \times 10^{-11})(2 \times 10^{30})}{(9 \times 10^{16})(7 \times 10^3)}$$
$$\approx 2 \times 10^{-1}$$

This result means that light emitted at 600 nm would be shifted to 720 nm by the time it reached an outside observer.

(b) Pulsars — Rotating Neutron Stars

In 1967, a large radio telescope was developed by Anthony Hewish in Cambridge, England, to study the scintillations of radio sources. Scintillation is the rapid twinkling of a radio source from density fluctuations in the interplanetary plasma (the solar wind) and in the interstellar medium; it is analogous to the twinkling of visible stars (from density fluctuations in the Earth's atmosphere). Almost immediately, weak, precisely periodic radio signals were detected.

Jocelyn Bell Burnell, then a graduate student in charge of preliminary data analysis, noticed a strange signal that suddenly disappeared only to reappear three months later. The Hewish group concentrated on this unusual signal and found radio pulses occurring at a regular rate, once every 1.33730113 s. Flushed with excitement, they searched the sky for any similar signals and discovered three more objects emitting radio pulses at different rates. They concluded that the objects must be natural phenomena

and named them *pulsating stars,* or *pulsars.* To date, a total of about 150 pulsars have been studied in detail. About an equal number have been recently discovered in special pulsar surveys, for a total of roughly 300.

For a given pulsar, the period between pulses repeats with very high accuracy, better than one part in 10^8. The amount of energy in a pulse, however, varies considerably; sometimes complete pulses are missing from the sequence. Although the intensity and shape vary from pulse to pulse, the average of many pulses from the same pulsar defines a unique shape (Figure 17–5). The average pulse typically lasts for a few tens of milliseconds, with no detectable radio emission between pulses. Individual pulses may be resolved into 20 to 30 subpulses of submillisecond duration, so that the primary pulses are actually the envelopes of these secondary pulses. Pulsars are most readily observed at low frequencies; for instance, the first discoveries were made at 81.5 MHz. The intensity of the pulses diminishes rapidly at higher frequencies, and the pulses become broader and more regular in shape.

For the well-studied pulsars, periods range from 1.6×10^{-3} to 4.0 s, with an average value of 0.65 s. (Only two pulsars are known so far to have millisecond pulses: one with a 1.6-ms interval, the other 6 ms.) In the cases where accurate radio observations have been made, periods have been noted to increase in regular fashion. The rates of change have typical values of about 10^{-8} s/year. Such small increases can be measured only with atomic clocks, whose stability

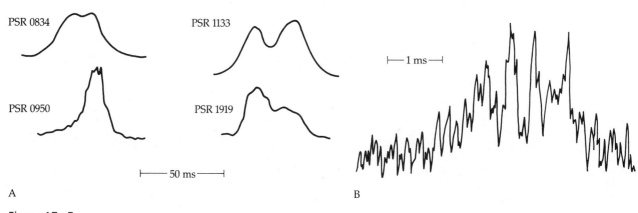

Figure 17–5

Pulsars. (A) A sample of the variety of pulse shapes. (B) Detailed structure within a pulse from PSR 1133. *(Adapted from a diagram by A. T. Moffet and R. D. Ekers)*

is better than 10^{-10} s/year. Note that, very roughly, the pulse period P divided by its rate of change with time dP/dt gives an estimate of a pulsar's age:

$$t \approx P/(dP/dt)$$

What we have done here is estimated the time for the pulse rate to decay from its present value to a very large interval. An example: the Crab pulsar (see below) has $P = 0.03$ s and $dP/dt = 4 \times 10^{-13}$ s/s, so that

$$\begin{aligned} t &\approx 0.03/(4 \times 10^{-13}) \\ &\approx 10^{11} \text{ s} \\ &\approx 10^4 \text{ years} \end{aligned}$$

Approximate distances to pulsars and some properties of the interstellar medium may be deduced directly from pulsar observations. A given pulse arrives at the Earth later as we look at lower frequencies. This phenomenon is called *velocity dispersion,* and it is due to a slowing down of the photon velocity by electrons in the line of sight to the pulsar (analogous to the index of refraction discussed in Section 8–1 and the lower propagation velocity of light in a material medium). Longer wavelengths are slowed down more, and from the observations, we may deduce the mean electron density in the line of sight. Conversely, if we know (or can estimate) the mean electron density, the distance to the pulsar follows immediately. Combined with the observation that most pulsars lie at low galactic latitudes, these data imply that pulsars are quite local (within a few kiloparsecs) and lie in the galactic disk.

Finally, we know that the plane of polarization of linearly polarized radiation (Section 8–1) is rotated when the radiation propagates through a magnetized plasma. This effect, known as *Faraday rotation,* depends upon (1) the mean electron density, (2) the mean magnetic field strength, (3) the square of the wavelength of the radiation, and (4) the distance traveled through the medium. Since pulsar bursts are strongly linearly polarized, we infer that the mean magnetic field in the galactic disk has a strength of about 10^{-10} T. In other words, for a given source, we can measure the angle through which the plane of polarization rotates as a function of wavelength. This gives a value for the product of the electron density and magnetic field strength integrated along the line of sight. Then if we can determine the electron density—and we can, from a measurement of the velocity dispersion—we can infer the mean compo-

nent of the magnetic field strength along the line of sight.

What mechanism keeps the precise clock of a pulsar? The accepted model is that of a rotating, magnetic neutron star known as the *lighthouse model.* The model has two key components: (1) the neutron star, whose great density and fast rotation insure a large amount of rotational energy and (2) a dipolar magnetic field that transforms the rotational energy to electromagnetic energy.

That neutron stars might possess extremely intense magnetic fields follows from the same conservation-of-flux argument applied earlier to white dwarfs. (Recall that observational evidence supports this argument; some white dwarfs have surface magnetic fields of roughly 10^2 T.) Imagine our Sun collapsed to the size of a neutron star 7 km in radius. Calculating the field strength from conservation of magnetic flux, we have

$$\begin{aligned} B_{ns} &= B_\odot (R_\odot/R_{ns})^2 \\ &= 10^6 \text{ T} \end{aligned}$$

More careful calculations indicate that the fields more typically have a strength of 10^8 T. The region close to the neutron star where the magnetic field directly and strongly affects the motions of charged particles is called the pulsar's *magnetosphere.* Here all the energy conversion takes place. The magnetic axis is tilted with respect to the rotational axis.

As the pulsar spins, its 10^8-T magnetic field induces an enormous electric field at its surface. This electric field pulls charged particles (mostly electrons) off the solid crust of iron nuclei and electrons. The electrons flow into the magnetosphere, where they are accelerated by the rotating magnetic field lines. The accelerated electrons emit synchrotron radiation in a tight beam more or less along the field lines.

You can now see how a pulsar emits regular pulses without actually pulsating. If the magnetic axis can fall within our line of sight, each time a pole swings around to view (like the spinning light of a lighthouse), we see a burst of synchrotron emission (Figure 17–6). The time between pulses is the rotation period. The duration of the pulses depends on the size of the radiating region. As the pulsar generates electromagnetic radiation, the torque from accelerating particles in its magnetic field slows down its rotation. As you saw, this slowdown is observed.

Here's a simple argument that shows that fast pulsars must have neutron-star densities. Assume

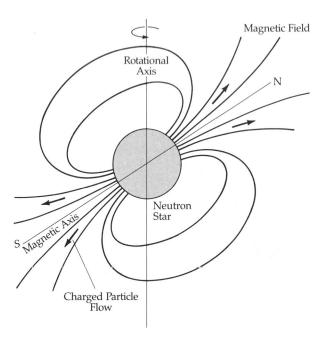

Figure 17–6

Model for a pulsar. A rapidly rotating, highly magnetic neutron star can emit synchrotron radiation along its dipolar axis.

that the clock mechanism is rotation. A sphere can rotate only at a speed such that the centripetal acceleration V^2/R at the equator is equal to or less than the gravitational acceleration GM/R^2:

$$V^2/R = GM/R^2$$
$$V = (GM/R)^{1/2}$$

where V is the equatorial velocity of the sphere, R its radius, and M its mass. The period of a rotating sphere is

$$P = 2\pi R/V$$

so that

$$P = 2\pi R/(GM/R)^{1/2}$$
$$= 2\pi R^{3/2}/(GM)^{1/2}$$

but

$$M = (4/3)\pi R^3 \rho$$
$$P = 2\pi R^{3/2}/[G(4/3)\pi\rho R^3]^{1/2}$$
$$P = (3.8 \times 10^5)/\rho^{1/2} \quad s \qquad \textbf{(17–4)}$$

with the density in kilograms per cubic meter. For a

period of, say, 2 ms,

$$2 \times 10^{-3}\ s = 3.8 \times 10^5/\rho^{1/2}$$
$$\rho^{1/2} = 1.9 \times 10^8$$
$$\rho \approx 4 \times 10^{16}\ kg/m^3$$

just the density of a neutron star.

(c) The Supernova Connection

Supernovae are the cataclysmic explosions of stars at the end of their lives (see next chapter for details). These explosions often involve the detonation of massive stars such that most of the mass is blown away. The blast drives the material in the core to extreme densities to create, perhaps, a neutron star. If this model is correct, we would expect to find neutron stars as pulsars associated with the remnants of supernovae. We do with certainty in two cases: the Crab Nebula, the most renowned of supernova remnants [Section 18–5(c)], and the Gum Nebula.

David H. Staelin and Edward C. Reifenstein discovered the Crab Nebula pulsar (Figure 17–7), called PSR 0531+21. (PSR stands for pulsar, and the numbers 0531+21 refer to its celestial coordinates in the sky—Appendix 10.) PSR 0531+21 has a pulse period of 0.033 s, or 30 pulses/s. The Crab pulsar was the first discovered to emit optical pulses as well as radio ones (Figure 17–8); the optical and radio pulses were found to have the same period. Observations of these visible pulses showed a smaller pulse between the main peaks (Figure 17–8B). The in-between pulse is usually called an *interpulse*. Remarkably, the star emitting these pulses was picked out by Walter Baade and R. Minkowski in 1942 as a possible candidate for the stellar remnant of the supernova. Although this star is now known to be the pulsar, astronomers had observed it for years without noticing the optical blinking; a flicker of 30 times/s would be masked in ordinary photographs. Special stroboscopic techniques were used to determine the period of the optical pulses.

The Crab pulsar is the only one so far observed to pulse in the infrared, radio, optical, X-ray, and gamma-ray regions of the spectrum (Figure 17–8). The total energy emitted in the pulses is about 10^{28} W. Another important feature of the Crab pulsar is that it was one of the first to exhibit a definite slowdown in pulse period, at a rate of about 4×10^{-13} s/s or 10^{-5} s/year.

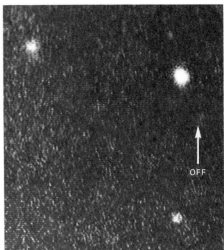

Figure 17–7

Optical pulses from the Crab pulsar. At left, we see the peak of an optical pulse; at right, offpeak. *(Lick Observatory)*

The discovery of the Crab pulsar seems to solve the energy problem of the Crab Nebula. Summing over all wavelengths, the Crab Nebula emits about 10^{31} W. If the pulsar is a rotating neutron star, its slowdown in period gives a change in rotational energy of about 5×10^{31} W. That's enough to power the nebula — if the rotational kinetic energy of the neutron star can somehow be converted to kinetic and radiative energy of the nebula. That's exactly what happens in the highly magnetic, rapidly rotating neutron-star model for pulsars. Let's look at this point in detail.

The rotational kinetic energy of the mass is

$$E_{\text{rot}} = (1/2)I\omega^2$$

where I is the moment of inertia and ω the rotational

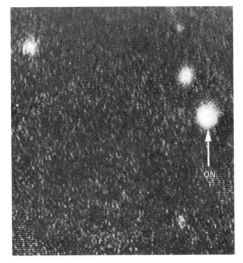

A X-ray

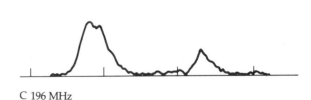

C 196 MHz

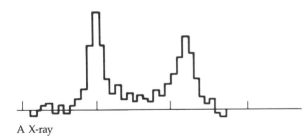

B Optical

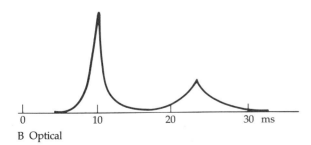

D 111 MHz

Figure 17–8

Crab pulses from radio waves to X-rays. Note how the height sharpness of the primary and secondary pulse changes with wavelength.

angular velocity, with

$$\omega = 2\pi/P$$

For a sphere of uniform density,

$$I = (2/5)MR^2$$

where R is the radius. Now suppose some process converts *all* of the rotational energy to radiative energy; then conservation of energy requires

$$dE_{\text{rad}}/dt + dE_{\text{rot}}/dt = 0$$

but

$$
\begin{aligned}
dE_{\text{rot}}/dt &= (1/2)d/dt(I\omega^2) \\
&= (1/2)(d/dt)(2/5)MR^2(2\pi/P)^2 \\
&= (4/5)\pi^2 MR^2(d/dt)(1/P^2) \\
&= -(8/5)\pi^2 MR^2 P^{-3}(dP/dt)
\end{aligned}
$$

Note, however, that

$$L = dE_{\text{rad}}/dt = -dE_{\text{rot}}/dt$$

so that

$$L = (8/5)\pi^2 MR^2 P^{-3}(dP/dt)$$

and

$$dP/dt = (5/8\pi^2)(LP^3/MR^2) \qquad \textbf{(17-5)}$$

which implies that the pulse rate should slow down

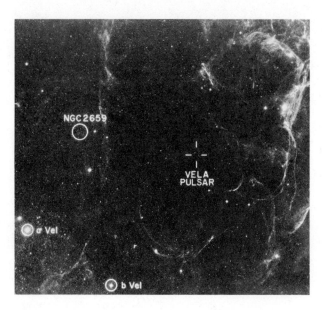

Figure 17–9

Gum Nebula, a supernova remnant. The Vela pulsar lies near its center. *(B. Bok and Steward Observatory)*

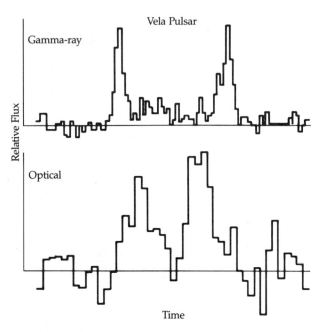

Figure 17–10

Optical and gamma-ray pulses from the Vela pulsar.

as the pulsar loses energy. Use 1 solar mass for M, 10 km for R, 10^{31} W for L, and 1 s for P to get, as a rough estimate,

$$
\begin{aligned}
dP/dt &= (5/8\pi^2)(10^{31})(1^3)/(2 \times 10^{30})(10^4)^2 \\
&= (5/16\pi^2)(10^{31}/10^{38}) \\
&\approx 10^{-8} \text{ s/s}
\end{aligned}
$$

For the Crab pulsar, $P \approx 0.03$ s, and so

$$dP/dt \approx 10^{-13} \text{ s/s}$$

which is the observed rate of slowdown.

If the Crab pulsar were the only one associated with a known supernova remnant, it might be a coincidence, but we know of another one: the pulsar in the constellation Vela, near the center of the Gum Nebula (Figure 17–9), called PSR 0833–45. Although the Vela pulsar was known in 1968, it was eight years before astronomers finally observed its optical pulses because they are very weak, with an average magnitude of only 25.2. The pulses come every 80 ms and have two peaks separated by about 22 ms. Gamma-ray telescopes have also detected pulses for Vela (Figure 17–10). This pulsar resembles the Crab pulsar in many ways. Both are rapid, both emit pulses over a wide range of the electromagnetic spectrum, and their gamma-ray pulse profiles are very similar. The period of the Vela pulsar also slows down but at the slightly different rate of about 1.3×10^{-13} s/s (4×10^{-6} s/year). So it, too, provides support for the lighthouse model of pulsars and the supernova–neutron-star connection.

The millisecond pulsars are quite remarkable in the context of this lighthouse model. The first one discovered, PSR 1937+214, spins at the phenomenal rate of 642 Hz! It has a very small spindown rate of 1.1×10^{-19} s/s.

17-3 Black Holes

A black hole is a region of spacetime in which gravity is so strong that nothing, not even light, can escape it. Contemporary physics indicates that once a certain minimum mass gets together in a small enough volume, it must eventually become a black hole—collapsing by its own gravity after all its nuclear fuel is exhausted. No known physical force can stop this self-swallowing of mass that makes a black hole. And that minimum mass is not large—about five times the Sun's mass. No material can withstand this final crushing point of matter. The volume will continue to decrease until it reaches zero; the density will increase until it becomes infinite. Neither of these events can be exactly true for a real object in this Universe. This theoretical collapse of a nonrotating mass to a singular point of zero volume and infinite density, called a *singularity*, marks a breakdown of the laws of physics as we know them.

(a) The Basic Physics of Black Holes

Here's a simple way to picture a black hole. Consider an escape velocity such that, when an object leaves with just the escape velocity, it will have zero velocity out at "infinity." There, its total energy ($KE + PE$) is

$$TE = (1/2)mv^2 - GmM/R = 0$$

Since total energy must be conserved, at the moment of launch we require

$$TE = 0 = (1/2)mV_{esc}^2 - GmM/R$$

$$(1/2)V_{esc}^2 = GM/R$$

$$V_{esc} = (2GM/R)^{1/2}$$

Now no object can travel faster than the speed of light, and so the maximum escape velocity is c. Then the equation for the black-hole radius is

$$R = 2GM/c^2 \qquad (17-6)$$

In units of solar masses for M,

$$R = 3M \text{ km}$$

This critical radius is called the *Schwarzschild radius*, after the German astrophysicist Karl Schwarzschild, who worked out this solution shortly after Einstein published his general theory of relativity. For the Sun, the Schwarzschild radius of 3 km would require compression to a density of roughly 10^{19} kg/m³— about the same density as the nucleus of an atom.

To examine the strange structure of spacetime around a black hole, let us take a theoretical journey into one. Start out from a spaceship orbiting a black hole of mass $10M_\odot$ at a distance of 1 AU. The ship orbits the black hole in accordance with Kepler's laws, as it would any ordinary mass. In fact, Kepler's third law and the spaceship's orbit permit you to measure the hole's mass. You will hop in with a laser light and digital watch to send signals back to the spaceship.

For a long time as you fall toward the black hole, nothing strange happens, but as you get closer, stronger and stronger tidal gravitational forces stretch you out from head to toe and squeeze you together at the shoulders. Near a black hole, tidal forces grow enormously. Ordinary human beings would be ripped apart at about 3000 km from a black hole of mass $10M_\odot$. You cross the Schwarzschild radius! But nothing new happens, no signs mark the edge of the black hole. The trip now swiftly ends. About 10^{-5} s after you cross the Schwarzschild radius, you crash into a singularity. Crushed to zero volume, you are destroyed.

What of the view from back in the spaceship? As you drop closer to the black hole, the light from your laser is redshifted. The light must do work against gravity, and so it loses energy and increases in wavelength—a gravitational redshift. The time between laser flashes increases because of the time-dilation effect predicted by general relativity. As you come closer to the Schwarzschild radius, the watches get more and more out of synchronization. In fact, a laser burst sent out just as you cross the Schwarzschild radius would take an *infinite* time to reach the spaceship—even though the light is moving at c! It also would suffer an *infinite* redshift. The fall seems to grow slower and slower to an outside observer as you get closer to the black hole. Time slows down so much that it seems to be frozen to a distant observer. The light gets more and more redshifted until it can no longer be detected. A black hole practices cosmic censorship; it prevents any outside observer from seeing you fall into it.

(b) The Structure of Spacetime Around a Black Hole

Let's look at the geometry of spacetime outside a black hole. To do so, we will have to examine a spacetime diagram for the geometry there. We get the appropriate spacetime map by solving Einstein's equations of general relativity to find the geometry of spacetime for an empty region of space surrounding a nonrotating, spherical mass (the simplest case). The main point is this: *Spacetime is not static, but dynamic.* You will also see that spacetime does even stranger things than have been described so far.

The spacetime diagram (Figure 17–11) has coordinates that are not space and time as you experience them. The horizontal axis has properties that are space-like and the vertical one has properties that are time-like, but they are not exactly the same as measured space and measured time. Past is at the bottom of the diagram, future at the top. Light follows a special path in this spacetime diagram; it moves at 45° with respect to the axes. Any object moving slower than light has a path between the time-like axis and the light path; a path between the light line and the space-like axis represents something moving faster than light, which is not normally possible.

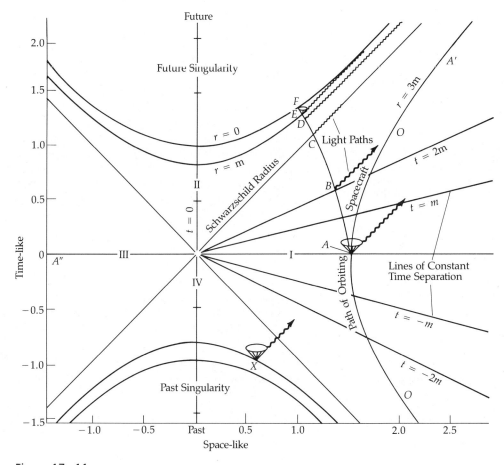

Figure 17–11

Spacetime diagram around a black hole. *(Adapted from a diagram by R. Ruffini and J. Wheeler)*

The diagram has been divided into four regions separated by *event horizons*. (An event horizon is another term to describe the Schwarzschild radius; it emphasizes the fact that any events taking place inside the Schwarzschild radius are cut off from outside view; they are beyond our visible horizon.) Note that a singularity exists both at the top (future) and bottom (past) of the diagram. Also drawn in the diagram is the path of a spaceship orbiting the black hole (line *O*) and that of a person jumping into the black hole from the spaceship (path *A* to *F*).

With these preliminaries in mind, look at the journey into a black hole, described above, in the spacetime diagram. As the spaceship orbits the black hole, it moves along line *O* from past to future (to the upper right). Your friend jumps out of the spaceship at *A*. Her laser signals are indicated by wavy lines; note that these lie at 45° with respect to the axes. The point at which the wavy line crosses line *O* is the point where and when you see it. The pulse emitted at *B* crosses line *O*, but the pulse emitted at *C*, that is, as your friend crosses the event horizon (Schwarzschild radius), does not intercept *O* until an infinite time has passed. All photons emitted after she crosses the event horizon (at *D* and *E*, for example) eventually get gobbled up in the singularity. At point *F*, your friend plunges into the singularity. You cannot see any events at *C* or beyond.

This example shows that region I is our region of spacetime, that is, our Universe outside the black hole. Region II is that part of spacetime within the Schwarzschild radius and containing the (predicted) singularity. What about region III? It's the mirror image of region I: another realm of spacetime outside of event horizons and singularities. At the bottom of the diagram, region IV, lies a singularity that is the mirror image of the one at the top, that is, a singularity in the past, or a time-reversed black hole. Notice that a permitted photon path (from *X*) can cross the event horizon into region I, our Universe. This we would see as light erupting from an event horizon, a phenomenon sometimes called a *white hole.*

Region III is inaccessible, and so we cannot demonstrate its existence. We are in region I. Suppose we tried to travel to region III along path *AA''*. To go that way means we must *travel faster than light* because such a path makes less than a 45° angle with the space-like axis. In fact, if we examine the figure, no path from region I to III requires less than light speed. Any slower-than-light path crashes into the top singularity. So we have no access to region III. Also, we cannot reach region IV because it lies in the past and we cannot travel back in time. We can get into region II, but we can't get out. Only in region I can we move in space at will.

Let's describe the inaccessibility another way. Imagine making a motion picture of the evolution of spacetime's geometry around a black hole. We can use the spacetime diagram to make such an imaginary movie. A horizontal slice across the diagram (such as *AA'*) gives the space geometry at a certain instant of time (Figure 17–12). Notice that slice *AA'* has a geometry that bridges two flat regions of spacetime. This geometric connection is called the

Figure 17–12

Evolution of spacetime near a black hole. (A) Take slices of spacetime at different times. These represent the evolution of the geometry. (B) Geometries corresponding to each of the time slices. (*Adapted from a diagram by C. Misner, K. Thorne, and J. Wheeler*)

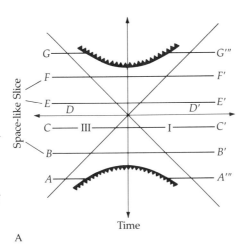

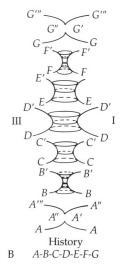

Einstein–Rosen bridge or *wormhole*. Region I (top) is our Universe; region III (bottom) is the mysterious part of spacetime. It looks like you could cross from region I to III along the surface of the wormhole, but you can't because the wormhole is not static and always open.

To see this point, make a movie by taking horizontal slices in the spacetime diagram (Figure 17–12B) from bottom to top. This procedure gives a history of the evolution of spacetime's geometry. Notice that the throat of the wormhole opens (connecting regions I and III), reaches a maximum expansion at AA', then closes again, disconnecting regions I and III. This opening and closing of the wormhole happens rapidly — so rapidly that if you tried to cross the bridge, it would shut before you got through, pinching you off into the singularity. Only by crossing faster than the speed of light could you avoid being caught in the pinchoff.

This evolution of spacetime geometry results from a special solution to Einstein's equations of general relativity, and it demonstrates one of the most profound insights of Einstein's theory: *spacetime is dynamic; it evolves*. This insight will be extremely fruitful when we look at the evolution of the Universe (Chapter 24). The solution also indicates one way to avoid the singularity — by making a black hole of *rotating* matter. The angular momentum gives spacetime around the black hole a different character that avoids the singularity problem.

This theoretical view of black holes will be examined in the light of observational evidence in the next chapter.

In summary: When stars die, they leave corpses behind. The kind of corpse is determined by the *mass at the time of death*, which is *less* than the star's main-sequence mass because of the mass lost as a red giant, and/or in a supernova, or in the formation of a planetary nebula. The three kinds of corpses are white dwarf (mass $< 1.4 M_\odot$), neutron star (mass between $1.4 M_\odot$ and $5 M_\odot$), and black hole (mass $> 5 M_\odot$). The masses on the main sequence will be higher.

Problems

1. **(a)** A white dwarf has an apparent magnitude $m_v = 8.5$ and parallax $\pi = 0.2''$. Its bolometric correction is -2.1 mag, and $T_{eff} = 28,000$ K. Calculate the radius of the star. Compare your value with the radius of the Earth.
   (b) A neutron star has $T_{eff} = 5 \times 10^5$ K and a radius of 10 km. What is its luminosity?
   (c) A protostellar cloud starts out with $T_{eff} = 15$ K and $R = (4 \times 10^4) R_\odot$. Determine $L/L_\odot$ and the wavelength of the peak of the Planck curve.

2. Calculate the kinetic energy ($mv^2/2$) for each of the following:
   (a) a nova outburst that accelerates a mass of $(10^{-5}) M_\odot$ to a velocity of 10^3 km/s
   (b) formation of a planetary nebula in which a mass of $0.1 M_\odot$ is accelerated to a velocity of 20 km/s
   (c) a supernova outburst that accelerates a mass of $1 M_\odot$ to a velocity of 4×10^3 km/s
   How many years would it take the Sun to radiate away these energies?

3. How hot would a cloud of material become falling into a black hole? If it emitted as a blackbody, at what wavelength would it peak?

4. Calculate the Schwarzschild radius for
   (a) the Earth
   (b) the Sun
   (c) a globular cluster
   (d) the Galaxy
   What trend do you notice?

5. Consider stars of mass $1 M_\odot$. Compute the mean mass density for the following:
   (a) our Sun ($R_\odot = 7 \times 10^5$ km)
   (b) a white dwarf ($R = 10^4$ km)
   (c) a neutron star ($R = 10$ km)
   Now consider a $^{12}_6 C$ nucleus of radius $r = 3 \times 10^{-15}$ m and compute its mean density. Discuss the significance of all these results!

6. The oldest white dwarfs were formed about 10^{10} years ago with initial temperatures of about 10^9 K. Determine the current temperature of an old white dwarf of maximum mass $1.4 M_\odot$, radius 7×10^6 m, and age 1×10^{10} years. Assume for simplicity that the density is constant throughout the star. What is the current wavelength of maximum intensity of this "white" dwarf? (*Hint:* Since the star cools by radiating as a blackbody, set $L = -$volume $\times$ number density of particles $\times k(dT/dt)$, where k is Boltzmann's constant. To solve this equation, separate the variables by putting all terms involving T on the left-hand side; then integrate.)

7. The speed of sound, given by $c_s = (5P/3\rho)^{1/2}$ for a nonrelativistic gas, where P is the pressure and ρ the

density, is the speed at which a star will pulsate once oscillations are generated. Determine the speed of sound and the period of pulsations $\approx R/c_s$ as a function of mass (in solar units) for a nonrelativistic white dwarf (equation of state $P \approx (3.2 \times 10^6)\rho^{5/3}$ in SI units). Assume constant density and pressure. How does this time scale compare with the periods of the fastest pulsars? Is it possible for pulsars to be pulsating white dwarfs?

8. The Crab Nebula pulsar radiates at a luminosity of about 1×10^{31} W and has a period of 0.033 s. If $M = 1.4 M_\odot$ and $R = 1.1 \times 10^4$ m, determine the rate at which its period is increasing (dP/dt). How many years will it take for the period to double its present value? (*Hint:* You must integrate after isolating all the terms involving P on the left-hand side for the latter calculation.)

9. Equation 17–3 describes the redshift of electromagnetic waves emitted near a massive, compact object. Since time is in many respects the inverse of frequency, we can express the time-dilation effect by the formula $\Delta t'/\Delta t_{obs} = \nu_f/\nu_i$, where $\Delta t'$ is a time interval between two events (for example, consecutive ticks of a clock) in the source's frame and Δt_{obs} is the time interval between the same two events as measured by the observer. Notice that a clock placed in a strong gravitational field ticks more slowly than normal as observed by a distant observer while a distant clock ticks more rapidly as observed by an observer in the gravitational field.
   (a) Describe how a distant observer would describe how the timing of events (time between events, how long events last, and so on) changes for an object falling into a black hole. Does the observer ever see the object cross the Schwarzschild radius? Comment.
   (b) Now describe the timing of distant events as observed by an observer falling into a black hole. What does the observer see as she or he crosses the Schwarzschild radius? Is there a paradox here? If so, can you resolve it?

10. A neutron decays into a proton, electron, and antineutrino via the weak nuclear interaction after about 15 min when it is outside the nucleus of an atom. Now imagine that a neutron is freed from its nucleus 3.00 km from the center of a black hole of mass $1 M_\odot$. How long will it take the neutron to decay as measured by a distant observer? (Use the expression for time dilation given in the previous problem.)

11. A star identical to the Sun is in a binary system with a black hole of mass M_H. Assume for simplicity that the density of the star is uniform and that the orbit is circular.
   (a) Use Equation 3–9 to obtain the minimum separation from the black hole the star must have in order *not* to be torn apart by tidal forces.
   (b) At what black-hole mass is this minimum separation less than the Schwarzschild radius? Black holes *more* massive than this can swallow a star whole!

12. The star VB8-B is thought to be a brown dwarf whose luminosity derives from gravitational contraction. Its mass is $0.05 M_\odot$, and its luminosity is $(3 \times 10^{-5})L_\odot$. If we assume that its luminosity has been constant (even when the star had a much larger radius), how long can a star of this type radiate before the contraction is halted by electron degeneracy pressure (when $R \approx (9 \times 10^6)M^{-1/3}$ m, where M is in solar units)?

13. A typical young neutron star has a radius of about 10^4 m and a temperature of roughly 10^4 K.
   (a) What is the blackbody luminosity of such a neutron star?
   (b) What is the maximum distance a neutron star with these properties could have and still be detected by its optical blackbody radiation? Assume a limiting magnitude of 25 for detection and a bolometric correction of zero. Comment.

14. A cloud of hot gas is in a Keplerian circular orbit of radius 4.0×10^4 m about a black hole of mass $10 M_\odot$. Plot the ratio of observed to rest frequency (ν_f/ν_i) versus time over one orbital period of an emission line radiated by the cloud, as seen by an observer who views the orbital plane edge-on. Include both the gravitational and normal Doppler effects and use the relativistic formulas.

15. A magnetic dipole rotating with a period P radiates with a luminosity that is proportional to P^{-4}.
   (a) Show that, long after its formation, a pulsar slows down according to the law $P \propto t^{1/2}$ (approximately).
   (b) Using the result from part (a), show that the luminosity of the pulsar decreases approximately as $L \propto t^{-2}$.

Variable
and
Violent
Stars

Chapter 16 discussed the structure and evolution of the majority of the observed stars, slowly changing *normal* stars. We alluded to the existence of rapidly evolving and spectacularly changing stars and mentioned that most normal stars pass through several such phases during their lifetimes. This chapter presents the observational and physical properties of such *variable stars*, which pass through intriguingly peculiar phases of stellar evolution. It illustrates the diverse phenomena of stellar astrophysics.

18–1 Naming Variable Stars

The variable stars of primary interest are the *intrinsic variables*, a term used to differentiate these stars from the geometric variables—extrinsic variables—(such as eclipsing binary stars). Intrinsic variables may be roughly divided into two categories: (a) *pulsating stars*, whose atmospheres undergo periodic expansion and contraction, and (b) *cataclysmic*, or *explosive*, *variables*, which exhibit sudden and dramatic changes. The pulsating stars include Cepheids, RR Lyrae stars, the irregular RV Tauri stars, and the long-period Mira stars. Novae, dwarf novae (U Geminorum stars), and supernovae are the cataclysmic variables. Variables that do not fit neatly into either category include flare stars, T Tauri stars, spectrum variables, and magnetic stars. Tables 18–1 to 18–3 list the characteristics of the most important types of variable stars, and Figure 18–1 indicates their generic positions in the H–R diagram.

A single star may pass through several stages of variability during its evolution. A star of one solar mass or less may initially be a T Tauri star before settling onto the main sequence; then, much later, it may pulsate as an RR Lyrae star after passing through the giant stage. More massive stars contract to the main sequence very rapidly, so that it is difficult to say whether any have been observed during this contraction phase. In later stages of evolution, however, these massive stars are observed as Cepheid variables, following the giant or supergiant stage. To become a white dwarf, a massive star must lose much of its mass, and it may do this as a red giant, a planetary nebula, or a supergiant.

To designate variable stars (both intrinsic and geometric), astronomers have devised a special system of nomenclature. Each variable star is given a set of capital letters followed by the genitive name of the constellation (Appendix 2) in which the variable occurs. The capital letters follow an alphabetical sequence based on the order of discovery: the first variable discovered in a given constellation is designated R, with the rest of the alphabet through Z used for successively discovered variables. After Z, double letters are used in the order: RR, RS to RZ, SS, ST to SZ, . . . , ZZ. Then we return to the beginning of the alphabet (the letter J is omitted to avoid confusion with I): AA, AB to AZ, BB, BC to BZ, . . . , QZ. If there are more variables within a constellation, we resort to numbers starting with V335 (V for variable), since the single and double letters have already accounted for 334 variables. Some examples of variable-star designations are (in sequence!) R Monocerotis, T Tauri, RR Lyrae, UV Ceti, AG Pegasi, BF Cygni, V378 Orionis, and V999 Sagitarii.

Until recently, not all novae were included in this system; each was simply designated by the constellation name and year of occurrence (such as Nova Aquilae 1918). Since 1925, novae have been given variable-star designations, such as RR Pictoris and DQ Herculis; today, even the earlier novae have been assigned such designations, so that Nova Aquilae 1918 is also known as V603 Aquilae.

Variable stars that are bright enough to have been given a proper name or a Greek-letter designa-

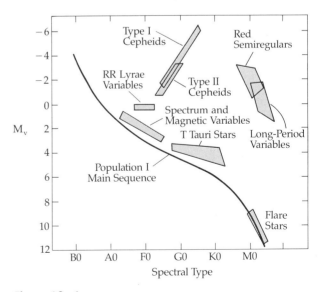

Figure 18–1

Locations of selected variable stars on the H–R diagram.

tion have not been renamed. Therefore, we have Beta Lyrae, Delta Cephei, and Algol, all of which are well-known variables and prototypes of their stellar classes.

18–2 Pulsating Stars

(a) Observations

The vast majority of known variable stars have been detected by observations of the fluctuating light. The most important of the pulsating stars are the *Cepheid variables.* In distinct contrast to the light curves of most eclipsing stars, the light curves of pulsating stars exhibit continual changes in brightness. The spectra of pulsating stars also vary periodically, corresponding to changes in stellar surface temperature that may range over an entire spectral class. The spectral lines show variable Doppler shifts, from which *radial velocity curves* (for the stellar atmosphere) may be deduced. From the periodic radial velocities, we find that variable stars alternately expand and contract; the changes in radii follow (Figure 18–2).

In general, the characteristics described in this section pertain to all pulsating stars. You are urged to study Table 18–1 with care, for it summarizes the distinguishing features and population characteristics of the several types of pulsating stars.

The Cepheids and RR Lyrae stars lie in a region of the H–R diagram called the *instability strip.* As low-

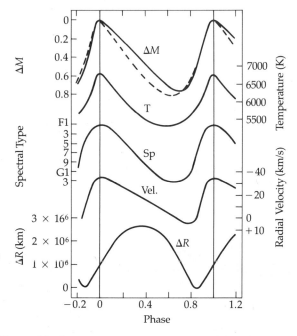

Figure 18–2

Observed properties of a pulsating star. Variations with phase of radial velocity curve for changes in magnitude (dashed curve for a constant radius), temperature, spectral type, radial velocity, and radius. (*Adapted from a diagram by W. Beckers*)

mass Population II stars ($0.5M_\odot$ to $0.7M_\odot$) traverse this strip during their core helium-burning phase, they become unstable and pulsate; these are the RR Lyrae stars. Population I stars of 3 to $18M_\odot$ also cross

Table 18–1   Pulsating Stars

Type	Prototype	$\overline{M}_v$	Spectral Class	Pulsation Period Range	Characteristic Period	Population
Classical Cepheids	δ Cephei	−0.5 to −6	F6 to K2	1^d to 50^d	5^d to 10^d	I
Population II Cepheids (W Virginis)	W Virginis	0 to −3	F2 to G6	2^d to 45^d	12^d to 28^d	II
RR Lyrae stars	RR Lyrae	0.5 to 1	A2 to F6	1.5^h to 24^h	0.5^d	II
Long-period variables (Mira)	o Ceti	1 to −2	M1 to M6	130^d to 500^d	270^d	Disk
RV Tauri stars	RV Tauri	−3	G, K	20^d to 150^d	75^d	II
Beta Canis Majoris stars	β Canis Majoris	−3	B1, B2	4^h to 6^h	5^h	I
Semiregular red variables	α Herculis	−1 to −3	K, M, R, N, S	100^d to 200^d	100^d	I and II
Dwarf Cepheids	δ Scuti	4 to 2	A to F	1^h to 3^h	2^h	I

the upper region of this strip during their phase of core helium burning, and they also pulsate, becoming the Cepheids.

(b) A Pulsation Mechanism

A star pulsates because it is not in hydrostatic equilibrium; the force of gravity acting on the outer mass of the star is not quite balanced by the interior pressure (Chapter 16). If a star expands as a result of increased gas pressure, the material density (and pressure) decrease until the point of hydrostatic equilibrium is reached and *overshot* (owing to the momentum of the expansion). Then gravity dominates, and the star starts to contract. The momentum of the infalling material carries the contraction beyond the equilibrium point. The pressure is again too high, and the cycle starts anew. Energy is dissipated during such pulsation (analogous to frictional losses), and eventually this loss of energy should result in a *damping* of the pulsations. The prevalence and regularity of pulsating stars imply that the dissipated energy is replenished in some way.

The rate at which energy is transported outward from the stellar interior can be altered by a damming process. We define *opacity* (Chapter 10) as the amount of radiative energy absorbed; therefore, a changing opacity will act as a valve. When the stellar atmosphere is transparent, radiation flows freely and the star is bright. When the opacity is greatest and radiation is prevented from escaping, the star is faint. If the star is compressed at the time of greatest opacity, the excess radiation is dammed up and exerts pressure on the outer layers of the star; this process provides the energy necessary to continue the pulsations. The atmospheres of pulsating stars have a zone in which the opacity increases because singly ionized helium absorbs ultraviolet radiation to become doubly ionized. The He^+ ionization region is cooler than the surrounding regions because energy normally used to heat the gas is used to ionize it. The helium ionization zone contributes to the instability of the stellar atmosphere and so perpetuates the pulsations.

Pulsating stars occupy well-defined areas of the H–R diagram (Figure 18–1); this observation can be explained in terms of the depth of the He^+ ionization zone. This depth depends upon the structure of the star, which in turn is a function of the star's stage of development. When the zone lies too deep, the valve action is insufficient to overcome damping. When the

zone is shallow, the damming action is inefficient and pulsations are not given the necessary impetus. The period–luminosity law (next section) is explicable in terms of the position in the H–R diagram at which the star becomes unstable to this damming valve mechanism.

Regardless of the pulsation mechanism, we can generally relate a star's period of pulsation to its average density. After maximum expansion, the star's layers free-fall inward. We consider this infall as a special case of orbital motion—along a straight line! So the gas obeys Kepler's law:

$$P^2/R^3 = 4\pi^2/GM$$

where P is the pulsation period, R the star's radius, and M its mass; hence

$$P^2 \propto R^3/M$$

but

$$M \propto \rho R^3$$

where ρ is the *average* density. Then

$$P^2 \propto R^3/\rho R^3$$
$$\propto 1/\rho$$

so that

$$P\rho^{1/2} = \text{constant}$$

Then the ratio of the pulsation periods for two different Cepheids is inversely proportional to the ratio of the square roots of their average densities:

$$P_A/P_B = (\rho_B/\rho_A)^{1/2}$$

which is observed to be approximately the case.

(c) The Period–Luminosity Relationship

Cepheids show an important connection between period and luminosity: the pulsation period of a Cepheid variable is directly related to its median luminosity. This relationship was first discovered from a study of the variables in the Magellanic Clouds, two small nearby companion galaxies to our Galaxy that are visible in the night sky of the southern hemisphere. To a good approximation, you can consider all stars in each Magellanic Cloud to be at the same distance. Henrietta Leavitt, working at Harvard in 1912, found that the brighter the median apparent luminosity, the longer the period of the Cepheid variable. Harlow Shapley recognized the importance of this *period–luminosity* (P–L) *relationship* and attempted to find the zero point, for then a knowledge of the period of a Cepheid would immediately indicate its luminosity (absolute magnitude). This calibration was difficult to perform because of the relative scarcity of Cepheids and their large distances. None are sufficiently near to allow a trigonometric parallax to be determined. Shapley had to depend upon the relatively inaccurate method of statistical parallaxes. His zero point was then used to find the distances to many other galaxies. These distances are revised as new and accurate data become available. Right now, some 20 stars whose distances are known reasonably well serve as the calibrators for the P–L relationship.

Further work showed that there are *two* types of Cepheids, each with its own separate, almost parallel P–L relationship (Figure 18–3). The classical Cepheids are the more luminous, of extreme Population I, and found in spiral arms. Population II Cepheids, also known as W Virginis stars after their prototype, are found in globular clusters and other Population II systems. Classical Cepheids have periods ranging from one to 50 days (typically five to ten days) and range from F6 to K2 in spectral class. Population II Cepheids vary in period from two to 45 days (typically 12 to 20 days) and range from F2 to G6 in spectral class. Population I and II Cepheids are both *regular,* or *periodic, variables;* their change in luminosity with time follows a regular cycle.

RR Lyrae stars are also periodic variables. These stars (named after their prototype, RR Lyrae, whose period is 13.6 h) vary in luminosity with periods of 1.5 to 24 h (typically 12 h). They are Population II, range in spectral class from A2 to F6, and have about 100 times the Sun's luminosity. About 5000 RR Lyrae stars are known. Note that RR Lyrae stars have a P–L relationship of sorts: they all have essentially the same luminosity (absolute magnitude of about 0.5) regardless of period.

(d) Long-Period Red Variables

Long-period variables exhibit a very large change in visible light because they are cool (about 2000 K) and so most of their radiation lies in the infrared. An increase of even 500 K in surface temperature will shift much more radiation into the visible wavelength band than appears to be warranted by an application of Stefan's law (Section 8–6). Although the total amount of radiation emitted by the star increases only slightly, the Planck spectral energy curve slides over into the visible band and dramatically increases the amount of visible radiation. In addition, molecules and dust can exist at these low temperatures, forming a veil over the star's surface; as the temperature increases, these molecules are dissociated and more radiation can penetrate the veil. The *red variables* have irregular cycles of light variation of up to a few magnitudes that last from 100 to 700 days. They contain both Population I and II stars of spectral classes K and M with luminosities roughly 100 times that of the Sun; these are red giants and supergiants. The long-period red variables fall in a region of the H–R diagram in which they undergo shell burning of helium. The cause of their irregular pulsations is yet unknown.

18–3 Nonpulsating Variables

Some curious variables are not pulsating stars; we briefly mention them here. They include *T Tauri stars,* which are thought to be pre-main-sequence, solar-mass stars; *flare stars,* which have arrived on the

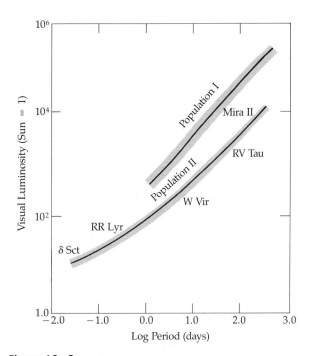

Figure 18–3

Period–luminosity relationship for Cepheids. Note that two relationships exist, one for Population I stars and one for Population II stars.

main sequence but exhibit stellar activity similar to that observed on the Sun; *magnetic variables,* which are in a late phase of evolution, and *RS Canum Venaticorum stars,* which are Sun-like stars evolving off the main sequence.

(a) T Tauri Stars

T Tauri stars are still in the process of contracting onto the main sequence. At this stage of stellar evolution, these low-mass ($0.2M_\odot$ to $2M_\odot$) stars have an extensive convection zone and surface magnetic activity (probably driven by convection) is rampant. Spectral activity includes emission in the hydrogen Balmer lines as well as emission from ionized calcium and other metals. Some T Tauri spectra also show the forbidden lines typical of gaseous nebulae, indicating the importance of the surrounding nebula. The underlying spectra of these stars are usually of spectral types F to M; a few have spectra resembling those of A0 or B8 stars. The continuum also varies, so that a total light variation occurs together with the sporadic appearance of emission lines. Large fluctuations occur in the ultraviolet; in fact, the ultraviolet is often excessively strong for the apparent spectral type. Recent observations with X-ray telescopes show that some T Tauri stars fluctuate rapidly and violently in X-rays, on the order of a factor of 10 in one day.

The position of T Tauri stars on the H–R diagram is just above and to the right of the main sequence, just where we expect to find pre-main-sequence stars (Figure 18–4). T Tauri and related objects radiate very strongly in the infrared region, an excess attributed to a surrounding dust cloud that absorbs much of the star's short-wavelength radiation and then re-emits it at infrared wavelengths. This dust may be either the remainder of the material from which the protostar formed or matter ejected from the star as it collapsed.

Infrared and radio observations indicate mass outflows (stellar winds) from T Tauri stars of roughly 10^{-7} to 10^{-8} solar mass per year. In fact, mass outflows appear associated with almost all protostellar and pre-main-sequence objects.

(b) Flare Stars

Solar flares are the most energetic and spectacular aspect of solar activity. There must be flares on other G stars, but the amount of energy radiated by even the largest solar flares amounts to little compared with the total stellar radiation. However, on a

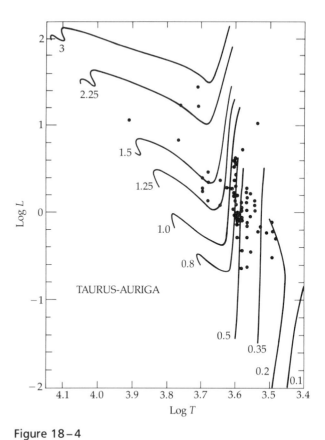

Figure 18–4

H–R diagram for T Tauri stars. Stars observed in the Taurus–Auriga region plotted with theoretical evolutionary tracks labeled by mass. *[M. Cohen and L. V. Kuhi, Astrophysical Journal Supplement Series 41:743 (1979)]*

dwarf M star, which radiates very much less than the Sun, a flare with the energy of a large solar flare would lead to a twofold increase in brightness! In fact, several cool main-sequence stars have flared at irregular intervals by brightening several magnitudes in a matter of seconds. The light curves are similar to those of solar flares, for the decline is far slower than the ascent (Figure 18–5). Joint observations by radio and optical observers have verified that some of these flare stars emit radio bursts simultaneously with the flares.

The total energies of flares from dwarf M stars range from 10^{21} to 10^{27} J for stars whose B-band luminosities range from 10^{21} to 10^{26} W. The B-band luminosity of the Sun—in the B filter of the *UBV* system—is about 2×10^{26} W. Although stellar flares are more energetic and have a faster rise time than solar flares, the underlying process is probably very

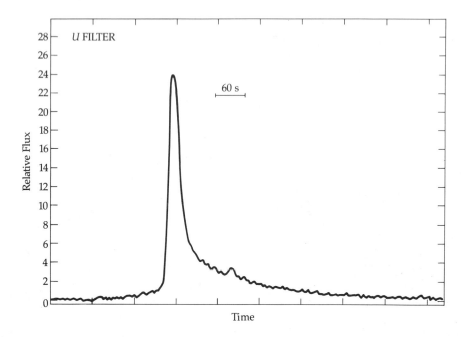

Figure 18-5

A stellar flare. *U*-band observations of a flare from the star YZ Cmi. Note the steep rise and more gradual decline. *[T. J. Moffett, Astrophysical Journal Supplement Series 29:1 (1974)]*

similar: a release of magnetically trapped particles into the corona. As with the Sun, most of the energy comes out in X-rays but much more energetically. For example, flares from dwarf M stars hit a maximum luminosity in X-rays of about 10^{24} W (compared with 10^{20} W for the Sun). It may be the case that all stars undergo a flaring phase during pre-main-sequence evolution. This is one interpretation of the X-ray fluctuations of T Tauri stars.

(c) Magnetic Variables

Stars that have the basic characteristics of A stars may show spectral peculiarities attributable to abundance abnormalities (Chapter 13). Some of these stars exhibit *variable spectra* such that the intensity of certain lines varies approximately periodically. Many of the peculiar and metal-lined A stars have strongly integrated magnetic fields; these fields range up to a few teslas, but most are from 0.01 to 0.1 T (recall that the strongest magnetic fields on the Sun are about 0.4 T, and these fields are restricted to very small areas in sunspots). Most observed stellar magnetic fields are variable, with some undergoing polarity reversals; sometimes the magnetic variability is coupled with spectrum variability. The light variations of magnetic stars and spectrum variables are very small, amounting to about ±10 percent.

One possible explanation for magnetic variability is that the magnetic axes are tilted with respect to the rotational axes (just as in the Earth). This model is referred to as the *oblique rotator*. Perhaps more simply, both the spectrum and magnetic variations are due to large spots on the stellar surface that are periodically brought into view by rotation.

(d) RS Canum Venaticorum Stars

In this tour of variable stars, we have encountered indications of stellar activity similar to solar magnetic activity but more energetic. For Sun-like stars, the best hints of hyperactive magnetic activity come from the *RS Canum Venaticorum* stars, or RS CVn stars for short.

Generally, the RS CVn stars are binaries. The typical orbital period is about seven days; however, periods range from 12 h to a few months. In almost all RS CVn systems, the stars are locked by tidal forces, so that the rotational period of each is almost, but not quite, equal to the orbital period. One star is about 1000 K hotter than the other; the hotter star is often main-sequence (luminosity class V), the cooler one a subgiant (luminosity class IV).

RS CVn stars are basically solar-type, with roughly the same size, mass, surface temperature, and chemical composition as the Sun. For instance, the hotter star of the pair is usually a spectral type G main-sequence star with a solar mass of material and a surface temperature near 6000 K. The cooler companion is spectral type K and has a mass 20 to 40 percent greater than the solar mass. The G star is about the size of the Sun; the K star is a few times larger.

The extraordinary aspect of RS CVn stars is that they are *hyperactive* relative to the Sun at radio, ultraviolet, and X-ray wavelengths. They have somewhat regular and sometimes sudden changes in their visual-light output over a few orbital periods and over a few years. They flare violently in the radio, H_α, optical, ultraviolet, and X-ray regions of the spectrum.

Observations with the VLA have shown that the average radio emission from RS CVn systems is about 10^{20} W. In contrast, the Sun's radio luminosity averages 10^{13} W. Radio flares from RS CVn stars peak at 10^{20} to 10^{21} W, which is 10^5 to 10^6 times stronger than the *strongest* solar radio flares. A few superflares have energies 10^7 times greater than any solar radio flares. The great outbursts are not common — one occurs in an RS CVn system only every few years. Radio observations over a range of wavelengths show that the flares' emissions are highly polarized with nonthermal spectra. These clues point to synchrotron radiation as the source of the flaring radio emissions.

Optically, a light curve of an RS CVn system

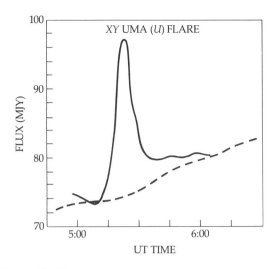

Figure 18–7

Flare from an RS CVn system. *U*-band observations of a flare from XY UMa in January 1982. (*Capilla Peak Observatory, University of New Mexico*)

exhibits an almost-regular wave even for noneclipsing systems. These waves are called *distortion waves* (Figure 18–6). They amount to 1 to 30 percent of the total light from the system. Their amplitude varies from as fast as a few months to over many years. These distortion waves are attributed to large (tens of percent of the total surface area) starspot regions concentrated in one or two groups. As the star rotates (in a period that is the same as the orbital period because the stars are close enough together to be tidally locked), we see spotted and unspotted regions that cause the fall and rise of the distortion wave. Such enormous active regions (at sunspot-cycle maximum, the spots cover at most 0.1 percent of the Sun's surface) indicate that these stars have stronger concentrations of magnetic fields than does the Sun.

RS CVn stars show optical flares, although few have been observed to date. One well-observed flare from XY UMa in January 1982 lasted for about 30 min (Figure 18–7). Its total energy output amounted to 10^{27} J — 10^3 times greater than a solar "white-light" flare.

What drives the hyperactivity of RS CVn stars? We know that magnetic fields do so for the Sun. From X-ray observations in particular, we now understand that the solar corona arises from *coronal loops* of magnetic flux tubes (Figure 18–8). The flux tubes, embedded in active regions (usually marked by sun-

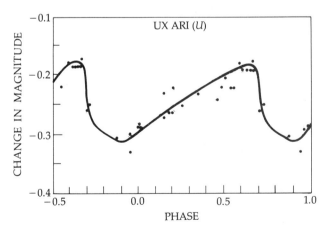

Figure 18–6

Light curve for an RS CVn binary, showing the variation of light at *U* band as a function of orbital phase for the system UX Ari. Note the large changes in this noneclipsing binary. (*Capilla Peak Observatory, University of New Mexico*)

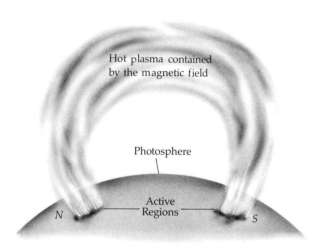

Figure 18–8

Coronal loops. Magnetic fields trap the hot coronal gas. An RS CVn star may have 1000 times as many loops as the Sun.

A loop model for the corona combined with X-ray observation for RS CVn stars results in 10^3 more loops than for the Sun—essentially a star completely covered with active regions. The temperature of the gas contained in these loops is a few *tens* of millions K, about ten times hotter than the Sun's coronal gas. Such very hot loops may reach lengths of *tens* of solar radii—larger than the RS CVn stars.

Although we do not yet know the details of how solar flares occur, we do realize that they erupt in the corona and flow down flux tubes to their bases in the photosphere in a violent rush of high-speed, charged particles. Coronal loops, active regions, flares, and strong X-ray emission are all associated phenomena on the Sun; the same view is supported for RS CVn systems, and similar processes may drive hyperactivity on young stars, such as the T Tauri class, and the flares from dwarf M stars.

spots) of opposite magnetic polarity, arch up to a few percent of the Sun's radius. Magnetic fields at their bases amount to 0.5 T; the temperature here is also 5000 K. Within their curving loops, magnetic fields of a few hundredths of a tesla contain and confine the hot coronal plasma at about 2 million K. The hyperactivity of RS CVn stars implies that strong magnetic fields, concentrated in active regions and coronal loops above them, drive the unusual behavior.

18–4 Extended Stellar Atmospheres: Mass Loss

(a) An Atmospheric Model

An extended gaseous envelope around a star (Table 18–2) is deduced from the line profiles of certain spectral lines (Figure 18–9). The portion of the *shell, or extended atmosphere*, seen projected against the star's photosphere produces a narrow ab-

Table 18–2   Extended-Atmosphere Stars

Type	Prototype or Example	$\overline{M}_v$	Spectral Class	Expansion Velocity (km/s)	Rotational Velocity (km/s)	Rate of Mass Loss ($M_\odot$/year)	Mass ($M_\odot$)	Population
Be	48 Per	−4	B	—	500 to 600	≤10^{-6}	≈10	I
Shell star	{ γ Cas	−4	B	50	500	≤10^{-7}	≈10	I
	{ Pleione		B					
Wolf–Rayet	{ HD 66811	−4 to −6.8	WN, WC	120 to 2500	—	10^{-6}	4 to 10	I
	{ HD 68273							
P Cygni	P Cyg	−6?	B	130	0	?	—	—
O and B supergiants	{ ζ Pup	−7	O and B	1000 to 1800	—	10^{-6}	10 to 30	I
	{ δ, ε, i Ori							
M supergiants	{ α Her	−2 to −8	M Ia to II	up to 26	—	5×10^{-6} to 5×10^{-9}	—	I
	{ α Ori							
Planetary nebulae	Ring nebula in Lyra	0 to +8	W stars? O,B	10 to 30	—	10^{-4} to 10^{-5}	≤1.4	II
Beta Lyrae	{ β Lyr	−6?	B	—	—	10^{-4}	≈5	—
	{ γ Cyg		B	130	0	?	—	—

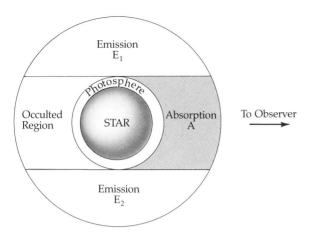

Figure 18-9

An extended-atmosphere model. Light from the star's photosphere is absorbed (region A) on its way to the observer and stimulates emission (regions E₁ and E₂) in the atmosphere seen side-on.

sorption line, and that part not projected against the disk (the annular region) is seen as an emission line. Normally, the emission is superimposed on the stellar photospheric absorption; the extent to which the underlying absorption profile is distorted by this emission line depends on the strength of the emission, which in turn is a function of the density of the stellar atmosphere. The atmosphere cannot be too dense, however, for then it would affect the continuum radiation as well as the spectral lines. The widths of the several components of the profile depend upon the motions of the contributing regions. For instance, if the atmosphere is turbulent, both the emission and the absorption features will be broad.

(b) Be and Shell Stars

If a star rotates more rapidly than its atmosphere, the underlying stellar absorption profile is widened more than the atmospheric emission line. This results from the fact that the Doppler shifts of the approaching and receding limbs of the photosphere are greater than those of the extended atmosphere (Figure 18-10A). The absorption due to the projected gas remains narrow because the motion is across the line of sight. The *B-emission stars (Be)* and *shell stars* both fit this model, but they differ in that the latter contain more material in the envelope. Sometimes net outward velocities are observed, indicative of expansion and possibly loss of material. The mass lost in this

manner is small relative to the mass of the star and to the mass the star must lose if it is to become a white dwarf. Observationally, massive main-sequence stars, giants, and especially supergiants are ejecting material at a substantial rate. The formation of a shell around the Be stars may result from the rapid rotation; the rotational velocities may be so high that material is spun off.

(c) Mass Loss from Giants and Supergiants

In a sense, we should include giants and supergiants in the category of stars with extended atmospheres. For instance, *mass loss* from M-type giants and supergiants indicates that the stellar atmospheres are expanding. The spectra of many M giants and supergiants have narrow absorption lines superimposed on and shifted with respect to the broad Ca II emission lines of the star itself (Figure 18-10B). These features are interpreted as being due to circumstellar material similar to the shells of the early-type stars discussed earlier. The Doppler velocities of the shifts are of the order of tens of kilometers per second. To convert these velocities to rates of mass loss, we must know the density of the material ejected. Estimates of the density lead to figures for the mass loss ranging from 10^{-6} to 10^{-8} solar mass per year. The more luminous the star and the later the spectral type, the higher the rate of mass loss.

At the hot end of the spectral sequence, O and B supergiants eject mass at high velocities (1000 to 2000 km/s). Spectra in the visible region do not show this phenomenon, for there are no suitable spectral lines. The material being ejected from the star has a very low density; hence, collisional excitation is negligible and most atoms and ions are in their lowest energy level (ground state). Owing to the high temperature, however, most of the gas is ionized; therefore, radiative excitation leading to an absorption line from the ground state (known as a *resonance line*) must originate from an ion. None of the ions produce such resonance lines in the visible spectral region. In the ultraviolet, however, there are several such transitions: from Si III and IV, C III and IV, and N V. The ejection velocities are far higher than for the cool supergiants, but the mass loss is only slightly greater because of differences in the gas densities. Ultraviolet spectra of main-sequence O and B stars seem to confirm the absence of expansion in that part of the H-R diagram.

Since both hot and cool supergiants are losing

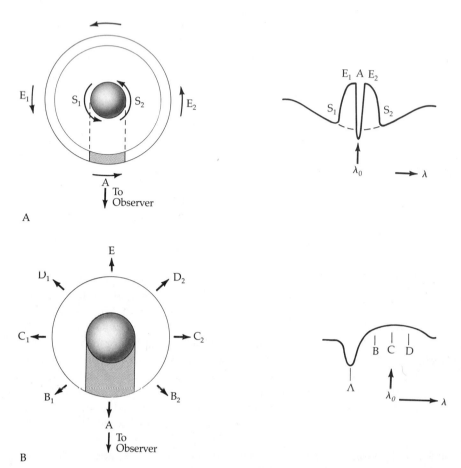

A

B

Figure 18–10

Expansion of an extended atmosphere. (A) From a rotating star, the line profile shows the absorption, A, of the undisplaced line center and the Doppler-shifted peak (E_1 and E_2). S_1 and S_2 are the absorption lines of the star Doppler-shifted to the blue and red by rotation. (B) From an expanding atmosphere, the line profile has absorption feature, A, displaced to the blue by the expansion toward the observer. The contribution to the emission, C, is at the undisplaced line center; B and D are Doppler-shifted from the expansion.

mass, we might expect similar behavior of the intermediate types, the A through K supergiants. Some evidence hints that these also eject matter but apparently rather more sporadically than those stars already discussed.

(d) Wolf–Rayet Stars

Wolf–Rayet stars are hot stars ($T_{\mathrm{eff}} \approx 30,000$ K) whose spectra show strong and wide emission lines of He I, He II, C III, C IV, N III, and N V. That they are young stars of Population I is deduced from their association with OB stars, in open clusters, in H II regions, and as binary companions to O or B stars. (Almost all WR stars are in binary systems.) The broad emission lines often have narrower absorption lines superimposed but displaced to the blue. (Such combined blue absorptions superimposed on wide emission lines are often referred to as *P Cygni line profiles* after the prototype star for which such pro-

files were first observed.) These observations fit an interpretation (Figure 18–10B) of a stellar atmosphere expanding at velocities of 1000 km/s or higher. Wolf–Rayet stars are helium-rich and hydrogen-deficient; carbon dominates the spectrum of some, the WC stars, and nitrogen that of others, the WN stars. Although the true nature of these stars is still not entirely clear, it seems that they represent a particular stage in a star's evolution and that the two types result from real differences in abundances, not from differences in excitation.

As you saw in Section 16–3(b), Wolf–Rayet stars may be initially massive stars with strong stellar winds that have lost their outer layers, revealing interiors with compositions greatly modified by nuclear reactions. A tantalizing hypothesis to explain the two sequences suggests that, in the WN stars, we see the products of CNO-cycle hydrogen burning, whereas WC stars have already undergone helium burning.

We may be seeing either the core of the star or, more likely, intermediate convective layers enriched with core products.

For WR stars in binary systems, we have determined their masses to be roughly 10 to 40 solar masses. Binary nature may be a basic characteristic of WR stars. Then the mass loss, rather than from a stellar wind, occurs by mass transfer in the red-giant phase of the initially more massive star, which must evolve more rapidly than its companion because of its greater mass. In such a case, the WR star is the remnant, consisting largely of helium and heavier elements. From an evolutionary approach, mass loss dramatically affects the evolution of WR stars [Section 16–3(b)]. Theoretical models, starting with a mass of $60M_\odot$, suggest that WR stars are in a core carbon-burning phase after having been a red supergiant and just before becoming a supernova.

(e) Planetary Nebulae

Planetary nebulae, so named because some resolve into disks reminiscent of planets when seen with a telescope, also fit an expanding-atmosphere model (Figure 18–11). In fact, the atmosphere is really a large shell, large enough and of sufficiently low density that most of the receding portion (E in Figure 18–10B) is visible and the spectral lines are doubled. The velocities of expansion, however, are only some tens of kilometers per second, far lower than those for WR stars.

A substantial fraction of all stars probably go through the planetary-nebula stage. The small number of observable planetaries is due to the short duration of this stage, which lasts only 50,000 years. The star develops into a planetary nebula when a nebula of material is ejected from the central star during the contraction that terminates the red-giant stage. The extended envelope includes so much of the former star that what is seen as the central star was formerly the core.

The central star is very hot and so radiates strongly in the ultraviolet; hence, the atoms and ions in the envelope *fluoresce* (absorb ultraviolet radiation to become ionized and emit at longer wavelengths upon recombination and cascade). The central stars generally have spectra that can be classified as type O or WR, but these stars are not identical to normal O and WR stars in either luminosity or mass. Both luminosity and mass are very difficult to establish because all planetaries are so far away that we cannot measure their trigonometric parallaxes. We can still make reasonable estimates on the basis of ionization processes operating between star and nebula, however. The stellar masses range from $0.5M_\odot$ to $1M_\odot$, and estimates of the mass contained within the envelopes range from $0.1M_\odot$ (or less) to $0.5M_\odot$. Apparently many planetary nuclei are underluminous for their spectral type, as might be expected for a star in the process of losing its mass as it changes from a giant to a white dwarf.

To ask what place the planetary nebulae occupy in the evolutionary scheme raises the problem of their population characteristics. The planetaries in our Galaxy are primarily concentrated in the disk and center, as the long-period variables and the RR Lyrae stars are. Their galactic orbits are elongated rather than circular, in contrast to the orbits of Population I objects. At least one is known to be a member of a globular cluster—that is, an extreme Population II object. It is probably safe to say that most planetary nebulae belong to the intermediate disk population, but isolated cases are found at both extremes, Populations I and II. Red variable stars, such as Miras, may well be their predecessors.

Figure 18–11

A planetary nebula. A shell of gas is expanding into space, forming the Helix Nebula. (*National Optical Astronomy Observatories*)

18-5 Cataclysmic Variables

This category includes stars that eject matter suddenly and violently, in contrast to the slow expansion and small mass loss of planetary nebulae. These eruptions are accompanied by enormous changes in luminosity, from a few magnitudes for the dwarf novae to over 20 magnitudes in the case of supernovae (Table 18–3). It is probable that all cataclysmic variables except the supernovae suffer several eruptions in their lifetimes. Repeated outbursts of dwarf novae and recurrent novae have been observed, with the interval between outbursts being a function of the amplitude of the eruption. If such a relationship extends to normal novae, their interburst interval must be of the order of 10,000 years.

(a) Novae

Nova is the Latin word meaning "new"; *novae* (plural) are stars that suddenly become visible in the sky where no star was seen before (Figure 18–12A). Actually, the star is usually too faint to be visible prior to its eruption, for a nova characteristically increases in brightness some ten magnitudes from the prenova stage to nova maximum. The rise to maximum brilliance is very rapid (Figure 18–13). The initial rise brings the star to within two magnitudes of the maximum in only two or three days, and the final increase in luminosity takes a day for fast novae and weeks for very slow novae. The decline from maximum is far more gradual; the time spent at maximum is relatively short, generally only a matter of days. Erratic light fluctuations of large amplitude may occur during the decline; occasionally a drop of many magnitudes is observed followed by a practically complete recovery some weeks later (Figure 18–13B). The length of time from the maximum to the final leveling off of the light curve (at approximately the prenova level) ranges from months for fast novae to years for slow ones.

Spectra obtained just prior to and at maximum light show that material is ejected from the star at velocities up to 2000 km/s. Complex changes in the spectrum occur during the development of the nova; several layers of ejected material are seen as the outer envelope becomes progressively more transparent and exposes the inner layers. The expanding material may change character, and its spectrum changes accordingly. After maximum, for instance, most nova spectra include the bright forbidden lines characteristic of emission nebulae. The velocities of ejection differ at different nova stages; some velocities are directly related to changes in the light curve, as if there were ejections subsequent to the initial eruption. One nova for which a spectrum was recorded prior to its outburst is Nova Aquilae 1918 (also known as V603 Aquilae). This spectrum resembles that of a hot blue star without any spectral lines, presumably a blue subdwarf. Most novae eventually return to such a state.

Some light fluctuations persist even at minimum. Superimposed on these rapid, erratic variations are others that are best interpreted as stellar eclipses.

Figure 18–12A

Nova Cygni 1975. These photographs were taken before (left) and during (right) the outburst. (*Lick Observatory*)

Table 18–3   Cataclysmic Variables

Type		Example	M_{max}	M_{min}	Δm	Energy per Outburst (J)
Supernovae I		Tycho's	−20			10^{44}
				?	>20	
Supernovae II			−18			10^{43}
Novae:	Fast	GK Per = Nova Per	−8.5 to −9.2	+2.7 to +4.8	11 to 13	6×10^{37}
	Slow	DQ Her	−5.5 to −7.4	+4.8 to +8.1	9 to 11	—
	Recurrent	T Cr B	−7.8	0	8	10^{37}
Dwarf novae	U Gem		+5.5	+9.5	4	6×10^{31}
	SS Cyg					

Type		Cycle Time	Mass Ejected per Cycle ($M_\odot$)	Velocity of Ejection (km/s)	Mass of Star ($M_\odot$)	Duration of Decline
Supernovae I		—	≤1	10,000	1	—
Supernovae II		—	?	10,000	≥4	—
Novae:	Fast	10^6?? years	10^{-5} to 10^{-3}	500 to 4000	1 to 5	0.5 to 2 years
	Slow	—	—	100 to 1500	0.02 to 0.3	3 to 9 years
	Recurrent	8 to 80 years	5×10^{-6}	60 to 400	2	<1 year
Dwarf novae		to 100 days	10^{-9}	—	≈0.4	Days to months

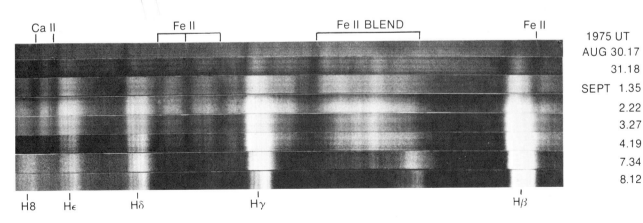

Figure 18–12B

Spectra or Nova Cygni 1975. The UT dates are given at the right. These eight spectra show the evolution of the emission lines, especially those of the Balmer series. Note how broad the emission lines appear. (B. Bohannon)

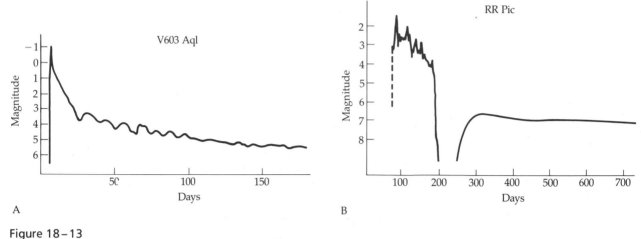

Figure 18–13

Light curves of novae. Note the rapid rise and more gradual decline, with fluctuations. (A) V603 Aql, a fast nova. (B) RR Pic, a slow nova. (*Adapted from a diagram by D. B. McLaughlin*)

Such novae clearly must be members of binary systems; in fact, the evidence implies that all novae are members of binaries. This conclusion suggests that the presence of a companion is a key condition for a star to become a nova. One binary model based upon this idea consists of a red giant or a star in the process of expanding into the red-giant phase and a hot subdwarf or white dwarf. As the red star extends, its atmospheric gaseous material crosses the Roche lobe to make a semidetached binary (Section 12–4). Hence, the gas from the red giant's atmosphere escapes and falls onto the white dwarf or forms a gaseous disk around it because of the conservation of angular momentum. The influx of hydrogen-rich gas from this accretion disk onto a degenerate star (Chapter 17) that has used up most of its hydrogen can cause further nuclear reactions at the stellar surface. This new supply of energy heats the outer envelope of the white dwarf so that it becomes nondegenerate, and the stellar atmosphere suddenly expands—a nova explosion. Repeated explosions may occur if more material flows from the red giant to the white dwarf.

Novae represent an advanced stage of the evolution of stars that happen to be members of close binary systems. The distribution of novae in our Galaxy corresponds to Population II. The discovery that several X-ray sources are close binary systems, in which the emission of X-rays is attributed to mass exchange, suggests a close link between novae and X-ray sources.

The high ejection velocities quoted earlier are based upon measurements of spectral-line profiles —P Cygni-type profiles resembling those of Wolf–Rayet stars. The ejected gas expands as a shell, and sometimes this shell becomes visible as a nebula surrounding the nova. Over the years, the nebula expands perceptibly, and its rate of expansion appears as a proper motion measurable in seconds of arc per year (Chapter 15). Spectra obtained during the same epoch give expansion velocities directly in kilometers per second. If we assume that the velocity of expansion is uniform in all directions, then the observed proper motion must correspond to the same velocity. The geometry of the expansion then allows us to find the distance to the nova. From Chapter 15, we have

$$d = v_r/4.74\mu'' \qquad (18–1)$$

where v_r is the radial velocity in km/s, 4.74 is a conversion factor that yields d in parsecs, and μ'' is the proper motion in arcsec/year. Nova Persei (GK Persei; Figure 18–14) is a good example of a nova with an expanding nebula. Observations show that the expansion of the shell increases about 0.5″/year and that the radial velocity of the shell is roughly 1100 km/s. Then the distance is

$$d = 1100/(4.74 \times 0.5) = 460 \text{ pc}$$

The importance of this method of determining the distances of novae is that it is both direct and unambiguous. In this way, we can establish the luminosities of novae at maximum and minimum, and from

Figure 18-14

Expanding shell from Nova Persei. The nova exploded in 1901; this photograph was taken in 1949. *(Palomar Observatory, California Institute of Technology)*

that information we can find the total amount of energy released in the explosion.

A nova occurred in August 1975: Nova Cygni, which peaked at magnitude 1.8 (Figure 18-15). Prenova photographs indicate that the star increased in luminosity at least 16 million times. Curiously, we have no evidence that Nova Cygni is a member of a binary system. It may be a rare instance when the infall of interstellar material triggers a nova; some studies indicate that an influx of a mere $10^{-10}M_\odot/$ year will do the trick in a few million years.

(b) Supernovae

Supernovae attain absolute magnitudes in the range from -16 to -20. Although supernovae are relatively rare in our Galaxy, some have been noted in historical records. From such records, especially the Chinese chronicles, we find, for example, that the supernova of A.D. 1054 reached an apparent magnitude of -6, even brighter than Venus at its brightest

(-4). Much of what we know about supernovae has been gleaned from studies of galaxies. In the case of small galaxies, the luminosity of a supernova may rival the total brightness of the galaxy. Like ordinary novae, supernovae exhibit a very rapid rise to maximum and then drop two or three magnitudes within a month before declining more gradually. The total energy output from any supernova is stupendous: 10^{44} J, or approximately as much energy as the Sun will produce in its entire lifetime of 10 billion years.

Two types of supernovae occur, differentiated by their spectra and by their light curves (Figure 18-16 and Table 18-3). Type I supernovae appear in both elliptical and spiral galaxies (Chapter 21), and Type II occur only in spirals (especially in the spiral arms). We deduce that Type I supernovae belong to Population II and Type II to Population I. At maximum brightness, a Type II supernova shows a nondescript spectrum; the only prominent line is an emission line at 656.3 nm, the H_α line. About a month later, the

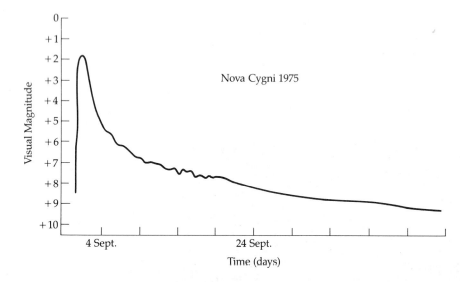

Figure 18–15

Light curve of Nova Cygni. Observations were made at *V* band. *[P. Young, H. Corwin, J. Bryan, and G. de Vaucouleurs, Astrophysical Journal 209:882 (1976)]*

spectrum has evolved to show more emission lines and a few weak absorption lines. In contrast, Type I supernovae exhibit messy spectra. At maximum light, broad emission lines appear along with some strong, dark lines, the combination having P Cygni-type profiles. Later, four emission lines from Fe II dominate the spectra, along with emission lines of Na I and Ca II. Basically, the difference between Type I and Type II spectra is that Type II show strong hydrogen lines and Type I do not—an indication that

Type I involve highly evolved, hydrogen-deficient stars.

Both types of supernova violently eject a large fraction of the original star's mass at speeds of 5000 to 10,000 km/s. Here we are dealing not with a surface phenomenon but with something far more fundamental—an explosion of the interior regions of a star. At maximum brightness, a supernova can become about as large as the Solar System. For the first months after maximum, the supernova's size in-

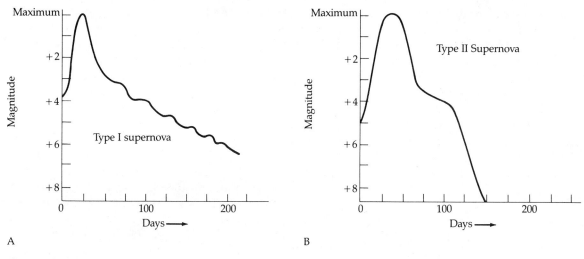

Figure 18–16

Supernovae light curves. (A) Type I; note the gradual decline. (B) Type II; note the sharp decline and shoulder.

creases dramatically. For example, one Type II supernova observed in 1970 had a radius at maximum of roughly 3×10^9 km; the star expanded at about 5000 km/s for 30 days, until it attained a radius of 2×10^{10} km, approximately 1 lightday in diameter. After that, the star's photosphere shrank. During the expansion, the photosphere decreased in blackbody temperature from 12,000 to 6000 K.

Type II supernovae are thought to be stars much more massive than the Sun (10 to 100 solar masses) —stars that live their normal lives as O and B stars. Current theoretical computer models for Type II supernovae suggest that the explosion occurs in the core of a red supergiant. Red supergiants are so large that they do not cool much when they expand to a size roughly that of the Solar System. And the atmosphere of a red supergiant has nearly constant density, so that a shock wave traveling through it moves at almost constant velocity and transmits energy efficiently to the star's surface. The models predict that, at peak brightness, the supernova should have a photospheric temperature of roughly 10,000 K and a surface speed of 5000 km/s.

The basic interior model for a Type II supernova involves the implosion of a stellar core to form a neutron star. A violent rebound from this sudden collapse produces the explosion that ejects the outer layers. The stars in which this process occurs fall in the mass range of $10 M_\odot$ to $100 M_\odot$. Such stars develop carbon–oxygen cores during their normal lives; the carbon fuses slowly to produce neon and magnesium. The core's density eventually reaches a level high enough that electron capture removes the electrons, whose degeneracy pressure has supported the core. It collapses; a neutron star forms. The infalling material rebounds off the neutron core to drive a shock wave outward; this triggers violent explosions in the unburned material in the outer layers.

Because Type I supernovae are associated with Population II, these stars are roughly the mass of the Sun. They are really a puzzle, for it is hard to see how a solar-mass star can detonate as violently as a supernova. One idea resembles that for binary novae. Imagine a binary system containing a white dwarf and a normal star in which the white dwarf has a mass very close to the maximum limit ($1.4 M_\odot$). If enough mass flows onto the white dwarf to push it over the limit, it will collapse violently to a neutron star. This collapse may release sufficient energy to make a supernova.

(c) The Crab Nebula — A Special Supernova Remnant

The Crab Nebula (Figure 18–17) is the nearest, most intriguing, and most spectacular *supernova remnant*. It was the first identified as such and the first connected to a pulsar with the central star, in 1968. Other supernova remnants share a few of the characteristics portrayed by the Crab Nebula, and so it serves as an important prototype.

The distance to the Crab Nebula has been established by the technique outlined for novae (expansion of the gaseous shell); it is about 2000 pc. We can also use the present rate of expansion of the nebula to extrapolate back to the time of outburst; the result corroborates identification with the A.D. 1054 supernova (if we account for some acceleration since the initial expansion). The expansion is not uniform in all directions, and so the central star was not unambiguously identified until the pulsar was identified optically.

In the visible region of the spectrum, the nebula presents varied aspects, depending upon whether it is photographed in the radiation from one of the emission lines (such as H_α), in the continuum radiation, or through a polarizing filter. The line radiation emanates from clearly defined filamentary features (Figure 18–17), where the gas density is enhanced over that of the rest of the nebula. Underlying the filaments and more concentrated at the central part of the nebula is the region emitting in the continuum. This region also possesses considerable structure, which can be described as vague wisps or fibers. The appearance and position of these wisps change with time because of motions of the gas or because of compression waves moving through the gas.

An important clue to the nature of the Crab Nebula is the fact that the continuum radiation is strongly linearly polarized (Figure 18–17). Another is that the nebula is a strong radio emitter, for it is identified with the radio source Taurus A (the first radio source to be discovered in the constellation Taurus). The wavelength dependence of the radiation in both the visible and the radio region differs greatly from a Planck blackbody curve: it is *nonthermal radiation*. These characteristics led I. S. Shklovsky to propose (in 1953) that *synchrotron radiation* is the source of both the optical and the radio continuum.

We have already discussed synchrotron radiation in Sections 4–6(c) and 6–1(c), but let us remind

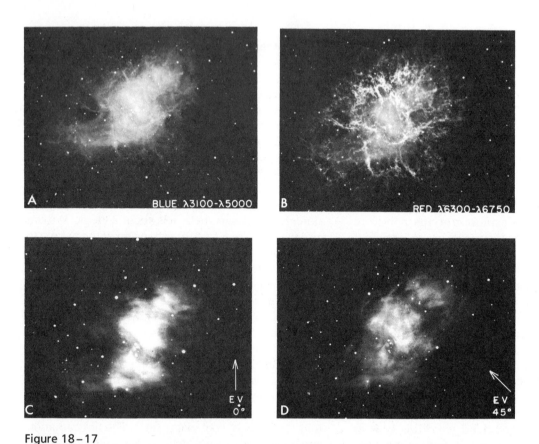

Figure 18–17

Crab Nebula. These photographs were taken in normal and polarized light. (A) Blue light. (B) Red light. (C) Polarized light, electric vector at 0° (arrow). (D) Polarized light, electric vector at 45° (arrow). *(Palomar Observatory, California Institute of Technology)*

you here of the basic points about it. When energetic electrons are accelerated by a magnetic field, they spiral around the magnetic field lines (Figure 18–18). This motion causes them to emit strongly polarized continuous radiation, whose intensity at a given frequency depends upon both the magnetic field strength and the energy of the electrons. The higher the mean electron energy, the higher the frequency at which the intensity is a maximum. Many radio sources have a synchrotron radio spectrum, but few simultaneously emit synchrotron radiation at the higher frequencies corresponding to visible radiation. The Crab Nebula is one of these few. In fact, synchrotron radiation also accounts for the X-ray emission that is observed from the Crab. The nebula as a whole, and not just the pulsar, generates X-rays.

The magnetic field strength in the nebula is estimated at 5×10^{-8} T. This field strength produces copious quantities of synchrotron radiation when the electrons are traveling at relativistic speeds (almost the speed of light), but where do these relativistic electrons come from? We know the following observational facts about the Crab Nebula: (1) we continue to see it approximately 930 years after the outburst, (2) the nebular gases exhibit accelerated motions, and (3) phenomenal amounts of synchrotron radiation continue to be emitted. Taken together, these facts demand that a strong energy source lies within the nebula. This energy source is none other than the Crab pulsar, which generates relativistic electrons and synchrotron radiation [Section 17–2(c)]. Recent radio observations show that the overall flux from

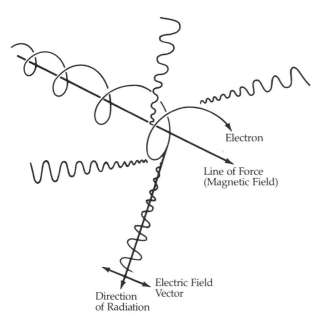

Figure 18–18

Synchrotron emission. A relativistic electron spirals along a magnetic line of force. The circular acceleration causes the electron to emit plane-polarized radiation.

the nebula has decreased at a rate of 0.17 percent per year (from 1968 to 1984), a value consistent with pulsar spindown time scales.

(d) Nucleosynthesis in Supernovae

As Chapter 16 pointed out, the most massive stars can fuse elements up to iron; heavier elements require reactions that absorb rather than produce energy. Elements heavier than iron are probably made in the supernova explosions of massive stars (Type II). Here is one scenario of that process.

A star with a mass greater than $10M_\odot$ to $20M_\odot$ will have a layered look at the end of its life: carbon, helium, and hydrogen shells at greater and greater distances from the iron core. This layering results from lack of convection and the outward drop in temperature. The iron core cannot support itself; it contracts and its temperature rises. At about 5 to 6 $\times$ 10^9 K, photodisintegration of iron gives $^{56}\text{Fe} + \gamma \rightarrow$ 13 $^4\text{He} + 4$ n—an *endothermic* reaction needing about 100 MeV. This robs the core of energy, and so it contracts more rapidly. Once the iron disintegrates,

$$^4\text{He} \longrightarrow 2\,\text{p} + 2\,\text{n}$$

$$\text{p} + \text{e}^- \longrightarrow \text{n} + \bar{\nu}$$

converts the core to a degenerate neutron gas. Meanwhile, the layers above the core fall rapidly inward and also heat up. They still have fuel for nucleosynthesis. This goes off explosively, blowing off the outer layers. This fusion results in a flood of energetic neutrons, which can be absorbed by heavy nuclei. Now the rapid and slow processes come into play, "rapid" and "slow" referring to how fast the process goes relative to beta decay:

$$\text{n} \longrightarrow \text{p} + \text{e}^- + \bar{\nu}$$

which takes about 17 min. In the rapid process, nuclei capture neutrons *faster* than beta decay; this builds up neutron-rich material. In the slow process, neutrons are captured more *slowly* than beta decay; proton-rich material results. The rapid process is usually abbreviated *r process* and the slow process, *s process.*

Here are the specific examples of this type of nucleosynthesis. Suppose starting with ^{56}Fe; then by the r process

$$^{56}\text{Fe} + \text{n} \longrightarrow {}^{57}\text{Fe}$$

$$^{57}\text{Fe} + \text{n} \longrightarrow {}^{58}\text{Fe}$$

$$^{58}\text{Fe} + \text{n} \longrightarrow {}^{59}\text{Fe}$$

$$^{59}\text{Fe} + \text{n} \longrightarrow {}^{60}\text{Fe}$$

$$^{60}\text{Fe} + \text{n} \longrightarrow {}^{61}\text{Fe}$$

Now, ^{61}Fe is stable for only about *6 min*, and so if no neutrons are captured in this time, then by the s process,

$$^{61}\text{Fe} \longrightarrow {}^{61}\text{Co} + \text{e}^- + \nu$$

Another example: transforming lead-206 to other isotopes of lead and finally to bismuth-209. Sock ^{206}Pb with a neutron; it becomes ^{207}Pb. Hit it with another neutron, and so on, up to ^{209}Pb. Each heavier isotope of lead is less stable than the previous one, and ^{209}Pb is so unstable that it decays rapidly (by beta decay) to ^{209}Bi (plus an electron and an antineutrino). Then the ^{209}Bi can absorb neutrons, build up to other bismuth isotopes, finally beta decay to the next element, and so on. In supernovae, time scales are short so that the r process is most effective in nucleosynthesis.

We have emphasized nucleosynthesis in supernovae, but red giants also manufacture some heavy elements. When a red giant undergoes helium shell burning, small numbers of neutrons are produced and add at low rates (s process) to iron to fuse into heavier elements. Although the s process is indeed slow, the red-giant stage lasts long enough (10^5 years or so) to synthesize an appreciable amount of heavy elements. Some are also made during red-giant helium flashes. The s process cannot synthesize the very heavy radioactive elements, however, for the neutron addition is so slow that such nuclei decay by fission before more neutrons can be added. In general, elements made by the s process complement those made in the r process to fill up the periodic table. Nucleosynthesis in red giants makes many of the elements heavier than iron and lighter than lead, but elements heavier than lead, such as uranium and thorium, are produced in Type II supernovae.

18–6 X-Ray Sources: Binary and Variable

X-ray telescopes have opened the high-energy Universe (photons with energies ranging from 1 to 100 keV) to our view. The Uhuru and Einstein X-ray instruments in Earth orbit have provided a wealth of data on galactic X-ray sources, many of which are variable and some of which are in binary systems. A few of these sources may contain black holes.

A number of binary X-ray sources have been seen (Table 18–4). Each is believed to be a main-sequence or post-main-sequence star with an X-ray

source. They have X-ray luminosities in the range from 10^{29} to 10^{31} W. Three of the sources (Hercules X-1, Centaurus X-3, and Small Magellanic Cloud X-1) have short-period X-ray pulses; they are X-ray pulsars. Five of the systems (Centaurus X-3, Small Magellanic Cloud X-1, Vela X-1, Circinus X-1, and Hercules X-1) exhibit X-ray eclipses; the X-ray source passes behind the normal star as we view the system. Using spectroscopic analysis of the light from the visible star (not the X-ray source), we can observe the changes in Doppler shift and so find the orbital periods. They are typically a few days. These short periods indicate that the orbits are only a few times larger than those of the primary stars. Then, if we can determine the separation of the two objects, we can ascertain, from Kepler's third law, the sum of the masses (normal star plus X-ray source). With an idea of the mass of the normal star from its luminosity (using the mass–luminosity law), we can also determine the mass of the X-ray source. And if that mass turns out to be large enough (greater than 5 solar masses, the upper limit for a neutron star), the X-ray source is likely a black hole!

(a) Cygnus X-1

The most likely candidate for a black hole is Cygnus X-1, a strong X-ray source in the constellation Cygnus. Cygnus X-1 emits about 4×10^{30} W in X-rays. Observations have shown that Cygnus X-1 flickers rapidly, in less than 0.001 s. This observation indicates that the flickering X-ray emitting region must be less than 0.001 lightseconds in diameter (less than 300 km). In 1971, radio astronomers discovered

Table 18–4   Some Binary X-Ray Sources

Name	Distance (kpc)	Binary Period (days)	Characteristics of X-Rays	Characteristics of Visible Star
Cygnus X-1	2.5	5.6	Vary in duration from 0.001 to 1 s	Blue supergiant of mass $\approx 20 M_\odot$
Centaurus X-3	5–10	2.087	Eclipses with 0.488-day duration; pulses every 4.84 s	Blue giant of mass $\approx 16 M_\odot$
Small Magellanic Cloud X-1	65	3.89	Eclipses with 0.6-day duration; pulses every 0.72 s	Blue supergiant of mass $\approx 25 M_\odot$
Vela X-1	1.4	8.95	Eclipses with 1.7-day duration; flares lasting a few hours	Blue supergiant of mass $\approx 25 M_\odot$
Hercules X-1	2–6	1.70	Eclipses with 0.24-day duration; pulses every 1.24 s	Companion HZ Her of mass $\approx 2 M_\odot$
Scorpio X-1	0.3–1.0	0.787	Vary in duration down to 1 s; no eclipses	Mass $< 2 M_\odot$

radio bursts from Cygnus X-1 and were able to pin down the location better than the X-ray astronomers could. In the most likely place for Cygnus X-1 lies an O supergiant, that is, a hot, massive star (Figure 18–19). It is called HDE (Henry Draper Extension) 226868 and is of spectral type O9.7 I, that is, a supergiant with a surface temperature of roughly 31,000 K.

Optical observations show that the dark lines in the spectrum of the blue supergiant go through periodic Doppler shifts in 5.6 days; the star orbits with the X-ray source (a massive but optically invisible companion) about a common center of mass every 5.6 days. The mass of Cygnus X-1 is hard to determine, however. We can observe the Doppler shift in the spectrum of the visible companion, but we cannot obtain the velocity of the X-ray source. And because Cygnus X-1 has not been found to eclipse, we do not know its orbital inclination. Hence, we cannot determine either individual mass.

We can make reasonable estimates, however. Blue supergiants typically have masses of 15 to 40 solar masses; take $20 M_\odot$ as typical. The orbital period and velocity of the supergiant give a relationship between the masses of the supergiant and the X-ray source, uncertain by the amount of orbital tilt. If the orbital tilt were 90°, the mass of the companion would be 4 to 5 solar masses, but the tilt cannot be so high because the star is not eclipsing. So the supergiant's orbital velocity must be higher than observed, and the mass of the companion must be higher [as expected from the mass function for binaries; Section 12–2(b)]. A study of the brightness of the supergiant indicates that it varies a little in 5.6 days. This fact leads to an inferred orbital tilt of 30°; then Cygnus X-1 has a mass possibly as great as 14 solar masses and most likely about 9 solar masses. If so, Cygnus X-1 must be a black hole, if the limit for a neutron star is 5 solar masses.

Note that the X-rays come *not* from the black hole itself but from an accretion disk of material around it. Mass from the blue supergiant, perhaps aided by a stellar wind, falls toward the black hole. Its angular momentum channels the infall into a disk, which is heated by tidal forces and the conversion of gravitational potential energy into thermal energy. Parts of this disk have temperatures of a few million kelvins; these generate the X-rays.

(b) Centaurus X-3

X-ray emission can arise not only from accretion disks around black holes but also from accretion onto neutron stars. An example is Centaurus X-3 (abbreviated Cen X-3). The Uhuru satellite showed that this X-ray source pulses every 4.84 s. Also, long-term observations have revealed that X-ray eclipses take place every 2.087 days and last about 0.5 day. So we know that the orbit of Cen X-3 is tilted so that its plane lies in our line of sight. A faint star at the X-ray source position varies in light with the same period as Cen X-3. The star turns out to be a blue giant about 25,000 lightyears away.

This information all falls into place with a simple model for the Cen X-3 binary system (Figure 18–20). Cen X-3 itself moves in an almost circular orbit around the blue giant at 415 km/s. Its orbit has a radius of about 11×10^6 km. At this close distance, mass flowing from the giant is picked up by the X-ray source. About every two days, the X-ray source orbits behind the giant as seen from the Earth and an X-ray eclipse occurs. These eclipses allow us to estimate the mass of Cen X-3: $0.6 M_\odot$ to $1.1 M_\odot$—a low-mass neutron star probably made in a supernova explosion.

The fact that Cen X-3 is an X-ray pulsar also supports a neutron star model, in analogy with the

Figure 18–19

The blue supergiant HDE 226868. This is the visible star about which Cygnus X-1 orbits. (*J. Kristian*)

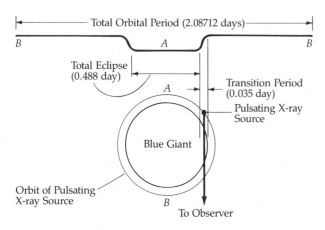

Figure 18–20

Model of emission from Centaurus X-3. *(Adapted from a diagram by H. Gurskey)*

model of radio pulsars as magnetic neutron stars. The X-ray pulses might arise from accreting matter channeled into the magnetic polar regions of the neutron star by the intense magnetic field.

(c) SS 433

A binary system with a neutron star can explain the bizarre radio, X-ray, and optical object called SS 433. That name comes from the fact that SS 433 is listed as entry 433 in a catalog of Milky Way objects showing strong emission lines compiled in 1977 by Bruce Stephenson and Nicholas Sanduleak. SS 433 at first appeared to be a nondescript fourteenth-magnitude star in the constellation Aquila. Radio observations showed that it lies near the center of a much larger radio source, called W 50, which may be a supernova remnant. The radio output of SS 433 varies by a factor of 2 to 3 over times of a few hours to days. X-ray observations revealed that the X-ray emission from SS 433 also varies.

These data prompted optical astronomers to take spectra of SS 433. A very strong H_α line appeared, surrounded by less intense but still rather strong lines. They were too strong to be lines emitted by oxygen or any of the heavier elements, and yet they were not at the correct wavelengths to be lines from hydrogen or helium. They turned out to be very highly redshifted and blueshifted lines of hydrogen and helium. Their displacements indicated radial velocities of more than 40,000 km/s, and the Doppler-shifted lines moved about 1 percent in a few days.

Long-term observations showed that the Doppler-shifted lines move periodically, with the red- and blueshifted lines swinging back and forth in a 164-day cycle (Figure 18–21). The maximum redshift value corresponds to a radial velocity of a little over 50,000 km/s. Very careful spectroscopic observations of the strongest emission lines, which at first seemed stationary, revealed a small Doppler shift (only 70 km/s, about 0.1 percent of the main moving lines) with a period of 13 days.

A model developed by Bruce Margon and his coworkers solves most of the puzzles presented by SS 433 (Figure 18–22): it is a binary system containing an ordinary (unseen) star and a neutron star orbiting every 13 days. The proximity of the two stars channels gas from the ordinary star onto an accretion disk around the neutron star. This gas heats up as it falls in and produces the strongest emission lines. Meanwhile, the accretion disk has two jets of material squirting out of it in opposite directions. The jet blasting away from us generates the highly redshifted lines; the one pointing toward us generates the blueshifted lines. The line of the jets is tilted about 20° to the rotational axis of the neutron star; that, in turn, inclines about 80° with respect to the line of sight from the Earth. Like the Earth's axis, the neutron

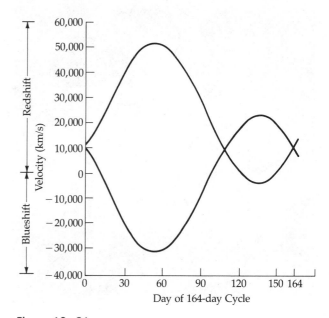

Figure 18–21

Cycle of Doppler shifts for SS 433. *(Based on observations by B. Margon)*

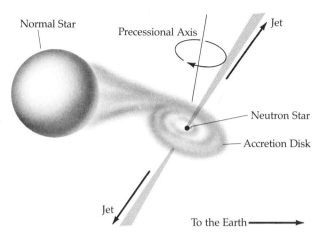

Figure 18–22

Model for SS 433. In this binary system, matter from a normal star accretes around a neutron star. Jets pour out along the axis of the accretion disk.

star's axis precesses in space, with a period of 164 days. As the spin axis slowly whirls in space, it carries the jets around with it. This variation causes the Doppler-shifted lines to fluctuate between maximum and minimum displacements as the inclination of the jets changes along the line of sight.

Radio observations with the VLA (Figure 18–23) imply that the jets consist of small blobs of ionized gas continuously blowing out of the source, like water from a hose. Incredibly, this material shoots out at 78,000 km/s, about 26 percent of *c*. The jets radiate by synchrotron emission — an idea backed up by the fact that they are highly polarized (15 to 20 percent). If so, they contain magnetic fields with a strength of about 10^{-6} to 10^{-7} T. X-ray observations of SS 433 confirm this model of relativistic jets. To date, SS 433 is the only *star* system observed to generate relativistic jets.

(d) X-Ray Bursters

X-ray bursters are set apart from other X-ray sources by their emission of brief but powerful bursts of X-rays (Figure 18–24). The bursts can occur in regular intervals of a few hours or a few days. Others fire off in a rapid sequence as a machine gun does, shooting off several thousand bursts in a day. A 10-s burst carries as much X-ray energy as the Sun gives off in a week at all wavelengths.

One X-ray burster is called 3U 1820-30 ("3U"

stands for the third Uhuru catalog; the other numbers refer to the object's coordinates on the sky). The source lies near the center of the globular cluster NGC 6624. In 1975, Jonathan Grindley and Herbert Gursky discovered a brief burst of X-rays from NGC 6624. Other X-ray satellites in operation at the time examined the source and delimited its position as being very close to the core of NGC 6624. X-ray astronomers at MIT then examined some new and old data to find bursts that lasted about 10 s. Remarkably, the bursts seemed to recur every 4.4 h.

Not all X-ray bursters are associated with globular clusters, and those that are do not always fall close to a cluster's core. To date, it appears likely that the neutron-star model best explains the properties of bursters: rise to peak in less than 1 s, duration of about 10 s, interval between bursts of about 2 h; blackbody effective temperatures of 3×10^7 K. The bursters appear concentrated near the galactic center; if they are 10 kpc from us, one burst has a luminosity at maximum of a few times 10^{31} W. An effective temperature of 3×10^7 K requires a blackbody of roughly 9-km radius to produce 2 to 3×10^{31} W. This hints that bursters are neutron stars.

Paul Joss has developed a model of helium-

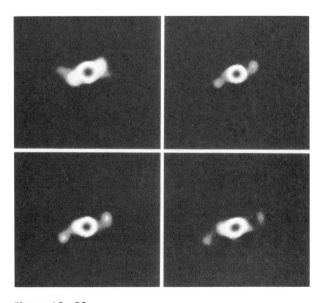

Figure 18–23

Radio observations of SS 433. The VLA made these observations over a four-month period. The material in the jets moved outward in this time. (*R. M. Hjellming and K. J. Johnston, National Radio Astronomy Observatory*)

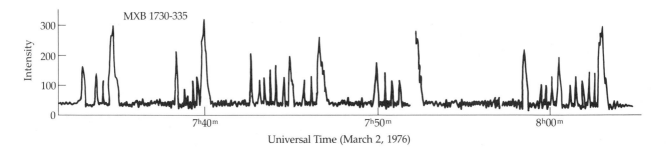

Figure 18–24

X-ray burster. The rapid burster MXB 1730-335 repeats a sequence about every 10 min.
[W. Lewin et al., Astrophysical Journal (Letters) 207:L95 (1976)]

burning flashes on an accreting neutron star that accounts for a good number of burster properties. It starts out with a neutron star accreting at a rate of roughly 10^{14} kg/s to build an envelope on it. As matter accretes, the temperature rises to 2×10^9 K at the base of the helium layer. The helium ignites in a flash, producing 2.5×10^{31} W just 0.2 s after the fusion reactions begin. With continuous accretion, the flashes can recur every 15 h. These calculated characteristics are much like those observed. The model does not rely on any special source for the accreted material; it may come from a binary companion. This model is similar to that for novae, the differences being that here the material falls onto a neutron star, not a white dwarf, and the temperatures are higher so that the energy comes out as X-rays rather than as visible light. In other words, bursters arise from thermonuclear flashes on neutron stars in binary systems. In a few cases, the spectra of possible companions to the X-ray sources suggest that a G8 III star is involved—similar to the metal-rich G giants found in some globular clusters.

(e) Binary Pulsars

Most stars in the Galaxy are members of binary or multiple-star systems. Even after one member of a binary becomes a supernova, the system usually remains intact. So a radio pulsar could exist in a binary system. The first one observed is called PSR 1913+16 and was discovered by Russell Hulse and Joseph Taylor in July 1974 during a search for new pulsars. This pulsar first attracted attention because its pulse period was only 0.059 s. When Hulse and Taylor reobserved PSR 1913+16 in September 1974, they found that its period went through a large cyclical

change in only 7.75 h. They recognized that such regular changes would naturally come about in a binary system consisting of the pulsar and a companion with an orbital period of 7.75 h. What was seen was a Doppler shift in the signal produced by the orbital motion of the system (Figure 18–25). When the pulsar is moving away from us, its pulses are spread out and come at longer intervals. When it is moving toward us, the pulses are pushed together and come at shorter intervals.

PSR 1913+16 lies about 5 kpc from us. Visual and X-ray observations have so far failed to detect

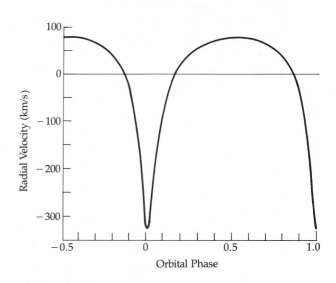

Figure 18–25

Doppler shift in the radial velocity of binary pulsar PSR 1913 + 16 from its motion about the center of mass.
[Adapted from a figure by R. Hulse and J. Taylor, Astrophysical Journal (Letters) 191:L59 (1975)]

either the pulsar or its companion. Radio observations alone indicate an orbital semimajor axis of only 7×10^5 km—a solar radius! The combined mass is 2.83 solar masses, and so if the pulsar has a mass of about 2 solar masses (a typical neutron star), its companion has a mass of about 0.8 solar mass. The companion might be a white dwarf.

Other binary pulsars have been discovered. One called PSR+065564 has a period of 24 h 41 min; its orbit is almost perfectly circular with a radius of only 750,000 km, just a bit larger than the Sun's radius. The companion to the neutron star must be small, perhaps also a white dwarf. In contrast to this system with a circular orbit, the binary pulsar PSR 2303+46 has a highly eccentric orbit: $e = 0.658$. The orbital period is 12.34 days. This system most likely contains two neutron stars of mass ratio around unity, and each star with a mass somewhat larger than $1M_\odot$.

Problems

1. In a given constellation, the following designations have been assigned to variable stars: V502, SU, II, V956, XY, and AK. List these stars in the order of their discovery.

2. In your own words, describe and correlate the fluctuation profiles of the typical pulsating star illustrated in Figure 18–2.

3. A Cepheid variable in a hypothetical galaxy is observed to pulsate with a period of ten days, and its mean apparent visual magnitude is 18. It is not known whether this is a Population I or Population II Cepheid.
   (a) What are the two possible distances to the galaxy (neglect interstellar absorption)?
   (b) What is the ratio of these distances?
   (c) Would this ratio change if we considered other galaxies? Explain.
   (d) Will this ratio differ for Cepheids of different periods?

4. If our telescope has a limiting magnitude of 22, what is the *maximum distance* to which we can see
   (a) RR Lyrae stars
   (b) classical Cepheids
   (c) W Virginis stars
   (d) ordinary novae
   (e) dwarf novae
   (f) supernovae
   How do these distances compare with the diameter of our Galaxy? (*Hint:* Consult Tables 18–1 to 18–3.)

5. The outburst of Nova Aquilae (V603 Aql) occurred in June 1918, at which time it attained a brightness of −1.1 mag. Spectra showed Doppler-shifted absorption lines corresponding to a velocity of 1700 km/s. By 1926, the star was surrounded by a faint shell 16″ of arc in diameter. Find the distance to Nova Aquilae in parsecs and the absolute magnitude at maximum.

6. If a star becomes a supernova, by what amount does its luminosity change if it originally had an absolute magnitude of 5.0? of 2.0? (The absolute visual magnitude of a supernova at maximum is about −15.0.)

7. Consult Chapter 10 to find the energy output in watts of the most energetic solar flares. Referring to Chapter 13, compare this with the typical energy outputs of stars of the following spectral types:
   (a) F
   (b) G
   (c) K
   (d) M
   By what factor is the luminosity of each star increased during such a flare event? Which stars would you consider to be *observable* flare stars?

8. Consider a rotating star with an extended atmosphere. If the stellar atmosphere is both rotating and expanding, draw the observed profile of a given spectral line (Figures 18–9 and 18–10) when the atmosphere rotates more slowly than the star.

9. A white dwarf in a binary system accretes enough material to increase its mass beyond the Chandrasekhar limit ($1.4M_\odot$) and collapse to the radius of a neutron star ($\approx 10^4$ m). Calculate the kinetic energy generated in such a collapse. Compare this with the value of approximately 10^{44} J relevant to a supernova explosion. Comment on the required efficiency of energy conversion.

10. A certain contact-binary system contains a red giant and a neutron star. The neutron star has a mass of $1M_\odot$ and a radius of 10^4 m. The system radiates 10^{31} W in X-rays. Determine the rate of mass flow (in solar masses per year) from the red giant to the neutron star required to produce this luminosity. Assume that half of the change in gravitational potential energy of an accreted gas particle is converted to X-rays and that the separation of the two stars is much greater than the radius of the neutron star.

11. Since the jets of SS 433 are oppositely directed, the point in Figure 18–21 at which the radial velocity

curves cross is also the point at which the jets are perpendicular to the line of sight. The velocity at this point is the relativistic transverse Doppler shift, with the equivalent radial velocity given by $1 + v/c = 1/[1 - (v_{jet}^2/c^2)]^{1/2}$.

(a) Determine the value of v_{jet}.

(b) What angle do the jets make to the line of sight at the point of maximum redshift and blueshift? (*Hint:* Use the relativistic Doppler formula and your answer to part a.)

12. Another suggested mechanism for generating a type II supernova is explosive nuclear "burning" of the heavier elements, especially silicon. One model for a massive red supergiant predicts the presence of a shell of mass $\approx 2M_\odot$ and containing mostly ^{28}Si deep within the interior. Once this shell reaches ignition temperature, calculations show that the entire shell undergoes fusion within a fraction of a second. For simplicity, assume that the shell has a mass of $2M_\odot$, with ^{4}He making up half the nuclei and ^{28}Si the other half. The reaction is ^{28}Si $+ ^4$He $\rightarrow ^{32}$S $+ \gamma$. Calculate the total energy released when the shell ignites, and compare this with the value of about 10^{44} J required for a supernova explosion. (The masses of the nuclei are 27.9769 amu for ^{28}Si, 31.9721 amu for ^{32}S, and 4.0026 amu for ^{4}He.)

13. Sketch the analogs to Figures 18–9 and 18–10 for the case of a cloud of hot gas falling in a spherical shell onto an even hotter star.

The Interstellar Medium and Star Birth

We have examined the general form of our Galaxy and the varieties of stars that populate it. Now to consider the content of the vast regions of space between the stars — the interstellar medium (ISM). Just as our own Solar System is pervaded by gas and plasma (the solar wind), magnetic fields, particles, and rocks, so also is interstellar space filled with gas, dust, magnetic fields, and particles. This chapter concentrates on the dust and gas in the galactic disk. Gas and dust make up the bulk of the interstellar medium, and here, from dense molecular clouds, new stars (and planets) are born.

19–1 Interstellar Dust

Interstellar dust cohabits the interstellar medium with the gas. On the average, one dust particle exists in every million cubic meters, but the dust amounts to about 1 percent of the total mass of interstellar mat-

ter, and it can cut out light from distant objects or from those enshrouded in dense clouds. Piercing the dust veil has been an important goal of radio and infrared astronomers in revealing the process of star birth.

(a) Dark Nebulae and the General Obscuration

Numerous dark patches stud the Milky Way; some appear only on photographs, but others, such as the Great Rift in Cygnus and the Coal Sack near the Southern Cross (Figure 19–1), are visible to the naked eye. These *dark nebulae* are opaque clouds obscuring the light of the stars behind them. In many cases, dark nebulae lie adjacent to or superimposed upon bright nebulae; an example is the famous Horsehead Nebula in Orion (Figure 19–2). Sometimes very small dark regions called *globules* overlie bright nebulae (Figure 19–3).

Far more difficult to detect is the *general obscura-*

Figure 19–1

The southern Milky Way. The Coal Sack is the dark region near the center. (*Harvard College Observatory*)

Figure 19–2

Horsehead Nebula. This dark cloud is in Orion. *(Palomar Observatory, California Institute of Technology)*

Figure 19–3

Dark globules. These small, dusty regions lie in the Rosette Nebula. *(Palomar Observatory, California Institute of Technology)*

tion caused by dust distributed more uniformly and thinly than in the dark clouds. We have already mentioned this obscuration in our discussion of observed and expected star counts (Chapter 14). The general absorption from dust requires that the equation for the distance modulus should be rewritten from

$$m - M = 5 \log d - 5 \qquad (19\text{-}1)$$

to

$$m - M = 5 \log d - 5 + A \qquad (19\text{-}2)$$

where A represents the total amount of absorption (in magnitudes) for the distance d from our Sun; m is the apparent magnitude, M the absolute. Note that stars appear to be fainter (larger m), and therefore farther away, whenever interstellar obscuration intervenes.

If this general interstellar absorption were truly uniform throughout the Galaxy, A could be expressed as a simple function of distance, $A = kd$. Observations clearly show, however, that the general absorption is patchy; the total amount of absorption between us and a star or cluster differs with direction in the sky and with the character of the intervening space. This patchiness should not surprise you: the dark nebulae have irregular appearances.

(b) Interstellar Reddening

The absorption A in Equation 19–2 depends upon wavelength. The interstellar dust between us and a star does not dim that star's light identically at all wavelengths; more light is scattered in the blue than in the red. As a result, the light from the star appears redder than in the absence of dust—hence the term *interstellar reddening*. The obscuration is primarily a form of scattering rather than absorption. The reddening arises from *selective scattering*, so that if equal numbers of red and blue photons are incident upon a dust cloud, a greater number of the blue photons are scattered out of the beam. Hence, a proportionately larger number of the red photons penetrate through the cloud and reach an observer (Figure 19–4). In a sense, the light is "deblued" rather than reddened.

Reddening must increase the color index observed for a star. We define *color excess* as the difference between the observed and the intrinsic color index [Section 11–4(b)]:

$$CE = CI \text{ (observed)} - CI \text{ (intrinsic)} \qquad (19\text{-}3)$$

The intrinsic color index depends upon the spectral

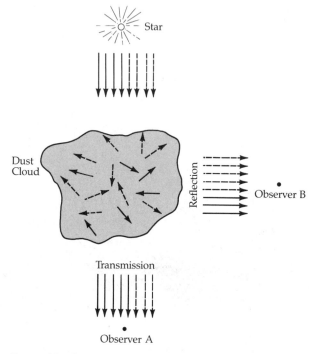

Figure 19–4

Scattering and reddening. Grains in a dust cloud scatter blue light more efficiently than red. So observer B sees a blue reflection nebula, and observer A sees reddened starlight.

type of the star, and it can be established from a spectrum. It is also possible for astronomers to determine the color excess without taking a spectrogram, by comparing two color indices, such as blue–visual $(B - V)$ and ultraviolet–blue $(U - B)$.

The wavelength dependence of the interstellar extinction (absorption) is found by comparing the brightnesses of stars of similar spectral type at a number of wavelengths. By such a comparison of stars reddened by different amounts, astronomers have found that the extinction is proportional to $1/\lambda$ in the visible region. Such data also established that, in most regions of the Galaxy, the absorption in visual magnitudes is three times the color excess; thus

$$A_v \approx 3(CE) \qquad (19\text{-}4)$$

We may use this value of A in Equation 19–2 to find the true distance to the star, providing the star itself does not have a peculiar spectral distribution or affect the nature of dust grains in its immediate vicinity.

As an example, consider a star of spectral type G0

V with $m_v = 13.0$ and $CI = 1.6$. The H–R diagram shows that a star of this spectral type has an intrinsic color index of 0.6, and $M_v = 5$. Therefore, from Equation 19–3, we have $CE = 1.0$ and from Equation 19–4, $A_v \approx 3.0$; now substituting in Equation 19–2

$$m_v - M_v = 5 \log d - 5 + A_v$$

we obtain

$$13.0 - 5.0 = 5 \log d - 5 + 3$$
$$\log d = 2$$
$$d = 100 \text{ pc}$$

Although the star appears to be 400 pc distant, its actual distance is only 100 pc. Or, we could say that the true distance modulus is

$$m_v - M_d = m_v - M_v - A_v = 5 \log(d/10)$$
$$= 13 - 5 - 3 = 5$$

Let's quantify the discussion a bit more. Rewrite Equation 19–2 as

$$m_\lambda - M_\lambda = 5 \log d - 5 + A_\lambda$$

where the wavelength (λ) subscript emphasizes that the amount of extinction depends on wavelength. Note that A_λ is the absorption along the line of sight, so that

$$A_\lambda = k_\lambda d$$

where k_λ is the absorption, or extinction, coefficient, which depends on the extinction cross section σ_λ and the number-density distribution of absorbing material n. On the average, $k_v = 1$ to 2 mag/kpc.

The amount of extinction can be directly related to the extinction characteristics of the dusty material. Recall that, for light of intensity $I(0)$ passing through a uniform slab of thickness L, absorption results in

$$I(L) = I(0) \exp(-\tau_\lambda)$$

where the optical depth

$$\tau_\lambda = \sigma_\lambda \int_0^L n(l)dl = \sigma_\lambda nL$$

so that over distance d in the interstellar medium,

$$\tau_\lambda = \sigma_\lambda \int_0^d n(r)dr = \sigma_\lambda nd$$

if $n(r)$, the number density of absorbing particles, is uniform along the line of sight. Now, the intensity ratio $I/I(0)$ is related to a difference in magnitudes

ΔM by

$$\Delta M_\lambda = -2.5 \log[I/I(0)]$$
$$= -2.5 \log[\exp(-\tau_\lambda)]$$
$$= 2.5(\log e)\tau_\lambda$$
$$= 2.5(0.434)\tau_\lambda = 1.086\tau_\lambda$$

Since this change in magnitude ΔM_λ is the absorption A_λ,

$$A_\lambda = 1.086\tau_\lambda$$
$$= 1.086\sigma_\lambda nd = k_\lambda d$$

where

$$\sigma_\lambda = \pi a^2 Q_\lambda$$

Here a is the radius of the dust particles and Q_λ is their relative *extinction coefficient*, which can be calculated in the laboratory from the optical properties of the appropriate materials. Note that the absorption in magnitudes approximately equals the optical depth.

(c) Interstellar Polarization

The light emitted by stars is basically unpolarized. However, observations show starlight to have a polarization of up to 10 percent at optical wavelengths. (See Section 8–1 for the nature and properties of polarized light.) The amount of polarization correlates directly with the amount of interstellar reddening: a large polarization is found for stars with large color excesses. Therefore, interstellar dust causes most of the optical polarization of starlight (although some stars do show intrinsic and often variable polarization).

We detect polarization by measuring the intensity of light transmitted through a polarizing filter or other polarization analyzer. Such a filter passes a maximum intensity $I_\parallel$ in one orientation and a minimum intensity $I_\perp$ when rotated through 90°. We define the *fractional polarization* of light by

$$FP = (I_\parallel - I_\perp)/(I_\parallel + I_\perp)$$

For completely unpolarized light, we have $I_\parallel = I_\perp$ and $FP = 0$, whereas for *linearly* polarized light, $I_\perp = 0$ and $FP = 1$.

Nonspherical particles can polarize light, for light vibrating parallel to the long axis of an elongated particle will be diminished more than that vibrating perpendicular to that axis. Hence, the discovery of interstellar polarization gives clues about the nature of *interstellar dust grains*. Also, even nonspherical particles cannot polarize light if they are oriented at random. On the average, the grains along the line of sight must have a preferential orientation for polarization to occur. Now, under certain conditions, even relatively weak magnetic fields can align particles. One model visualizes elongated particles spinning with their short axes aligned along the magnetic field. So we can use polarization data to map the magnetic field of the Galaxy as seen from the Sun. Polarization will be strong and ordered when the magnetic field is perpendicular to the line of sight and weak and random when we look along the field (down a magnetic flux tube). Observations of interstellar polarization indicate that the galactic magnetic field (on the average) lies along spiral arms. As discussed in Chapter 20, radio interstellar polarization confirms the interpretation that the optical polarization is related to the magnetic field.

(d) Reflection Nebulae

When a dust cloud lies to one side of a star rather than between the observer and the star, it scatters light from the star toward the observer. This is the same scattering phenomenon that is responsible for interstellar reddening, but instead of viewing the light that filters through the dust, we see the light that is scattered out of the star-to-cloud direction (Figure 19–4)—a *reflection nebula* (Figure 19–5). Each of the dust particles scatters a bit of the starlight toward us. Since the particles scatter blue light more effectively than red, reflection nebulae appear *bluer* than the incident starlight they scatter. Another type of bright nebula, which appears reddish, is an *emission nebula*, which consists primarily of gas excited by a hot star [Section 19–2(b)]. Although we distinguish between these two types of bright nebulae, they are often found in close proximity to one another; a reflection nebula may lie adjacent to an emission nebula, as occurs in the constellation Orion. This illustrates that dust and gas are usually closely mingled in the interstellar medium (Section 19–3).

One of the key observational features of reflection nebulae is that the light from them is highly polarized—often as much as 20 to 30 percent. Now, the light emitted by stars is basically unpolarized. When scattered by small particles, light is selectively plane-polarized. The polarization occurs because light is a transverse wave, and so small particles selectively scatter light perpendicular to the direction of

Figure 19–5

Reflection nebula. A dust cloud around the star Merope in the Pleiades. *(National Optical Astronomy Observatories)*

travel of the incident light. This polarization differs from that of the general interstellar polarization in that it results from reflection rather than transmission of light. It, too, depends on the nature of the medium but also on the angles of the incident starlight and the line of observation. So again we may deduce some of the properties of the dust grains around stars by observing the polarization of the reflected starlight.

(e) The Nature of the Interstellar Grains

The observed effects of the interstellar grains—interstellar reddening, extinction, reflection, and polarization—give clues to the nature of the particles involved. Although both theorists and experimentalists have worked diligently to decipher the data, they have not yet found a complete solution. The possibilities that can explain most of the observations include the following:

1. Elongated, perhaps needle-shaped, dirty-ice grains
2. Grains of graphite (carbon)
3. Particles with graphite cores and ice mantles
4. Large, complex molecules similar to hydro-carbons
5. Silicate particles

The strength of the interstellar obscuration and the characteristics of the reddening require solid particles rather than molecules, atoms, or electrons. Interstellar polarization requires *nonspherical* particles that can be aligned by a magnetic field; pure ice particles are excluded because they are not magnetic. Graphite or graphite-core particles could fit, for carbon in the form of graphite readily forms into highly flattened plates or flakes. This nonspherical attribute is not unique to graphite, however, for ice and ice-like materials also tend to form flat crystals. There is strong support for silicate particles and for a *mixture* of silicates and either alternative 2 or 3.

A key clue comes from the average *interstellar extinction curve* (Figure 19–6) in the visible and ultraviolet. Note that the curve rises in the visible, has a bump in the ultraviolet (about 0.2 μm), and then, after a slight dip, rises again into the far ultraviolet (the data are limited at very short wavelengths by our observational techniques). No one type or size of grain can fit the extinction curve; it must be a composite resulting from the interstellar mixture. Calculations show that the bump and the rise in the ultraviolet must be caused by very small particles, 0.005 to 0.02 μm in radius. The rise in the visible can be the result of larger grains, 0.05 to 0.2 μm in radius. Note that dust that absorbs ultraviolet will heat up and emit, in equilibrium, in the infrared. This infrared emission has been observed on a large scale from throughout the Milky Way, especially by the IRAS satellite.

The bump at 0.2 μm can be explained by bare, small-radius (≈ 0.02 μm) graphite (pure carbon) particles. The bonds between the carbon atoms resonate and absorb at this wavelength. The rise in the ultraviolet must also come from very small particles; silicate particles with a radius of 0.005 to 0.01 μm can play this role. For the visible region of the spectrum, larger particles are needed; their radius must be about

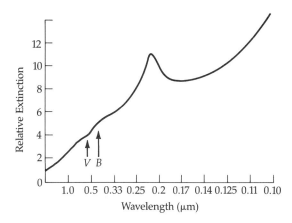

Figure 19–6

Interstellar extinction curve. This curve is an average one over many directions in the sky; note the peak near 0.2 μm.

0.2 μm. Such particles cannot be composed entirely of silicates, graphite, or pure iron; to account for how much extinction is seen requires some of the abundant hydrogen in the icy materials. To account for the shape and amount of the interstellar extinction curve, astronomers have developed *core–mantle grain models*. The small core, about 0.05 μm in radius, can consist of silicates, iron, or graphite; silicates are plausible. The mantles are made of icy materials, likely some composite of all possible kinds. Note that these grains are much smaller than the dust in your house; in fact, they are smaller than the particles in tobacco smoke.

Infrared observations bolster the idea that silicates and ices (at least water ice) make up part of the interstellar grains. They show absorption bands at 9.7 and 3.07 μm (Figure 19–7). Silicates in terrestrial rocks, meteorites, and lunar rocks have absorption bands at about 10 μm; these involve changes in the energy of vibration in the Si–O bonds. Silicates also have another, but weaker, absorbing band at 18 μm that involves the energy of bending the O–Si–O bonds. This absorption feature has been seen in a few other sources, strengthening the identification with silicates. The band at 3.07 μm likely occurs from water ice, but the amount of water ice in the grain is not enough to account for all of the extinction. Probably present are other icy substances that have not yet been positively identified because their infrared bands are much weaker than that of water ice.

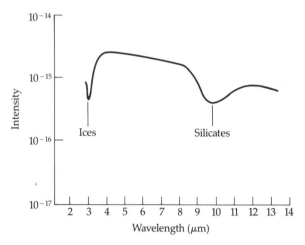

Figure 19–7

Infrared spectrum showing absorption features. This spectrum is from the Becklin–Neugebauer source in the Orion Nebula. The absorption bands are probably from ices and silicates. *[F. Gillett and W. Forrest, Astrophysical Journal 179:483 (1973)]*

What is the source of interstellar grains? Basically, they come in two types, ices and denser materials, and so they must have at least two different sites of origin. Ices solidify at a few hundred kelvins, the denser materials at a few thousand kelvins. The heavier grains are probably made in the atmospheres of cool supergiants. We know such stars are blowing mass into space at rates of about 10^{-6} solar mass a year. The surfaces of these M stars have temperatures of only 2500 K or less. As gaseous material streams outward from them, its temperature drops and solids can condense out of the vapor. The spectra of some supergiants show the 9.7-μm silicate feature, indicating that such dust exists around them. In a rarer class of stars, in which carbon is somewhat more abundant than oxygen, graphite-like grains and particles made of silicon carbide can form in the outflowing material. The infrared spectra of these stars indicate a surrounding cloud of carbon particles.

What about the ices that make up grain materials (or perhaps entire grains)? These likely condense on cores in the deep interiors of dense molecular clouds. Here the temperatures are low and the gas densities high, and so bare grains can grow crusts of ices. A core may have to grow a mantle once every 10^8 years or so, since grains will lose their mantles when in an environment where temperatures range above a few hundred kelvins.

19–2 Interstellar Gas

In addition to dust grains, interstellar space contains gas. This section investigates the physical and observational properties of the many species of *interstellar gas*, which makes up most of the interstellar medium. While the dust between the stars makes its presence known by its broad spectral influence upon the continuous light from distant stars, the interstellar gas produces its own characteristic emission and absorption line spectra. The temperature and density of the gas determine these characteristic spectral features. In general, the gas is essentially transparent over a wide spectral range despite the fact that the total mass of the gas in our Galaxy is greater than the total mass of the dust by a factor of about 100.

If there is so much more gas than dust, why is it the dust that is responsible for interstellar obscuration? The number density of dust grains is vastly smaller than the number density of the gas: roughly 10^{12} to 1. To answer the original question, we must consider the *absorption coefficient* of each material; this parameter is analogous to opacity. Dust particles interact very strongly with visible light over a broad range of wavelengths, but the interstellar gas cannot interact with visible light (hydrogen and helium in their ground states cannot absorb visible-light photons). Therefore, at visible wavelengths, the absorption coefficient per particle is astronomically greater for the dust than for the gas.

(a) Interstellar Optical Absorption Lines

Some stars have in their spectra absorption lines that are quite out of character with the spectral class. For instance, many B stars exhibit sharp, sometimes multiple lines of Ca II. Some spectroscopic binaries show particular spectral lines that remain fixed in wavelength while the rest of the spectral lines shift periodically to the red or blue in response to the binary stellar motions (Figure 19–8A). Clearly, these absorption lines must originate in the interstellar medium. Multiple lines arise when there are several absorbing clouds along the line of sight (Figure 19–8B). Optical absorption lines, identified as interstellar in

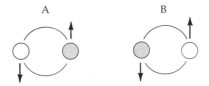

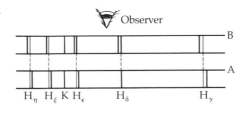

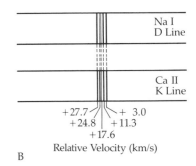

Figure 19–8

Optical interstellar absorption lines. (A) Lines from cool clouds appear in absorption toward a binary system, but they do not show any Doppler shift. (B) The lines have several Doppler-shifted components, each from a cloud with a different radial velocity.

origin, include those from Ca I, Ca II, Ti I, Ti II, Na I, and the molecules CN and CH.

These absorption lines are sharp because thermal Doppler broadening is negligible at the low temperatures that characterize some parts of the interstellar medium. The intensity of a line depends upon the amount of gas lying between the star and the observer; if the gas is distributed uniformly through space, the intensities of interstellar absorption lines depend directly upon the path length traversed by the starlight. Low gas density plays a role in preventing ions from recombining into neutral atoms after photoionization. Sufficiently energetic photons and cosmic rays will occasionally encounter and ionize the widespread gas atoms and molecules. In order to recombine, an ion must capture an electron, but at typical interstellar densities the chance of such a capture is very small.

(b) Emission Nebulae: H II Regions

Hydrogen Line Emission

Among the most spectacular objects to be photographed with telescopes are the *emission nebulae*—clouds of gas caused to shine by the intense radiation from a hot star. Hot O and B stars emit tremendous amounts of ultraviolet radiation; such energetic photons, with wavelengths less than 91.2 nm, ionize any hydrogen atom they encounter. If such a hot star is surrounded by a cloud of gas, the hydrogen atoms close to the star will be ionized and form an *H II region*, or emission nebula. Away from the star, the energetic photons have been used up for ionization; eventually none are available to ionize the hydrogen and the H II region terminates (neutral hydrogen H I prevails). Let's now examine the physical processes that determine the structure of an H II region in greater detail.

The hydrogen gas in interstellar space is extremely dilute and cold. Most of the gas is H I (neutral hydrogen) in the ground state because collisional excitation is rare. Therefore, only photons whose wavelengths are less than or equal to 91.2 nm can ionize the gas to H II; note that 91.2 nm corresponds to the Lyman continuum limit (ionization potential) of hydrogen. Imagine a hot star in the midst of this cool H I gas with $T_{\text{eff}} \gtrsim 20{,}000$ K, which the Planck spectral curves imply will produce ample ultraviolet radiation ($\lambda \gtrsim 91.2$ nm). If the gas density is reasonably uniform, the ultraviolet radiation from the central star ionizes all of the hydrogen in a roughly spherical volume of space; we term this region the *Strömgren sphere*. (Because the interstellar gas is clumpy, H II regions are rarely spherical.) Equilibrium is established when the rate of recombination (H II + e⁻ → H I) equals the rate of photoionization; the H II region is maintained by the continual reionization of recombined H I atoms due to the flux of ultraviolet photons from the central star.

The outer boundary of an H II region arises from several factors. At greater distances from the star, the inverse-square law diminishes the flux of ultraviolet photons and ionization of the recombined H I atoms is no longer possible. So the ratio of H I to H II rises sharply with increasing distance from the star, and the material quickly becomes opaque to the Lyman continuum, giving rise to the sharp boundary. In addition, most of the H II recombines to an excited state of the neutral H I; the atom then quickly cascades to the ground state, emitting several low-energy ($\lambda >$ 91.2 nm) photons in the process. Because the H I atoms spend so little time in the excited states, practically all of these low-energy photons (as well as the star's photons with $\lambda > 91.2$ nm) escape from the H II region. So the H II region *fluoresces* by converting the stellar ultraviolet radiation to lower-energy photons, with the bulk of the radiation escaping as the visible (longer-wavelength) Balmer lines. The H II region is therefore spectacularly visible as a bright reddish *emission nebula*—reddish because most of the Balmer radiation is in H_α. Note that the ionized gas within the H II region is hot (about 10,000 K) because the photons with energies above 13.6 eV expel freed electrons with kinetic energy that they share with the ions by collisions.

Although some radiation short of the Lyman continuum limit does leak out, the spectrum of the central star (as seen through the gas cloud) exhibits severe depletion at ultraviolet wavelengths. The strongest optical emission lines from the nebula itself are the hydrogen Balmer lines, especially H_α. Radio line emission at centimeter wavelengths has been observed from very-low-energy electronic transitions between very high excitation levels of H I, such as from level $n = 110$ to $n = 109$ and from $n = 105$ to $n = 104$. The ionized hydrogen (H II) has no electrons, and so it cannot radiate spectral lines; nevertheless, radio continuum radiation emanates from the H II region as a result of free–free transitions (next section). Optical fluorescence lines of helium are also strong in the spectra of emission nebulae; together with the radio recombination lines of helium (arising from transitions between high-excitation levels), these lines permit us to (1) study the excitation mechanisms operating in H II regions, (2) investigate the elemental abundances (especially He/H) of the interstellar medium, and (3) probe the spiral structure of our Galaxy.

Continuous Radio Emission

The ultraviolet photons from a hot star liberate electrons from hydrogen atoms in the gaseous nebula

surrounding the star. These electrons then move freely through the gas, sometimes recombining with ions and sometimes exciting atoms or ions (leading to the emission of forbidden lines), but more often interacting with ions in a *free-free transition.* A free electron travels past an ion in a hyperbolic orbit of a given energy. This orbit can be altered by the quantum-mechanical absorption or emission of a photon with any amount of energy. When an assembly of electrons and ions (a plasma) is involved, the individual free-free emissions add up to a continuum; because the characteristic kinetic energies are small, this continuum radiation occurs predominantly at infrared and radio wavelengths. In short, an H II region is a source of radio emission characterized by the mean energy of the electrons, by the temperature of the gas. To distinguish this emission from synchrotron radiation, we use the term *thermal radio emission* or *thermal bremsstrahlung.*

Observations at radio wavelengths of thermal bremsstrahlung from H II regions provide a diagnostic of their physical conditions. The emission falls into two regimes: optically thick and optically thin con-

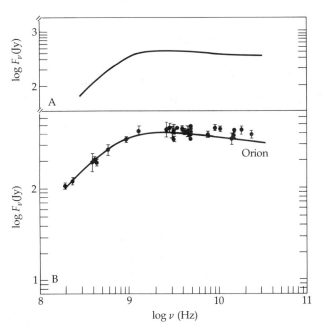

Figure 19–9

Continuous emission from an H II region. (A) Typical spectrum of free–free emission from an H II region. (B) Microwave spectrum for the Orion Nebula. (*Adapted from a diagram by E. Chaisson*)

nected by a turnover region (Figure 19–9A). In the optically thin regime, the flux is a power law:

$$F_\nu \propto \nu^{-\alpha}$$

where $\alpha \approx 0.1$ for a thermal source. On a log-log plot (Figure 19–19A), α is just the slope of the spectrum. For the Orion Nebula, for instance, the spectrum turns over at about 1 GHz and α does equal 0.1. In the optically thick regime at low frequencies, the spectrum rises as ν^2, the same as for blackbody radiation at low frequencies.

The turnover frequency gives a direct measurement of the average electron density in an H II region; typically, it averages about $10^9/\text{m}^3$. Measurements in the optically thin regime (at high frequencies) also allow us to compute the size of the H II region, the mass of ionized hydrogen, and the number of ionizing photons per second from the embedded stars and hence to infer the number and spectral type of the stars creating the H II region. These values can be found even if the region is obscured optically by dust surrounding it and along the line of sight.

(c) Supernova Remnants

Material ejected from supernovae certainly becomes part of the interstellar medium. Moreover, the ejected matter sweeps up any surrounding gas and dust as it expands; this produces a shock wave that excites and ionizes the gas, which then becomes visible as an emission nebula. X-rays emitted by supernovae are also instrumental in ionizing nearby gas. Supernova remnants are radio emitters because of their synchrotron radiation. The Loop Nebula in Cygnus is such a remnant (Figure 19–10). Note that it looks spherical — a shell produced by the interaction between the interstellar medium and a supernova shock wave.

A similar nebula in the southern sky — the Gum Nebula (Figure 19–11) — extends over 50° in the sky. The Gum Nebula has a diameter of about 700 pc, its closest edge being only 100 pc from the Sun. This nebula was created by the pulse of ultraviolet radiation and X-rays generated by a supernova 20,000 years ago. An X-ray source named Vela X lies almost in the nebula's center. It is a prime suspect as the supernova site. The discovery of a pulsar near the location of the Vela X source supports its nature as a supernova remnant.

About 150 galactic supernova remnants are cata-

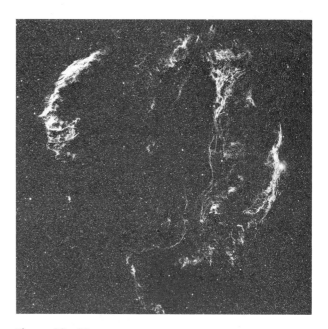

Figure 19–10

Supernova remnant in Cygnus. The Loop Nebula in red light. *(Palomar Observatory, California Institute of Technology)*

loged, and more than 30 have been observed with the Einstein X-Ray Observatory. The huge shock waves plow through the interstellar gas and heat it to temperatures of at least a few million kelvins in the zone just behind the wave. This gas emits X-rays because it has such a high temperature. The X-ray pictures of Type I remnants, such as Tycho's Supernova (Figure 19–12) typically show symmetrical shells with variations in brightness around their rims — a possible indication of the patchy structure of the interstellar medium.

(d) Planetary Nebulae

Planetary nebulae differ from H II regions in that they are more compact and of higher surface brightness and have a different exciting source. When seen through a telescope, a planetary nebula appears as a round, greenish disk that superficially resembles a planet — hence the name. Closer examination reveals that the nebula is excited by a very hot central star. Gas densities in the nebulae surrounding these stars are higher than in H II regions; hence, collisions between electrons, atoms, and ions occur more fre-

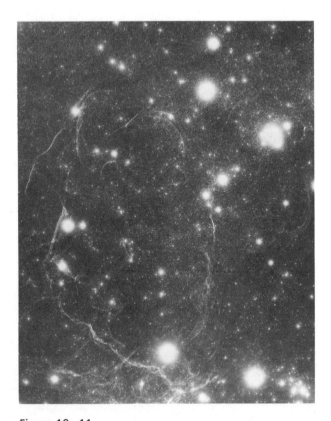

Figure 19–11

Vela supernova remnant. A part of the Gum Nebula, these filaments are moving through the interstellar medium at thousands of kilometers per second. *(Royal Observatory, Edinburgh)*

2×10^6 K prevail, whereas in planetary nebulae, the temperature is of the order of 20,000 K or less. Forbidden lines occur in both cases because of the low gas density; when an atom is excited to a metastable level, the chance of collisional de-excitation from that level is slight, so that the atom may remain in that level long enough to make the forbidden (low-probability) transition to the ground level.

The forbidden nebular lines of [O III] at 500.7 nm and 495.9 nm correspond to the calculated energy differences between the metastable level and two of the three closely spaced ground levels of O III (Figure 19–13A). (The square-bracket notation signifies forbidden transitions.) These lines give planetaries their greenish appearance. A similar pair of forbidden lines arises from [O II], except that in this case the metastable level is double and the ground level is single (Figure 19–13B).

Ions such as O II, O III, Ne III, and N II act as cooling agents in gaseous nebulae. Hydrogen atoms require large amounts of energy to become excited (10.15 eV for the first excited state), but most of the free electrons in the nebula do not have this much

quently. Collisional excitation and de-excitation are therefore significant, so that spectra of planetaries differ in important ways from those of H II regions.

Although the spectral lines of hydrogen and helium are quite pronounced in the spectra of planetary nebulae, the strongest lines are those of O III, O II, and Ne III. These lines do not correspond to ordinary electronic transitions but instead arise from electronic transitions from an excited metastable state to the ground state, resulting in *forbidden lines.* This situation resembles coronal forbidden lines (Chapter 10) except that in the solar corona the forbidden lines arise from ions that have lost nine or more electrons whereas in planetary nebulae only one or two electrons have been removed. The physical reason for this difference is the temperatures characterizing the two situations: in the solar corona, temperatures of

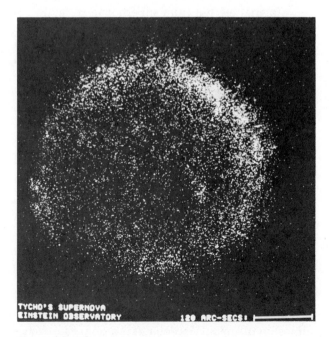

Figure 19–12

X-ray image of Tycho's Supernova remnant. Note the spherical shape. *(P. Gorenstein and F. Seward, Center for Astrophysics)*

Figure 19-13

Forbidden transitions. The metastable state is M, and the ground state is G. (A) For O III, M is a single level and G has three sublevels. Two forbidden lines are seen. (B) For O II, M has two sublevels and G is single. Again, two lines are seen.

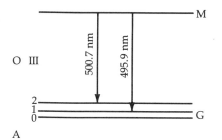

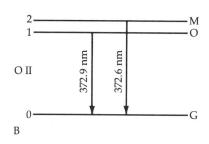

kinetic energy. The cooling-agent ions, however, all have energy levels near 2 or 3 eV; when an electron collides with one of these ions, it gives up part of its kinetic energy to excite the ion to one of these low (metastable) levels. Within a minute or two (in contrast to the 10^{-8} s for ordinary levels), the ion gives up this energy by emitting a forbidden line that escapes the nebula (Figure 19-14). Collisions must be rare; otherwise, de-excitation would prevent the occurrence of forbidden transitions. On the other hand, collisions must be frequent enough that collisional excitation is fairly common. So these ions extract energy from the electrons, and because the kinetic energy of the electrons is a measure of the temperature of the nebula, the result is a lower temperature for the planetary nebula (or H II region).

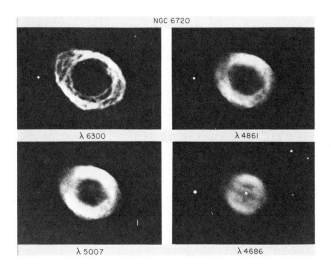

Figure 19-14

Emission from a planetary nebula. Upper left: H_α; upper right: H_β; lower left: forbidden O III; lower right: He II. *(National Optical Astronomy Observatories)*

(e) Interstellar Radio Lines

The Neutral-Hydrogen Line at 21 cm

Where the interstellar gas is cold, hydrogen is neutral and in its ground state. This ground state has two levels separated by a very small energy difference. The reason for this phenomenon lies in the fact that both the proton and the electron have an intrinsic spin. You are aware that a moving charge produces a magnetic field. Because both the proton and the electron are charged particles, their spin motion generates a dipolar magnetic field (like the field of a tiny bar magnet) that we can characterize by the term *magnetic moment.* The magnetic moment of a spinning particle is represented by a *vector,* which is proportional to the vector angular momentum (Chapter 1) of the particle.

Two possible ground-state configurations of the neutral hydrogen atom exist (Figure 19-15). In one configuration, the magnetic-moment vectors of the proton and electron are parallel, or aligned; since vectors add (Mathematical Appendix), a high state of magnetic energy is present. Just as two parallel bar magnets will repel one another, so, too, will the proton and electron be less tightly bound to each other in their mutual orbits. If the magnetic-moment vectors are antiparallel, or opposed, we have the second configuration, which is characterized by less magnetic energy and a more tightly bound orbit. So the aligned state lies at a slightly higher energy than the opposed state; we refer to this effect as the *hyperfine splitting* of the ground state of the hydrogen atom. A spontaneous transition from the higher hyperfine state to the lower one can occur, accompanied by a relative spin flip of the electron (from aligned to opposed) and the simultaneous emission of a very-low-energy photon. This emission produces the 21-cm radio spectral line of neutral hydrogen at a frequency of 1.420406 GHz.

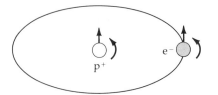
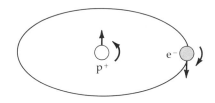

Figure 19–15

Spin alignment for hydrogen. The spins of the proton and electron can be either aligned (left) or antiparallel (right), which is the lower energy state.

When hydrogen atoms collide in the interstellar medium, they generally exchange their electrons; this collisional transfer is the chief mode of changing the hyperfine states of these atoms. If the spin of the newly acquired electron has the same orientation as the old, no change in energy level occurs; otherwise, there is a change in level (either up or down). In other words, collisions may result either in no change, in excitation, or in de-excitation. A change in either direction takes place about once every 400 years for a given interstellar hydrogen atom. On the other hand, an atom in the excited hyperfine state makes a spontaneous downward transition followed by the emission of a 21-cm line quantum only once every few million years (on the average) because this transition is strongly forbidden. Numerous collisional excitations and de-excitations (of the hyperfine levels) occur in the interim. Eventually, an equilibrium is established with an aligned-to-opposed ratio of 3:1, but over distances on the order of a kiloparsec, an enormous number of hydrogen atoms lie along a line of sight in spite of the exceedingly low gas densities in our Galaxy. Among these atoms, enough downward radiative transitions occur to produce a detectable 21-cm spectral line. The profile of the 21-cm line often has Doppler-shifted several peaks; this indicates that the gas is concentrated in discrete regions, such as spiral arms, rather than distributed smoothly throughout the Galaxy.

Molecular Lines

Interstellar molecules range from simple molecules like CO, CN, and OH to such complex organic molecules as formaldehyde (H_2CO) and methanol (CH_3OH), all found by searching for spectral lines at radio wavelengths. These molecules (Table 19–1) let us probe dense clouds of gas and dust, some of which contain protostars. The study of these molecules will eventually lead to a better understanding of the chemistry of the interstellar medium. For many people, however, the most exciting aspect of the molecules concerns their implications with respect to life outside the Solar System. Molecules of H_2O, NH_3, HCN, H_2CO, HC_3N, and HNCO are used in laboratory experiments to synthesize amino acids and nucleotides, the building blocks of life. The fact that these molecules exist in interstellar space indicates that their formation does not require biological conditions.

Although grains make up a very small fraction of the total interstellar medium, they influence the form of the gas. Grains are probably the sites of molecule formation for some of the simpler molecules. Their surfaces act as catalysts by allowing atoms (or simple molecules) to stick to them so that there is time for a second atom to land, interact, and form a molecule that then evaporates back into the gas. Dust grains also shield molecules from dissociating ultraviolet radiation, thus letting the molecule population build up within a cloud.

The first molecule to be detected by radio was the hydroxyl radical, OH, in 1963, after the characteristic spectral frequencies had been firmly established in the laboratory. Four transitions near a wavelength of 18 cm (frequencies of 1612, 1665, 1667, and 1721 MHz) occur because of the splitting of the ground level of the OH molecule. Molecules appear to be connected with dust because OH, H_2CO, and CO lines are fairly widespread and found in large dust clouds. Many of these dense clouds lie in the direction of and are connected to H II regions; the Orion Nebula is a prime example of an H II region containing many dense condensations. The number densities in such clouds are estimated as 10^9 to 10^{12} H_2 molecules/m³; other molecules are, of course, far less abundant though more readily observed. The cloud temperatures are low, usually 10 to 30 K and sometimes as high as 100 K. (Note that we cannot observe H_2 directly by radio because it emits no lines in that wavelength range. Instead, we observe CO and assume that it acts as a good tracer of H_2.)

Most of the radio molecular lines are emission

Table 19–1   Some Interstellar Molecules Observed to Date

Complexity	Inorganic		Organic	
Diatomic	H_2	hydrogen	CH	methylidyne radical
	HD	deuterized hydrogen	CH^+	methylidyne ion
	OH	hydroxyl radical	CN	cyanogen radical
	SiO	silicon monoxide	CO	carbon monoxide
	SiS	silicon monosulfide	CO^+	carbon monoxide ion
	NS	nitrogen monosulfide	CS	carbon monosulfide
	SO	sulfur monoxide	C_2	carbon
	NO	nitric oxide		
Triatomic	H_2O	water	CCH	ethynyl radical
	HDO	heavy water	HCN	hydrogen cyanide
	N_2H^+	imidyl ion	HNC	hydrogen isocyanide
	H_2S	hydrogen sulfide	DCN	deuterium cyanide
	SO_2	sulfur dioxide	DNC	deuterium isocyanide
	HNO	nitroxyl	HCO	formyl radical
4-atomic	NH_3	ammonia	H_2CO	formaldehyde
			HNCO	hydrocyanic acid
			H_2CS	thioformaldehyde
			HC_2H	acetylene
5-atomic			CH_4	methane
			H_2NCN	cyanamide
			HCOOH	formic acid
			HC_3N	cyanoacetylene
			H_2C_2O	ketene
6-atomic			CH_3OH	methyl alcohol
			CH_3CN	methyl cyanide
			$HCONH_2$	formamide
7-atomic			CH_3NH_2	methylamine
			HC_5N	cyanodiacetylene
8-atomic			$HCOOCH_3$	methyl formate
			CH_3C_3N	methyl cyanoacetylene
9-atomic			CH_3CH_2OH	ethyl alcohol
			HC_7N	cyanotriacetylene
11-atomic			HC_9N	cyanotetraacetylene
13-atomic			$HC_{11}N$	cyanopentaacetylene

lines from rotational transitions [Section 8–3(c)]. Emission requires that the molecules be excited above the ground state by some mechanism. For example, OH lines appear in emission from parts of bright H II regions. Some, but by no means all, OH sources show strong H_2O emission as well. The H_2O emission is variable, with intensity changes occurring in periods of months or days. Although also variable, the OH radiation changes far less erratically. Superimposed on some H II regions are several groups of OH emission regions separated by distances of only a few astronomical units.

The emissions from these small OH and H_2O regions far exceed that expected from thermal excita-tion by collisions, which would require temperatures as high as 10^{13} K. The energy levels of the molecules are apparently subject to population inversion, by which we mean that more molecules are in the upper levels than in lower levels; hence, the Boltzmann equation [Section 8–4(a)] is violated and thermal equilibrium does not exist. A *maser action* (molecular laser) is responsible for these inversions. Some mechanism (and several have been proposed but none agreed upon) amplifies the energy so as to pump the molecules into the appropriate excited state. Atoms or molecules in a gas are excited to some particular energy state and then stimulated to fall to a lower energy state at a more rapid rate than normal.

We illustrate the maser process with a hypothetical three-level molecular laser (Figure 19–16), with levels numbered 1, 2, and 3 in increasing energy. A molecule in the ground state, 1, is excited to the highest level, 3, either by colliding with another particle or by absorbing radiation. This process is called *pumping* a maser or laser. Assume that from level 3 the most probable transition when the molecule gives off a photon is to level 2, which is relatively stable; the probability of dropping to level 1 is relatively small. The pumping puts many molecules into level 2. Imagine that a photon with energy equal to the difference between levels 1 and 2 comes close to such an excited molecule. It triggers the molecule's drop from level 2 to 1, sending off another photon, which has the same energy as the original photon and is moving in the same direction. The electromagnetic field of the photon promotes emission of a photon equal in energy to the incoming photon. This process is called *stimulated emission.* The two photons now can stimulate two more molecules to radiate. The resulting four photons can trigger four more molecules to radiate, making a total of eight photons, and so on. The chain reaction amplifies the original photon millions of times as the photons travel through the gas. As a result, a maser's light is intense, narrowly directed, and at a single frequency.

Consider this maser process in OH (Figure 19–17). This radical is pumped, probably by infrared photons, to level 5. The possibilities for transitions to lower levels are such that the natural decay in energy leaves most molecules in level 3. The molecules can then be stimulated to drop to level 1 and emit a 1665-MHz photon; another possible, but less likely, drop is to level 2, with the emission of a 1612-MHz photon.

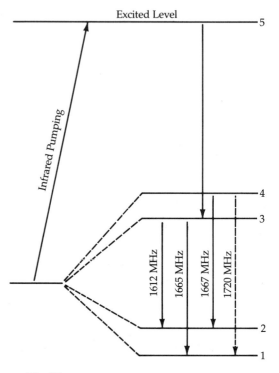

Figure 19–17

Energy-level diagram for maser transitions of OH.

This, in fact, is what is seen in sources such as the Orion Nebula. Other cosmic masers are water at 22,235 MHz (1.35 cm) and silicon monoxide at 43,122 MHz (6.95 mm) and 86,243 MHz (3.47 mm).

Many hydroxyl and water interstellar masers tend to be found near giant molecular clouds. The regions of maser emission are extremely compact—only a few tens of astronomical units across—and very dense. They turn out to be signposts of incipient star birth.

(f) Intercloud Gas

Optical observations of interstellar absorption lines and 21-cm data indicate that a large fraction of the interstellar gas consists of cool clouds along with denser molecular clouds. The H I clouds have diameters up to a few tens of parsecs, temperatures around 100 K, and densities of 10^6 atoms/m^3. Filling the space between these clouds is an ionized gas, some of which is very hot. Radio observations indicate one partially ionized component at 10,000 K with a mean density of roughly 10^4 ions/m^3. Ultraviolet and X-ray

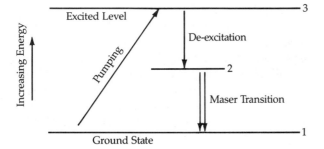

Figure 19–16

Energy-level diagram for a three-level maser.

observations reveal a hotter component at 10^6 K— called the *coronal interstellar gas*. It appears to occupy the largest volume of local interstellar space.

(g) The Evolution of the Interstellar Gas

Driven by star birth and death, the interstellar gas evolves in various forms. Supernovae play an important role in interstellar gas dynamics. Recall that a supernova blasts a tremendous amount of energy (about 10^{44} J) and material (1 to 50 solar masses) into space. The material blown off by a supernova expands as a shell into the interstellar medium. The expanding shell compresses and heats up the interstellar gas; behind the shell the gas is left hot and rarefied. It is so hot, in fact, that not only is all the hydrogen ionized, but also very highly ionized species such as O VI are formed. The supernova shells are large structures, a few of them hundreds of parsecs (up to 3 kpc) in diameter (Figure 19-18). As these sweep through the interstellar gas, they heat it to about 50,000 K and thin it out to a low density. In the process, they distort and destroy existing cool interstellar clouds. These large, expanding shells may power the evolution of a large fraction of the interstellar gas.

Consider the interstellar medium as containing several major components, which may evolve from one to another as the imbedded stars go through their various life stages:

1. *H II regions.* Zones of glowing, ionized hydrogen surrounding young, hot stars (spectral types O and B); contain a minor amount of the interstellar gas, perhaps 10 million solar masses total in the Galaxy; temperature about 10^4 K; density about 5×10^3 ions/m³.
2. *H I regions/diffuse neutral clouds.* Clouds of cool, neutral hydrogen roughly 5 pc in diameter and each containing about 50 solar masses of material; total mass in Galaxy may be 3 billion solar masses; temperature about 100 K; density about 5×10^7 atoms/m³.
3. *Molecular clouds.* Small to huge, containing mostly molecular hydrogen (H_2); total mass of a few billion solar masses; temperature as low as 10 K; density about 10^9 molecules/m³ or greater. Although they occupy less than 1 percent of the interstellar space, they contain a substantial portion of the matter that constitutes the interstellar medium. Stars form out of the dense molecular clouds, some fragments of which develop into H II regions.
4. *Intercloud medium.* A relatively hot gas composed largely of neutral hydrogen (and therefore observable at 21 cm) plus about 20 percent

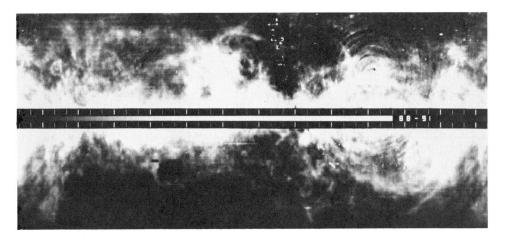

Figure 19-18

Filaments and shells of H I in the interstellar medium. This 21-cm map shows the large-scale structure of the interstellar gas, probably from supernova explosions. The bar through the center marks the galactic plane; the view extends 60° above and below the plane. (*C. Heiles*)

ionized gas, including electrons (observable in the radio continuum). This gas surrounds the cooler interstellar clouds and fills about 20 percent of the volume; temperature from 5000 to 10,000 K; density about 3×10^5 atoms/m³ and 5×10^4 electrons/m³.

5. *Coronal gas.* A very hot (10^6 K), low-density ($< 10^4$ particles/m³), ionized gas that permeates the rest of interstellar space and occupies well over half of it, perhaps as much as 70 percent.

19–3 Star Formation

We now turn to the interstellar medium as a star-forming factory to tie together some pieces of the observational picture. We divide the topic into massive (ten or more solar masses) and solar-mass star births. Massive protostars have greater luminosities than solar-mass ones, and once they reach the main sequence, massive stars ionize the gas around them. The ionized gas is detectable by radio telescopes. Since this action takes place cloaked by dust, only infrared and radio observations allow inspection of stellar wombs.

(a) The Birth of Massive Stars

Before we present specific cases, let's outline the hallmarks, in the radio and infrared, of the birth of a massive star. First, stars condense from molecular clouds (visible by their emission at millimeter wavelengths). Second, the free-fall collapse, at its early phases, heats the dust to low temperatures, roughly 30 to 50 K. This dust emits infrared radiation that peaks at roughly 10 μm. Third, as the protostar forms, the temperature of the interior dust reaches about 1000 K and therefore it emits with a peak at 3 μm. The exterior dust is cooler, still about 100 K. So the spectrum shows the combination of two blackbody peaks, one near 3 μm and the other near 30 μm. Fourth, as the protostar reaches the main sequence, it ionizes the hydrogen gas and a compact H II region develops, most easily observable at microwave wavelengths. Fifth, the hot, ionized gas expands. The dust, pushed outward, is cooled; it emits with a peak in the far infrared, but less intensely. When the dust has been pretty much dissipated, the H II region emits a weak continuous spectrum of radio waves. Finally,

the H II region expands enough to blow off its dusty cloak, and the star appears to optical view.

With this scenario in mind, consider the Orion Nebula (Figure 19–19). The H II region around the Trapezium marks the oldest (most evolved) part of the region. The Trapezium cluster consists of a few hundred stars within 1 lightyear of each other; it is the O and B stars of this group that ionize the gas. These massive stars are no more than 1 million years old. Various observations show the gas flow here to be chaotic and swift, with velocities as high as 100 km/s. In an evolutionary sense, the molecular cloud core that lies behind the Orion Nebula is the youngest part of the region.

Where is star birth happening? We see the embryonic material (the molecular cloud) and the end results (the Trapezium cluster). The most likely place for protostars lies between the two, in the infrared cluster associated with the Becklin–Neugebauer object and the Kleinmann–Low nebula. High-resolution infrared observations indicate that at least five sources lie in the cluster, with separations of a few thousand astronomical units. These may be protostars along with the Becklin–Neugebauer object itself. Its observed characteristics match those expected from a massive protostar in its PMS evolution.

Infrared astronomers have observed the Becklin-Neugebauer object carefully at an infrared line at 4.05 μm, called the *Brackett alpha line*, which arises from a transition from level 5 to level 4 in hydrogen. From the viewpoint of protostellar evolution, this line can arise from recombination within a very small, newly formed H II region around a massive star approaching the main sequence. Observations to date indicate that the emission does come from a compact H II region. So the Becklin–Neugebauer object, embedded in the molecular cloud, is, in this interpretation, a B0 star just reaching the main sequence, surrounded by dust and gas that it has just begun to ionize. It is less than 1 million years old.

The situation in Orion and other regions where H II regions abut giant molecular clouds supports a scenario of sequential star formation within them. Massive star formation begins at one end of the giant molecular cloud (Figure 19–20). (Such clouds tend to be elongated and cigar-shaped.) A small group of about ten O and B stars forms. They evolve to the main sequence. Their ultraviolet radiation then dissociates hydrogen molecules around them and ionizes the gas. The H II region, since it is hot, expands,

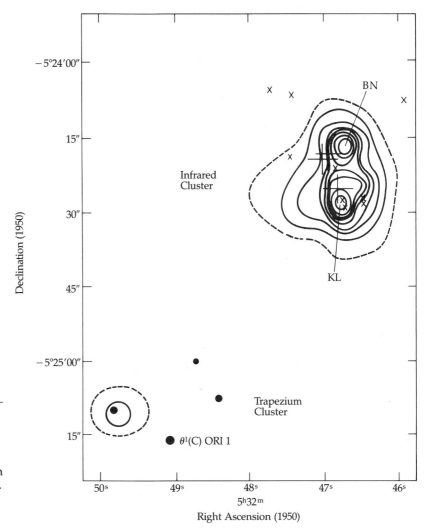

Figure 19–19

Infrared emission from the Orion Nebula region. The infrared cluster has two strong sources: the Kleinmann–Low (KL) source and the Becklin–Neugebauer (BN) object. The Trapezium marks the visible part of the H II region. *(Adapted from a diagram by E. Becklin, G. Neugebauer, and C. Wynn-Williams)*

pushing a shock wave into the molecular cloud. The gas behind the shock wave is compressed to densities sufficient to start gravitational collapse. A new group of O and B stars is born about 1 million years after the previous one. The process repeats. Small groups of massive stars are born in a sequence of bursts across the molecular cloud.

This model predicts that the fossil remnants of a molecular cloud will be a string of small groups of O and B stars, 10 to 30 pc apart and more or less in the same space as the parent molecular cloud. Such loose groupings of O and B stars, called *OB associations,* must be very young. Since these stars do not live very long (a few tens of millions of years), the OB associations themselves cannot be more than 10^7 years old.

Many OB associations, which are 30 to 200 pc across, consist of small clusters of stars called *OB subgroups.* These subgroups contain four to 20 stars (average about ten). Within an association, OB subgroups are lined up in an evolutionary sequence. The most spread out, oldest subgroup lies at one end and the most compact, youngest subgroup at the other. For example, Orion contains a large association that has four OB subgroups; the smallest and youngest subgroup (estimated to be 2 million years old) is the Trapezium cluster. Recall that the Trapezium adjoins the southern Orion molecular cloud. Here we see the signposts of massive star formation.

Once star formation starts at one end of a molecular cloud, it propagates through the cloud in a chain

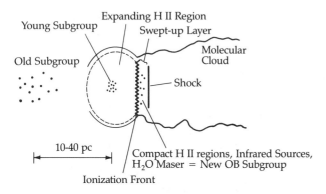

Figure 19-20

Sequential model of star formation from a giant molecular cloud. *(Adapted from a diagram by B. Elmegreen and C. Lada)*

reaction. But what starts the first burst of star formation? The answer to that question is not yet known. Perhaps it starts from the collision of molecular clouds or, more likely, from the blast wave of a supernova remnant crushing into an end of a molecular cloud. This idea is aesthetically appealing: a supernova signals the death of a massive star; then the mark of its death ignites the birth of other massive stars.

Note also in this model that giant molecular clouds do not last long once star birth begins—only a few tens of millions of years. Since we see many molecular clouds now, they must form rapidly in order to balance their rapid destruction. Where and how these clouds form most likely has to do with the spiral structure of our Galaxy (Chapter 20).

(b) The Birth of Solar-Mass Stars

The observational picture for the formation of stars like the Sun is skimpy. It does seem clear that, like massive stars, solar-mass stars are also born from molecular clouds. The questions are in which clouds and how? To date, however, we have no well-confirmed observation of a solar-mass protostar—and not for lack of trying!

Solar-mass stars may form within *dark clouds*, which are interstellar clouds containing enough dust to blot out the light of stars within and behind them. We now know that dark clouds are one type of molecular cloud. They typically have temperatures of 10 K, densities about 10^9 atoms/m³, and masses from a few

tens to a few hundreds of solar masses. Infrared observations reveal a few tens to about 100 candidates for solar-mass protostars within well-examined dark clouds. These form from fragments throughout the cloud rather than at the edges as massive stars do.

The births of massive stars or perhaps the death of one in a supernova sweeps away the gas and dust to reveal the stars. In this picture, most star birth takes place in dark, massive clouds, out of which OB associations form. Thus the Sun may have been born in an OB association such as the one we described in Orion, perhaps blasted clean by a supernova.

An alternate idea is that solar-mass stars form out of small, isolated molecular clouds, not giant ones. These clouds are no more than a few parsecs across, contain at most 1000 solar masses of gas, and are not near any giant molecular clouds. An example of such

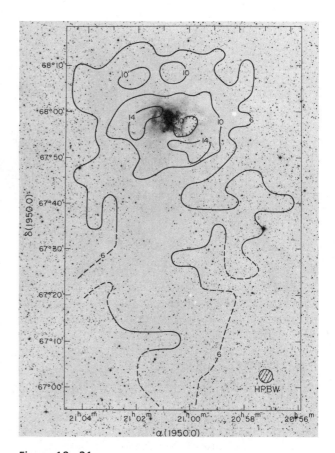

Figure 19-21

Carbon monoxide map of NGC 7023. *(D. Elmegreen and B. Elmegreen)*

a small cloud is NGC 7023 (Figure 19–21), studied at millimeter wavelengths. This cloud lies some 430 pc from the Sun and about 1000 pc above the galactic plane. No large cloud is within 80 pc of it. It has overall dimensions of only 4 by 9 pc and contains 300 solar masses of material in its densest region and at most 1000 solar masses overall.

Inspecting the carbon monoxide map of Figure 19–21, you can see that the cloud is very clumpy. For example, the region to the east of the central star is much denser than that to the west, where you find a hole that allows you to look through the cloud. The region to the east contains a small, weak radio continuum source—an indication of a young, perhaps B-type star embedded in the clump next to the visible star. Various observations show that the molecular cloud contains a number of relatively young stars spread throughout it. So it appears that star forma-

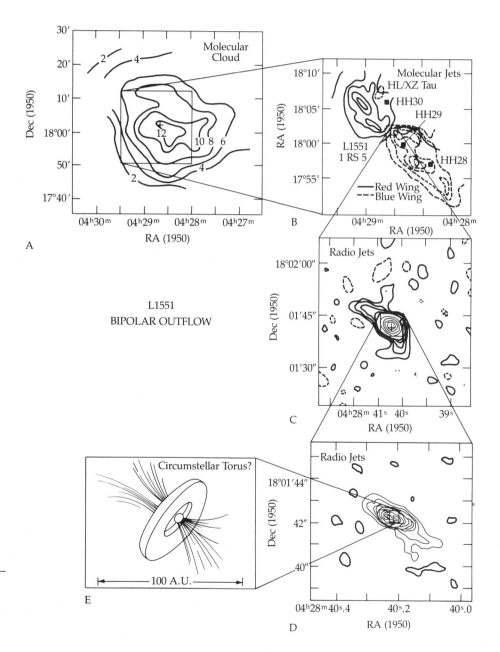

Figure 19–22

Bipolar outflow from L 1551.
(Adapted from a diagram by R. L. Snell, J. Bally, S. E. Strom, and K. M. Strom)

tion has taken place and continues in NGC 7023 in many places, not just at the edges, as appears to be the case with the formation of massive stars from giant molecular clouds.

From observations so far, we are beginning to suspect that, in general, stars are born from molecular clouds; massive stars from massive clouds and less massive stars (the majority of those in the Galaxy) from less massive clouds.

(c) Molecular Outflows and Star Birth

Observations of molecules around protostars have revealed high-speed (30 to 100 km/s) flows of gas. Doppler-shift measurements show that these flows tend to be bipolar: two streams moving in opposite directions. The flows carry considerable mass and can span a few parsecs; so enormous amounts of energy push them along. The exact source of that energy and the origin of the flows are a mystery at the moment. Such outflows do appear associated with the birth of massive stars.

A simple model to explain the outflows envisions a young massive star putting out a strong stellar wind. Surrounding the star is a dense disk or torus of gas and dust. This disk would naturally channel the flow of the stellar wind so that it streams out along the thin axis of the disk, making two streams. When these two streams push enough material outward, two opposing lobes of gas should form.

One of the cleanest examples of this collimated bipolar flow is associated with a dark cloud called L 1551, which is embedded in a large molecular cloud (Figure 19–22A). An infrared source named IRS 5 lies at the center of a bipolar CO outflow (Figure 19–22B): it has a luminosity of a few tens that of the Sun. VLA observations reveal a radio source coincident with IRS 5 (Figure 19–22C). Curiously, high-resolution radio maps show two jets (Figure 19–22D) aligned with the axis of the molecular bipolar flow. These observations, combined with CCD images, strongly suggest that a disk or torus surrounds the protostar to collimate the outflow (Figure 19–22E), which may arise from a very strong stellar wind from the protostar.

The discovery of these two-sided flows strongly hints that disks of material typically form around massive stars during their formation. It is from such disks that planetary systems might form. So we have a clue that the nebular model for planetary formation might actually operate elsewhere in the Galaxy.

Problems

1. An A0 V star has an apparent visual magnitude of 12.5 and an apparent blue magnitude of 13.3.
   (a) What is the color excess for this star?
   (b) What is the visual absorption in front of this star?
   (c) Calculate the distance of the star (in parsecs).
   (d) What error would have been introduced if you had neglected interstellar absorption?

2. (a) How does interstellar reddening alter the Planck spectral energy curve of a star? Sketch approximate curves for an A star (10,000 K) in the visible part of the spectrum, both with and without reddening.
   (b) What effect would interstellar reddening have on the color–magnitude (H–R) diagram of a star cluster? Draw a diagram to support your answer.

3. What are the observational clues to the nature of the interstellar dust grains, and how have these clues been interpreted in terms of models?

4. Several galactic nebulae have been photographed in color. Two different types occur, red and blue.
   (a) Explain the physics of these two types of nebulae.
   (b) Briefly describe the spectra you would expect to observe for these two types of nebulae.

5. Why are none of the hydrogen Balmer lines seen as interstellar absorption lines, even though hydrogen is the most abundant element in the Universe? (*Hint:* Recall the energy level diagram of hydrogen.)

6. (a) Name the two factors that primarily determine the size of an H II region.
   (b) Explain the physical basis for your answer to part (a) in terms of the Strömgren sphere.
   (c) The brightness of an H II region depends only on the gas density of the nebula. Explain this phenomenon in terms of what you know about the hydrogen atom. (*Hint:* Remember the Saha equation.)

7. Forbidden lines are observed both in the solar corona and in gaseous nebulae, but they are not the same forbidden lines.
   (a) How do the lines differ in the two cases?

(b) Why are nebular forbidden lines not emitted by the solar corona? What about coronal forbidden lines for nebulae?

8. Somewhere in our Galaxy resides a cloud of neutral hydrogen gas with a radius of 10 pc. The gas density is 10^7 atoms/m^3.

   (a) How many 21-cm photons does the cloud emit every second?

   (b) If the cloud is 100 pc from the Sun, what is the energy flux of this radiation (in W/m^2) at the Sun?

9. What are the energies (in electron volts) of the photons that characterize

   (a) the Lyman continuum limit (91.2 nm)

   (b) the nebular line of [O III] at 500.7 nm

   (c) the neutral hydrogen line (21 cm)

   (d) the ammonia (NH$_3$) emission (1 cm)

   (e) the hydrogen Balmer line, H$_\alpha$ (656.3 nm)

10. Using Bohr's formula for the wavelengths of hydrogen lines, calculate the wavelength and frequency of the radio recombination line with upper level 93 and lower level 92.

11. A pure hydrogen gas cloud of number density $n = 10^7$ atoms/m^3 surrounds an O star that generates 10^{49} photons/s at wavelengths shorter than 91.2 nm. The rate at which recombinations occur is $\alpha = (2 \times 10^{-19})n^2$/m$^3 \cdot$ s.

    (a) Balance the number of ionizations with the number of recombinations to determine the Strömgren radius of the resultant H II region.

    (b) The Sun produces about 5×10^{23} photons per second with $\lambda < 91.2$ nm. Calculate its Strömgren radius in astronomical units for an interplanetary medium density of 10^9 atoms/m^3.

12. Approximate the Galaxy as a uniform disk of constant thickness. Show that the optical depth of extinction by interstellar dust should approximately obey the law $\tau_\lambda \propto \csc b$ (except for small b), where b is galactic latitude. At what galactic latitudes would an astronomer want to observe other galaxies without having to worry much about extinction?

13. An open cluster of stars is found to contain main-sequence O5 stars with observed color indices $(B - V)$ of 0.4. These O stars are observed to have apparent magnitudes of 10.0.

    (a) Use Table A4–3 to calculate the distance to the cluster. Include the effects of extinction by dust.

    (b) Determine the apparent magnitude of a G0 main-sequence star in the cluster.

14. Consider an extended region filled with neutral hydrogen gas at a density of nm_H and temperature T, where m_H is the mass of the hydrogen atom. Now consider a small spherical volume of radius R inside this region where the density is slightly higher than in the surrounding region.

    (a) Show that the radius of this volume must satisfy the condition $R \gtrsim (3kT/2\pi Gnm_H^2)^{1/2}$ in order for gravitational collapse of the denser region to occur. (*Hint:* Obtain an expression for gravitational pressure by integrating Equation 16–1 over r.)

    (b) Obtain a lower limit for the mass of the collapsing volume.

    (c) For the general interstellar medium, we can take the approximate values $n = 10^6$/m^3 and $T = 100$ K. Obtain a numerical value for the mass limit in solar masses.

    (d) Is a collapsing region of the general interstellar medium likely to produce a single star like the Sun? If not, how would a region which *could* form a single, solar-type star differ from the general interstellar medium?

15. A star count is made on a photograph that contains a dark nebula. Ten times fewer stars are counted in front of the cloud than are found on a region away from the cloud with the same solid angle. For ease of calculation, assume that the dark cloud is completely opaque, that the stars are uniformly distributed in space, that all stars have absolute magnitudes of 5.0, and that the limiting magnitude of the photograph is 15.0. Calculate an approximate distance to the cloud.

The
Evolution
of
Our
Galaxy

We have so far described the physical characteristics of two main parts of the mass of the Galaxy: the stars and the interstellar medium. This chapter ties them together in the grand design of the structure of the Milky Way Galaxy. You will see that our Galaxy has a spiral layout, with much irregularity imposed on the overall pattern. This spiral imprint must evolve, and its evolution connects to the evolution of the stars, gas, and dust contained within.

20–1 The Structure of Our Galaxy from Radio Studies

To map out the Galaxy's structure requires a technique that distinguishes the paths of spiral arms from the bulk of the interstellar medium. The traditional mapping technique uses the 21-cm line of H I. As you will see, this technique has serious limitations. Recently, the mapping of molecular clouds in CO has complemented 21-cm observations. In fact, observational evidence so far indicates that molecular clouds outline spiral arms more tightly than do H I clouds.

(a) 21-cm Data and the Spiral Structure

The hyperfine transition from neutral hydrogen (H I) at the radio wavelength of 21 cm, combined with radial velocity variations from differential galactic rotation, allows us to deduce the *spiral-arm*

structure in the galactic plane. If the galactic rotation curve is known (and that is a crucial *if!*), the distances to concentrations of neutral hydrogen may be found from the observed Doppler-shifted 21-cm line profiles. Implicit in these distance determinations are two assumptions: (1) differential galactic rotation and (2) circular galactic orbits for the gas near the galactic plane.

Because interstellar absorption is insignificant at the 21-cm wavelength, the line emission is observable throughout our Galaxy. Hence, we can probe galactic regions far beyond the solar neighborhood. The 21-cm line profile for a given line of sight exhibits several Doppler-shifted peaks that are fairly narrow and well defined. The concentration of hydrogen into spiral arms produces the observed forms of the line profiles (Figure 20–1). Line profiles seen near the galactic longitude *l* have Doppler peaks that shift in a concerted fashion as *l* varies. Each peak characterizes a spiral arm intersected by the line of sight. If we interpret the Doppler shift in terms of the radial velocity of that section of the arm and apply the rotation formulas of Chapter 15, we find the distance to the arm; since we have assumed circular orbits, the distance is uncertain to the extent that asymmetric motions occur. More realistic results are obtained by including modifications to circular motions.

Let's illustrate this procedure by interpreting one 21-cm profile (Figure 20–2). This profile corresponds to the line of sight at $l = 48°$, and it consists of three

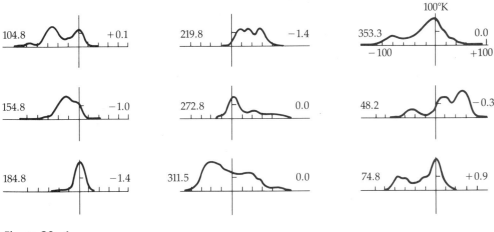

Figure 20–1

Line profiles at 21 cm in the galactic plane. These emissions come from regions spanning the galactic equator. *(F. J. Kerr and G. Westerhout)*

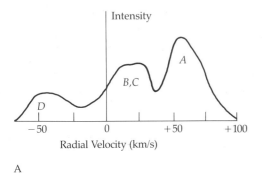

A

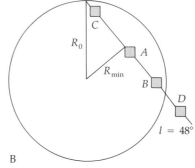

B

Figure 20–2

Line profiles and Doppler shifts. (A) Line profiles from a number of H I clouds at longitude 48°. (B) The line-of-sight geometry for the profiles in A.

Doppler peaks at radial velocities of 55, 15, and −50 km/s. If we denote the Sun's distance from the galactic center by R_0 and note that positive radial velocities correspond to recession, then two of the hydrogen clouds are receding and one is approaching. From the rotation formulas (Chapter 15), we find that recession corresponds to $R < R_0$ and that an approaching cloud must lie at $R > R_0$. Remembering the approximate nature of Equation 15–16, we compute the maximum radial velocity as 57 km/s; therefore cloud A, with a radial velocity of 55 km/s, must lie very close to the tangent point. Cloud D, which is approaching us (−50 km/s), lies at $R > R_0$, as indicated in the figure. There is a hint of double structure (clouds B and C) in the 15-km/s peak; latitude scans imply that cloud B lies beyond the tangent point (near $R = R_0$) and cloud C is very close to the Sun.

By combining 21-cm data from both the northern and southern hemispheres, we can construct a schematic picture of the neutral hydrogen distribution in the spiral arms of our Galaxy (Figure 20–3). The spiral structure is poorly determined near $l = 0°$ and $l = 180°$—that is, toward the galactic center and in the diametrically opposed *(anticenter)* direction. Circularly orbiting hydrogen clouds in those directions should exhibit no radial velocity; hence we cannot determine the distances to such clouds, for the line profile is a single peak at 21 cm. A distance ambiguity exists for hydrogen clouds that are closer to the galactic center than is the Sun. Because the maximum radial velocity of recession occurs when the line of sight passes closest to the galactic center (tangent point), a cloud closer to the Sun than the tangent point may have the same (lower) recession speed as a cloud beyond the tangent point. The ambiguity is difficult to resolve with certainty.

(b) The Galactic Latitude Distribution of Hydrogen

Neutral hydrogen is concentrated in the galactic plane. If we define the thickness of the gas layer as the distance from the galactic plane to the half-density point (where the number density falls to half the value found at the galactic plane, or galactic latitude $b = 0°$), then the thickness of the hydrogen layer is observed to range from 80 to greater than 250 pc. The smaller value refers to the region between the Sun and the galactic center. The thickness increases to 250 pc at the spiral arms near the Sun ($R = R_0$), flares out to several hundred parsecs for $R > R_0$, and reaches almost 2 kpc at $R = 30$ kpc. The galactic latitude distribution of neutral hydrogen (Figure 20–4) has the gas layer very flat (near $b = 0°$) in the region $R < R_0$, while beyond $R = R_0$, the layer is bent in opposite directions (relative to $b = 0°$) at $l = 180°$. This is the *galactic warp*, with a peak-to-peak displacement of some 3 kpc.

Another way to view the H I distribution is to examine its volume density. Outside the solar circle ($R = R_0$), the density peaks between 12 and 14 kpc and then drops rapidly beyond $R = 20$ kpc (Figure 20–5A). Within the solar circle, the distribution is constant from $R = 4$ kpc. Most of the H I gas lies outside the solar circle—at least 80 percent of the total gas mass. The CO distribution does not in general follow that for H I. Observations indicate that the CO is densest 6 kpc from the Galaxy's center (Figure 20–5B). Outward from 8 kpc, the density of CO drops but the H I density stays roughly the same. Inside 4 kpc, the CO density also decreases but not as rapidly. The CO layer has a thickness of 125 pc. Remember, the CO indicates the presence of H_2, molec-

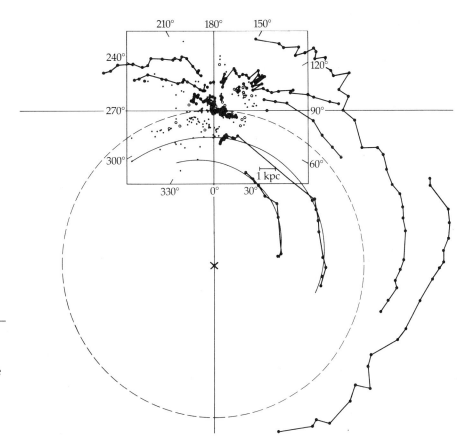

Figure 20–3

Galactic structure from 21-cm observations. The lines connecting the dots are from radio data. The unconnected dots within the box are optical observations of spiral-arm tracers, such as H II regions. (*Adapted from a diagram by H. Weaver*)

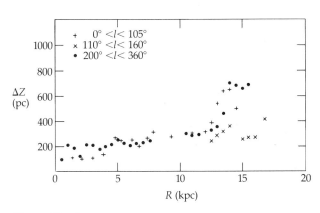

Figure 20–4

Warping of gas in the disk. From 21-cm observations, we see that the gas spreads out near the edge of the disk to a height of about 600 pc. (*Adapted from a diagram by H. Van Worden*)

ular hydrogen. Within the Sun's distance from the Galaxy's center, about 93 percent of the hydrogen exists as H_2. In contrast, outside of the solar circle, the hydrogen takes the form of H I.

(c) The Galactic Center Region

If the gas motions were perfectly circular, there would be no observed radial velocities (Doppler-shifted line profiles) near $l = 0°$. The distances to such gas clouds would be indeterminate because the expected 21-cm line profile is a single intense peak at zero radial velocity. Nevertheless, the gas structure in the region may be approximated by extrapolating the features seen at longitudes flanking $l = 0°$.

The 21-cm line profiles observed within a few degrees of the galactic center are extremely complex, largely because of peculiar geometric and velocity perturbations. The major feature seen is a sharp peak at -50 km/s, which reveals material moving out-

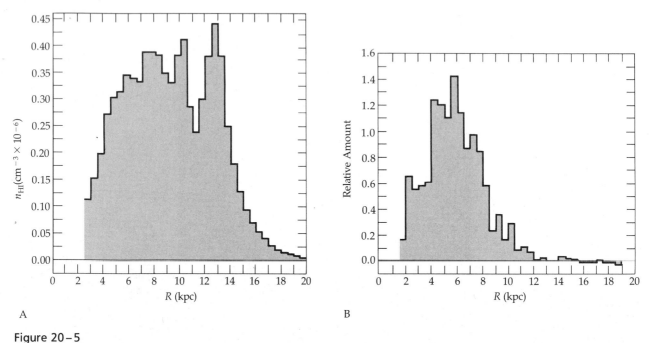

Figure 20–5

Gas distribution in the disk. (A) Volume density of H I as a function of distance.
(B) Relative density distribution of CO. *(From a diagram by M. Gordon and B. Burton)*

ward from the galactic center; this makes up the *expanding,* or *3-kpc, arm.* Within 2000 pc of the center lies a thin gaseous disk tilted 24° with respect to the galactic plane. The disk has a thickness of 100 pc and a maximum rotational velocity of 360 km/s. It contains 10^7 solar masses of H I. It may have a central bar. In one respect, the molecular distribution here follows that of H I; it shows the same tilted disk structure in the innermost part of the Galaxy. The disk contains 10^9 to 10^{10} solar masses of H_2. The total mass, determined from the rotation of matter outside the disk, is only a few times 10^{10} solar masses. So this inner part of the Galaxy may be exceptionally rich in gas, rather than mostly stars.

(d) High-Velocity Hydrogen Clouds

Chapter 15 showed that stars that move in higher eccentric orbits about the galactic center appear to us as high-velocity stars. The 21-cm profiles at high galactic latitudes ($|b| \geq 10°$) reveal hydrogen structures far from the galactic plane. This gas is distinctly distributed in the form of discrete clouds that fall into

three velocity categories: (a) the *high-velocity clouds* with velocities (relative to the Sun) of 70 to 120 km/s, (b) the *intermediate-velocity clouds* with velocities in the range 30 to 70 km/s; and (c) the *very-high-velocity* clouds with velocities greater than 120 km/s in the negative galactic latitudes between $l = 0°$ and $l = 180°$. Nearly all of these clouds exhibit radial velocities of approach (negative), and they appear to be concentrated near $l = 120°$ and $b = 40°$ (they occur in the regions $l = 60$ to 200° and $b = 10$ to 80°).

Interpretations of this fast-moving gas have been the subjects of intense debate, which has yet to be resolved. The most prominent feature of the high-velocity H I is called the *Magellanic Stream.* It envelops the Magellanic Clouds (two companion galaxies to the Milky Way; Chapter 22) and runs as a long filament close to the southern galactic pole and then into the galactic plane near $l = 90°$. About 30° from the pole, the streams break into cloud fragments—the tip of which may appear as the high-velocity clouds mentioned above. Most models view the stream as a thread of gas pulled from the Magellanic Clouds by a tidal interaction with our Galaxy.

20–2 The Distribution of Stars and Gas in Our Galaxy

(a) Spiral Arms: Extreme Population I Objects

Neutral hydrogen and carbon monoxide emissions enable astronomers to trace out roughly the spiral-arm structure over a large portion of our Galaxy, but spiral arms also include all other objects of *extreme Population I*. The objects belonging to the various populations, and their characteristics, are summarized in Table 20–1. Because gas and dust are found together, both occur in the spiral arms of our Galaxy. The newborn T Tauri stars are usually surrounded by the gas and dust from which they formed. These stars are not very luminous, and even when they occur in groups, they cannot be seen to very great distances. We know that they are good examples of extreme Population I, but they are not good spiral-arm indicators because they have low luminosities.

Massive stars evolve very rapidly and may be seen as main-sequence O or B stars; they, too, may still be enveloped by the gas and dust from which they formed. The ultraviolet radiation from such stars ionizes the surrounding gas, rendering it visible as an H II region (or emission nebula; Chapter 19). *H II regions* are usually luminous, for they characteristically include several bright, hot stars whose ultraviolet radiation is converted to visible light. H II regions and the OB associations exciting them are extreme Population I objects and are indicators of spiral structure (Figure 20–6). The radio recombination lines (Chapter 19) from H II regions allow us to determine their radial velocities and also their distances. Such velocities can then be used in the same manner as the 21-cm line velocities to delineate the spiral structure, making possible a more complete comparison between the distributions of H I and H II regions. The spiral-arm patterns delineated by H I and H II are roughly similar, but a marked difference appears in the large-scale distribution, for the ionized hydrogen attains its greatest concentration closer to the galactic

Table 20–1 Characteristics of Stellar Populations

	Population Group				
	Extreme Population I	Older Population I	Disk Population II	Intermediate Population II	Halo Population II
Typical objects	Interstellar dust and gas O and B stars Supergiants T Tauri stars Young open clusters Classical Cepheids O associations H II regions	Sun Strong-line stars A stars Me dwarfs Giants Older open clusters	Weak-line stars Planetary nebulae Galactic bulge Novae RR Lyrae stars $(P < 0.4$ day$)$	High-velocity stars $(Z > 30$ km/s$)$ Long-period variables $(P < 250$ days$)$	Globular clusters Extremely metal-poor stars (subdwarfs) RR Lyrae stars $(P > 0.4$ day$)$ Population II Cepheids
Characteristics					
$\langle |z| \rangle$, pc	120	160	400	700	2000
$\langle |Z| \rangle$, km/s	8	10	17	25	75
Distribution	Extremely patchy in spiral arms	Patchy	Smooth	Smooth	Smooth
Age (10^9 years)	<0.1	0.1 to 10	3 to 10	≈ 10	≥ 10
Brightest stars (M_{vis})	-8	-5	-3	-3	-3
Concentration to galactic center	None	Little	Considerable	Strong	Strong
Galactic orbits	Circular	Almost circular	Slightly eccentric	Eccentric	Highly eccentric

Adapted from A. Blaauw.

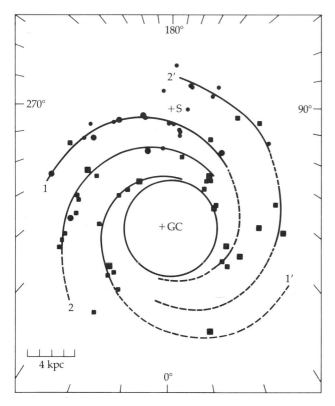

Figure 20 – 6

Spiral structure based on Population I objects. Indicated are the positions of H II regions from optical (circles) and radio (squares) observations. *(Adapted from a diagram by Y. M. Georgelin, Y. P. Georgelin, and J.-P. Sivan)*

Disk and intermediate Population II stars constitute most of the total mass of our Galaxy, and they form a fairly thick layer, or disk. The *Disk Population* is intermediate between Populations I and II. From the $\langle|z|\rangle$ entry in Table 20 – 1, we see that representative disk stars may lie quite far from the galactic plane. This quantity $\langle|z|\rangle$ is the mean stellar distance from the plane, and the related parameter $\langle|Z|\rangle$ is the mean of the stellar velocity components perpendicular to the galactic plane. The greater the perpendicular velocity component of a star, the greater the likelihood of that star's being far from the galactic plane. Contrary to Population I, Disk Population objects are distributed fairly smoothly and show no spiral structure.

Intermediate Population II stars may lie even farther from the galactic plane; their mean distance is 700 pc. They may also have highly eccentric orbits because many of them are high-velocity stars. Disk Population objects show a slight concentration at the galactic center, but Population II stars are strongly concentrated at the center. We are dealing with highly evolved stars, which are very old, when we speak of Population II. These stars probably formed at a place and time of scarce heavy elements.

The *metal abundance* of stars is an indicator of the degree of heavy-element enrichment of the gas out of which they formed. This metal abundance is not a smooth function of population, although this was thought to be the case earlier. Extreme Halo Population II stars (see below) do have a low metal abundance, as do some intermediate Population II stars, but stars in the nucleus and some disk stars are metal-rich, perhaps because of an accelerated birth rate of massive stars early in the life of the Galaxy. *Metal rich* is a relative term. The metal abundance of ordinary Population II stars is about 1 percent of the Sun's, and for metal-rich Population II stars, this value may go up to 10 percent or higher.

Enveloping the disk and spiral arms is the *galactic halo*, which extends far above the plane but is still concentrated at the galactic center. Objects that lie in this domain are called *Halo Population II*; these include the globular clusters and RR Lyrae stars with periods greater than 0.4 day. The orbits of these objects are highly eccentric, with large velocity components perpendicular to the galactic plane. For instance, the globular clusters form a sphere around the Galaxy's center. Their elliptical orbits bring them out to extreme distances of 10 to 12 kpc from the Gal-

center than does the neutral hydrogen. Because early-type stars are extreme Population I, the *young open clusters* that contain these stars must also be of this population. In these clusters, few stars have evolved away from the upper main sequence.

(b) Stellar Populations: Galactic Disk and Halo

Our Sun belongs to an Old Population I. Although they are not strictly confined to the spiral arms, Old Population I objects still lie fairly close to the galactic plane and have an inhomogeneous distribution throughout the Galaxy. Old Population I (Table 20 – 1) includes clusters whose upper-main-sequence stars have evolved to the giant and variable evolutionary stage. The older open clusters are prototypes of this population.

axy's nucleus. The clusters orbit at speeds of about 100 to 150 km/s, diving into and shooting out of the disk. These passages have helped to wipe globular clusters clear of any gas and dust they once had.

The halo also contains gas, but much less than the disk. The neutral hydrogen shows up as the high-velocity clouds. Otherwise, the halo gas is ionized and extends to greater distances above and below the galactic plane than does the H I. We have observed the ionized gas in two forms: hot (80,000 K) and cool (10,000 to 20,000 K). The hot component has its greatest density at z from 1 to 3 kpc and can be seen as far out as 10 kpc. In contrast, the cool component has been seen only up to 2 kpc. The cool component contains about ten times more mass than the hot, but together they amount to only 1 or 2 percent of the gaseous mass in the disk.

The halo may also contain other objects, currently unobservable. Low-mass main-sequence stars would be very hard to detect. Smaller objects similar to planets and asteroids may also exist. Some astronomers have even suggested numerous low-mass black holes. Recent observations of the rotation curves of a few other spiral galaxies imply that they may have extensive, massive halos of faint red stars.

(c) The Central Bulge and Galactic Nucleus

Wide-angle pictures of our Galaxy show that it has a *central bulge* similar to that of other spiral galaxies. This bulge, which is about 2 kpc in radius, is amorphous. It contains a mixture of heavy-element-enriched stars, especially late-type M giants, ordinary Population I K giants, and a few metal-poor objects such as the RR Lyrae stars. At the very center of the bulge is the *galactic nucleus*, which is analogous to the stellar-like nucleus in M 31, the Andromeda galaxy. Spectra of the Andromeda galaxy's nucleus show that it, too, consists of metal-rich giants as well as a large number of low-mass dwarfs. Infrared observations confirm that such stars reside in the Galaxy's nucleus; these and radio data also suggest the presence of very young M and O supergiants. Gas exists as molecular clouds and H II regions. Motions of gas here suggest a high concentration of mass at the center, perhaps a black hole. In the very center of the Galaxy lies a radio source less the 140 AU in diameter.

The continuous radio emission from the nucleus shows that an intense radio source lies right in the direction to the center. It is called *Sagittarius A (Sgr A)*. Clustered around Sgr A and lying more or less along the galactic equator is a string of radio sources (Figure 20–7). When investigated at different radio wavelengths, these sources appear to have characteristics of H II regions. The total extent of this region is about 90 by 260 pc; and the ultraviolet energy output from the OB stars needed to keep the region ionized is at least 2×10^{33} W (5×10^6 solar luminosities).

Sgr A emits both thermal (Sgr A East) and nonthermal radiation. Within Sgr A West is a nonthermal point-like radio source less than 0.1″ in diameter that may mark the core of the Galaxy. The ionized gas here, which amounts to a few million solar masses of material, seems to be rotating at about a few hundred kilometers per second. The thermal emission of Sgr A West, from the inner 3 pc of the Galaxy, shows that the bulk of the radio emission lies along a ridge-like source at the galactic center (Figure 20–8).

Radio hydrogen recombination lines from the center are very broad. One physical process that commonly broadens spectral lines is the Doppler shift resulting from the rotation of a mass of gas. If Doppler-broadened, the line widths imply rotational speeds of 150 km/s at a distance of 2 pc from the Galaxy's center. CO observations indicate that this molecular cloud may contain as much as a million solar masses of material.

The galactic center region emits strongly at 2.2 μm (Figure 20–9). The most intense part of this emission coincides with Sgr A. The source of this radiation is simply the combined 2.2-μm emissions from all the old Population I stars (probably mostly from K giants) that inhabit the galactic nucleus. The region just around Sgr A is packed with 2.2-μm sources (Figure 20–9). One of these coincides with Sgr A West. This infrared cluster coincides with the ridge of radio emission seen in Figure 20–8. Observations of the same region at 10 μm show the infrared emission from dust that is heated by the radiation from old Population I stars and from high-luminosity O stars; the condensations in the 10-μm map are probably the locations of newly formed O stars. These regions have diameters of less than a few parsecs, the same size as small H II regions. The combined luminosity from them in the range from 2 to 20 μm is roughly 1 million times that of the Sun. Far-infrared observations have found that Sgr A emits more intensely at 40 to 300 μm than at 10 μm, about 100 million times the Sun's luminosity. This emission,

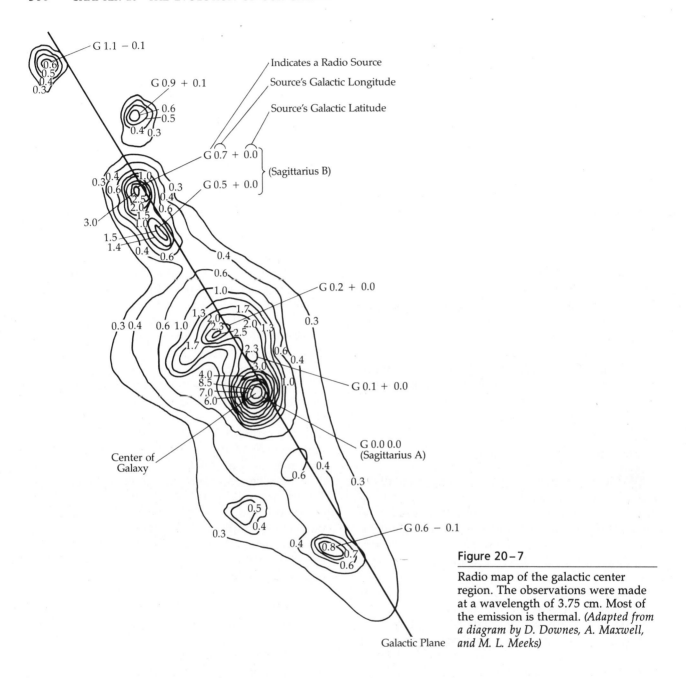

Figure 20–7

Radio map of the galactic center region. The observations were made at a wavelength of 3.75 cm. Most of the emission is thermal. *(Adapted from a diagram by D. Downes, A. Maxwell, and M. L. Meeks)*

too, probably comes from dust heated by O and B stars.

The infrared emission line at 12.8 μm produced by singly ionized neon (Ne II) comes from ionized gas in the nucleus, and as for radio recombination lines, information about the core can be inferred from the Doppler shift of the line. These observations show that the ionized gas concentrates in a region about

1.5 pc in radius. The Doppler shifts indicate rotational motions of around 200 km/s. These ionized zones are small, less than 0.5 pc in diameter, and contain a few solar masses of ionized material. They appear to orbit around the galactic center on an axis tilted 45° to the main rotational axis of the Galaxy. So the galactic core contains, within the Sgr A molecular cloud complex, a disk of rotating ionized gas.

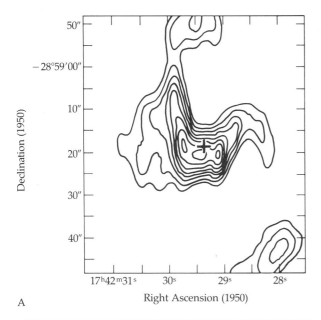

A

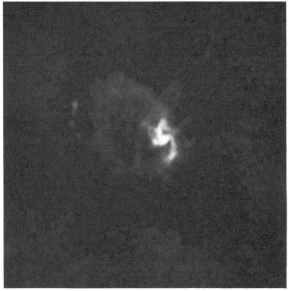

B

Figure 20–8

High-resolution maps of Sgr A. (A) Map at 5 GHz made with the VLA; the cross indicates the position of Sgr A West, the point-like nonthermal source. *[R. Brown, K. Johnston, and K. Lo, Astrophysical Journal 250:155 (1981)]* (B) Sgr A at 6 cm, a VLA map. *(R. D. Ekers, U. J. Schwarz, and W. M. Goss)*

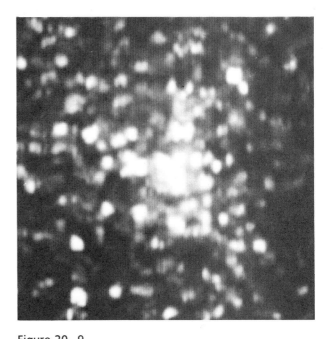

Figure 20–9

Galactic center at 2.2 μm. The region covered is about 5° square. *(D. A. Allen, A. R. Hyland, and T. J. Jones)*

The Einstein X-Ray Observatory has made an X-ray image of the galactic center region. It shows modest X-ray emission at wavelengths below 0.6 nm within 100 pc of the galactic center. This emission consists of a complex of weak sources (10^{27} to 10^{28} W) embedded in a weaker halo of diffuse X-ray emission (Figure 20–10). The discrete sources lie along a ridge just south of the galactic equator, in the same location as the cluster of infrared sources. The diffuse emission may come from hot, coronal gas.

Finally, gamma rays from the nucleus have been detected. In particular, a line has been observed with an energy of 511 keV and a luminosity of 10^{31} W, which is attributed to electron–positron annihilation. The change in luminosity over time leads to the conclusion that the nucleus has a diameter of less than 1 lightyear. If this interpretation is correct, there is a black hole at the center of the Galaxy!

(d) The Distribution of Mass in the Galaxy

Keep in mind that the neutral and ionized hydrogen gas represents only a small fraction of the total mass of the Galaxy—only 5 to 10 percent. On the

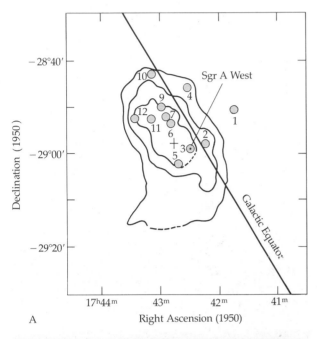

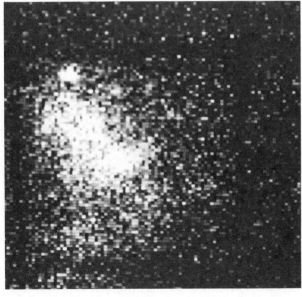

Figure 20–10

X-ray emission from the galactic center. (A) This schematic map shows discrete sources (circles) and diffuse emission (contour lines). The cross marks the center of the diffuse emission. (B) A computer-enhanced image of the X-ray emission. [M. Watson, R. Willingale, J. Grindley, and P. Hertz, Astrophysical Journal 250:142 (1981)]

average, approximately 1 percent of all the hydrogen is ionized, but the percentage varies with distance from the galactic center. For instance, there is a ring of ionized gas beyond the galactic bulge at about 4 kpc, and in this region the concentration of ionized hydrogen is about 10 percent.

The total mass of the Galaxy can be determined, to a first approximation, on the assumption that the Sun moves in a circular Keplerian orbit about a point mass; such calculations lead to a value of $(1.5 \times 10^{11})M_\odot$. However, recent rotation curves show a rise in the rotational velocity to 300 km/s at 20 kpc. This rotation curve implies a mass of $(3.4 \times 10^{11})M_\odot$ interior to 20 kpc. So at least as much mass lies exterior to the solar circle as interior to it. The Galaxy has a massive halo of nonluminous matter, the form of which is currently unknown.

The nucleus may hold the densest concentration of mass in the Galaxy (Figure 20–11). Recall that radio and infrared line observations show rapid rotational motions near the Galaxy's core. The rotational velocities increase closer to the core, where they are so high that a huge mass is needed to hold all that rapidly moving gas together. To account for the rapid rotation requires a mass in the core of several million solar masses—all lumped together in a region only 0.04 pc in diameter. An approximate calculation shows this point. The infrared line observations imply rotational velocities of about 200 km/s at a radius of 10^{16} m. Apply the argument used for pulsars, in which a rotating spherical mass just holds together by it own gravity:

$$V_{eq} = (GM/R)^{1/2}$$

where V_{eq} is the velocity of the object's equator. Solve this equation for the mass:

$$M = RV_{eq}^2/G$$

$$= (10^{16} \text{ m})(200 \times 10^3 \text{ m/s})^2/ \\ (7 \times 10^{-11} \text{ N} \cdot \text{m}^2/\text{kg}^2)$$

$$= 6 \times 10^{36} \text{ kg} \\ = 3 \times 10^6 \text{ solar masses}$$

What form might this mass have? One possibility is that it is locked up in a black hole. Indirect support for this idea comes from observations that a few other galaxies may have a similar mass concentration in their nuclear regions.

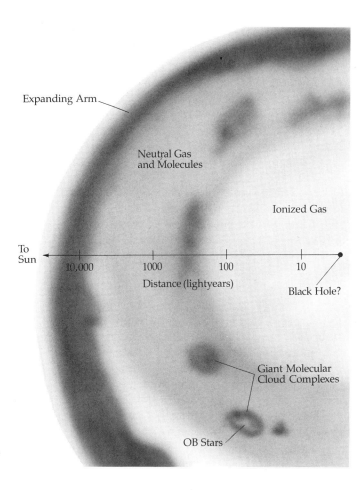

Expanding Arm

Neutral Gas
and Molecules

Ionized Gas

To
Sun

10,000   1000   100   10

Distance (lightyears)

Black Hole?

Giant Molecular
Cloud Complexes

OB Stars

Figure 20–11

Model of the central region of the Galaxy.

20–3 Evolution of the Galaxy's Structure

Why do galaxies such as ours exhibit a spiral structure? How stable are the spiral arms? Spiral structure may arise from an instability, or perturbation, so that some density irregularity is pulled into a spiral form by the differential galactic rotation. The difficulty with this explanation is that such a feature is expected to persist for only a short time (about 5×10^8 years) before being pulled apart again by the differential rotation. An additional problem is that the initial spiral-like distortion would not extend over the entire galaxy but would occupy only a small part of it. Yet we observe spiral structure throughout the entire plane of a galaxy, and such spiral galaxies are sufficiently common to suggest that they are in fact stable.

One very promising approach to this problem is the *density-wave model* developed by C. C. Lin and Frank Shu. The spiral structure of a galaxy is regarded as a *wave pattern* resulting from gravitational instabilities. The wave extends far beyond the initial localized perturbation. The presence of a density wave means that the distribution of mass is nonuniform; therefore, the gravitational potential varies over the galactic disk. Stars and gas are concentrated in the regions where the gravitational potential is low, and these mass concentrations in turn influence the orbits of other stars and clouds of gas. The density wave is a self-sustained phenomenon and is stable. The spiral pattern produced by the density wave is not tied to the matter but instead moves through it. According to this idea, the angular velocity of the pattern may differ appreciably (by a factor of one-half) from that of the material. Hence, stars formed from the gas and dust concentrated at the arms eventually migrate out of the arms. Gas concentrated near the potential minimum largely defines the spiral arms; this is ever-changing gas, for some is consumed in star formation and some is ejected from stars by some mode of mass loss. Stars traveling in orbits that differ greatly from circular orbits come under rapidly varying gravitational attractions. Density waves certainly influence their motions, but not systematically, and therefore no structure will persist for these eccentric-orbit stars.

The density-wave model assumes that a two-armed, spiral density wave sweeps through the galactic plane, but the model does not yet explain the wave's origin. Once formed, it persists for a fairly long time. The gas in the disk piles up at the back of the wave. The buildup of pressure and density heats up the gas suddenly so that a shock wave forms along the front of the density wave. This shock may initiate the collapse of the clouds to form giant molecular cloud complexes, which in turn form young stars and H II regions. Such squeezing also helps to make dust out of the gas, and a thin dust lane forms along the shock front. The compression of the interstellar medium by the density wave forms the features associated with a spiral arm. During the short lifetimes of the newly formed O and B stars, the density wave moves only a short distance. So these stars, while they last, clearly mark the spiral arm. As the density wave moves on, it provokes the formation of more stars. These take the place of the ones that have rapidly faded out. So the spiral arms persist by a continual destruction and creation maintained by the density wave.

How well does the density-wave model describe the observed spiral structure? First, it outlines the grand scheme of the two-armed spiral pattern that we observe in other galaxies and probably in our own. Second, it explains the persistence of the spiral arms in the face of galactic rotation. Third, it predicts the general features of a spiral arm. So the density-wave model succeeds fairly well in explaining the prominent features of spiral structure. However, the model fails on a number of points. It does not explain the origin of the density waves. Nor does it clearly work out what keeps them going. As the density waves ripple through the interstellar medium, they lose energy and should dissipate in about 1 billion years. As evidenced by the abundance of spiral galaxies, however, they must last longer than this. Some

mechanism must keep supplying energy to maintain them.

We will now try to place the Galaxy's present structure in the context of its history. The crucial clues come from the chemical composition of galactic material and its dynamics. The process of galactic evolution links the chemistry with the dynamics.

We stated earlier that the chemical compositions of Population I and Population II stars differ considerably in their abundances of heavy elements. In general, Population II stars contain about 1 percent of the metal abundance of Population I stars. However, we do not find a simple division of metal abundances into just two groups. Rather, we find a range of abundances, from about 3 to less than 0.1 percent for the mass ratio of metals to hydrogen. So, though the division into two populations is a useful first approximation, there is really a continuous range of populations. When objects are cataloged by metal abundance, a striking correlation emerges: *the lower the metal abundance of an object, the greater its height from the Galaxy's disk.*

Interpretation of this key observation relies on a basic concept of the recycling of the interstellar medium. First, stars are born from clouds in this medium. Their atmospheric elemental abundances reflect that of the gas from which they formed. Second, stars' orbital motions about the Galaxy are inherited from their parent gas and dust clouds. Third, massive stars evolve quickly and spew back into the interstellar medium material enriched with heavy elements. So as long as new stars, especially massive ones, are born, the abundance of heavy elements in the interstellar medium of the disk gradually increases as the Galaxy ages. Observations show that the youngest objects (highest heavy-element abundances) hug close to the disk and the oldest objects (lowest heavy-element abundances) range far from it. Other objects fall between these extremes.

We can estimate the Galaxy's age by finding the oldest stars in the halo. A comparison of theoretical models for globular cluster stars with their H–R diagrams indicates an oldest age of 17 billion years. Because globular clusters contain the oldest stars associated with the Galaxy, the halo marks the fossil remains of the Galaxy's birth. Within it, globulars orbit the Galaxy on extremely elongated elliptical paths, moving slowly through the halo at the outer extremes of their orbits and only briefly whipping in and around the nucleus. These stars exhibit the motions of the cloud from which they were formed. So the Galaxy must have been born from an initially huge gas cloud—at least 100 kpc in radius.

Consider a tremendous cloud of gas roughly twice as big as the Galaxy's halo today. This proto-Galaxy cloud was probably turbulent, swirling around with random churning currents. Slowly at first, the cloud's self-gravity pulled it together, with its central regions getting denser faster than its outer parts. Throughout the cloud, turbulent eddies of different sizes formed, broke up, and died away. Eventually, the eddies became dense enough to contain sufficient mass to hold themselves together. These might be hundreds of parsecs in size—incipient globular clusters. This process happened 17 billion years ago, and it took place quickly—no more than 2 billion years elapsed during the formation of halo stars.

Not all of the gas was consumed in this burst of globular cluster formation. As the material contracted further, it fell slowly into a disk. Because the original cloud had a little spin, conservation of angular momentum required that it spin faster around its rotational axis as it contracted. As the disk formed, its density increased and more stars formed. Each burst of star birth left behind representative stars at different distances from the present disk. Finally, the remaining gas and dust settled into the narrow layer we see today. Somehow density waves appeared and drove the formation of spiral arms. During this time, massive stars manufactured heavy elements and blew some back into the interstellar medium. So as stars were born in succession, each later type had a greater abundance of heavy elements. That enrichment continues today in the disk of the Galaxy.

There is one problem with this model. During the quick formation of halo stars, a rapid enrichment of metals occurred. Globular clusters all appear to have the same age and so must have been formed at roughly the same time (within a span of 2 billion years). Yet, all globulars do not have the same metal abundances. For example, some globular cluster stars have 100 times more iron than others. How did this difference come about in only 2 billion years? A related curiosity is that disk stars—the oldest of which match the ages of globular clusters—show a much more gradual metal enrichment.

20–4 Cosmic Rays and Galactic Magnetic Fields

(a) Observations of Cosmic Rays

Before we can discuss the role of *cosmic rays* in our Galaxy, we must examine their characteristics. They are not rays at all but rather high-energy charged particles. These particles may be nuclei of atoms whose electrons have been stripped away, or they may be electrons or even positrons. They are called cosmic rays when they travel at essentially the speed of light; in general, they possess large kinetic energies—up to 10^{20} eV. As you saw in Chapter 10, our Sun ejects low-energy (tens to hundreds of MeV) cosmic rays in flares. The Sun also modulates those cosmic rays coming from outside the Solar System; the interplanetary magnetic field and the solar wind severely distort the orbits of particles with energies of less than 10^9 eV (1 GeV). The amount of modulation varies with solar activity, and the modulation obscures the intrinsic properties of low-energy galactic cosmic rays. We do know, however, that the number of particles increases rapidly with decreasing energy. Most of the observational data on extra-Solar-System cosmic rays pertain to particles having energies in excess of 10^9 eV. Energies as high as 10^{20} eV have been observed for single particles, but these are extremely rare.

The chemical composition of cosmic rays gives us information on both their source and their journey through space. As in stars, hydrogen nuclei (protons) are by far the most abundant component, constituting about 90 percent of all cosmic-ray nuclei. Another 9 percent are helium nuclei, and the remainder is divided among heavier elements. Much interest centers on the so-called *light nuclei*—lithium, beryllium, and boron. These nuclei are virtually absent in stellar atmospheres, where their abundance is 10^{-7} that of helium, while among cosmic rays the light-nuclei-to-helium ratio is about 1:100. The much greater proportion of such nuclei among cosmic rays is attributed to the fact that the original heavy cosmic rays collide with interstellar matter. During such collisions, the heavy nuclei break up into lighter ones; this process is termed *spallation*. Observations of the light, and therefore *secondary*, nuclei enable us to estimate the amount of material traversed by the original cosmic rays. The density of the interstellar medium and the lifetime of cosmic rays are inexorably linked. Current data suggest an average density of interstellar matter roughly 10^5 atoms/m³ and a mean lifetime of cosmic rays from a few million to some tens of millions of years.

From what directions do cosmic rays come? Observationally, they appear to come more or less uniformly from all directions; they are *isotropic*. This phenomenon is not necessarily due to a random spatial distribution of the cosmic-ray sources but probably arises because galactic magnetic fields deflect cosmic rays to such an extent that they cannot travel in straight paths. In fact, these high-energy particles travel along the magnetic lines of force in spiral paths, whose size is determined by both the magnetic field strength and the particle energy.

(b) The Source and Acceleration of Cosmic Rays

If cosmic rays arrive at the Solar System isotropically, how can we locate and identify their sources? One clue is the high energy of the particles; we should look for highly energetic phenomena. We have more direct evidence concerning the sources, however, for one component of cosmic rays is electrons, which represent about 1 percent of all cosmic-ray particles. Cosmic-ray electrons have been observed directly. The fact that positrons (positive electrons) are only one-tenth as numerous indicates that these electrons are primary particles and not secondaries as the light nuclei are. This conclusion is based on the fact that, because of charge conservation, more positrons than electrons are produced when primary cosmic-ray protons collide with interstellar atoms.

We already know that high-energy electrons are observable from a distance, for they emit synchrotron radiation at radio frequencies when they move in a magnetic field. As you saw in Chapter 18, supernova remnants, such as the Crab Nebula, are strong sources of synchrotron radiation from relativistic electrons. Moreover, supernovae generate energy on a scale sufficient to produce cosmic rays. Other sources of cosmic rays may be rotating neutron stars, such as those observed as pulsars.

The lifetime of cosmic rays in the Galaxy is limited; some escape from the Galaxy altogether, and others are consumed by interactions with the interstellar medium. The supply of cosmic rays must be steady and continuous if the energy density and the total energy in them are to remain more or less constant. Supernovae can indeed maintain the supply if supernova explosions occur every 50 years in the Galaxy. Statistics based on the observations of supernovae in other galaxies suggest that this is a reasonable estimate of the frequency of supernova outbursts. A possible additional source of cosmic rays is the galactic nucleus, which appears to be a strong source of synchrotron radiation. The highest-energy cosmic rays (over 10^{18} eV) may well come from outside our Galaxy — that is, from the extragalactic radio galaxies and quasars that are observed to generate such enormous amounts of energy (Chapter 23).

Charged particles must be powerfully accelerated if they are to attain the energies or velocities observed for cosmic rays. Most of this acceleration occurs in the shock wave from a supernova explosion as it drives through the interstellar medium. Regions in interstellar space where magnetic field lines converge are very efficient in accelerating charged particles to produce cosmic rays.

(c) The Galactic Magnetic Field

The existence of a *galactic magnetic field* is proved by the observation of the Faraday rotation of radiation from radio sources; such radio sources include galactic sources such as pulsars (Chapter 17) and extragalactic sources. *Faraday rotation* is the rotation of the plane of polarization as linearly polarized radiation passes through a magnetized plasma [Section 17–2(b)]. If the electron density and the distance traversed are known, we can solve for the mean magnetic field strength. The best current estimates of the galactic magnetic field strength are about 0.5 nT. Before the observations of Faraday rotation, a galactic magnetic field had already been surmised to exist, both on the basis of interstellar polarization observations (Chapter 19) and from cosmic-ray data.

Another probe of the galactic magnetic field involves the Zeeman effect in 21-cm absorption lines by interstellar clouds along the line of sight to strong radio sources. For hydrogen in the ground state, the magnetic field splits the line into three components, and the frequency difference between two of these (of opposite circular polarization) is directly proportional to the magnetic field strength along the line of sight:

$$\Delta v = (2.80 \times 10^2)B \quad (\text{Hz})$$

where the field strength is in teslas. Such observations are consistent with those from Faraday rotation and range from about 0.1 to 1.0 nT.

The orientation of the magnetic field with respect to the spiral structure is primarily along the axes of the spiral arms. Superimposed on this mean field is a local field in our vicinity, which dominates most of the observations, making it difficult to ascertain the true nature of the general field. Locally, within

500 pc of the Sun, the field seems to have a turbulent, possibly helical form, with the mean magnetic axis along the spiral arm. Because cosmic rays travel in spiral paths about magnetic field lines, they are closely tied to the galactic magnetic field. Where the field lines converge, the particles are accelerated, as occurs in the Earth's magnetosphere for charged particles from the Sun.

Problems

1. Shown below is a sketch of the distribution of the neutral hydrogen maxima in the spiral arms of our Galaxy; the position of our Sun is denoted by $\odot$. Draw the 21-cm line profiles you would expect to observe in the directions $l = 50°$, $110°$, and $230°$; label the points on these profiles corresponding to the spiral arms. (Do not make any detailed calculations of radial velocity; just make certain that the signs of the velocities and the relative positions of the peaks are correct.)

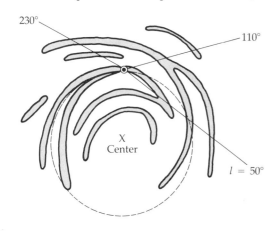

2. Name three important physical characteristics that distinguish Population I stars from Population II stars. Explain these differences in terms of the evolution of our Galaxy.

3. The objects that constitute our Galaxy are usually divided into five major population classes.
   (a) What are these classes?
   (b) Give an example of an object from each class.
   (c) Draw an edge-on view of our Galaxy, indicating the spatial distribution of each of the five classes. Label your diagram carefully.

4. (a) What would be the apparent magnitude of a star like our Sun if it were at the distance of the galactic center (8.5 kpc) from the Sun?

   (b) The Milky Way Galaxy contains 4×10^{11} stars. If we assume that all of these stars are like our Sun ($M_b = 4.7$), what is the absolute magnitude of the whole Galaxy? Compare this result with the apparent magnitude of the Sun.

5. When a particular globular cluster is at its farthest point from the galactic center (apogalacticon), its distance from the center is 10^4 pc. What is its period of galactic revolution? What assumptions must you make to arrive at a unique answer? Can you give any physical justification for your assumptions? (*Hint:* 1 pc = 2×10^5 AU. Assume that the mass of the Galaxy is $10^{12} M_\odot$.)

6. We can deduce the average time between stellar collisions by considering the figure below. If we have identical stars of radius R scattered through space with a mean number density N (stars per unit volume), then a star moving with speed V will sweep out the volume $\pi R^2 V$ per unit time. The average number of stars in this volume is $\pi R^2 VN$, so that in the time $T = 1/\pi R^2 VN$, the star will collide with one other star (on the average)! The average distance between each collision is just $L = VT = 1/\pi R^2 N$; this is called the *mean free path*. In each of the following situations, compute the mean collisional time and the mean free path for
   (a) the solar neighborhood, where $V = 20$ km/s and $N = 0.1/\text{pc}^3$; consider stars of radius $R = R_\odot$
   (b) a galactic nucleus, where $V = 1000$ km/s and where there are 10^9 stars (of radius $R = 10R_\odot$) within a sphere of radius 5 pc. Comment briefly on your results.

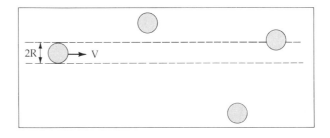

7. The 21-cm line of H I gas has a rest frequency of 1.420406 GHz. A cloud at $b = 0°$, $l = 15°$ has an observed 21-cm emission line frequency of 1.420123 GHz. Use Figure 15–10 to determine, approximately, *two* possible distances to the cloud. Assume only circular motion about the galactic center.

8. The following maximum velocities [relative to the Local Standard of Rest (LSR)] of H I gas are observed using the 21-cm emission along the lines of sight corresponding to the given galactic longitudes: *(i)* 123 km/s, $l = 15°$; *(ii)* 95 km/s, $l = 30°$; *(iii)* 64 km/s, $l = 45°$; *(iv)* 29 km/s, $l = 60°$; *(v)* 7.5 km/s, $l = 75°$. Assume circular motions along the galactic center.

   (a) Compute and draw a rough plot of the rotation curve of the inner part of the Galaxy ($R < R_0$) using these data.

   (b) Using the approximation that the mass distribution in the inner part of the Galaxy is spherically symmetric, determine the mass interior to the Sun's orbit, in solar masses.

   (c) Using the same approximation, how must the mass density depend on R between about 6 and 8.5 kpc from the galactic center? (Give a rough proportionality.)

9. A bright radio source has several 21-cm absorption lines caused by neutral hydrogen clouds along the line of sight. Discuss qualitatively how observations of the Doppler shifts of the lines can be used to estimate a lower limit to the distance to the radio source. Under what circumstances would this limit *not* possess the ambiguity caused by two distances along the line of sight having the same radial velocity?

10. Expanding gas rings are thought by some astronomers to be evidence of violent explosions near the center of the Galaxy.

    (a) Calculate the kinetic energy of the 3-kpc arm, which is thought to contain about $10^8 M_\odot$ of gas expanding at about 50 km/s.

    (b) Astronomers also believe that there is a ring of molecular gas of mass $10^7 M_\odot$ expanding at 150 km/s at a distance of 200 pc from the galactic center. Calculate its kinetic energy.

    (c) Compare the above energies with that of a single supernova explosion. Comment.

11. Speculate on how one might indirectly observe cosmic-ray electrons in our Galaxy. Do the same for cosmic-ray protons. (*Hint:* Ask how such high-energy particles might produce radiation.) Such measurements indicate that cosmic rays are present throughout the Galaxy and even in the halo!

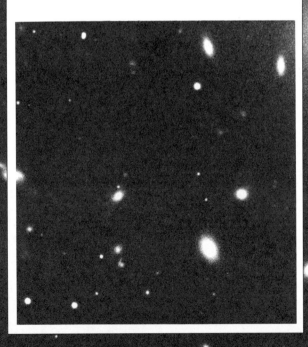

PART FOUR

The Universe

Galaxies Beyond the Milky Way

Galaxies are the grandest pieces of the Universe. They are the largest objects containing stars, gas, and dust that can still fit in the view of a telescope. Our Milky Way Galaxy contains 10^{11} stars; in the space beyond it, the other galaxies we can observe have even more stars. When we perceive the Universe, we usually picture it as a universe of galaxies. Yet, galaxies were recognized but recently (in 1924 for sure) as vast assemblages of stars. This chapter deals with ordinary galaxies in terms of their physical properties, influenced by the distribution of mass within, the relative amounts of gas and dust, and the types of stars they contain.

21–1 The Classification of Galaxies

Edwin Hubble pioneered the study of galaxies based simply on appearance. Most galaxies may be separated into three major categories: *elliptical*, *spiral*, and *irregular*. We illustrate several galaxies here; you are urged to browse through *The Hubble Atlas of Galaxies* for many more examples. Astronomers now consider this scheme too simple, but it still serves, with the addition of one more type—*peculiar galaxies*—as the basis for classification.

(a) The Classification Scheme

The *elliptical galaxies* (designated by E) have the shape of an oblate spheroid (perhaps—there is no consensus as to whether they are oblate, prolate, or triaxial; we'll stick with oblate); they appear in the sky as luminous elliptical disks (Figure 21–1). The distribution of light is smooth, with the surface brightness decreasing outward from the center. Elliptical galaxies are classified according to the elongation of the apparent projected image; that is, if a and b are the major and minor axes of the apparent ellipse, then $10(a-b)/a$ expresses the observed ellipticity. The true ellipticity cannot be found because the orientation of a particular galaxy cannot be determined. So an E0 galaxy appears circular, while increasingly elliptical ones are given designations from E1 to E7 (the latter is the flattest observed). The E galaxies have *no* axis of rotation; their stars follow orbits with a variety of inclinations. Statistical studies lead to the conclusion that the true ellipticities of these galaxies are represented fairly uniformly from E0 to E7. Some clusters of galaxies contain exceptionally large ellipti-

Figure 21–1

An elliptical galaxy. This example, M 32, is type E2 and companion to the Andromeda galaxy. *(National Optical Astronomy Observatories)*

cal galaxies, the supergiant ellipticals. An important subclass of these is the *cD giant* or *supergiant galaxies*, which have elliptical nuclei and extended envelopes. Their diameters range up to a few megaparsecs (10^6 pc). Most ellipticals, however, are *dwarf ellipticals*, only a few kiloparsecs in diameter.

The *spiral galaxies* are divided into the *normal* (designated S) and the *barred* (SB) spirals. Both types have spiral-shaped arms, with two arms generally placed symmetrically about the center of the axis of rotation. In the normal spirals (Figure 21–2A), the arms emerge directly from the nucleus; in the barred spirals, a bar of material cuts through the center (Figure 21–2B) and the arms originate from the ends of the bar. Both types are classified according to how tightly the arms are wound, how patchy they are, and the relative size of the nucleus. Normal spirals of type Sa have smooth, ill-defined arms that are tightly wound about the nucleus; in fact, the arms form almost a circular pattern. The intermediate Sb galaxies have more open arms, which are often partly re-

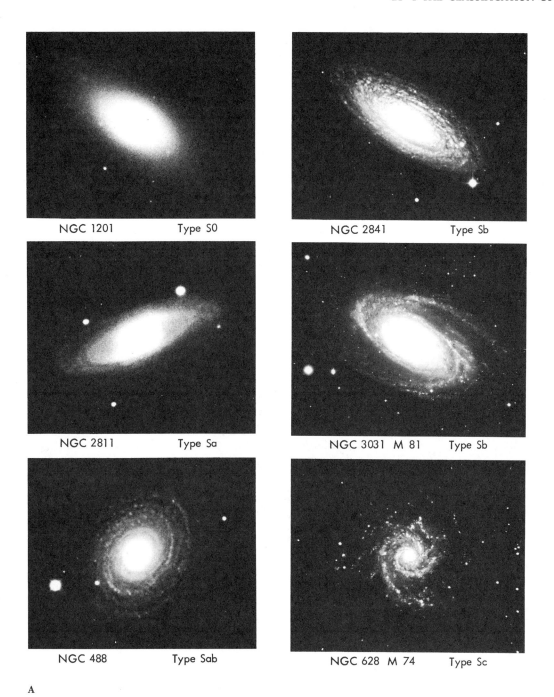

NGC 1201   Type S0

NGC 2841   Type Sb

NGC 2811   Type Sa

NGC 3031 M 81   Type Sb

NGC 488   Type Sab

NGC 628 M 74   Type Sc

A

Figure 21–2

Types of spiral galaxies. (A) Normal *(above).* (B) Barred *(see next page). (Palomar Observatory, California Institute of Technology)*

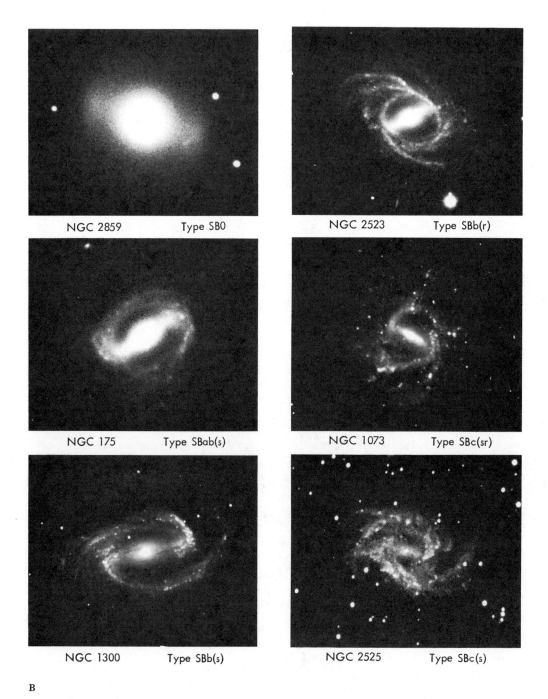

NGC 2859   Type SB0

NGC 2523   Type SBb(r)

NGC 175   Type SBab(s)

NGC 1073   Type SBc(sr)

NGC 1300   Type SBb(s)

NGC 2525   Type SBc(s)

B

Figure 21–2   *(Continued)*

solved into patches of H II regions and Population I stellar associations. The nuclei in Sc galaxies are usually quite small, and the spiral arms are extended and well resolved into clumps of stars. Both old and young populations coexist in spiral galaxies, but the proportion of young Population I objects increases from Sa to Sc. Barred spirals exhibit a parallel sequence of types: SBa, SBb, and SBc.

The degree to which the spiral arms are developed is related to the luminosity of a galaxy. In analogy to the stellar luminosity classes, those for galaxies are I, II, III, IV, and V, with I the most luminous and V the least. An ScI galaxy, then, is a very luminous spiral with a small nucleus and extended, well-resolved arms. These galactic luminosity classes also relate to mass, so that for the same galaxy type, class I galaxies are the most massive and class V the least.

One peculiar type of galaxy is given the S classification of spirals even though spiral structure is not usually seen. These *S0 galaxies* are intermediate between the ellipticals, E7, and the true spirals, Sa. They are flatter than the E7s and also differ from ellipticals in having a thin disk as well as a lenticular (lens-shaped) nucleus. In many respects, S0 galaxies resemble the true spirals but do not contain Population I objects.

Other galaxies fall into the *irregular class* since they show no symmetrical or regular structure; nevertheless, even these galaxies may be divided into two distinct groups. Those of type *Irr I* have resolved

O and B stars and H II regions and clearly a large Population I component. In some Irr I galaxies, incipient spiral arms are visible. The classification *Irr II* is quite ambiguous and may include galaxies that are simply peculiar. Primarily, however, these galaxies are amorphous and do not resolve into stars. Such galaxies do show marked absorption by interstellar dust, and gaseous emission is also observed. The peculiar galaxy M 82 is an Irr II (Figure 21-3); it is remarkable in that dusty material extensively blocks out light from its stars in a fashion that makes it appear to be exploding; it is *not*.

Finally, we note that not all galaxies fall into one of the three Hubble classes, and those that do not are generally called *peculiar galaxies*. Good examples are the bizarre ring galaxies (Figure 21-4) and other galaxies that may have undergone tidal disruption with another galaxy.

(b) Implications of the Classification Scheme

Do not think of the classification scheme as an evolutionary sequence! The form a galaxy takes depends upon its *angular momentum*; the greater the angular momentum, the more flattened the galaxy. In elliptical galaxies, the condensation of gas into stars was efficient and therefore rapid, which led to a more or less spherical distribution of stars and a high concentration of stars at the nucleus. Conversely, in spiral galaxies, star formation occurred more slowly and a dual type of distribution resulted. The slowly rotating system contains stars distributed spherically, and the rapidly rotating part is a flattened, disk-like system containing stars, dust, and gas. The dust and gas in the plane become subjected to the density waves that create spiral arms. The turbulence and magnetic fields may also play important roles in controlling the final shape of a galaxy. The total mass of the system *cannot* be an important factor in determining the type, for elliptical galaxies range from dwarf systems (with masses similar to that of a large globular cluster) to the supergiant systems that dominate galaxy clusters. These giant ellipticals are far larger and more massive than the largest spirals. Spiral galaxies exhibit a relatively small range of masses (Table 21-1).

About 77 percent of *observed* galaxies are spirals, 20 percent are ellipticals, and 3 percent are irregulars. This sample is dominated by the luminous spirals, however, which are visible at very great distances.

Figure 21-3

The irregular galaxy M 82. *(Lick Observatory)*

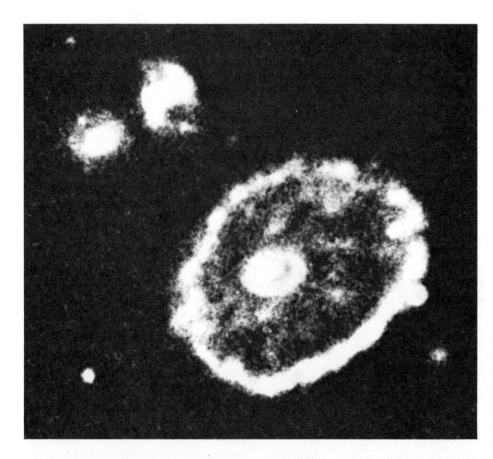

Figure 21–4

A ring galaxy. Called the Cartwheel Galaxy, this object may be the result of the collision between a large spiral (now the ring) and two other galaxies. *(Royal Observatory, Edinburgh)*

The relative numbers in a given volume of space are quite different. A survey of the region of space out to 9.1 Mpc shows that only 33 percent of the galaxies in this volume are spirals, 13 percent are ellipticals, and 54 percent are irregulars. Many of the irregulars are small galaxies of fairly low luminosity, as are dwarf ellipticals.

Dwarf galaxies may be either ellipticals or irregular I. In either case, they are small and often sparsely populated with stars; therefore, they are faint, of low

Table 21–1   The Fundamental Characteristics of Galaxies

	Ellipticals		Spirals		Irregulars I
Mass ($M_\odot$)	10^5 to 10^{13}		10^9 to 4×10^{11}		10^8 to 3×10^{10}
Absolute magnitude	-9 to -23		-15 to -21		-13 to -18
Luminosity ($L_\odot$)	3×10^5 to 10^{10}		10^8 to 2×10^{10}		10^7 to 10^9
M/L ($M_\odot/L_\odot = 1$)	100		2 to 20		1
Diameter (kpc)	1 to 200		5 to 50		1 to 10
Population content	II and old I		I in arms, II and old I overall		I, some II
Presence of dust	Almost none		Yes		Yes
		Sa	Sb	Sc, Sd	
$M_{\text{H I}}/M_\text{T}$(%)	0	2 ± 2	5 ± 2	10 ± 2	22 ± 4
Spectral type	K	K	F to K	A to F	A to F

surface brightness, and hard to detect. Dwarf ellipticals represent the low-mass end of the ellipticals and differ from globular clusters only in having about ten times a globular cluster's diameter. The dwarf ellipticals are extreme Population II. In contrast, dwarf irregulars contain large amounts of neutral hydrogen, and their stellar content also represents Population I.

21–2 Characteristics of Galaxies

Many of the important characteristics of galaxies, such as their structure and their population characteristics, are inherent in their classifications (Table 21–1). From the stellar population content, we can conclude that elliptical galaxies contain primarily old stars and spiral galaxies have a mixture of old (in their disks and haloes) and young stars (in their spiral arms). Irr I galaxies contain a high proportion of young stars.

(a) Integrated Colors

The spectrum of a galaxy arises from all of its stars, with the radiation contribution from the brightest hot stars competing with the light from the fainter (but far more numerous) cool stars. The composite spectrum measures the relative contributions of population types to the total light. In a gross way, we can use a galaxy's color to infer its stellar component. A direct correlation exists between a galaxy type and its color. Ellipticals tend to be much redder than spirals, and spirals redder than irregulars. Within the spiral group, the galaxies appear redder as their nuclear bulges grow larger and their spiral arms become less extensive. One way to describe a galaxy's color is to specify the stellar spectral class whose color resembles that of the galaxy. The elliptical and Sa galaxies have the same color as K stars, Sb galaxies resemble stars of classes F to K, and Sc and Irr galaxies have the same color as classes A to F.

The progression of color from the bluer irregulars to the redder ellipticals reflects a trend in the composition of the galaxies' populations. In general terms, an old Population I predominates in ellipticals and a much younger Population I stands out in irregulars. The mixture in the spirals is determined by the size of the nucleus (old Population I) relative to that of the spiral arms (young Population I). (Population II is probably a minor contributor in all large galaxies,

existing mainly in the globular clusters and galactic halo.)

(b) Masses

Spectra can also be used to determine the masses of galaxies. This method involves finding the velocity or rotation curve as a function of distance from the center of the galaxy; hence, the method applies only to galaxies that are near enough and bright enough to allow us to obtain spectra at several points. A model of the mass distribution is then fitted to the velocity curve, and the total mass is calculated, as was done for our Galaxy. A slit is placed across the galaxy's diameter, and Doppler shifts are measured at points along it. Knowing the tilt of the galaxy to our line of sight, we can translate the radial velocities to rotational ones.

Recent observations show a remarkable feature of the rotation curves for spiral galaxies (Figure 21–5). They rise steeply near the nuclear region and then flatten out at large distances from the nucleus, but they do *not* decrease, as they would if the mass were concentrated at the center and the velocity followed Kepler's law. This fact implies that a large fraction of the mass lies not in the interior regions but in the halo. If the rotation curve of our Galaxy resembles that for other spirals and so continues flat out to 60 kpc, then its total mass is 7×10^{11} solar masses!

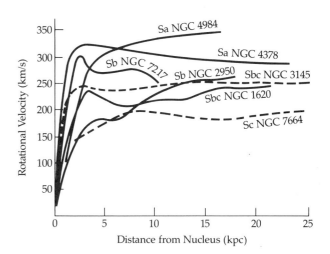

Figure 21–5

Rotation curves for spiral galaxies. Note how all flatten out at large distances from their centers. [*V. Rubin, W. K. Ford Jr., and N. Thonnard, Astrophysical Journal (Letters) 225:L107 (1978)*]

For binary galaxies, we again use the Doppler shift. Just as with visual binary stars, we can apply Newton's form of Kepler's third law to determine the mass if we know the distance, the angular size of the orbit, the period, and the position of the center of mass. However, galaxies revolve too slowly for us to see their orbits, periods, and relative centers of mass. So we cannot find the individual masses of binary galaxies. All we can measure are the radial velocities and the separations; we do not know what part of the orbit the galaxies are on or what the inclination is, and so we also do not know what the true orbital velocities are. If we examine a large sample of galaxies, however, and assume that their orbits are nearly circular and randomly oriented to our line of sight, we can estimate from these data the average masses of the galaxies sampled. An investigation of 279 binary systems, mostly spirals, used the Doppler shift of the 21-cm line to determine radial velocities. It gave an average mass of 10^{12} solar masses for these spirals (for $H_0 = 75$ km/s $\cdot$ Mpc—Section 21–3).

Another method employs observations of clusters of galaxies and uses the *virial theorem* to determine masses from the motions of individual galaxies relative to the whole cluster. This theorem states that, on the average, the magnitude of the total gravitational potential energy of a bound cluster equals twice the total kinetic energy:

$$2\langle KE \rangle = -\langle PE \rangle \text{ or } \langle mv^2/2 \rangle = \langle GMm/2r \rangle$$

(21–1)

where m is the mass of an individual galaxy, v is its speed, M is the mass of the whole cluster, and r is the distance from the center of the cluster. The *mean velocity* can be estimated from the velocities of the individual members, and the *mean distance* (from the center of the cluster) is deduced by averaging the measured distances (with a correction for projection effects). We can then use average values in Equation 21–1 to find the total mass because the individual masses cancel out. The total mass can be divided by the number of known or estimated members to find the mean individual masses. This procedure yields considerably higher masses for galaxies than do other methods. Hence, clusters may contain a considerable amount of invisible material (Chapter 22).

(c) Neutral-Hydrogen Content

Neutral-hydrogen 21-cm radiation gives us much information about the H I content of nearby galaxies, including an estimate of the total amount of hydrogen in the galaxy, the ratio of the mass of the hydrogen gas to the total mass of the galaxy, the distribution of the neutral hydrogen in the system, the rotation curves as a function of distance from the center of the galaxy and the derivation of its mass, as discussed earlier, and the radial velocity of the galaxy. The most complete data, of course, are for galaxies that may be resolved by radio telescopes.

As you might expect, the total amount of hydrogen in a galaxy is primarily a function of galaxy size. On the other hand, the ratio of the hydrogen mass to the total galaxy mass ($M_{\text{H I}}/M_{\text{T}}$) depends on galaxy type (Table 21–1). The percentage of the total mass that is in the form of neutral hydrogen is relevant to our ideas concerning evolution within galaxies. The less hydrogen relative to stars, the more original gas there was that must have already been condensed into stars. Data for a sample of spiral and irregular galaxies indicate that the H I mass makes up a small fraction of the total mass, only 3 percent for lenticulars and 22 percent for irregulars. So Irr galaxies have more H I gas, relative to their total mass, than do Sa galaxies.

The present rate of star formation depends on both the amount of hydrogen available and its density. Sc spirals and the irregular galaxies are not necessarily younger than the Sa spirals, but their development has been different. We note, for instance, that Population II stars have been observed in irregular galaxies, even though these galaxies have high $M_{\text{H I}}/M_{\text{T}}$ ratios.

For Sc spirals and irregulars, the extent of the hydrogen in many cases is almost double the optical size of the galaxy. For example, studies of the Magellanic Clouds (companion galaxies to the Milky Way) show that there is hydrogen between these two galaxies, both as a bridge to the Milky Way and as a common envelope surrounding both galaxies.

(d) Sizes

Once you know the distance to a galaxy, you can determine its diameter from a measurement of its angular diameter. The hitch here is that the definition of the "edge" of a galaxy is more or less arbitrary; different definitions result in different diameters. No matter how the angular diameter is defined, to find the linear diameter relies on the relationship $\alpha_{\text{rad}} = s/d$, where α_{rad} is the angular diameter in radians, s is the linear diameter, and d is the distance (both in the same units). In general, astronomers judge the edge of a galaxy by using some limiting level of observed brightness. The intensity contour of this designated level—called an *isophotal level*—is drawn around the image of the galaxy. This isophotal level then determines the galaxy's apparent angular size. Such a procedure is relatively easy to carry out today using CCD pictures and computerized image processing.

Dwarf ellipticals and small irregulars tend to be the smallest galaxies, some only 3000 pc in diameter. The typical diameter of galaxies of all types is about 15 kpc. Giant ellipticals can range up to 60 kpc across. The very largest galaxies are the supergiant ellipticals, often called *cD galaxies*. These can have diameters up to 2 Mpc—greater than the distance from our Galaxy to the Andromeda galaxy.

(e) Luminosities

If we know distances along with fluxes (apparent magnitudes) for galaxies, we can calculate luminosities (absolute magnitudes). One problem—the same as with size determinations—is how to be sure to have measured the *total* flux rather than some fraction of it. In addition, we have to apply corrections for dust obscuration within our Galaxy and in the galaxy being observed.

Absolute magnitudes range from -8 (2×10^5 solar luminosities) for dwarf ellipticals to -25 (10^{12} solar luminosities) for supergiant (cD) ellipticals. Our Galaxy, viewed from the outside, would have an absolute magnitude of roughly -21 (2.5×10^{10} solar luminosities).

(f) Mass–Luminosity Ratios

Divide the total mass of a galaxy by its luminosity to calculate its *mass–luminosity ratio,* which is an indication of the average energy output per unit mass from the galaxy. It is usually expressed in units of solar masses and solar luminosities. Recent determinations using binary galaxy masses give 35 for the average value of M/L for spiral galaxies and about twice as much for giant ellipticals and lenticulars. (For comparison, M/L for stars in the Sun's neighborhood is about 1.)

Ellipticals have a large value of M/L because they contain a greater percentage of low-mass stars with low light output—main-sequence stars of spectral class M. If true, this extra abundance of M stars means that ellipticals should be redder in overall color than spirals—and they are. Other possibilities are neutron stars, black holes (including perhaps a giant one in the nucleus), and dark interstellar matter, which contribute to M but not to L.

One other trend relates to total luminosity and H I mass. Along the Hubble sequence from S0 to Irr, the ratio of hydrogen mass to optical luminosity increases. So the more H I gas a galaxy has, the greater its light output per unit mass. This relationship probably arises from the greater rate of ongoing star birth in spirals with more gas and dust.

Note that the sequence from ellipticals to irregulars marks a sequence from galaxies in which star birth ceased long ago (ellipticals, with old Population I, some Population II, and no gas and dust) to those in which star birth is now carried on very actively (irregulars, with Population I stars and considerable gas and dust). The trend provides a clue to the evolution of galaxies.

21–3 The Distance Scale and the Hubble Constant

How far away are galaxies? This is *the* crucial question to answer to infer some of their physical properties—and to undertake cosmological research. The galaxies are grouped (Chapter 22) and give the structure of luminous matter in the Universe. They also serve as markers for the Universe's expansion—a key cosmological discovery of this century.

(a) Distances to Galaxies

Cosmology involves the *distance scale*—the scale of astronomical distances. Let's see how this scale is established in a stepwise fashion. Distances to even the nearest galaxies are far too great to be measured directly, and so we must use indirect methods related to the brightness of objects and the inverse-square law of radiation intensity. The goal here is to calibrate the most luminous objects possible, those that can be seen at the greatest distances.

The calibration process starts locally in the Solar System and progresses by steps to span our Galaxy (Table 21–2 and Figure 21–6), the galaxies nearest to ours [called the Local Group; Section 22–1(b)], and more distant galaxies. Objects that are brighter than absolute magnitude zero can be used as distance indicators for galaxies within the Local Group. Galactic-cluster main-sequence fitting is not useful even for nearby galaxies, but this method allows calibration of bright stars, such as OB stars and classical Cepheids. Although the methods listed in the second

Table 21-2   Distance Indicators

Object	M_v	Population	Method	Basis of Calibration
Nearby stars	$\geq 0^m$	Disk	Trigonometric parallax	Radar determination of AU
Galactic clusters	—	I	Main-sequence fitting	Trigonometric parallaxes Moving clusters Main sequence of nearby stars
Lower main-sequence stars (A–M)	$\geq 0^m$	Disk	Spectroscopic parallax	Trigonometric parallaxes Galactic-cluster C–M diagrams
O and B stars	0^m to -6^m	I	Spectroscopic parallax	Galactic-cluster C–M diagrams Statistical parallaxes
Supergiants	$-6^m; -7^m$	I	Spectroscopic parallax	Galactic-cluster C–M diagrams
RR Lyrae stars	$+0^m5$	II	Period from light curve	Statistical parallaxes Globular clusters
Classical Cepheids	-0^m5 to -6^m	I	P–L law	Statistical parallaxes Galactic-cluster C–M diagrams
W Virginis stars (Population II Cepheids)	0^m to -3^m	II	P–L law	Statistical parallaxes Globular-cluster C–M diagrams
Globular clusters	-5^m to -9^m	II	Integrated magnitude	RR Lyrae stars C–M diagrams
Novae	-8^m	Disk	Maximum light	Expansion rate of shell
H II regions	-9^m	I	Angular size	Nearby galaxies
Supernovae	-16^m to -20^m	I and II	Maximum light	Nearby galaxies
Brightest galaxies in clusters of galaxies	-21^m	I and II	Integrated magnitude (a) Brightest galaxy (b) Fifth brightest (c) Mean of ten brightest	Nearby galaxies
Galaxies	—	I and II	Hubble constant of redshift (expansion of Universe)	Doppler shifts of nearest clusters of galaxies

C–M = color–magnitude (=H–R); P–L = period–luminosity.
Statistical parallaxes are averaged over a group of stars, using one component of their proper motions, chosen so that they are *not* affected by solar motion and so that their peculiar velocities are random.

part of Table 21–2 permit one to find distances to galaxies with large distance moduli, these distance indicators are substantially less reliable than those in the first part of the table. The power of the former lies in the very great brightness of the objects and in the subsequently large limiting distance moduli. The bright objects can give distances to galaxies well beyond the Local Group to other large clusters of galaxies, such as the Virgo cluster. Statistics from these clusters tell us about the luminosities (absolute magnitudes) of the brightest galaxies in a cluster and about the luminosity function within a cluster. Although there is some variation from cluster to cluster, the statistics are sufficiently good to give a first ap-

proximation of distance. Distinct groupings appear at certain luminosities; for instance, the brightest galaxy in a large cluster is nearly always a giant elliptical.

What kind of galaxies serve as useful standards? In the work of Allan Sandage and Gustav Tammann, the candidates are supergiant spiral galaxies with small nuclei and spread-out spiral arms, that is, Sc I galaxies. (M 101 is such a galaxy.) Sandage and Tammann have calibrated the absolute magnitude of such galaxies as about -21.2, or 25×10^9 times the Sun's luminosity. Seen in distant clusters, these galaxies are relatively easy to identify because of their high luminosity and distinctive shape. With contemporary telescopes, they serve as standard candles to dis-

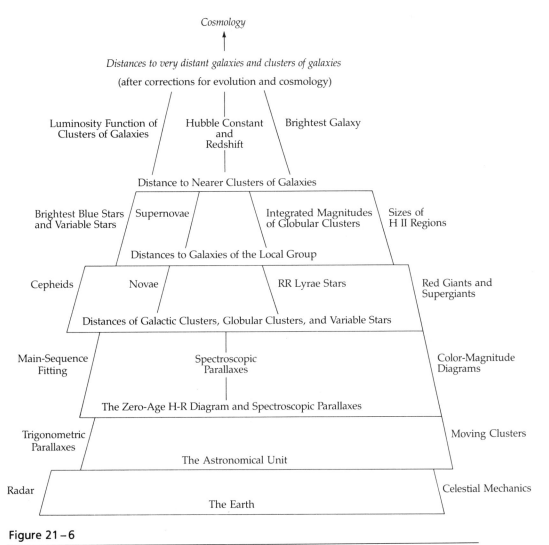

Figure 21–6

The distance-scale pyramid. (*Adapted from a diagram by P. W. Hodge*)

tances of roughly 400 Mpc. Giant elliptical galaxies are also used. Again, their luminosities must be calibrated by reference to some nearby galaxy or cluster whose distance is known by other techniques. Note that the lower levels of the distance pyramid (Figure 21–6) affect the outcome most strongly.

(b) Redshift and Distance

Galaxies move relative to one another. The radial velocities of other galaxies relative to our Galaxy may be found by studying Doppler-shifted features in the spectra of these galaxies (Figure 21–7). At visible wavelengths, we may use Doppler-shifted absorp-

tion lines (H and K of Ca II), and at radio wavelengths, the 21-cm emission line of neutral hydrogen is useful.

In 1912, Vesto M. Slipher discovered that most of the galaxies he observed had redshifted spectral features; that is, most galaxies are receding from our Galaxy. Edwin Hubble and Milton Humason later established that the amount of redshift increases with the faintness of the galaxy; the correlation is known as *Hubble's law* (Figure 21–8). Consider a given galaxy at a distance r. If this galaxy emits a spectral line of wavelength λ_0 and we detect this line at a (greater) wavelength λ, then the *redshift z* is defined as

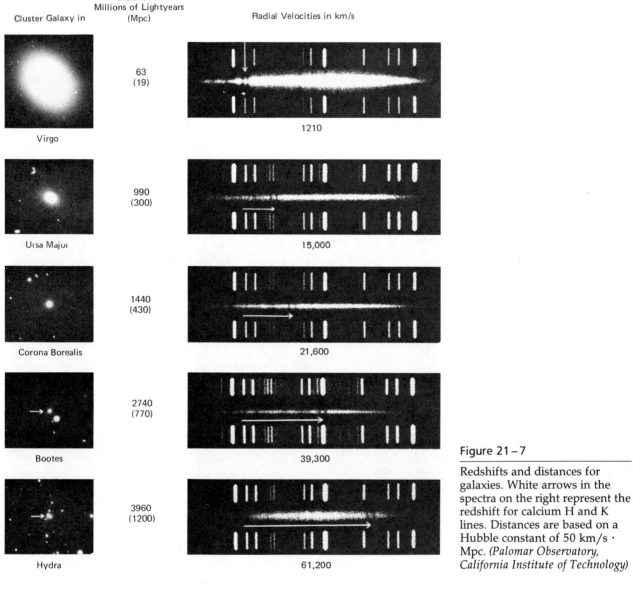

Cluster Galaxy in	Distance in Millions of Lightyears (Mpc)	Radial Velocities in km/s

Virgo — 63 (19) — 1210

Ursa Major — 990 (300) — 15,000

Corona Borealis — 1440 (430) — 21,600

Bootes — 2740 (770) — 39,300

Hydra — 3960 (1200) — 61,200

Figure 21–7

Redshifts and distances for galaxies. White arrows in the spectra on the right represent the redshift for calcium H and K lines. Distances are based on a Hubble constant of 50 km/s · Mpc. *(Palomar Observatory, California Institute of Technology)*

$$z = (\lambda - \lambda_0)/\lambda_0 = \Delta\lambda/\lambda_0 \qquad (21-2)$$

If this change in wavelength is interpreted as a *velocity* Doppler shift, then the *speed of recession* of the observed galaxy is

$$v = (\Delta\lambda/\lambda_0)c = zc \qquad (21-3)$$

where c is the speed of light. (When $v \geq c/10$, we must use the *relativistic* expression for the relationship between Doppler shift and recession speed.) By using the apparent magnitude of the galaxy to measure its relative distance, Hubble discovered the cor-

relation

$$zc = Hr \qquad (21-4a)$$

where H is the *Hubble constant*; comparing Equation 21–4a with Equation 21–3, we find an alternative form of Hubble's law:

$$v = Hr \qquad (21-4b)$$

Since apparent magnitude goes as $\log r$, we may illustrate Hubble's linear relationship between redshift and distance (Equation 21–4a) by plotting $\log z$

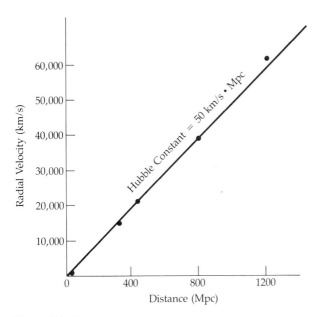

Figure 21-8

Hubble's law. Plotted here are the galaxies shown in Figure 21-7.

versus the apparent magnitude of the galaxy. The problem in finding H is to find the distance r accurately because redshifts (z) can be measured well. So what is the value of H? Basically, we don't know yet to within a factor of 2! Estimates in the last ten years range from 50 to 100 km/s · Mpc (if megaparsecs are used as the distance unit). Let's examine two techniques that give results at either end of this range. (Compare them in light of Table 21-2.)

The Sandage–Tammann Process to Obtain H

Here are the steps in a contemporary procedure followed by Allan Sandage and Gustav Tammann that yields a value of 50 km/s · Mpc (±10 percent) for H:

1. Measure the astronomical unit using radar reflection from Venus and then measure the distances to nearby stars using trigonometric parallax with the astronomical unit as the baseline.
2. Determine the distance to the Hyades, a galactic cluster, using the moving-cluster method. Check for consistency by comparing brightnesses of Hyades stars with those of nearby stars with distances known by trigonometric parallax.
3. Find the distances to Cepheids in our Galaxy by searching out galactic clusters that contain Cepheids, comparing the brightness of the stars in these clusters with those of stars in the Hyades cluster, and so finding the distances to the clusters with Cepheids. This calibrates the Cepheid period–luminosity relationship.
4. Use Cepheids to determine the distances to nearby galaxies by the period–luminosity relationship.
5. Measure the angular sizes of H II regions in these nearby galaxies. Find that the sizes of H II regions in spiral and irregular galaxies depend on the luminosity of the galaxy. Use nearby galaxies to calibrate the relationship between size of H II region and luminosity.
6. Extend this calibration to Sc I galaxies, for which the nearest, M 101, has its distance determined by different methods as a check. We now have the sizes of H II regions in Sc I galaxies.
7. Use the size–luminosity relationship for H II regions to find the distances of galaxies (about 60) to the limit of this method. Now we know the absolute magnitudes of these galaxies and their relationship to luminosity classes.
8. Look at Sc I galaxies, the objects we can see distinctly at the greatest distances. Use the absolute-magnitude calibration (step 7) to find their distances. Measure their redshifts to get their radial velocities; divide their radial velocities by their distances to get the Hubble constant.

Sandage and Tammann's result rests on *five* distance indicators. These indicators apparently include systematic errors that are not obvious, as the next technique shows.

Note that this procedure relies strongly on Cepheids as distance indicators. When the Hubble Space Telescope is launched, astronomers will be able to resolve Cepheids in galaxies about ten times farther away than the present ground-based limit. This telescope will increase direct measurements of distances to galaxies by a factor of 10 — and the number of galaxies by a factor of 1000.

The Aaronson–Huchra–Mould Way to H

Marc Aaronson, John Huchra, and Jeremy Mould have developed a new technique — independent of the Sandage–Tammann steps — that gives a value of $H = 90$ km/s · Mpc (with a claimed uncertainty of 5 percent). Let's briefly look at their procedure, which relies on radio and infrared observations to infer the luminosities of galaxies.

The technique rests on a relationship, derived by R. Brent Tully and J. Richard Fisher, between the absolute magnitudes (in blue light) of spiral galaxies and the spreads in frequency of their 21-cm H I emissions. The wider the 21-cm line, the greater a galaxy's luminosity. The 21-cm emission comes from the neutral gas in a spiral galaxy's disk, which is broadened by rotational motion. Most galaxies have maximum rotational velocities of 100 to 300 km/s. So the Tully–Fisher relationship is really one between rotational velocity and luminosity of spiral galaxies. Note that the rotational velocity at a given distance from the center relates to the mass of a galaxy, and the luminosities of the stars in a galaxy are also related to their masses. So for one type of galaxy, if the galaxies have more or less the same mix of stars of different masses and if the degree of concentration of mass toward the center is more or less the same, we might expect the total luminosity to be related to the rotational velocity. The relationship shows relatively little variation from one galaxy to the next, at least among galaxies of the same type.

If properly calibrated, the Tully–Fisher relationship provides a galaxy's luminosity from a measurement of the width of its 21-cm line. The main problem with the method, as originally developed, lay in its use of blue magnitudes. Dust scatters blue light well, so that blue-light observations of another galaxy suffer from an unknown amount of extinction produced by dust in our Galaxy and in the disk of the other galaxy. Aaronson, Huchra, and Mould noted that infrared light suffers virtually no extinction from galactic or extragalactic dust. So they made infrared observations of nearby galaxies (such as M 31 and M 33) whose distances are known from standard candles, such as the Cepheids. This gives the absolute infrared magnitudes (essentially the infrared luminosities at a wavelength of 1.6 μm). These infrared absolute magnitudes connect well with the widths of the 21-cm lines of local galaxies. Hence, we can infer the infrared luminosities of galaxies from the widths of their 21-cm lines (Figure 21–9). Compare these with their infrared fluxes, and we can infer the distances and so a value for H. This technique yields a "best" result of 90 km/s · Mpc for observations of 11 clusters of galaxies (Figure 21–10).

Who's right? Both the Sandage–Tammann method and the Aaronson–Huchra–Mould method

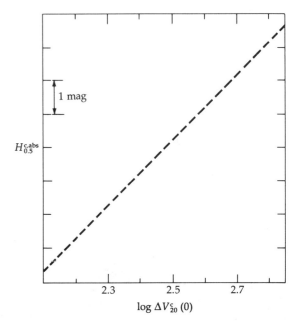

Figure 21–9

The infrared Tully–Fisher relationship. Plotted here are infrared absolute magnitude versus rotational velocity for more than 200 spiral galaxies. The line is the average of all values. *(Adapted from a diagram by M. Aaronson)*

probably contain unknown systematic errors, but we can say confidently that H lies in the range 50 to 100 km/s · Mpc. Several other investigations also give values in this range. A very recent one using supernovae as standard candles concluded that H is 68 to 76 km/s · Mpc (Figure 21–11).

Redshift, Distance, and the Age of the Universe

If H is known, we can infer a galaxy's distance from its redshift by using

$$r = zc/H \qquad \textbf{(21–4c)}$$

Then, for example, if $z = 0.1$,

$$r = (0.1)(3 \times 10^5 \text{ km/s})/(50 \text{ km/s} \cdot \text{Mpc})$$
$$= 600 \text{ Mpc}$$

or 300 Mpc if $H = 100$ km/s · Mpc. The percentage uncertainty in the distance equals the percentage uncertainty in H (a factor of 2 right now).

We emphasize that the simple relationship of Equation 21–4c holds only for nearby galaxies with small z. For values of z greater than about 0.8, cosmo-

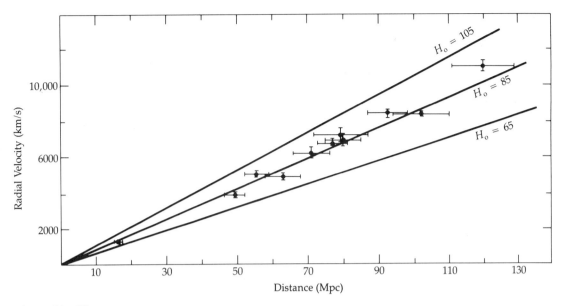

Figure 21–10

Hubble diagram for 11 galaxy clusters. These results are based on the infrared Tully–Fisher relationship. Shown is a range of values for the Hubble constant. *(Adapted from a diagram by M. Aaronson)*

Figure 21–11

Hubble constant versus year of measurement. Astronomers are beginning to come close to the actual value. *(Adapted from a diagram by D. N. Schramm)*

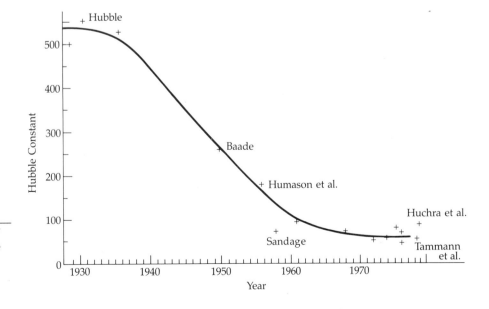

logical effects become important for converting red-shifts to distances. For a flat Universe, the proper relationship is

$$r = cz(1 + z/2)/H(1 + z)^2 \qquad \textbf{(21–4d)}$$

Now to compare Hubble's law, Equation 21–4b, with the distance–time relationship for motion at a constant speed. This relationship is

$$d = Vt$$

so that

$$t = d/V$$

Compared this with

$$1/H = d/V$$

which implies that

$$t = 1/H \qquad \textbf{(21–5)}$$

This t is the time since the expansion started, an "age" of the Universe—it is commonly called the *Hubble time*. Using a value of 50 km/s · Mpc for H and conversion factors to get everything in the same units, we obtain the value of $t = 1/H$ in years:

$$
\begin{aligned}
t = 1/H &= 1/(50 \text{ km/s} \cdot \text{Mpc}) \\
&= [1/(50 \text{ km/s} \cdot \text{Mpc})] \times (10^6 \text{ pc/Mpc}) \\
&\quad \times (3 \times 10^{13} \text{ km/pc}) \\
&= 6 \times 10^{17} \text{ s} \\
&= (6 \times 10^{17} \text{ s})/(3 \times 10^7 \text{ s/year}) \\
&= 2 \times 10^{10} \text{ years}
\end{aligned}
$$

Note that for $H = 100$ km/s · Mpc, t is only 10^{10} years, less than the inferred age of globular clusters.

(c) The Physical Meaning of the Cosmic Expansion

The Hubble time just calculated is the time in the past when all galaxies, if no acceleration has occurred in their motions, were jammed together at the beginning of the expansion. We call this event the *Big Bang*. Note that the Big Bang took place at a certain time, not at a certain place, because all parts of the Universe were close together at the Big Bang. Since then, the Universe has been expanding—but *not* into "empty" space: space *itself* expands as time passes. Galaxies simply serve as luminous markers of this expansion. Hence, the redshifts of galaxies, although commonly called Doppler shifts, are not really so. They arise from the different distances of our cosmic markers—galaxies—at different times in the history of the Universe.

Neither does the observed expansion imply that we lie at the center of the Universe. If the expansion is uniform, then an observer on another galaxy would observe the same Hubble's law, albeit with different units.

Finally, recognize that H is not really a constant; it must change with time because of the gravitational effects the galaxies have on each other. In general, H decreases as the Universe ages. The values of H mentioned in this chapter are for *now*, our epoch; they are usually written as H_0.

Problems

1. An approximate distance for the separation between our Galaxy and the Magellanic Clouds is 50 kpc. What will be the observed apparent magnitude for the following stars in the Magellanic Clouds:
   (a) an RR Lyrae variable with a period of 0.5 day
   (b) a classical Cepheid with a pulsation period of 100 days
   (c) a Population II Cepheid with a ten-day period

2. Elliptical galaxies are designated by Ex, where $x = 10(a - b)/a$ and the quantities a and b are the major and minor axes of the apparent elliptical disk of the galaxy. Draw a reasonably accurate picture of the observed shapes of E0, E2, E4, and E7 galaxies.

3. Our Galaxy and the Andromeda galaxy (M 31) are by far the most massive members of the Local Group. If these two giant galaxies form a *binary system* and move about one another in circular orbits, then calculate
   (a) the distance to the center of mass of the system from our Galaxy
   (b) the orbital period
   Perform the same computations for the orbit of the pair M 32 and M 31.

3. What methods of distance determination are most useful for finding the distance to
   (a) the Pleiades
   (b) a globular cluster in our Galaxy
   (c) the Large Magellanic Cloud
   (d) M 31, the Andromeda galaxy
   (e) the Virgo cluster of galaxies
   (f) the Hercules cluster of galaxies

(g) our Sun

(h) the nucleus of our Galaxy

5. Use the rotation curves in Figure 21–5 to calculate the masses of NGC 7664 and NGC 4378.

6. The Sun orbits around the Galaxy at a radius of about 8.5 kpc. In a mere 110 million years from now, the Sun will be on the other side of the Galaxy. The nearby galaxies will appear to shift in the sky relative to more distant background galaxies—a galactic parallax. What then would be the number of parsecs in a "gal-sec," the distance of a galaxy whose galactic parallax equals 1.0"?

7. The Ca II K stellar absorption line has a rest wavelength of 393.3 nm. In a particular galaxy, the Ca II K line is observed at a wavelength of 410.0 nm. What is the distance to the galaxy, assuming

(a) $H = 50$ km/s $\cdot$ Mpc

(b) $H = 100$ km/s $\cdot$ Mpc

8. How would the derived age of the Universe (Equation 21–5) change if

(a) the expansion is decelerating (H decreasing with time)

(b) the expansion is accelerating (H increasing with time)

(c) What observations could be made to determine if either of these possibilities is correct?

9. Most of the nearest galaxies do not obey Hubble's law. Explain why.

10. Space-based telescopes allow much higher resolution than ground-based optical telescopes. Describe how the availability of better resolution will refine the distance scale. Be specific.

Large-Scale Structure in the Universe

We so far have only hinted at the large-scale (many megaparsecs) layout of matter in the Universe. Recent observations have revolutionized that picture, changing it from one in which galaxies congregated in spherical clusters imbedded in a uniform background of galaxies to one in which chain-like superclusters snake among vast voids of space. This chapter deals with that structure, which may be the undistorted fossil of an earlier stage in the Universe's evolution.

22–1 Clusters of Galaxies

Most galaxies—maybe all—are members of some type of cluster. The Milky Way Galaxy is one of the dominant members of the Local Group of galaxies, which in turn is part of a local supercluster that also includes the rich Virgo cluster.

(a) Types of Clusters

There are different ways of defining categories of galaxy clusters, but the simplest is that of George Abell, who separates rich clusters into *regulars* and *irregulars*. ("Rich" means that more than 50 galaxies

fall within two magnitudes of the brightest galaxy and lie within a radius of 2 Mpc.) *Regular clusters* tend to be giant systems with spherical symmetry and a high degree of central condensation; they frequently contain many thousands of member galaxies, of which perhaps 1000 are brighter than absolute magnitude − 15. Almost all members of regular clusters are either elliptical or S0 galaxies, whereas *irregular clusters* contain a mixture of all types of galaxies. Among the irregular clusters of galaxies are included (1) small groups, such as our own Local Group, (2) loose aggregates of subgroups with several centers of condensation, and (3) fairly large but diffuse clusters. The *luminosity function for galaxies* in clusters (the number of galaxies per integrated magnitude interval) indicates that there is a great preponderance of *faint* galaxies.

(b) The Local Group of Galaxies

Our Milky Way Galaxy and the Andromeda galaxy (M 31) dominate the small group of galaxies referred to as the Local Group, which contains at least 20 members (Table 22–1). The other members are fainter and less massive and to some extent seem to be concentrated around one or the other of the two largest spiral galaxies. A wide range of galaxy types is

Table 22–1   Selected Members of the Local Group

Name	R.A. (1950)	Dec. (1950)	Type	M_v	Distance (kpc)	$m - M$	Mass ($M_\odot$)	Relative Velocity (km/s)
M 31 = NGC 224	00^{h}40^m	+41.0°	Sb	−21.1	690	24.6	3×10^{11}	−267
Galaxy	17 42	−29.0	Sb or Sc	−21	8.5	14.7	4×10^{11}	—
M 33 = NGC 598	01 31	+30.4	Sc	−18.9	690	24.6	4×10^{10}	−190
LMC*	05 24	−69.8	Irr I	−18.5	50	18.6	6×10^9	+275
SMC*	00 51	−73.2	Irr I	−16.8	60	19.1	1.5×10^9	+163
NGC 205	00 38	+41.4	E6p	−16.4	690	24.6	—	−239
M 32 = NGC 221	00 40	+40.6	E2	−16.4	690	24.6	2×10^9	−220
NGC 6822	19 42	−14.9	Irr I	−15.7	460	24.2	1.4×10^9	−34
NGC 185	00 36	+48.1	dE0	−15.1	690	24.5	—	−270
NGC 147	00 30	+48.2	dE4	−14.8	690	24.5	—	—
IC 1613	01 02	+01.8	Irr I	−14.8	740	24.5	4×10^8	−235
Fornax	02 37	−34.7	dE3	−13.0	188	21.4	2×10^7	−73
Sculptor	00 57	−34.0	dE3	−11.7	84	19.7	3×10^6	—
Leo I	10 06	+12.6	dE3	−11.0	220	21.8	3×10^6	—
Leo II	11 11	+22.4	dE0	−9.4	220	21.8	10^6	—
Ursa Minor	15 08	+67.3	dE6	−8.8	67	19.5	10^5	—
Draco	17 19	+58.0	dE3	−8.6	67	19.6	10^5	—

* LMC = Large Magellanic Cloud; SMC = Small Magellanic Cloud.

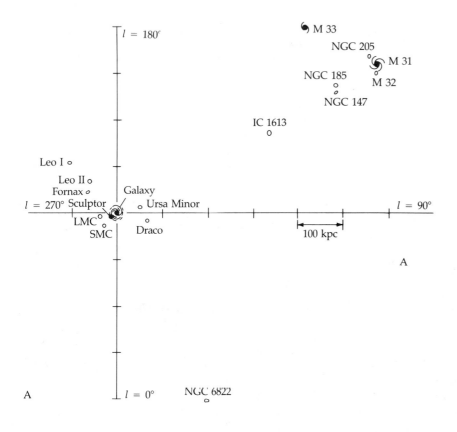

A

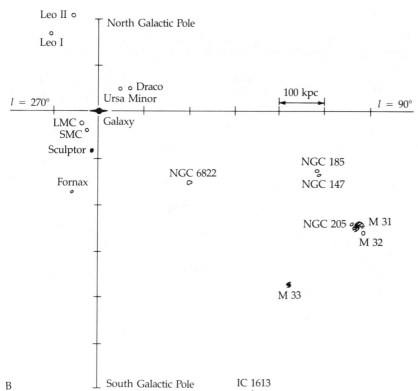

B

Figure 22–1

The Local Group. (A) A top view projected on the plane of our Galaxy. (B) A side view. Note the concentrations around our Galaxy and M 31.

included in the group, from the three major spirals (our Galaxy, M 31, and M 33) to the dwarf ellipticals and irregulars. Most of the galaxies are dwarfs with low mass and luminosity. The Local Group spans about 1 Mpc along its largest dimension (Figure 22–1).

The importance of the Local Group lies not only in the fact that it is the closest cluster but also in the fact that the study of its individual galaxies enables us to learn a great deal about the characteristics of galaxies; we may then use this knowledge to extrapolate to the more distant galaxies. Local Group members also serve as critical calibrators for the distance scale. Finally, we point out that, in comparison with other clusters (such as Virgo), the Local Group does not contain a large number of galaxies, nor are many of these very massive.

(c) Other Clusters of Galaxies

Other clusters range from compact groups to rather loose collections. The Fornax cluster, one relatively close to us, contains many types of galaxies even though the total number is only 16. The huge Coma cluster spreads over at least 7 Mpc of space and contains thousands of galaxies. From these observa-

tions, we know that a typical cluster contains about 100 galaxies brighter than $M = -16$ and is separated by tens of millions of lightyears from its neighboring clusters. Under the Abell classification, the *Coma cluster* (Figure 22–2) is a regular cluster. Two large, bright elliptical galaxies lie near the center, about which the others seem to concentrate; other rich, regular clusters, such as A 2199, are dominated by cD galaxies, which are supergiant, elliptically shaped galaxies with extensive halos. Examples of irregular clusters are the Local Group, the *Hercules cluster* (Figure 22–3), and the *Virgo cluster*. Of the 205 brightest galaxies in the Virgo cluster, the four brightest are giant ellipticals, but ellipticals make up only 19 percent of the total, whereas in the spirals, this figure is 68 percent. The Virgo cluster covers about 7° in the sky, which with its distance of 15.7 Mpc, implies that its diameter is some 3 Mpc.

(d) Clusters and the Galaxian Luminosity Function

The local stars have a luminosity function; that is, a certain number of stars can be found in a given range of luminosity or absolute magnitude. The basic trend is that there are many fewer extremely lumi-

Figure 22–2

Central region of the Coma cluster. *(National Optical Astronomy Observatories)*

Figure 22–3

Central region of the Hercules cluster. *(Palomar Observatory, California Institute of Technology)*

nous (OB) stars than low-luminosity (M V) stars [Section 14–2(a)]. A similar hierarchy exists for galaxies in clusters. In the Local Group, only three galaxies are most luminous (the Milky Way, M 31, and M 33); most galaxies in the Local Group are dwarf, low-luminosity galaxies. So galaxies have a luminosity function somewhat like that of stars.

Clusters provide a direct way of determining the luminosity function of galaxies, for you can see a wide range all at once. You can count the number of galaxies in specific apparent-magnitude ranges, from the brightest to the faintest. Because every galaxy in a cluster lies at about the same distance, a plot of number versus apparent magnitude translates into a plot of number versus absolute magnitude once you know the distance to the cluster. The main drawback here is that dwarf galaxies are undercounted because they are so faint at great distances. Abell determined the luminosity functions of a few clusters and found that the number of bright galaxies falls off very rap-

idly as the luminosity increases (Figure 22–4). This behavior tells us that there are many more low-luminosity galaxies in a cluster than high-luminosity ones.

Let's examine the concept of luminosity function in a bit more detail. We do so because if we know the luminosity function for clusters, then we can determine (1) the (unobserved) dwarf galaxy populations in clusters, (2) the mass distributions in clusters, and (3) the distances to clusters. A basic luminosity function gives the number of galaxies per luminosity (or magnitude) interval per unit volume of space. The exact expression of the function can take many forms. Paul Schechter has proposed an analytical form that seems useful and appropriate to a wide variety of samples of galaxies:

$$\Phi(L)dL = \Phi^*(L/L^*)^\alpha \exp(-L/L^*) \, d(L/L^*)$$

where $\Phi(L)$ is the density of galaxies in the range from L to $L + dL$, Φ^* is the normalization parameter,

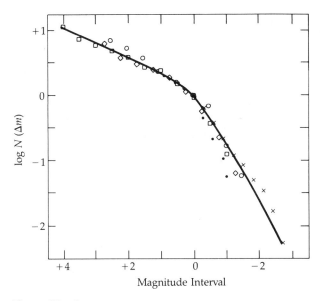

Figure 22–4

Luminosity function for galaxies. Number of galaxies in a magnitude interval versus magnitude interval, with an arbitrary zero. *(Adapted from a diagram by G. Abell)*

L^* is a characteristic luminosity at which the slope of the function changes rapidly (at $L_B = -19.4$), and α is the slope of a plot of $\log \Phi$ versus $\log L$ (for $L < L^*$). Typically, for rich clusters, $\alpha = -\frac{5}{4}$ and

$$\Phi^* = 0.005(H_0/50)^3/\text{Mpc}^3$$

See Figure 22–5 for a graph of such a function.

It's difficult to estimate the masses of these clusters. Not all the material in them can be seen, and so adding up all the observable galaxies gives a lower limit to a cluster's mass. On the other hand, if the cluster is assumed to be gravitationally bound, the motions of its members establish an upper limit on its mass. (The actual value lies between these two limits.) Masses range from 10^9 to 10^{15} solar masses. However, it is not possible to tell if all clusters are bound and stable or if they are unstable and expanding. If they are *not* stable, the mass estimates fall into the lower end of the range. A higher value brings up the *dark-matter* problem; if most of the mass is nonluminous, in what form does it reside within a cluster?

The result rests heavily on the application of the virial theorem:

$$[mv^2/2] = [GMm/2r] \qquad (22-1)$$

where m is the mass of an individual galaxy, v its

speed, r its distance from the center of the cluster, and M the total mass of the cluster. Applying this idea, astronomers come up with mass estimates that are an order of magnitude *higher* than that observed by summing up the masses of luminous galaxies. That is, the ratio M/L for clusters ranges from 200 to 400; in contrast, spiral galaxies have M/L ratios of about 100 (if they have massive halos), ellipticals around 10. Hence, the dark matter does not reside within the galaxies.

(e) Galactic Cannibalism?

A remarkable fact about clusters is that — relative to the sizes of the galaxies in them — the galaxies are spaced pretty close together. Compare galactic spacings with those of planets and stars. In the Solar System, planets are spaced out about 10^5 times their diameters. In the Galaxy, stars are spaced about 10^6 times their diameters. In a cluster, however, the spacing is only 100 times a typical galaxy's diameter.

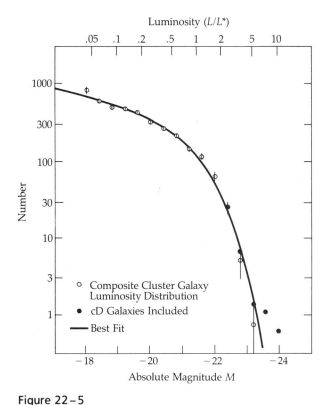

Figure 22–5

Schechter luminosity function. *[P. Schechter, Astrophysical Journal 203:297 (1976)]*

Now, if in relative terms galaxies are so close, consider the additional fact that the most massive galaxies (supergiant ellipticals) are at least 10^7 times more massive than the least massive ones (irregulars and dwarf ellipticals). Tidal forces could cause the largest galaxies to disrupt the smaller ones strongly enough to destroy their structure and then pull the pieces in. This devouring of a smaller galaxy by a larger one has been called *galactic cannibalism.*

What observations support this idea? Some point to supergiant elliptical galaxies (cD galaxies) as a special class of galaxies. Their peculiar properties include (1) extensive halos, up to 1 Mpc in diameter, (2) multiple nuclei (sometimes), and (3) location at the center of clusters. These observed properties plus theoretical calculations of the motions of these galaxies in clusters suggest that the cD galaxies result from galactic cannibalism, that is, from close encounters at the centers of clusters or the infall of material tidally stripped from other cluster members. Dynamical friction may also play a role. These processes all assume that the growing galaxies lie in the center of the gravitational potential well of a cluster, so that material freed from other galaxies congregates there.

Some observations support this notion: (1) cD

Figure 22–7

The galaxies M 82 (top) and M 81 (bottom). The activity of M 82 may be the result of a tidal interaction. (*National Optical Astronomy Observatories*)

galaxies *do* lie at the centers of clusters—within 200 kpc, which is about the diameter of a cD galaxy, and (2) photometry of the inner parts of cD galaxies indicates that they have the same properties as E galaxies. This observation backs up the scenario that cD galaxies were once E galaxies that added material to create their extensive halos.

At least 50 percent of cD galaxies have more than one nucleus. Might these be the leftovers from cannibalism? Recent observations suggest so. Dopplershift data indicate that the nuclei within a cD move at relative speeds of about 1000 km/s. In contrast, the stars within a cD galaxy move in orbits at about 300 km/s. Hence, the nuclei do not share the stellar dynamics in a cD galaxy. They are moving much faster, and their velocities carry them far away from the center of the galaxy. So the nuclei may be normal elliptical galaxies that have come close to and passed through the cD galaxy. Dynamical friction has decreased their orbital energies so that they are now bound to the cD galaxy.

Although galaxies may not actually merge, they certainly undergo close encounters and interact by tidal forces. Such interactions would have some gen-

Figure 22–6

Tidally interacting galaxies. NGC 4038 and 4039 are interacting by tidal forces, which result in the streaming tails. (*Palomar Observatory, California Institute of Technology*)

eral effects. First, as illustrated by the Earth's tidal bulges, matter would be pulled out in bulges on both sides of each galaxy. Second, because galaxies rotate, their material would conserve angular momentum after a tidal encounter and move off in arc-shaped streams. So we expect that tidal bridges may join two tidally interacting galaxies and tails may flow away from each in opposite directions. Have such interacting galaxies been seen? Many galaxies with peculiar shapes—those that do not fall into the standard Hubble form— show some of the characteristics of tidal interactions. An excellent example is the pair NGC 4038 and 4039 (Figure 22–6). Here is visible a bridge of material between the galaxies and tails heading off in opposite directions. Computer simulations of such an encounter show that similar structures should result from the gravitational interactions. The peculiar galaxy M 82 likely had its star formation triggered by a tidal interaction with M 81 (Figure 22–7).

22–2 Superclusters

Does the Universe have a higher level of organization than clusters of galaxies? Are there clusters of clusters—*superclusters?* For some years, astronomers were extremely skeptical about the reality of superclusters. Before observations drove home that reality, the standard picture was that of more or less spherical clusters embedded in a very uniform distribution of background, noncluster galaxies. That view has recently and rapidly changed to one in which superclusters have a strung-out, filamentary structure some 100 Mpc long. Between them lie vast voids, empty of luminous matter such as galaxies. The superclusters may be interconnected—the fundamental network of the Universe.

Some early workers gathered strong hints about superclustering. Donald Shane and Carl Wirtanen of Lick Observatory counted galaxies brighter than

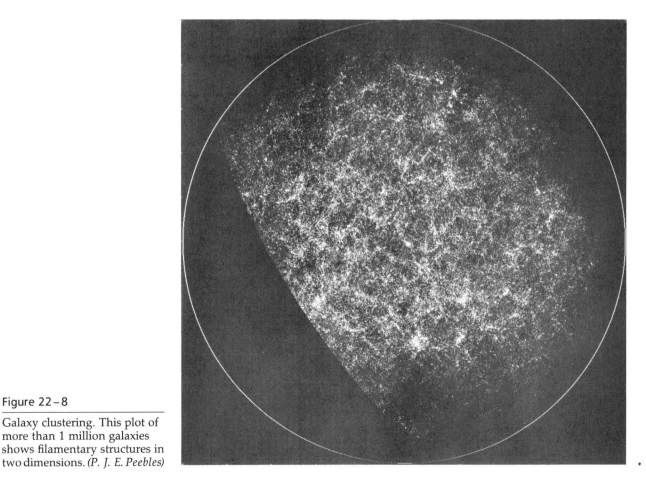

Figure 22–8

Galaxy clustering. This plot of more than 1 million galaxies shows filamentary structures in two dimensions. (*P. J. E. Peebles*)

magnitude 19 in $\frac{1}{6}°$ squares in the northern sky — some million galaxies in a survey that took 12 years. The map (Figure 22–8) shows that clusters of galaxies interconnect in a chain-like fashion — the explosive imprint of the Big Bang. (But note that this map is a two-dimensional projection of three-dimensional structure, so that the chains may not be real.) In the 1930s, Clyde Tombaugh (the discoverer of Pluto) made the first map of the Perseus supercluster (discussed below) in a survey of 2000 galaxies. Tombaugh showed the result to Edwin Hubble, who refused to believe that galaxies could be grouped in superclusters!

Both these pioneering efforts lacked an essential element — the third dimension, depth in space. We now use redshifts as the indicator of radial distance — a useful process even if the value of H_0 is not well known, for redshifts can be scaled by H_0. A simple way to do so is to define

$$h = H_0/(100 \text{ km/s} \cdot \text{Mpc})$$

and use h rather than H_0 to specify distances. Or, just use the redshifts z as a stand-in for distance. Modern work has relied on the observation of redshifts for large numbers of galaxies.

Much early work on superclustering was done by Gerard de Vaucouleurs. He found evidence for a *Local Supercluster* that has a diameter of 30 Mpc. It contains the Local Group and Virgo cluster, among others, for a total mass of 10^{15} solar masses. The Local Supercluster appears to be flattened (about 2 Mpc thick), which may imply that it is rotating, but it may simply have been formed this way. The center of mass lies in or near the Virgo cluster (which contains about 20 percent of all the galaxies in the Local Supercluster); our Galaxy, which lies at the edge of the Supercluster, moves at about 500 km/s about the apparent axis of rotation of the Supercluster.

Tully and Fisher have completed a three-dimensional map (Figure 22–9) of the Local Supercluster. (This project involved measuring the positions and redshifts of 2200 galaxies and took nine years to complete.) They discovered a rich, convoluted structure that breaks into two main clouds with streamers

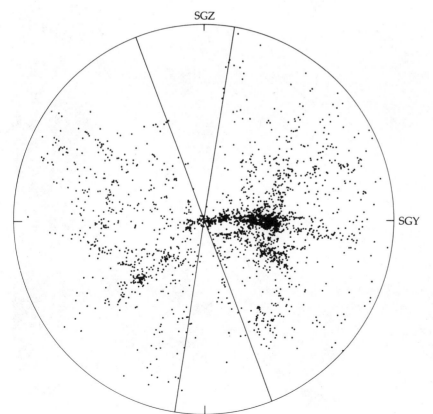

Figure 22–9

The Local Supercluster. This map is a top view that roughly corresponds to the plane of our Galaxy. Each dot represents one galaxy. The wedges indicate regions of the sky obscured by dust in our Galaxy. *[R. B. Tully, Astrophysical Journal 257:389 (1983)]*

—thin, cigar-shaped clouds—emerging above and below the central plane. Most of the Supercluster is empty space: 98 percent of the visible galaxies are contained in just 11 clouds that fill a mere 5 percent of the overall volume. Yet, the clouds do delineate a disk structure, with a width about 10 times its thickness —a cosmic pancake of clusters of galaxies. Note that distinct voids show up—regions empty of galaxies. These voids play a crucial role in theories of super-clustering.

The other well-studied superclusters are those in Perseus, Hercules, and Coma. These groupings show up most clearly in wedge diagrams, in which position on the sky (in one dimension) is plotted against red-shift (distance), with our Galaxy at the vertex of the wedge. This kind of map reveals filaments and voids visually. For example, the Coma supercluster (Figure 22–10) shows that two rich clusters (Coma = Abell 1656 and Abell 1367) link together. In front of them lies a void and a chain of smaller clusters. Behind

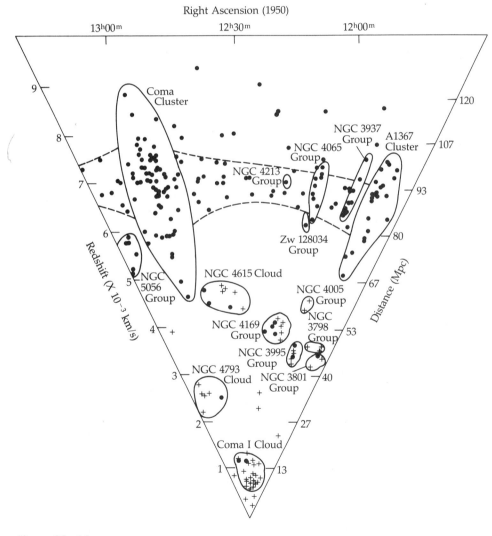

Figure 22–10

The Coma supercluster. This wedge diagram shows that the clusters in the supercluster are connected by bridges of galaxies. Note the void in front of the supercluster. *[S. A. Gregory and L. A. Thompson, Astrophysical Journal 222:784 (1978)]*

them appears a larger void. The Hercules supercluster (Figure 22–11) covers a volume of 60,000 Mpc³. It centers on a radial velocity of 11,000 km/s or 220 Mpc (if $H_0 = 50$ km/s · Mpc). The rich clusters Hercules (=Abell 2151) and Abell 2199 are connected in a band 100 Mpc thick. In the foreground appears a void also 100 Mpc deep. Generally, more giant elliptical galaxies are in the central regions of superclusters and more spiral galaxies are in the connecting filaments.

Overall, superclusters represent the largest-scale structures we can see in the Universe. Because the times it takes for clusters to move across their superclusters are greater than the age of the Universe, superclusters are not well mixed (compared with, for instance, the stars in globular clusters). Consider, for example, the Local Supercluster, which has a major axis of 40 Mpc and a velocity distribution of a few hundred kilometers per second. The crossing time is roughly

$$t_c \approx (40 \times 10^6 \text{ pc})(3 \times 10^{14} \text{ km/pc})/(300 \text{ km/s})$$
$$\approx (120 \times 10^{20} \text{ km})/(300 \text{ km/s})$$
$$\approx 4 \times 10^{19} \text{ s}$$
$$\approx 10^{12} \text{ years}$$

compared with 10^{10} years for the age of the Universe. This fact leads to an important inference: superclus-

ters may be the fossils of the time of galaxy formation. They may even contain information about primordial density fluctuations in the early Universe (Chapter 24).

This discussion of superclusters raises a fundamental question: how do we recognize a supercluster when the galaxies appear without depth on the sky? One way to tackle this issue is by a statistical approach that smooths out fluctuations that result from different distances to galaxies. We want to examine to what extent a clump of galaxies stands out relative to the average over the whole sky (or a large part of it). A *two-point correlation function* $Z(r)$ is defined by

$$n(r) = [1 + Z(r)]\langle n \rangle$$

where $n(r)$ is the space number density of galaxies within a distance r from any given galaxy and $\langle n \rangle$ is the average density of galaxies in the entire sky or in the region being studied. Then $Z(r)$ gives the probability of finding two galaxies separated by distance r or less. Generally, $Z(r)$ has the form

$$Z(r) = (r/r_0)^{-g}$$

where r_0, the *clustering length,* is the distance over which the density of galaxies is twice the general average. Over small regions of the sky, $g = 1.77$ and $r_0 \approx 10$ Mpc.

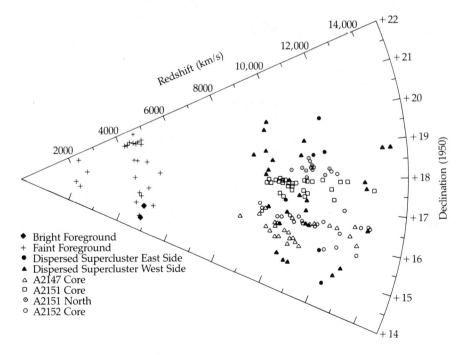

Bright Foreground
+ Faint Foreground
● Dispersed Supercluster East Side
▲ Dispersed Supercluster West Side
△ A2147 Core
□ A2151 Core
⊙ A2151 North
○ A2152 Core

Figure 22–11

The Hercules supercluster. This wedge diagram shows the large void between the supercluster and foreground galaxies. *[M. Tarenghi, W. G. Tifft, G. Chincarini, H. J. Rood, and L. A. Thompson, Astrophysical Journal 234:793 (1979)]*

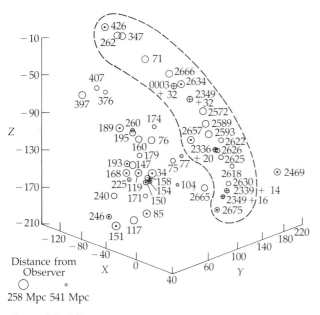

Figure 22–12

Perseus–Pegasus supercluster filament. Circles show the location of Abell clusters projected against the sky. The dashed contour outlines the filament. (*J. Burns and D. Batuski*)

Recently, Jack Burns and David Batuski of the University of New Mexico prepared a catalog of supercluster candidates, using rich clusters as a tracer. One of these candidates (Figure 22–12) stretches more than 300 Mpc (for $H_0 = 75$ km/s · Mpc) in the Perseus–Pegasus region of the sky.

22–3 Intergalactic Matter

Is intergalactic space empty, or is there an intergalactic medium similar to the interstellar medium? If an intergalactic medium is present, it may contain both gas and dust. The gas (probably hydrogen) may be neutral or ionized. We can look for the intergalactic medium in two locations: *between* clusters of galaxies and *within* clusters.

Consider the first possible location. Let's examine the possibility of intergalactic dust. Such dust, if it resembles the interstellar dust in our Galaxy, would extinguish and redden the light from distant galaxies. This extinction and reddening effect has been searched for but not found. It is less than 4×10^{-4}

magnitudes per megaparsec. So intergalactic space cannot contain very much dust; the density must be less than 4×10^{-30} kg/m³. In other words, intergalactic dust is optically thin over the distances between clusters of galaxies ($\tau_d < 1$).

How to detect neutral hydrogen? Hydrogen atoms absorb ultraviolet radiation well, especially at 121.6 nm, the Lyman alpha absorption. Such ultraviolet absorption has been sought in the spectra of distant objects at both small and large redshifts. It has not been detected. This lack of ultraviolet absorption implies that neutral hydrogen cannot have a density greater than about 10^{-9} atom/m³. So if hydrogen is there, it must be ionized because H II is far more transparent than H I. These observations probe the Universe's past, and so they also imply that any intergalactic gas must have remained highly ionized for most of the history of the Universe.

These arguments leave ionized hydrogen (H II) as the most likely candidate for the intergalactic medium. Because intergalactic material would not have a high density, ionized hydrogen would take a very long time to find an electron and recombine. Unfortunately, detecting a low-density ionized gas is a tough job. If it is hot (a few tens of millions of kelvins), you can expect X-ray or ultraviolet emission. X-ray observations of local superclusters show 15 sources that are probably clustered in seven superclusters. The sources consist of spots centered in rich clusters; this implies that hot gas in the superclusters is highly clumped.

Recent X-ray observations back up this idea (Figure 22–13). To date, at least 40 clusters of galaxies are known to emit X-rays. The X-ray luminosities of clusters range from 10^{36} to 10^{38} W. The sizes of the X-ray-emitting cores range from 50 kpc to 1.5 Mpc. The richer clusters tend to be the more luminous in X-rays. A reasonably confirmed model for this X-ray emission is that it comes from hot, ionized gas. This model requires typical temperatures of 10 to 100 million K and densities of about 1000 ions/m³ to explain the X-ray observations. So we have reasonable evidence of intergalactic gas in clusters, about equal to the amount of mass in the galaxies themselves. We have no evidence so far for much gas *between* clusters.

Finally, we note that for plasmas with temperatures between 10^4 and 10^7 K, trace elements dominate the emission and produce emission lines, which can be used to deduce the physical properties of the

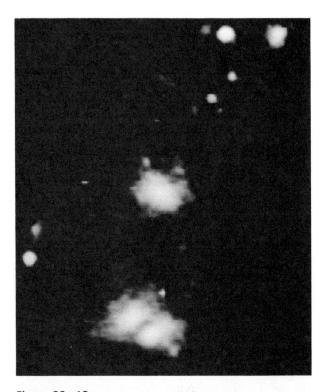

Figure 22–13

X-ray emission from superclusters. The cluster at the center is about 6 Mpc from the double cluster below it. (*C. Jones and W. Foreman*)

emitting gas. In particular, inner-shell transitions of neon, silicon, iron, sulfur, and argon can be very strong. X-ray spectroscopy of the emission of the gas from clusters shows highly ionized iron lines in almost all cases; theoretical models of the emission requires iron abundances (relative to hydrogen) about half that of the Sun. Hence, the intergalactic gas must have been processed through stars and then removed from the galaxies (perhaps by supernova explosions).

22–4 The Missing (?) Mass

As you will find out in detail in Chapter 24, Einstein's general theory of relativity predicts that for a certain average density of the Universe—around 5×10^{-27} kg/m³ for $H_0 = 50$ km/s · Mpc—the geometry of the cosmos is transitional between open and closed. In comparison, the known density of *lu-*

minous matter amounts to about 10 percent of this critical value. If for other reasons you believe that the Universe has a closed geometry, then the difference between this known density and the critical one raises the issue of the *missing mass*. (Of course, the mass is not missing if the Universe is open.) Let's presume that the mass is missing and see where it might hide. (Note that clusters within the Universe may have missing mass even if the Universe is open.)

A useful quantity to use in examining the missing mass is the mass-to-light ratio M/L, where both are in solar units. In the Solar System, $M/L \approx 1$. For a closed universe, $M/L \approx 700$ (for $H_0 = 50$ km/s · Mpc). For spiral galaxies with flat rotation curves, $M/L = 100$ to 300; for ellipticals, $M/L \approx 10$. For a cD galaxy, M/L increases with distance from the core and can exceed 100. Hence, a cluster containing just these types of galaxies would not have a density sufficient to close the Universe. However, clusters initiated the missing-mass problem, for masses inferred from the virial theorem (correct *only if* a cluster is gravitationally bound) give $M/L \approx$ a few hundred. For the core regions of clusters, a typical value of M/L is 300.

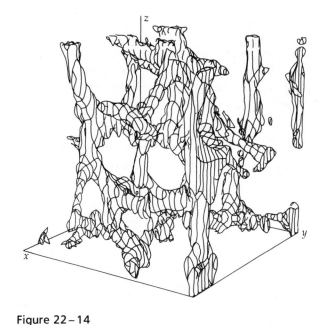

Figure 22–14

Simulation of superclustering. This theoretical calculation shows the clumping of dark matter, which ends up in filaments. (*Adapted from a diagram by J. Centrella and A. Melott*)

How much does hot gas in a cluster contribute to the value of M/L? For dynamically evolved clusters, the hot gas has relaxed into a spherically symmetrical distribution, usually centered on a cD galaxy that marks the bottom of the gravitational potential well of the cluster. For these, the X-ray emission traces out the total mass distribution, and the mass of hot gas can be estimated from the X-ray luminosity (if the distance to the cluster is known). Such an analysis gives $M_{gas}/M_{total} \approx 0.1$ and $M_{gas}/L \approx 10$; hence, the hot gas does *not* contribute enough to account for the missing mass in clusters.

That leaves superclusters as the hideouts for the dark matter. Unfortunately, because they are unrelaxed, we cannot apply the virial theorem. Counting galaxies will not work. If superclusters are bound by gravity, however, it will take a considerable amount of dark matter to do so. Basically, dark matter can come in two forms: baryonic (ordinary heavy parti-

cles, such as protons and neutrons) and nonbaryonic (everything else, such as photons and neutrinos). Baryonic dark matter could reside in quantity in planets, black dwarfs, neutron stars, or black holes, but none of these appears sufficient to bind clusters or superclusters. That leaves the subatomic leptons, such as neutrinos (which may have mass) and exotic types (such as magnetic monopoles).

Computer simulations of the clumping of matter after the Big Bang support the idea of nonbaryonic dark matter. For instance, massive neutrinos would clump into pancake-like shapes hundreds of millions of parsecs in diameter. At their boundaries, knots and strings of ordinary matter can condense (Figure 22–14). Hence, galaxies would form in long chains (superclusters?) with large, less-dense, pancake-shaped areas (voids?) between them. Baryonic matter of its own accord does not gather in such a distribution, as shown by theoretical numerical calculations.

Problems

1. Calculate the crossing time for the Hercules supercluster and compare it with the age of the Universe ($\approx 15 \times 10^9$ years).

2. **(a)** Demonstrate that a low-density hydrogen gas at 10^7 K produces X-rays. What is the peak wavelength of emission?
   (b) Compare the mass of the hot gas in a cluster with the mass contained in galaxies.

3. We state in the chapter that intergalactic space must have a dust density of less than 4×10^{-30} kg/m³. Assume that ten times this amount really exists. What would the interstellar extinction be, in magnitudes per megaparsec? How would that affect our ability to see the Virgo cluster?

4. How close does a spiral galaxy like the Milky Way have to get to a cD galaxy to be tidally disrupted?

5. Estimate the free-fall time for a cluster of galaxies and compare your result with the age of the Universe ($\approx 15 \times 10^9$ years). What do you conclude?

6. An approximation to the luminosity function of galaxies in rich clusters is $\Phi(L) = \Phi^*(L/L^*)^{-5/4}$ for $L < L^*$ and $\Phi(L) = 0$ for $L > L^*$, where Φ^* is $0.005/\text{Mpc}^3$ for $H = 50$ km/s · Mpc, $L^* \approx (10^{10})L_\odot$, and $L_{min} \approx$

$(10^3)L_\odot$. The number of galaxies per cubic megaparsec is

$$N_{gal} = \int_{L_{min}}^{L^*} \Phi(L)dL$$

and the total luminosity per cubic megaparsec is

$$l_{gal} = \int_{L_{min}}^{L^*} \Phi(L)L \, dL$$

   (a) Show that most galaxies have luminosities near the low end of the range.
   (b) Show that most of the luminosity of a cluster is produced in the higher-mass galaxies.

7. Only the radial velocity is observable for galaxies that are members of clusters. By integrating to obtain the mean value, determine how $\langle v^2 \rangle$ is related to $\langle v_r^2 \rangle$.

8. Discuss how the existence of superclusters affects the Hubble diagram. What are the dangers in deriving the Hubble constant using galaxies within the Local (Virgo) Supercluster?

9. Discuss what happens to the kinetic energy of a galaxy that is cannibalized by a central cD galaxy. Relate this to the extended haloes of cD galaxies.

Active
Galaxies
and
Quasars

We turn now from the placid world of ordinary galaxies to the explosive realm of active ones. As technology advanced the power of radio astronomy, we discovered that many (most?) distant galaxies emit radio waves with luminosities much greater than those of our Galaxy and others like it. X-ray telescopes then revealed that strong radio emitters usually produced high-energy photons with large total energies. This output is largely nonthermal, so that it cannot be explained simply as the combined emission of a group of stars.

This chapter examines the power and the glory of active galaxies and quasars in the distant Universe — galaxies that may contain supermassive black holes in their cores to supply their enormous energy outputs.

23–1 Active Galaxies

Galaxies that have high luminosities and nonthermal spectra (in whole or in part) make up the class of *active galaxies.* The term *active* is used in contrast to *normal,* with our Galaxy taken as a typically normal spiral galaxy. As a group, active galaxies show all or most of the following characteristics: (1) high luminosity, greater than 10^{37} W; (2) nonthermal emission, with excessive ultraviolet, infrared, radio, and X-ray flux (compared with normal galaxies); (3) a small region of rapid variability (a few lightyears across at most); (4) high contrast of brightness in the nucleus and large-scale structures; (5) explosive appearance or jet-like protuberances; and (6) broad emission lines (sometimes). The nucleus of the Milky Way Galaxy has some of these characteristics, but it does not generate as much energy in total (some 10^{35} W) as do the nuclei of active galaxies.

(a) Radio Galaxies

A majority of the spiral galaxies of apparent visual magnitude 11 or brighter radiate in the radio continuum; the typical amount of this radiation appears to be at about 10^{33} W. A few extragalactic radio sources are exceptionally strong radio emitters, with some of them generating energies in excess of 10^{37} W. We use the term *radio galaxy* for those galaxies with a radio luminosity greater than 10^{33} W. This is not to say that the "weak" sources are all normal, but rather

that normal galaxies are relatively weak radio sources.

Radio galaxies fall into two types: *compact* and *extended. Extended* means that the radio emission is larger than the optical image of the galaxy; *compact* means it is the same size or smaller. Compact radio galaxies often display *very* small (usually nuclear) radio sources, often no more than a few lightyears in diameter. Extended radio sources, in contrast, sometimes show a double structure of two gigantic lobes separated by distances of megaparsecs and symmetrically placed on opposite sides of the nucleus. In such sources, often called *classical doubles,* the two radio components are well separated from the optical galaxy and are much larger. As a rule, these sources lie on a line, with the galaxy at about the center of the radio-emission pattern. Typically, the nucleus is also a radio source.

M 87 is a typical radio galaxy (Figure 23–1). A giant elliptical galaxy, it dominates the Virgo cluster of galaxies and so lies about 20 Mpc away. One radio

Figure 23–1

The active elliptical galaxy M 87, a nearby radio galaxy. *(National Optical Astronomy Observatories)*

source only 1.5 light*months* in diameter appears in the core of M 87 along with a group of other compact radio sources. Poking out from the core, an optically visible jet (Figure 23–2A) extends over a length of some 6000 lightyears. This jet has a luminosity of roughly 10^{34} W; its emission is polarized. A detailed photograph shows that the jet contains at least six blobs of material, each no more than a few tens of lightyears across (Figure 23–2B). Over 22 years, the blobs have changed significantly in intensity and polarization.

M 87 also emits X-rays with about 50 times more energy than its optical emission, about 5×10^{35} W in X-rays from the whole galaxy. The jet itself is also

known to emit X-rays. High-resolution observations with the Einstein Observatory show that the jet contains knots that emit X-rays strongly. The VLA has mapped the M 87 jet in detail and confirmed that its radio emission coincides with the optical and X-ray emissions and extends into one radio lobe (Figure 23–3). So the jet overall emits over a wide range of frequencies, from radio to X-rays, and each knot of the jet generates this same spectrum of energies. Recent infrared observations have demonstrated a break in the spectrum's slope at about 600 nm. The emission is nonthermal both before and after the break.

In general, the flux, F, of nonthermal emission

A

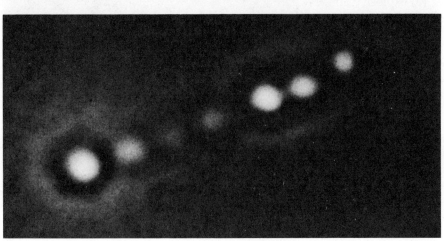

B

Figure 23–2

The jet of M 87. (A) Photograph of galaxy and jet exposed for maximum detail (negative print). (B) Computer-processed image to bring out the structure in the jet. *(H. Arp and J. Lorre)*

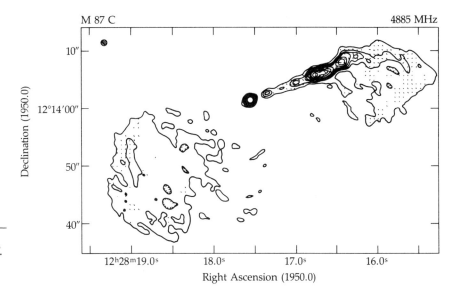

Figure 23-3

Radio emission from M 87. A VLA map made at 5 GHz; note the jet. *(F. Owen and P. Hardee, National Radio Astronomy Observatory)*

has the spectral form

$$F(v) = F_0 v^{-\alpha}$$

so that

$$\log F(v) = -\alpha \log v + \text{constant}$$

Hence, on a log–log plot of the spectrum, the *spectral index* α is simply the slope. For the M 87 knots, $\alpha \approx$ 0.6 before the break (in the radio) but 1.7 after the break (in the optical). For synchrotron emission from relativistic electrons, such a sharp change in spectral index probably results from a sudden change in the electron energy distribution (of the accelerated electrons). The average magnetic field strength in the knots is about 10^{-7} T.

Other elliptical galaxies also possess nuclear jets. In fact, radio jets are common — almost all radio galaxies in the lowest luminosity range have them.

Extended radio galaxies are commonly double, with the lobes lined up with the galaxy's center. These radio clouds are huge: most are 50 to 1000 kpc in diameter. When classified by structure, extended radio galaxies fall into three main groups: (1) classical doubles (example, Cygnus A), with highest luminosities, lobes aligned through center of galaxy, and bright hot spots at ends; (2) wide-angle tails, or bent tails (example, Centaurus A), with intermediate luminosities, a bend through the nucleus, and tail-like protrusions; (3) narrow-tail sources (example, NGC 1265), with lowest luminosities, U shapes, and rap-idly moving galaxies in a cluster. (Most extended radio galaxies fall into this last group.)

Cygnus A, one of the strongest radio sources in the sky and one of the first discovered, provides an excellent example of the classical-double structure typical of an extended radio galaxy. Its radio output, 10^{38} W, comes from two gigantic lobes set on opposite sides of the optical galaxy (Figure 23–4). Each lobe has a diameter of 17 kpc, and they extend roughly 50 kpc away from the central galaxy. Each lobe contains a cloud of energetic electrons and magnetic fields that stores the 10^{53} J needed to account for the radio luminosity lasting 10^7 to 10^9 years. The central galaxy of Cygnus A is an elliptical with a dust lane down its middle (Figure 23–4). It has an active nuclear region (perhaps with two nuclei), with a spectrum showing emission lines and a synchrotron continuum. Beyond 8 kpc from the center, the spectrum is just that of a mix of stars.

The VLA has produced a remarkable radio image of Cyg A (Figure 23–5) that shows a needle-like jet connecting the elliptical galaxy to one of its lobes. The radio map details a wispy structure in the lobes with small, bright spots of emission. Bright patches appear along the jet, which clearly holds its structure through most of the lobe. The filamentary structure of the lobes implies that the emitting gas occupies only 3 to 30 percent of their overall volumes. Within the lobes lie "hot spots" of intense radio emission.

Centaurus A (NGC 5128) is another extended

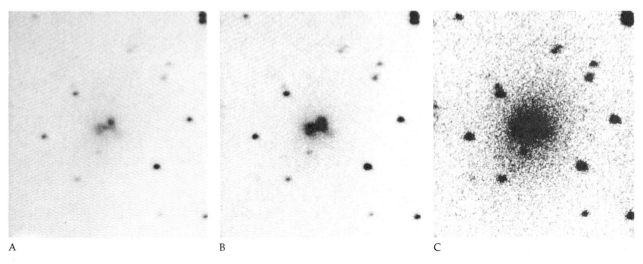

Figure 23–4

CCD images of Cygnus A. Made under conditions of excellent seeing, these images show (A) brighter features in the core, (B) lower intensity levels resembling previous photographs, and (C) lowest intensity showing the galaxy's halo. *(L. A. Thompson)*

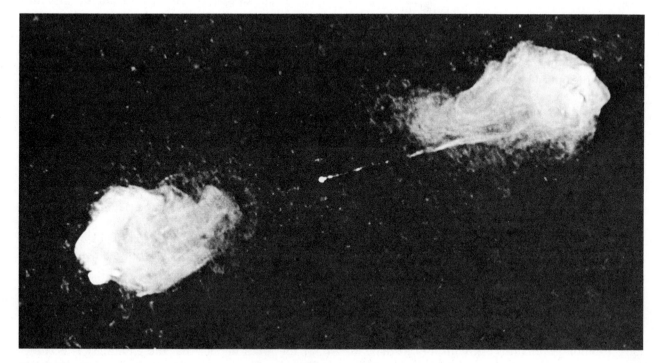

Figure 23–5

Radio map of Cygnus A. This extraordinary image shows the filamentary structure in the lobes and the thin jet to the right lobe. Made map with the VLA. *(R. A. Perley, J. W. Dreher, and J. J. Cowan, National Radio Astronomy Observatory)*

radio source, somewhat similar to Cygnus A. It is an E2 galaxy bisected by an irregular dust lane. At a distance of 4 Mpc, Centaurus A is the closest active galaxy; it almost outshines M 33 visually. Viewed with a radio telescope, Centaurus A has two huge outer lobes, 200 kpc and 400 kpc in diameter. The lobes form a bend through the nucleus and have tail-like protrusions. Closer in, another pair of radio lobes sits on the edges of the optical galaxy; each lobe is about 10 kpc in diameter. The inner and outer lobes have almost but not exactly the same alignment.

The nucleus of Centaurus A has a direct connection to one of the inner radio lobes. The Einstein X-Ray Observatory detected an X-ray-emitting jet streaming northeast from the nucleus and consisting of at least seven distinct blobs. A VLA map at a wavelength of 20 cm shows radio emission along a jet that extends to one of the nuclear radio lobes (Figure 23–6). The jet has a blob-like structure that coincides with the X-ray emitting blobs. So Centaurus A and M 87 look similar in that both have nuclear jets that emit radio waves and X-rays. New high-resolution (24 pc by 7 pc beam size) VLA observations of Cen A

Figure 23–6

Nuclear region of Centaurus A. A VLA map showing the nuclear emission and the inner lobes with a jet to the left-hand lobe is superimposed on an optical photograph. (*J. Burns, National Radio Astronomy Observatory*)

have revealed new structural features in the inner 700 pc of the jet (Figure 23–7). The radio emission shows limb brightening on alternating sides—in contrast to the center brightening of the M 87 jet. Thin filaments of radio emission emanate from the knots; they appear to point "downstream" in the direction of presumed matter flow along the jet (from the nucleus out to the lobe).

Many radio galaxies have emission in the form of lobes or streams that extend far beyond the visible galaxy. The lobes may be a few million lightyears apart and thousands of lightyears across. The vexing problem with these extended radio lobes is the vast amount of energy they contain: a typical lobe luminosity is 10^{36} to 10^{37} W, whereas the visible elliptical galaxy with which it is associated may emit only 10^{35} W. If the emission is synchrotron radiation, the lobes must be energy reservoirs of tangled magnetic fields (strengths about 1 nT) and high-speed electrons. Then a typical lobe contains more than 10^{52} J. The jets are the channels through which the nucleus conducts particles and energy to the lobes. The lobes and jets emit nonthermally, which suggests that the synchrotron process is operating. The nucleus provides the high-energy electrons. These are expelled either as a fairly constant beam of particles or as a sequence of ionized blobs thrown out along a magnetic field so that half the particles fly in one direction and the other half fly in the opposite direction. If the nuclear machine is active and stable, extended lobes of ionized material build up at the end of the jets. Repeated bursts from the nucleus can account for variability. The jets imply that high-speed electrons are channeled, perhaps in bursts, from the nucleus into the circumgalactic medium, where they pile up to form a lobe.

Extended radio galaxies show a bending sequence (Figure 23–8) from linear classical doubles to nuclear emission bunched up at one end of a tail. This sequence strongly implies that clusters of galaxies contain a hot, ionized gas. Imagine that a galaxy moving rapidly through this medium shoots out material (high-speed electrons, for instance) in a jet. The material flowing out of the galaxy is decelerated by the intracluster medium, and the moving galaxy leaves it behind. As the galaxy travels along, it leaves behind a radio-visible trail—a fossil record of where it's been (Figure 23-9). The bending sequence shows that a dense gas exists in clusters. The discovery of head–tail galaxies prompted the acceptance of intra-

Figure 23–7

High-resolution image of the Cen A jet. This map shows the nucleus and all 700 pc of the jet. Note the limb brightening on alternating sides. *(J. Burns, D. Clarke, E. D. Feigelson, and E. J. Schreier, National Radio Astronomy Observatory)*

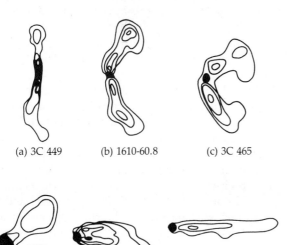

(a) 3C 449   (b) 1610-60.8   (c) 3C 465

(d) IC 708   (e) 3C 83.18   (f) IC 310

Figure 23–8

Bending sequence for radio galaxies. The position of the tails is influenced by the speed of the galaxy through the local intergalactic medium and by the density of that medium. *(Adapted from a diagram by G. Miley)*

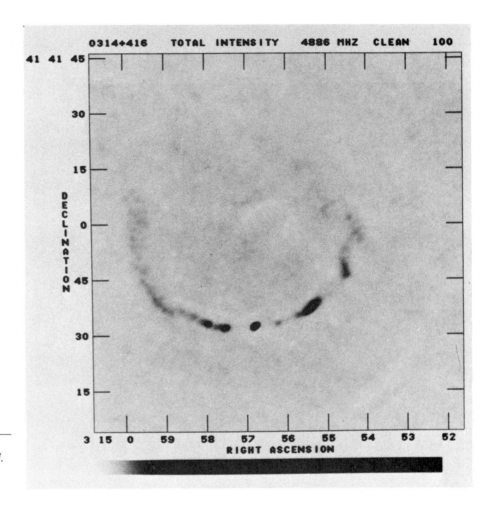

Figure 23–9

VLA map of the head–tail galaxy NGC 1265. *(F. Owen, J. Burns, L. Rudnick, National Radio Astronomy Observatory)*

cluster gas *before* X-ray observations confirmed its existence (at a density of 10^{-24} to 10^{-27} kg/m³) and showed it to be very hot (about 10^7 K).

To sum up: as more extensive observations are made of radio galaxies, it appears that at least 50 percent of the classical doubles (which also have relatively high luminosities) show jets, which tend to be one-sided. About 80 to 90 percent of lower-luminosity radio galaxies exhibit jets—perhaps *all* of them do. The jets are typically two-sided, clumpy, and well-aligned for distances up to kiloparsecs. The physics of these cosmic radio jets is a critical, unsolved problem in modern astrophysics.

(b) Seyfert Galaxies

In 1943, Carl Seyfert identified six spiral galaxies with unusual, broad emission lines in their spectra. When viewed optically, such galaxies have unusually

bright nuclei (Figure 23–10). They are now called *Seyfert galaxies;* some 90 are known to date. Seyfert galaxies, which are close by, provide a possible clue to the nature of active galaxies—a clue that comes from their broad emission lines and their partially non-thermal, synchrotron continuum.

We interpret the widths of the emission lines as coming from Doppler shifts produced by motions in the emitting gas. Observing the galaxy, we see some blueshifted photons from gas moving toward us, some redshifted photons from gas moving away in the other direction, and some unshifted radiation from gas at rest with respect to the average velocity of the galaxy. Taken together, the multitude of shifted lines blend into one broad line. Other galaxies have emission lines with Doppler widths equivalent to a few hundred kilometers per second, but the lines from a Seyfert galaxy have Doppler widths equiva-

Figure 23–10

The Seyfert galaxy NGC 1295 in Perseus. *(National Optical Astronomy Observatories)*

lent to a few *thousand* kilometers per second, 10 times broader (Figure 23–11). Note that these widths cannot arise simply from thermal broadening; that would require temperatures of about 10^8 K, which would ionize the gas so highly that it would not produce the observed emission lines.

Seyfert galaxies are almost always spirals. A detailed survey of 80 Seyferts finds that only 5 to 10 percent might be ellipticals. (The small angular size of some Seyferts makes it hard to classify them.) Compare this with the fact that extended radio galaxies are all ellipticals. Overall, about 1 percent of all spiral galaxies (ordinary and barred) are Seyferts. Perhaps all spiral galaxies go through a Seyfert phase, with only 1 percent active at any time.

Seyferts fall into two classes, based on their spectra. Class 1 Seyferts have strong, wide Balmer lines but narrow forbidden lines of other elements, such as those from [O III]. Class 2 Seyferts have Balmer lines and [O III] lines of the same width, both a bit wider than the [O III] lines of Class 1 Seyferts. The use of the terms *wide* and *narrow* here is relative. For example, the average Balmer line width for 40 Seyfert 1 galaxies is 3400 km/s. Their other lines have Doppler widths of only 200 to 500 km/s. In some Seyfert 1

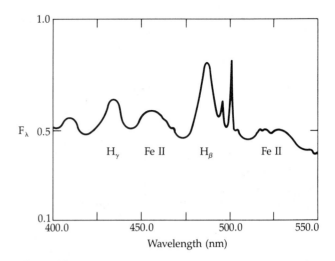

Figure 23–11

Broad emission lines from a Seyert galaxy. This spectrum shows the very broad emission lines in the Seyfert galaxy 1 Markarian 376. *(D. Osterbrock, Lick Observatory)*

galaxies, the Balmer lines have Doppler widths as great as 10^4 km/s. In contrast, the lines in Seyfert 2 galaxies are only 500 to 1000 km/s wide.

As in H II regions, the Balmer lines in a Seyfert are produced by recombination of hydrogen ions with electrons, which then cascade to lower levels. What makes the lines so broad? The simplest explanation is filaments or clouds of gas moving at speeds of a few thousand kilometers per second. To explain the Balmer emission from a Seyfert 1 requires a few tens to thousands of solar masses of ionized gas in the nucleus at densities of 10^{13} to 10^{15} ions/m³ moving at such speeds. The narrow line emission then might come from a large, surrounding volume of lower-density gas moving at lower speeds.

What is the energy source in the nucleus that generates the synchrotron emission, ionizes the gas, and drives its motions? Also in the nucleus must reside a source of high-energy electrons and gas. Observations indicate that the nuclei of Seyferts are small, only a few lightyears in diameter. Gas moving at 10^4 km/s would flow across such a tiny nucleus in only 100 years. So the gas must be replaced as it flows out.

To sum up, Seyferts are mostly spiral galaxies with extraordinary features. The most prominent are (1) extremely small and bright nuclei; (2) nuclear spectra that show emission lines not usually seen in the spectra of spiral galaxies; (3) very wide emission lines, implying gas motions of 500 to 4000 km/s; and (4) compact, low-luminosity radio sources within. (This last is a feature of only some Seyferts.)

In addition, the continuous spectra of Seyferts have a combination of stellar, nonthermal, and infra-red (from dust) radiation. The total energy output is 10^{37} to 10^{38} W. Their luminosities vary, but only by small amounts, over time spans of a few days to a few months. Seyferts tend *not* to be strong radio sources; sensitive radio surveys have detected only about half of the Seyferts examined.

A recent survey of spiral galaxies reached a curious conclusion with respect to Seyferts: most of them tend to be in close, binary galactic systems. Tidal interactions may then induce the Seyfert phenomena for some short period of time.

(c) BL Lacertae Objects

One other type of active galaxy is named after its prototype, BL Lacertae, and so these galaxies are called *BL Lac objects*. As a group, the BL Lac objects have most of the following characteristics: (1) rapid variability at radio, infrared, and visual wavelengths, (2) *no* emission lines, (3) nonthermal continuous radiation (Figure 23–12) with most of the energy emitted

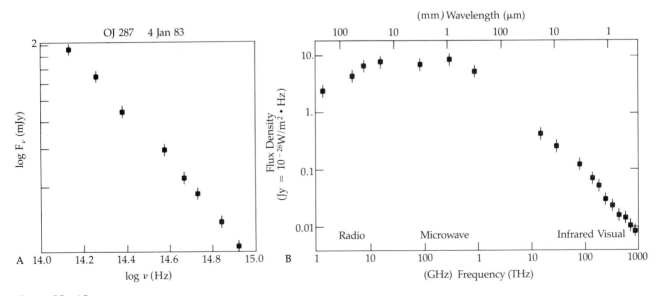

Figure 23–12

Spectra of OJ 287. (A) Infrared–optical continuum. *(P. Smith)* (B) Radio–optical spectrum in March 1983. *(T. Balonek)*

in the infrared, and (4) strong and rapidly varying polarization. Also, BL Lac objects generally have a star-like appearance—structure is rarely visible. The radio emission from many BL Lac objects is compact or only slightly extended. The extended radio structure is weak (only a few percent of the total) in contrast to the intense emission from the nucleus.

The greatest difference between BL Lac objects and other active galaxies is that the emissions of the former vary so frequently and erratically (Figure 23–13). For example, BL Lac itself, recorded on a long sequence of Harvard Observatory photographs, fluctuates between visual magnitudes 14 and 16 with occasional bursts to brighter than 13. These fluctuations mean that BL Lac's optical emissions vary by a factor of 20 times or so. Observers have noted night-to-night luminosity variations of 10 to 30 percent. A few BL Lac objects have changed their luminosities by as much as a factor of 100.

A great puzzle about the BL Lac objects is that their energy variations take place in objects that show almost no emission lines in their spectra. As discussed above, the standard model for active galaxies pictures synchrotron emission output in the ultraviolet (and even the electrons themselves) as ionizing any gas near the nucleus and producing emission lines through recombination. But where are the BL Lac's emission lines if they are powered the same way? For a few BL Lac objects that have weak continuum outputs, the emission lines *are* visible and are probably powered by nonthermal radiation from the nuclear source. The hint here is that the emission lines are usually present but often submerged in the continuum emission.

Some 40 BL Lac objects have been classified to date. They may actually not all be the same kind of beast. A few are possibly the nuclei of galaxies. Some, like BL Lac itself, have a faint surrounding "fuzz" that might be a galaxy. Others look point-like without a hint of enveloping material. A number of BL Lac objects are found in clusters of galaxies—indirect evidence that they are also galaxies.

Finally, we must note that we do not have good distance determinations for very many BL Lac objects. A redshift of 0.07 appears in the weak absorption features of the nebulosity around BL Lac, which corresponds to a radial velocity of 2.1×10^4 km/s. The nebulosity has a spectrum like that of a luminous elliptical galaxy. If the redshift is cosmological (from the expansion of the universe), this radial velocity corresponds to a distance of 430 Mpc (for $H_0 = 50$ km/s $\cdot$ Mpc).

(d) Starburst Galaxies

Appearances can be deceiving in astronomy— especially for distant objects of which we get a two-dimensional view projected on the sky. The nearby galaxy M 82 (NGC 3034) makes this point most dramatically, for it was once believed to be an exploding galaxy because of its optical appearance (Figure 21–3). An irregular II galaxy, M 82 is a radio source but not a strong one, radiating only 10^{32} W. Photographs of this galaxy in H_α show filaments extending above the plane of the galaxy to a distance of 3 to 4 kpc. Spectra of this galaxy also show material expanding outward from the center at velocity differences of about 100 km/s.

The nuclear region of M 82 is much larger than that for our Galaxy and a strong infrared emitter: about 10^{37} W. Star clusters and giant H II regions dominate the nuclear region; within it lies a small, nonthermal source. Clumps and filaments of dust here block out the optical view, a fact that misled

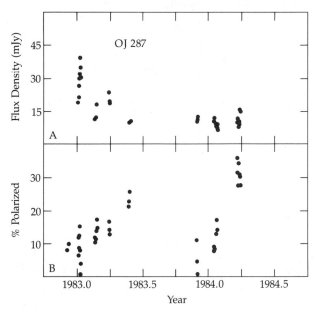

Figure 23–13

Variability of OJ 287. (A) Intensity variations; note the large outburst at the beginning of 1983. (B) Polarization variations of the linear polarization. *(P. Smith)*

people into believing that the galaxy's appearance resulted from explosive, high-velocity outflows.

Recent X-ray observations show that the nucleus has a clumpy structure with a total power of 3×10^{33} W in soft X-rays. The most likely candidates for the emission source are massive Population I objects, such as X-ray binaries (like Cyg X-1), OB stars, and supernova remnants. These and other observations lead to the interpretation that the nuclear region underwent a huge burst of star formation 10^7 to 10^8 years ago and that this burst is continuing today. M 82 lies close to the spiral galaxy M 81 (Figure 22–7). Tidal interactions between them could trigger the star formation in M 82, which is rich in the gas and dust needed to give birth to stars.

Other strange-looking galaxies may also have undergone a starburst episode initiated by tidal interactions; hence the name *starburst galaxies.* Seyferts may be a part of the starburst process, in which an active galaxy is driven to action by an *external* (gravitational?) influence.

23–2 Quasars: Discovery and Description

In the late 1950s, radio astronomers compiled catalogs replete with radio sources that were not identified with any familiar visible objects. Hunting for possible associations of radio and optical sources, Thomas Matthews and Allan Sandage in 1960 discovered a faint, 16th-magnitude star-like object (hence the name *quasi-stellar object,* or *quasar*) at the position of radio object 3C 48. (3C means the Third Cambridge Catalog.) This object had a spectrum of broad emission lines that could not be identified, and it emitted more ultraviolet light than an ordinary main-sequence star.

3C 48 remained a unique object until 1963, when the strong radio source 3C 273 was identified with a 13th-magnitude star-like object (Figure 23–14). The emission lines of 3C 273 were just as puzzling as the emission lines from 3C 48: they coincided with no known atomic lines. What was happening?

(a) Spectral Characteristics

The first quasi-stellar object whose spectral redshift was identified was the radio source 3C 273. The emission lines in the spectrum of 3C 273 show a

Figure 23–14

The quasar 3C 273. Note the jet. (*Palomar Observatory, California Institute of Technology*)

familiar regularity reminiscent of the hydrogen Balmer lines, but they seem to be greatly displaced to the red of the normal Balmer lines (Figure 23–15). Maarten Schmidt did identify them as the Balmer series, and he calculated a redshift of $z = 0.158$ (where z is $\Delta\lambda/\lambda_0$). With the acceptance of the concept of large values of z, other lines could be identified. Most of the emission features, all of which are *broad* lines, are similar to those found in gaseous nebulae. In addition to the hydrogen Balmer lines, there are forbidden lines, such as [O II], [O III], [Ne III], and [Ne V], and the permitted lines of Mg II, C III and C IV.

The spectra of most quasi-stellar objects show very broad emission lines similar to those found in the spectrum of 3C 273. The outstanding feature of these lines is the fact that they are all very greatly shifted to the red, so that the redshift corresponds to values of z from 0.06 to 3.53. For example, in several cases the Lyman line, normally 121.6 nm, is shifted into the visible part of the spectrum.

Note that redshifts greater than unity require the use of the *relativistic Doppler-shift* formula rather than the classical one,

$$z = \Delta\lambda/\lambda_0 = v/c$$

which for $z > 1$ would imply that the source is mov-

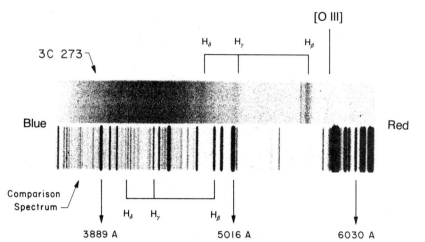

Figure 23–15

Spectrum of 3C 273. The comparison spectrum below serves as a wavelength standard. *(M. Schmidt)*

ing faster than the speed of light! One form of the relativistic Doppler shift is

$$z = \Delta\lambda/\lambda_0 = [(1 + v/c)/(1 - v/c)]^{1/2} - 1$$
$$\text{(23-1)}$$

where $\Delta\lambda = \lambda - \lambda_0$, λ_0 is the original (laboratory) wavelength, v the radial velocity, and c the speed of light. Note that for $v/c \ll 1$, this equation reduces to the classical result.

Now to use the relativistic Doppler shift to find v for quasars. Say $z = 2$; then

$$z = 2 = \Delta\lambda/\lambda_0 = [(1 + v/c)/(1 - v/c)]^{1/2} - 1$$

so that

$$(1 + v/c)/(1 - v/c) = 3^2 = 9$$
$$v/c = 8/10 = 0.8$$

Once identified, the emission lines from quasars can be analyzed in the same way as those from H II regions. This process indicates that a strong flux of ultraviolet and X-ray photons ionizes a low-density, transparent gas. Within that gas are clouds or filaments moving at high speeds—about 1000 km/s —to explain the width of the emission lines. Part of this broadening arises from the fact that the ions and atoms in the filaments are hot. However, the emission lines from quasars are 1000 times broader than the width predicted from the broadening effect of temperature. Most of the broadening, then, comes from the motions of the clouds or filaments themselves at relative velocities from 1000 to 1500 km/s.

Most (all?) quasars with emission-line redshifts greater than 2.2 also have strong absorption lines in

their spectra; those with lesser redshifts typically do not have absorption lines. The absorption-line redshifts are always less than or equal to the emission-line redshift. These absorption lines are much narrower than the emission lines. Generally they are identified with ionized states of such common elements as carbon, silicon, and nitrogen. Sometimes they show more than one redshift. For example, the spectrum of the quasar PHL 938 has an emission redshift of 1.955 and absorption redshifts of 1.949, 1.945, and 0.613.

These absorption lines may be produced by clouds of gas in or close to the quasar at temperatures cooler than those of the emission-line regions. If the clouds are moving outward from the quasar, their velocity relative to us, and hence their redshift, are less than those of the quasar. They might be produced by intervening (and otherwise invisible) intergalactic clouds of gas. Another possibility is a halo of gas around an intervening galaxy. The intervening clouds or halos have smaller redshifts because they are closer and so not moving away as fast as expected from the Hubble expansion. None of these possibilities has been well confirmed, but the discovery of the double quasar (see below) supports the third model.

(b) Continuous Emission

In general, when observed over a span of wavelengths from radio to X-rays, the continuous emission from quasars is nonthermal, so that a plot of log (flux) versus log (frequency) results in a straight line. Comparing spectral indices α, we find two distinct categories: flat and steep, with the division at $\alpha \approx 0.5$. The spectral indices range from 0.0 to 1.6. So synchrotron emission is most likely the source of the continuous radiation.

Quasars also split into two groups when their polarization is examined. Most quasars have low polarizations (less than 3 percent). Only 1 percent or so of optically bright quasars have polarizations greater than 3 percent (up to 35 percent). The highly polarized quasars are compact radio sources, have flat radio spectra and steep optical ones, and exhibit rapid (days to years), large-amplitude variability at optical wavelengths. Hence, high-polarization quasars share many characteristics with BL Lac objects.

(c) Optical Appearance

As suggested by the name, the optical appearance of quasars is stellar; that is, their angular diameters are less than 1″. Some, however, have faint nebulosities associated with them. The optical variations of several quasars have been well established; in some cases, changes of as much as 1 to 2 magnitudes over a few months have been observed. Some of these variations are long-term, with time scales of years; in other cases, the brightness changes have been very rapid, with small changes (of the order of a few percent) occurring in a matter of 15 min. Other changes occur more slowly, over a matter of several days or a month or two. One phenomenal case, 3C 446, changed its luminosity by a factor of 20 within a year and continued to show sizable variations thereafter. Quasars that vary appreciably at radio frequencies are also optical variables. A strong correlation exists between optical variability and polarization. In fact, the brightness changes are often accompanied by a variation in polarization, indicating that it is the polarized component that is responsible for the light fluctuations. About 20 percent of known quasars have rapid (days to weeks) variations at optical wavelengths (and also in the radio region).

Optical variations are of great importance in setting limits on the size of quasars. If an object varies with a period τ, the radius of the object must be equal to or smaller than τ times the speed of light—$R \leq c\tau$. Outbursts at different places within the source would average out to only slight overall changes if this limit did not hold. A total variation by any appreciable factor requires that the outbursts be synchronized; this means that the signal from one region must travel the distance to all other regions within the period of variation. Moreover, if an object with radius larger than $c\tau$ varied as a whole, the light travel time would smooth out the time variations. If $c\tau$ is 1 lightmonth but the object is 1 lightyear in diameter, radiation from the farthest point is delayed by 1 year relative to that from the side nearest the observer, masking the monthly variations.

The fact that the variations in quasars have been observed to take place with time scales of less than 1 month (even days) suggests that the radius of the object must be about 1 lightday, or 10^{13} m. The entire quasar need not participate in the variation, but the fact that variations by a factor of 2 and even more have been observed does mean that a very sizable portion of the total radiation from the object participates in this variation; so the region from which this radiation arises is restricted by this size limitation.

23-3 Problems with Quasars

The main feature of quasars is their very large redshifts. The most natural explanation of these redshifts is a cosmological one: quasars participate in the Universe's expansion. If so, their enormous redshifts indicate that they are very far from us and must expend vast amounts of energy. For example, the redshift of $0.16c$ for 3C 273, if due to the expansion of the Universe, implies a distance of 770 Mpc for $H_0 = 50$ km/s · Mpc. At this distance, to appear at its observed apparent magnitude of 13, 3C 273 must emit about 10^{40} W, or about 40 times as much as the most luminous galaxies. A typical quasar produces about 1000 times as much power as an ordinary spiral galaxy, most of it emitted in the infrared.

Not only do quasars emit energy at enormous rates, but that energy comes from relatively small regions of space in the quasar centers—from lighthours or lightdays to no more than a few lightyears in diameter. Two pieces of evidence point to small energy-emitting volumes. First, the variation of light output over days to years. The size of the region that emits the energy can be no more than the light travel time across it, and so the regions cannot be larger than a few lightyears. Second, VLBI observations have revealed in some quasars radio structures that are no more than a few tens of lightyears in diameter (for a cosmological redshift).

This then defines the energy problem for quasars: how to generate about 100 times the energy of a galaxy in a region only a few lightyears across!

(a) Energy Sources

Almost all of a quasar's continuous spectrum comes from synchrotron emission: high-speed electrons gyrating in a magnetic field. As these electrons emit electromagnetic radiation, they lose energy and move more slowly. So they emit radiation with lower and lower energy. This loss of speedy electrons implies that the supply of high-energy ones must be replenished at least about every year or so. The central energy source of a quasar must yearly blast out clouds of high-energy electrons containing a total of at least 10^{43} J. The rest of the quasar acts like a transformation machine, trapping the energy of electrons and converting it to other forms. What energy source lies at the heart of a quasar?

The most developed quasar model to date involves supermassive black holes, objects of mass about 10^7 to 10^9 solar masses. This model originates from that for binary X-ray sources (Chapter 17), in which material from a normal star forms an accretion disk around a black hole before the material falls into it. In the quasar model, a supermassive black hole in a dense galactic nucleus is fueled by the tidal disruption of passing stars. The stellar material forms an accretion disk and radiates as it spirals into the black hole, powering the quasar. Outflows of ionized gas may be generated perpendicular to the spin axis of the disk; these may be visible as jets from the nucleus [Section 23–3(d)].

The model calculations show that luminosities of 10^{12} solar luminosities, about that of bright quasars, are possible with an infall of 1 solar mass or less of material per year. One aspect of this model that works is that the supermassive black hole can easily generate the level of quasar luminosity in a region of space only a few lightyears in diameter (the Schwarzschild radius of a $10^8 M_\odot$ black hole is only 3×10^8 km, or about 2 AU). And it does the energy conversion (from gravitational to radiative) with high efficiency.

You can see this quantitatively by recalling that the potential energy of mass m brought in from infinity to distance R next to mass M is

$$PE = -GMm/R$$

so for a 1-kg mass moved to the Schwarzschild radius of a $10^8 M_\odot$ black hole, $PE \approx 4 \times 10^{16}$ J/kg. Then to provide $10^{12} L_\odot = 4 \times 10^{38}$ J/s requires the infall of

$$L/E = (4 \times 10^{38} \text{ J/s})/(4 \times 10^{16} \text{ J/kg})$$
$$\approx 10^{22} \text{ kg/s} \approx 10^{29} \text{ kg/year}$$

with the assumption that the conversion from potential energy to light will be 100 percent efficient. (It won't be, so the amount of mass will be greater.)

(b) Superluminal(?) Motions

VLBI radio observations have also presented a new puzzle with several quasars. Parts of these objects appear to be moving with speeds greater than c! This apparent faster-than-light motion is called *superluminal motion*.

A good example is 3C 273. Recall that it has an optical jet about 100 pc long. Its radio emission comes mainly from the body of the quasar (called 3C 273B) and from near the end of the jet (called 3C 273A). So 3C 273 has a core with a radio jet protruding from it. A series of radio observations from 1977 to 1980 showed that a knot in the source has steadily moved away from the central peak (Figure 23–16). The total separation has increased in the three years by 2 milliarcseconds. The expansion seems to have happened at a constant rate over this time. Now, if 3C 273 is at a distance given by its redshift, the observed angular separation rate corresponds to a transverse velocity of almost *six times the speed of light!*

Sources that show superluminal motions all have a common feature: a radio structure consisting of a strong central source with a weaker jet out one side. So they resemble nearby radio galaxies with single jets, such as Centaurus A. These jets are thought to be electrons flowing outward at close to the speed of light from a nuclear source; they may be relativistic jets. The same process may occur in the superluminal sources. In 3C 273, the jet is pointing almost directly at us (tilted only about 10° away from a perfect alignment). The moving knot is a blob of material streaming out along the jet. The knot is *not* moving faster than the speed of light, however—it only *appears* to be. The apparent superluminal speed is an optical illusion caused by the almost head-on orientation of the relativistic jet and the finite speed of light.

To understand this effect, consider a jet emitting blobs of material moving at close to c. Suppose the jet opens at some small angle, say 8° with respect to our line of sight (Figure 23–17). Imagine that a blob ejected by the nucleus (point N) gets to point A in 101 years. Suppose the light emitted from the blob when it was at N reaches point B after 100 years. The separation between A and B is 14 lightyears (for an 8° angle), but the light at B is one year ahead of that emitted by the blob when it reaches A. (It has taken

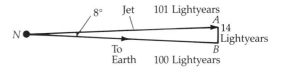

Figure 23-17

Geometry for superluminal motions.

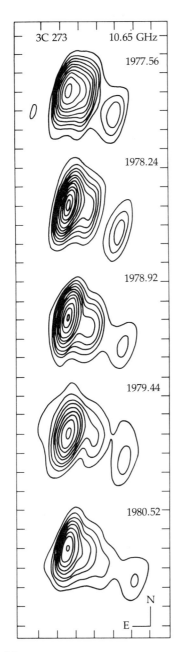

Figure 23-16

Motions in 3C 273. High-resolution radio maps at 10.65 GHz show a blob of material moving away from the nucleus. *(T. J. Pearson, S. C. Unwin, M. H. Cohen, R. P. Linfield, A. C. S. Readhead, G. A. Seielstad, R. S. Simon, and R. C. Walker, Owens Valley Radio Observatory and the National Radio Astronomy Observatory)*

100 years for the light from N to reach B, 101 years for the blob to reach A.)

Many years later, the light that was at B reaches us; only one year later, that emitted at A reaches us. The source seems to have moved from B to A — 14 lightyears — in only one year. It seems to have a transverse speed of $14c$. Yet no such physical motion has occurred. The superluminal velocity is only an apparent motion. And the smaller the angle of the jet to our line of sight, the more superluminal the motion will appear.

The main point is that bulk relativistic linear motion aligned close to the line of sight provides a simple explanation for rapid flux density outbursts, the superluminal motions, and the strength of the emission in the jets. Physically, this model requires that the speeds in the beam be very close to light ($0.99c$ or greater) and that the Lorentz factor γ, defined by

$$\gamma = (1 - \beta^2)^{-1/2}$$

where $\beta = v/c$, is on the order of 10. For $\beta = 1$, for small angles (θ) to the line of sight, the apparent transverse velocity is

$$v = 2c/\sin \theta$$

so for $\theta = 8°$, as in the example above, the apparent transverse velocity is increased by a factor of 7. Also, relativistic effects concentrate the flux in a narrow forward beam — an effect known as *relativistic beaming* — that boosts the observed flux. For a beam that is head-on with $\gamma > 1$, the flux is increased in proportion to γ^3, so that if $\gamma = 10$, the flux goes up by 10^3 relative to an unbeamed signal.

(c) Double Quasars and Gravitational Lenses

Another optical illusion related to quasars derives from the bending of light in strong gravitational fields — an effect predicted by general relativity and confirmed in the Solar System by the bending of starlight as rays pass close to the Sun. Large masses can image light, usually imperfectly. Recently, observations of quasars have confirmed this gravitational-lens effect on a cosmological scale.

Quasars are rarely close together. Two called 0957+561A and 0957+561B are only 6 arcsec apart. Even more surprising, their spectra are nearly identical (in line positions and intensities), and their red-shifts are the same: $z = 1.41$. It turns out that these quasars are not twins but rather optical images of the *same* quasar made by an intervening galaxy. Photographs made on a night of exceptionally good seeing show that quasar B has a little piece of "fuzz" sticking out of it (Figure 23–18A). This fuzz turns out to be

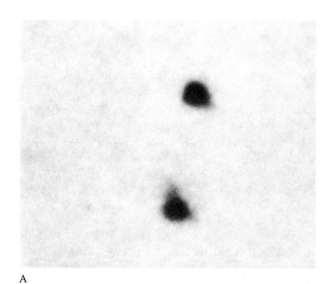

A

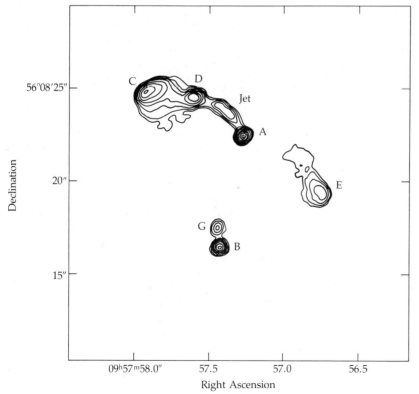

B

Figure 23–18

The double quasar. (A) Optical photograph on a night of excellent seeing; note the fuzz sticking up from the lower image. *(A. Stockton)* (B) Radio map at 6 cm done with the VLA. The double quasar (A and B) is seen; note that A has a jet flowing into two lobes (C and D). G has properties similar to those of an active elliptical galaxy. *(D. H. Roberts, P. E. Greenfield, J. N. Hewitt, B. F. Burke, and A. K. Dupree, National Radio Astronomy Observatory)*

the poorly resolved image of a faint galaxy—the gravitational lens! Since the galaxy is an extended mass, however, it acts like an imperfect lens and produces a complex pattern of up to three images. By a quirk of placement, we see two of the three images formed by a gravitational lens: an intervening, elliptical (probably cD, with mass $\approx 10^{12}M_\odot$) galaxy between us and the quasar (Figure 23–18B).

This discovery has three important implications. One, it provides another confirmation of general relativity. Two, it proves in this case that the quasar is more distant than the galaxy, and so the quasar's redshift is cosmological. Three, cool gas around the galaxy creates the quasar's absorption line spectrum; this situation may well be the case of other quasars, too.

(d) Quasars Compared with Active Galaxies

Quasars and active galaxies share some observed characteristics. First, consider the radio galaxies. Most have only absorption lines in their spectra. Those that have emission lines come in two types. One type has *narrow* emission lines. Cygnus A is a good example; it has strong emission in the lines of [O III] at 500.7 nm and [N II] at 658.3 nm and other lines from such ionic states as [S II], [Ne V], and [Fe X]. These lines are all narrow; the Doppler widths are roughly 500 km/s. Other narrow-line radio galaxies have emission-line widths of 400 to 800 km/s. Narrow-line radio galaxies make up about two-thirds of the total. The other one-third have broad lines of hydrogen and helium, some as wide as 10^4 km/s. Other emission lines for these galaxies tend to be narrow. Quasars with low redshifts have optical spectra that resemble those of the broad-line radio galaxies in terms of the emission lines present, their widths, and the shape of the optical continuous spectrum. For instance, the hydrogen and helium lines of such quasars have widths of 3000 to 6000 km/s.

Quasars also resemble Seyferts in their emission-line spectra. The Seyfert 1 galaxies, just like the broad-line radio galaxies, look like low-redshift quasars in terms of their emission spectra. So the physical conditions in the regions producing the spectra may be basically the same in Seyfert 1 galaxies and in low-redshift quasars. Note that if Seyferts were more distant than they are, we would see *not* the galaxy but only the bright core. A distant Seyfert would look a lot like a quasar.

Comparing quasars and BL Lac objects, we note a crucial difference: BL Lac objects do *not* have strong emission lines. However, in both BL Lacs and quasars, the nonthermal nature of the continuous emission stands out more clearly than in other active galaxies. BL Lac objects and about 15 percent of radio-bright quasars show wide variations in optical output over periods of days, weeks, and months. These swings in luminosity often occur very abruptly.

One other possible connection relates to radio structure. Quasars have high luminosities and tend to exhibit symmetrical double structures. The higher-luminosity active galaxies have similar radio shapes. So the physical processes responsible for both may be the same. The nuclear jets imply that this structure is somehow tied to violent activity in the nuclei of luminous active galaxies and quasars.

The observations suggest that quasars are somewhat physically similar to active galaxies. One popular idea views the two as objects at different stages of evolution, that is, the quasar phenomenon signals very violent activity in the nucleus of a galaxy at a very early stage in its life. So quasars may be hyperactive galaxies. This evolutionary connection is a nice idea and implies that the energy sources in quasars relate directly to those in active galaxies. If the sources are black holes, then all (or most) galaxies need to have supermassive black holes in their nuclei.

If we accept a black-hole model, then the nucleus of a quasar should show an intense, point-like source of light (from a concentration of stars around the black hole), stars orbiting the center should have high velocities, and emission lines with high Doppler shifts might be visible from the infalling matter. Observations of such effects have been reported for a high-luminosity point in the nucleus of M 87, including higher velocities for stars in the nucleus than for those outside. A model consistent with these observations is one of a $(5 \times 10^9)M_\odot$ black hole hiding in the inner 100 pc of the nucleus. (An alternative model uses a dense concentration of ordinary stars.)

Given that a quasar may simply be the hyperactive nucleus of a very distant galaxy, what about the rest of the galaxy? At cosmological distances, the disk of a quasar's parent galaxy would be too small and faint to see easily. For example, if the Andromeda galaxy were placed at a distance of 900 Mpc —equivalent to a redshift of 0.2—its angular diameter would be a mere 4″. Remember that the Earth's atmosphere limits seeing to 1″ at best. So a distant galaxy would make an image that is difficult to distinguish from that of a faint star.

Recent CCD observations of the quasar 1059+730 ($z = 0.089$) have noted that the bright nucleus is surrounded by an elongated, fuzzy structure about 9" by 16" in size that resembles an edge-on disk. In May 1983, a stellar object with an apparent visual magnitude of 19.6 appeared within the fuzz. This outburst was most likely a supernova explosion. If so, then 1059+730 lies at a cosmological distance as inferred from its redshift and the fuzz surrounding it must contain stars—the stellar disk of the underlying galaxy.

We might also look for a quasar in a cluster of galaxies—a good sign that we're not looking at individual stars. An example is the quasar 3C 206, which is surrounded by some 200 faint galaxies (Figure 23–19). The quasar's redshift is 0.206; that of a nearby pair of galaxies is 0.203 (and the same is presumed for the cluster). This result argues in favor of the quasar's residing in the cluster.

Other observations are beginning to reveal that quasars are surrounded by very faint envelopes that may well be the hard-to-detect disks of galaxies. For example, 3C 273, the closest of the high-luminosity quasars, has a fuzzy appearance in computer-processed photographs. The faint fuzz has emission lines in its spectrum with the same redshift as the quasar. Another good example is Markarian 1014, a low-red-shift (0.163) quasar whose fuzz has been examined spectroscopically. Stellar absorption lines in the spectra show a mixture of early- and late-type stars with the same redshift as the quasar. The luminous matter surrounding it appears asymmetric and spiral-like and has a linear diameter of about 100 kpc. This seems to be a galaxy with a "quasar" nucleus.

(e) Noncosmological Redshifts?

Our discussion has presumed that the redshifts of quasars are cosmological—in other words, that they arise from the expansion of the Universe—but a small, persistent band of astronomers have argued that all or part of the redshifts come from noncosmological causes. Such effects require that quasars be closer so that the energy problem is alleviated somewhat.

The proponents of noncosmological redshifts have looked for evidence that some quasars are associated with galaxies—in the sense of proximity on the sky—and then searched for a possible *physical* connection between the galaxy and the quasar. The arguments rely heavily on statistical inference to demonstrate that the chances of proximity are small. (On the average, two quasars appear per square degree of sky down to a limiting magnitude of $V = 19$.)

We give two representative examples, both from the work of Halton Arp. In one, a quasar-like object ($z = 0.044$) lies in *front* of an elliptical galaxy ($z = 0.009$), based on the fact that a dark ring rims the quasar—material perhaps absorbing light from the galaxy. Arp concludes that the object was ejected by the galaxy. He finds another curious case in the cluster of galaxies Abell 1367, where two quasars lie close to (within about 1') a galaxy—an association that Arp argues has only 7×10^{-6} probability of happening by chance. A physical connection has yet to be established in either case.

Frankly, we both believe that the evidence to date more strongly supports a cosmological rather than a noncosmological interpretation, and most other astronomers do, too.

Figure 23–19

Quasar 3C 206 in a cluster of galaxies. *(S. Wyckoff, P. Wehinger, H. Spinrad, and A. Boksenberg)*

23–4 Astrophysical Jets

High-resolution radio observations in recent years have revealed that long, narrow flows of material—loosely called *astrophysical jets*—are a common and colorful phenomenon. Within the Galaxy, we see bipolar flows from protostars (Chapter 19), a possible outflow from the Galaxy's nucleus (Chapter 20), the relativistic flow from SS 433 (Chapter 18), and surprising flows from the Crab Nebula and some planetary nebulae. This chapter describes the nuclear jets from active galaxies and quasars. These jets range in size from a few parsecs to megaparsecs across, with lengths typically 100 times the diameters. In just about all cases (such as Cen A and M 87), the jets show internal knots, the emission from which is likely synchrotron radiation from electrons in magnetic fields. In most cases, the knotty beam ends in an amorphous, wispy lobe that surrounds a hot spot. Some jets twist and bend before they reach the limits of their flows.

The observed properties of astrophysical jets appear similar to those of supersonic gaseous jets that have been studied in laboratories and in theory for well over a century, pioneered by Ernst Mach and others. A key property of a supersonic flow is the *Mach number M*, which is the ratio of the flow speed v to the local sound velocity s:

$$M = v/s$$

where

$$s = (\gamma P/\rho)^{1/2}$$

and γ is the adiabatic index, which equals $\frac{5}{3}$ for an ideal monotonic gas. When $M > 1$, the flow is supersonic, that is, greater than the local sound speed.

Supersonic flows generate shocks (such as the sonic boom created by a jet airplane that crosses Mach 1 in our atmosphere). Laboratory experiments show that artificial jets with high Mach numbers produce internal, reflected shocks and, as a result, high-pressure knots. Astrophysical jets exhibit knotty structures of higher pressures, temperatures, and

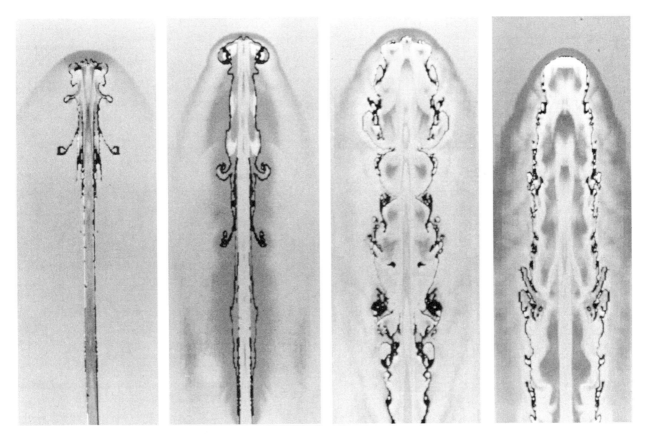

Figure 23–20

Shock wave in front of a jet. These theoretical calculations show the envelope of a shock around the front of a supersonic jet. (*M. Norman and K.-H. Winkler*)

densities that may very well be caused by internal shocks. These shocks do not disrupt the beam's flow, which does end in a strong shock at the head of the beam. The termination is surrounded by a cocoon of low-density material (the radio lobe).

Michael Norman, Larry Smarr, and Karl-Heinz Winkler have used a supercomputer (Cray-1) at Los Alamos National Laboratory to investigate the properties of computational supersonic jets and to compare them with the observed astrophysical ones. We will outline their basic results here for jets whose internal pressures initially just match those of the surrounding media. (Indirect evidence suggests that astrophysical jets are kept collimated by the external gas pressures of the surrounding media.)

As a jet penetrates the ambient medium, it drives a shock before it (Figure 23–20). The discontinuity forms at the boundary of the jet and ambient gas. A shock forms behind the contact boundary; here the

gas in the beam stops and is deflected around in a return flow that creates a cocoon around the jet. This cocoon contains vortices within itself and waves along its surface. The extent of the cocoon depends on the Mach number and density of the jet's gas relative to the medium. Greater Mach numbers generate larger cocoons, and lower-density jets also produce relatively large cocoons. Note that the cocoon works to dissipate the flow of the jets. In real astrophysical jets, shocks within the cocoons might generate the synchrotron emission that is observed. Eventually, the flow becomes unstable and develops kinks and wiggles far out from the source. Here the jet also hits a high-density piece of the surrounding medium that brings the flow to an end. Overall, the model of knotty supersonic flows describes some of the key features of astrophysical jets, and more computations are in progress to gain a better match between theoretical models and the observed properties of jets.

Problems

1. The special theory of relativity says that no material object can move faster than the speed of light. The classical Doppler formula says that when the redshift z is greater than unity, we have $v > c$ (which is impossible).
   (a) Referring to Chapter 8, derive the *exact relativistic relation* between v and z. Express your result in the form $v = f(z)$.
   (b) Make a table with three columns: column one for z, where you should enter five exemplary values of z from 0 to 3.0; the second column headed $(v/c)_{cla}$, where you compute the classical *nonrelativistic* result for your five values of z; and the last column headed $(v/c)_{rel}$, where you use your formula from part (a).

2. Assume that the optical galaxy associated with the radio source Centaurus A is the same size as our Galaxy and draw a scale diagram of the radio and optical emission regions of Centaurus A. Clearly indicate the dimensions and the relative positions of the various components.

3. The radio galaxy Cygnus A has an observed radio *flux density* of 2.18×10^{-23} W/m² · Hz at a frequency of 10^3 MHz. (Note that the unit of bandwidth $\Delta\nu$ is 1 Hz.) The observed redshift of the galaxy is $\Delta\lambda/\lambda_0 = z = 0.170$.
   (a) If the radiation is received at 10^3 MHz, at what *(rest)* frequency was it emitted by Cygnus A?

   (b) What is the distance to Cygnus A? (Use a Hubble constant of $H_0 = 50$ km/s · Mpc.)
   (c) What is the *radio luminosity* (W/Hz) of this radio source at 10^3 MHz?
   (d) To find the total radio luminosity of Cygnus A, we must multiply the result of part (c) by the *bandwidth* $\Delta\nu$ of our detector. Assume $\Delta\nu = 10^4$ Hz and compute the energy radiated per second at radio frequencies.
   (e) What is the minimum mass of hydrogen (in solar masses) that must be converted to helium during each second to provide this luminosity?
   (f) If Cygnus A continues to radiate at this rate for 10^8 years, how many solar masses of hydrogen must be converted to He? Express this result in terms of the mass of our Galaxy ($\approx 10^{12}M_\odot$).

4. 3C 9 is a quasi-stellar object that has a redshift of 2.0 and an apparent visual magnitude of 18.2. Answer the following questions using the cosmological interpretation of the redshift.
   (a) What is the speed of recession?
   (b) What is the distance to 3C 9?
   (c) What is the intrinsic luminosity relative to that of our Galaxy?
   (d) What is the maximum size of the emitting region if 3C 9 exhibits luminosity variations on a time scale of two months?

5. (a) Quasar 3C 273 has a redshift of 0.16. What is its distance?
   (b) The V magnitude of 3C 273 is 12.8. What is its flux density at V band? Its luminosity at V band?
   (c) The fuzz around 3C 273 has a diameter of 15″. What is its linear size?
   (d) The absolute magnitude of the fuzz is -25. What is its luminosity?

6. The quasar PKS 1402+044 has a redshift of 3.2. What is its distance? Note that $z > 1$!

7. Observations of the radio jet in Cen A indicate that the spectral index is about 0.5. The 20-cm flux density of the strongest blob of the jet is 2.3 Jy (janskies, 1 Jy = 10^{-26} W/m$^2 \cdot$ Hz). If this emission is synchrotron, what should the flux density be at 2.2 μm?

8. (a) How close must a star pass a black hole of mass $10^6 M_\odot$ to be disrupted tidally?
   (b) What *yearly* rate of matter infall is needed to power a quasar at 10^{39} W if a black hole of mass $10^6 M_\odot$ lies in its core?

9. In Cen A, the radio jet and the innermost lobe are separated by a distance of 4 arc-min.
   (a) What is their physical separation?
   (b) How long would it take relativistic electrons to travel from the jet to the lobe?

10. The quasar 1059+730 has a redshift of 0.089.
    (a) What is the distance to the quasar if the redshift is cosmological? What is the range, given the uncertainty in H_0?

    (b) The fuzz around the quasar has an angular size of 9″ $\times$ 16″. What is its physical size? How does this compare with the size of a typical spiral galaxy?
    (c) The supernova observed in 1059+730 had an apparent V magnitude of 19.6. Use this value to estimate the distance to the quasar. Compare this value with your results from (a).

11. The argument is often made that the dimensions of an object with variable brightness cannot exceed the speed of light times the time scale of the variations. Critically examine this argument by considering
    (a) two nonspherical geometries
    (b) special relativistic effects (note that observed time intervals are inversely proportional to observed frequencies)

12. The quasar NRAO 140 ($z = 1.26$) has a compact radio component that moves at an angular speed of 0.15 milliarcseconds/year. What is that apparent velocity of the component as a percentage of c? (Multiply your answer by $1 + z$ to correct for relativistic time dilation. Use a Hubble constant of 50 km/s $\cdot$ Mpc.)

13. A quasar has an emission line, identified as Ly$_\alpha$ of hydrogen ($\lambda_0 = 121.6$ nm), that is observed at 581.2 nm. Calculate the redshift and distance to the quasar for a Hubble constant of 50 km/s $\cdot$ Mpc. How fast is the quasar moving away from us?

Cosmology: The Big Bang and Beyond

Our *Universe* is the total of all that exists or has existed, both in space and in time. In this chapter, we first present key observational characteristics of the Universe. We then consider present theories of our Universe and its evolution. We end by sketching the current consensus view of the evolution of the Universe: the standard Big Bang model and its connections to elementary particles.

24-1 Fundamental Observations

(a) The Physical Structure of the Universe

This entire book deals with the detailed characteristics of those entities that compose our Universe. On the largest scale, our Universe is filled with numerous superclusters of galaxies. To the limits of the largest telescopes, we can see about 10^9 galaxies. Between the superclusters are vast voids empty of bulk luminous matter. The structure of superclusters is one of filamentary strings. Hence, the large-scale distribution of galaxies is very *inhomogeneous.* Following Hubble's law, the clusters of galaxies (*not* the galaxies within clusters) are moving away from each other. The average density of luminous matter in clusters amounts to 4×10^{-28} kg/m^3.

(b) Size and Distance Scale

Chapter 21 describes the bootstrap process by which increasingly larger astronomical distances are measured. Once we know the angle subtended by a given object and the distance to it, we can determine its diameter. Here we give some perspective on the typical dimensions found in our astronomical Universe (Figure 24-1).

You already have an intuitive feeling of how large the Earth is relative to a person and of the relative dimensions of our Solar System. So let's begin with the stars. If stars are imagined to be the size of cherries, then we must place cherries an average of 1 km apart to mimic the mean stellar distribution in our Galaxy. On this scale, our Galaxy would be about 20,000 km in diameter. Now shrink our Galaxy to the size of a flat doorknob (about 4 cm across). The Andromeda galaxy (M 31) is a similar doorknob about 1 m away; this is essentially the size of the Local Group of galaxies. Also, this scale represents the mean distribution of galaxies in our Universe, except that in clusters of galaxies there may be thousands of these doorknobs about 20 cm apart in a volume of diameter 10 m or so. On this same scale, the most distant optical galaxy is 1 km away, while the most distant quasar is about 2 km away. The greatest distance from which light can reach us now in our Universe is about 3 km away because such light has been traveling for essentially the lifetime of the Universe. Interplanetary and interstellar space is incredibly empty, but the extragalactic regions are fairly densely populated with galaxies. No wonder we frequently have difficulty comprehending cosmic dimensions, when they are a factor of 10^{26} larger than we are!

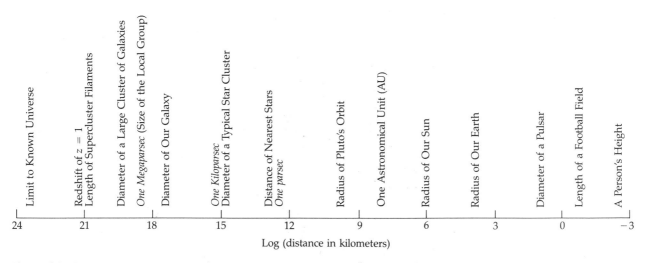

Figure 24-1

A cosmic size scale.

(c) Cosmic Expansion and the Age of the Universe

Chapter 21 discusses Slipher's and Hubble's discovery that distant galaxies appear to be moving away with recession speeds v directly proportional to their distances r—Hubble's law:

$$v = Hr \qquad \textbf{(24-1)}$$

where the Hubble constant is $H_0 \approx 50$ to 100 km/s · Mpc. (We will use 50.) A more correct statement is that all *clusters* of galaxies recede in accord with Hubble's law. Because most individual galaxies are gravitationally bound to clusters, they are not free to move apart—it is only the clusters that separate relative to each other. So the entire Universe of clusters is expanding. Once we have obtained the recession speed of a galaxy (or cluster) by measuring its redshift (Chapter 21), we can find its distance with Equation 24–1; hence, the distance scale of the Universe is related to the Hubble constant. The simplest way to obtain the approximate size of the Universe is to compute r from Equation 24–1 when $v = c$; that is, assume that the farthest galaxy we can see is moving at the speed of light. Since $H_0^{-1} \approx 2 \times 10^{10}$ years, we obtain $r_{horizon} = cH_0^{-1} \approx 2 \times 10^{26}$ m (note that this is the number used in Figure 24–1). We label this distance the *horizon*, for it is the maximum scale possible (covered by light) since the time the cosmic expansion began.

Note that H_0^{-1} is related to the *age of the Universe* because it has the dimension of years. Consider (Figure 24–2) a plot of distance r to an arbitrary galaxy as a function of time. If the Universe has always been expanding at a constant rate ($\tau = 1/H_0$), then Equation 24–1 tells us that H_0^{-1} years ago, $r = 0$; therefore, all the galaxies (and the Universe itself) must have originated in a titanic explosion about 20 billion years ago! Astronomers call the beginning of the expansion of the Universe the *Big Bang*. On the other hand, every bit of matter (and energy) in the Universe gravitationally attracts every other bit; so the expansion could not be free-coasting (the solid line in Figure 24–2); rather, it must be a *deceleration* (the dashed curve). To illustrate this point, consider a rock thrown vertically upward from the Earth's surface. This rock would move at a constant speed if there were *no* gravitational attraction from the Earth, but you are well aware that the rock slows down and

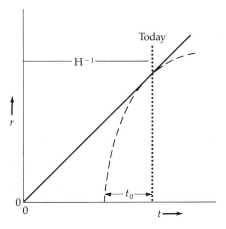

Figure 24–2

Age of the Universe. This diagram shows the distance to a certain galaxy as a function of time. The solid line assumes the galaxy moves at a constant speed; the dashed line is for a decelerating galaxy.

eventually falls back to the Earth. If you could give the rock a parabolic or hyperbolic velocity—an escape speed (Chapter 1), it would never return to the Earth but it would slow down nonetheless. Hence, the actual age of the Universe must be $t_0 < H_0^{-1}$.

Because H_0^{-1} is a good order-of-magnitude estimate of the age of the Universe (at least an upper limit), we should check whether the contents of the Universe are less than 2×10^{10} years old. Lead–uranium dating gives the greatest age on the Earth as about 4.0×10^9 years. Some rock pieces in the Apollo lunar surface samples have an age of 4.6×10^9 years. This finding corroborates the dating of meteorites, most of which have ages of about 4.6×10^9 years; this figure represents the age of our Solar System. To find older objects, we must look to the stars. In particular, the ages of globular clusters may be deduced from theoretical considerations by using the main-sequence turnoff method (Chapter 16). This method indicates an age for the oldest stars of about 15×10^9 years with an uncertainty of 2×10^9 years. When we consider the factor-of-2 uncertainty in the value of H_0, H_0^{-1} is a reasonable estimate of the age of our Universe, but note, however, that if $H_0 \approx 100$ km/s · Mpc, then $H_0^{-1} \approx 10 \times 10^9$ years, which presents problems with the ages of Population II globular cluster stars—they seem to be older than the Universe!

24-2 Models of the Universe

Cosmology means the science of the cosmos, or Universe. We will not consider experimental cosmology here, for we have already discussed the observational data; rather we'll concentrate on *theoretical* models of the Universe.

We will focus on *relativistic* cosmology — models that rely on Einstein's theory of relativity. With the appearance of Einstein's special (1905) and general (1915) theories of relativity, theoretical cosmology became a true science. In the special theory, Einstein united space and time in a four-dimensional spacetime continuum and deduced the correct kinematic theory for all objects. The general theory extended the special theory to include all gravitational phenomena; it is a theory of gravitation. In its geometry, spacetime is curved and the trajectories of objects are determined by this curvature; the curvature, in turn, is produced by the material and energy content of the Universe. In contrast to Newton's view, Einstein saw gravity not as a force but as a manifestation of the curvature of spacetime.

(a) Einstein's Theory of General Relativity

In 1905, Einstein had the insight that space and time must be considered together as one entity; with his general theory of relativity in 1915, he replaced the force of gravitation by a coupling between this geometric spacetime and the material content of the Universe. In Einstein's view, all events in the Universe involve space and time together. When anything in the world takes place, it happens in *both* space and time. He called such a happening an *event.* You mark an event by noting where (space) and when (time) it took place. These events make up points in *spacetime,* a four-dimensional system consisting of three dimensions of space and one of time. The general theory of relativity examines the geometry of spacetime and finds that, for accelerated observers, spacetime is curved — not flat.

Einstein noted that Newton defined mass by two different operations: Newton's second law and the law of gravitation. Suppose you apply a known force to an object and measure its acceleration. When the force and acceleration are known, Newton's second law gives the object's mass — its *inertial mass.* Now take the same object and weigh it. Weight is a force, the amount of gravitational force acting on the object. Mass measured this way is called *gravitational mass.* Newton believed that an object's inertial mass and its gravitational mass were the same. He knew this from Galileo's experiments with falling bodies as well from his own careful experiments. These results demonstrated that, near the Earth, all masses fall with the same acceleration.

Experimentally, this equality of gravitational and inertial mass holds true to a very high degree of accuracy. To the limits of sensitivity of the testing methods, no difference has ever been detected. The best experiment to date, done by V. B. Braginsky and V. I. Panov at Moscow University, found the inertial and gravitational masses of gold and platinum to be the same within one part in 10^{12}. Einstein felt that the equality of gravitational and inertial mass was no accident. He saw it as a fundamental fact about the Universe and gave it a special place in the general theory as the *principle of equivalence.* Here is Einstein's own example of the principle of equivalence. Imagine that you are on the Earth in a spacecraft with no windows. You drop objects in the spacecraft, measure their accelerations, and find that they all fall with the same acceleration, 9.8 m/s^2. Now suppose that, without your knowledge, you and the spacecraft are instantly transported out into space and constantly accelerated at 9.8 m/s^2. As you continue your experiment, there is no change in the acceleration of the falling objects. It remains 9.8 m/s^2. Only by being able to look out a window could you tell that you had left the Earth. You cannot, by your experiments, distinguish between gravitational mass and inertial mass.

The principle of equivalence provides a way to cancel out gravity locally. Put yourself in an elevator in a tall building. Let the elevator free-fall. You find yourself weightless; gravity has vanished! Transport yourself into space far away from any large masses. Your condition is the same — you are weightless without gravity. You might object that in this example of free fall, gravity will reappear quite dramatically when the elevator hits the ground. Imagine, however, a long tunnel drilled through the Earth, so that the elevator never hits the ground. Then the elevator would free-fall back and forth from one side of the Earth to the other, completing a cycle in about 84 min. Throughout this swing, no acceleration would be felt inside the elevator, even though it's passing through the Earth!

Note that if we dropped two test particles—one from each pole—through the tunnel, they would execute simple harmonic motion with a period of 84 min. Relative to each other, their paths are curved (Figure 24–3). So this experiment demonstrates that *spacetime within the Earth is curved*. In Einstein's general theory of relativity, the distribution of mass (and energy) determines the geometry of spacetime. A massive object produces a curvature of nearby spacetime. And that curvature shows itself by accelerated motion, which Newton would say was caused by gravitational forces. In other words, spacetime curvature is proportional to mass–energy density. Therefore, cosmological models may be constructed by making reasonable estimates of the type and amount of material content of the Universe; the equations of general relativity then yield the *evolutionary* behavior of the model.

(b) Newtonian Cosmology

Before we develop the equations of motion for a relativistic Universe, we will do so for a Newtonian model. These equations have a direct analogy to the relativistic ones, and they are physically a bit easier to intuit. Assume here that the Universe is infinite and homogeneous, that the speed of light is infinite, and that, as a consequence, a universal time ("absolute time") applies to all observers.

Newton proved that only the matter *interior* to some point affects the motion at that point *if the distribution of matter is homogeneous* (which we assume). Consider the matter in the Universe to be a noninteracting gas (so that pressure = 0); then the equation of motion for a test particle is

$$d^2R/dt^2 = -GM(R)/R^2 \quad \text{(spherical symmetry!)} \quad \textbf{(24–2)}$$

where $M(R)$ is the mass interior to R:

$$M(R) = 4\pi \int_0^R \rho(r)r^2 dr$$
$$= (4/3)\pi\rho R^3$$

since $\rho(r) = \rho = $ constant. Multiply Equation 24–2 by dR/dt to get

$$(dR/dt)(d^2R/dt^2) = -[GM(R)/R^2](dR/dt) \quad \textbf{(24–3)}$$

Integrate this equation with respect to t:

$$\int_0^t (dR/dt)(d^2R/dt^2) + \int_0^t [GM(R)/R^2](dR/dt) = 0$$

To do this integration, note that

$$d/dt[(dR/dt)^2/2] = 2[(dR/dt)/2](d^2R/dt^2)$$
$$= (dR/dt)(d^2R/dt^2)$$

so that we get

$$(dR/dT)^2/2 - GM(R)/R = k = \text{constant}$$

but $M(R) = (4/3)\pi\rho R^3$, and so

$$(dR/dt)^2/2 - (4/3)\pi G\rho R^2 = k$$

Dividing this equation by R^2, we have

$$[(dR/dt)/R]^2 - (8\pi/3)G\rho = 2k/R^2$$
$$[(dR/dt)/R]^2 = 8\pi G\rho/3 + 2k/R^2$$
$$(dR/dt)/R = (8\pi G\rho/3 + 2k/R^2)^{1/2} \quad \textbf{(24–4)}$$

What does this equation mean? Note that $R \geq 0$. So if we start out (initial condition) with $(dR/dt)/R > 0$ (expansion), then

$k = 0$ means $(dR/dt)/R$ is always greater than 0
$k > 0$ means $(dR/dt)/R$ is always greater than 0
$k < 0$ means $(dR/dt)/R$ eventually equals 0 and expansion "turns around"

Equation 24–4 in essence is the energy form of the escape-velocity equation. Recall that

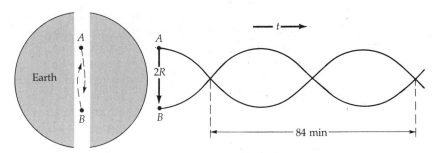

Figure 24–3

Motion of two masses falling through the Earth. One oscillation takes 84 min.

$$V_{esc} = (2GM/R)^{1/2}$$

so that

$$V_{esc}/R = [(8/3)\pi G\rho]^{1/2}$$

which is Equation 24-4 for the case $k = 0$ (total energy equals zero at infinity). The Universe would collapse if its initial expansion did not provide the material in the Universe with escape velocity.

(c) Relativistic Cosmology

Now assume that the Universe is isotropic and homogeneous and that the speed of light is finite and constant (no universal time). Consider observers at rest relative to local matter so that spacetime locally is flat.

To say that the Universe expands is to say that the measured (light travel time) distance between observers increases with time. If the expansion is both isotropic and homogeneous, then

$$R(t) = \text{measured distance} = R_0 a(t) \qquad (24-5)$$
$$= \text{(constant depending on choice of}$$
$$\text{observers)} \times \text{(universal function of time)}$$

Note than an expansion obeying this equation preserves the angles and shape of a triangle. (That's just what it means to have an isotropic and homogeneous expansion.) Take the derivative of Equation 24-5 with respect to time:

$$dR/dt = \text{rate of expansion} = R_0\, da/dt$$

but

$$R_0 = R(t)/a(t)$$

from Equation 24-5, and so (with $\dot{a} = da/dt$)

$$dR/dt = (R/a)\dot{a} \qquad \text{or} \qquad (dR/dt)/R = \dot{a}/a$$

Recall Hubble's law, however:

$$V = HR$$

from which you can see that

$$H = \dot{a}/a$$

Note that $\dot{a}$ and a are both functions of t, and so H is also a function of time. (It's *not* a constant.)

Again assume that the cosmological gas (galaxies, radiation, and so on) does not interact so that pressure is zero. Take the Newtonian result (Equation 24-4) and substitute $R(t) = R_0 a(t)$ for R:

$$\dot{a}/a = H = [(8\pi G/3)\rho + (2k/a^2 R_0^2)]^{1/2} \qquad (24-6)$$

Here ρ now stands for the total mass–energy density (which must include the $E = mc^2$ contribution from radiation). As shown for the Newtonian case, the value for k determines whether the expansion continues or reverses. The same holds true here, so that k can be considered a geometric term: $k = 0$, flat; $k > 0$, hyperbolic; $k < 0$, closed. For relativistic cosmology, the second term on the right-hand side of Equation 24-6 can be called the *curvature term*. The flat ($k = 0$) case marks the transition between an open Universe and a closed one. Set $k = 0$ in Equation 24-6 to get

$$H = [(8\pi G/3)\rho_c]^{1/2} \qquad (24-7)$$

where H and ρ_c are measured at some time ($H = H_0$, $\rho_c = \rho_{0c}$ at present epoch) and ρ_c is the mean mass–energy density of the Universe. *So a measurement of H_0 tells us the critical density needed to close the Universe:*

$$\rho_{0c} = 3H_0^2/8\pi G \qquad (24-8)$$

Take $H_0 = 50$ km/s $\cdot$ Mpc, and then $\rho_{0c} = 5 \times 10^{-27}$ kg/m³.

Note that we have three generic models of an expanding Universe. We term these types the *closed, flat,* and *hyperbolic* models. Each model is characterized by a function $R(t)$, which represents the "radius" of the Universe; since t is "cosmic time," the models evolve (Figure 24-4). All three models are expanding

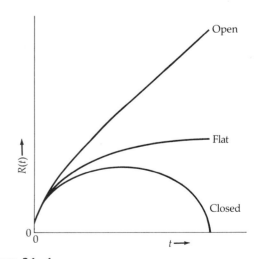

Figure 24-4

Dynamics of cosmological models. For isotropic and homogeneous models, distances vary with time depending on the overall geometry (open, flat, or closed).

at present. They all suffer deceleration, with the *closed* model decelerating fastest and the hyperbolic model decelerating least. All three models emerge from a singularity (at $t = 0$) for an initial creation, where $R = 0$. Because the volume of the Universe goes as R^3, the entire material content was crushed to infinite density (and therefore had a very high temperature) at the creation—the Big Bang. Since the slope of each curve represents the Hubble rate H, we see that the present epoch (t_0) is greatest for the hyperbolic model and smallest for the closed model. Note that only the closed model reaches a maximum size, and then recollapses to zero volume in a finite time—the Big Crunch!

The flat model marks the critical point between the hyperbolic and closed models. Its properties are quite simple. The model evolves with $R(t) \propto t^{2/3}$, so that the Hubble rate goes as $H = (2/3)t$. If we denote the present by the subscript 0, we find

$$\text{Age of Universe} \equiv t_0 = (2/3)H_0^{-1}$$
$$\approx 13 \times 10^9 \text{ years} \qquad \textbf{(24-9)}$$

$$\text{Mass density} \equiv \rho_0 \approx 5 \times 10^{-27} \text{ kg/m}^3$$

Note that the critical values listed in Equation 24-9 are uncertain by about a factor of 2 for t_0 (since H_0 is uncertain by that amount) and by a factor of 4 for ρ_0. We see that t_0 is in reasonable agreement with the observed ages of objects in our Universe; ρ_0 is the critical mass density, which determines whether our Universe is *open* or *closed*, and it is the number responsible for the current search to find the missing mass in intergalactic space.

The hyperbolic model is also simple in its consequences:

$$(2/3)H_0^{-1} \leq t_0$$
$$\rho_0 \leq 5 \times 10^{-27} \text{ kg/m}^3 \qquad \textbf{(24-10)}$$

As t_0 goes to its maximum value (2×10^{10} years), $\rho_0 \to 0$. Because the age of the Universe is somewhat greater in this model and because ρ_0 can be matched to the mass density of the observed *luminous* matter, many astronomers are inclined to accept this model. (MZ does.)

The *closed* model is the most complicated; its properties are given by

$$t_0 \leq (2/3)H_0^{-1}$$
$$\rho_0 \geq 5 \times 10^{-27} \text{ kg/m}^3 \qquad \textbf{(24-11)}$$

To many people, it is aesthetically satisfying that this

model stops expanding and recontracts; this implies a finite lifetime for the Universe and permits the possibility of a rebirth after $R \to 0$. Hence, the Universe may be cyclic, pulsating from one collapsed state to the next along repeating, identical curves. Note that the recollapse of the closed model implies that galaxies will approach us and appear blueshifted after we pass the maximum of expansion.

24-3 The Primeval Fireball

In 1964, Arno Penzias and Robert Wilson, scientists with the Bell Telephone Laboratories in New Jersey, began a sensitive study of the radio emission from the Milky Way. They detected an annoying excess radiation in their special low-noise radio antenna. They had tuned their radio receiver to 7.35 cm (4080 MHz), where the radio noise from the Galaxy is very small. Still they picked up the static. They further discovered that the noise did not change in intensity with direction in the sky, time of day, or season. The excess noise had an intensity equivalent to that of a blackbody at 3.5 K. What could it be?

At the same time, a group at Princeton was pondering the consequences of the expansion of the Universe from a *hot* dense state. Photons from such a time would be highly redshifted by now and would permeate the cosmos. Also, if the early Universe were so dense that it was opaque to the photons, then they would have had a blackbody spectral distribution. Subsequent redshifting by expansion would lower the temperature of the blackbody spectrum but not change its shape.

Penzias and Wilson got in touch with the Princeton group and concluded that their excess noise could very well be redshifted radiation from a hot Big Bang. That conclusion was solidified when the Princeton group and others confirmed the existence of the background radiation and found that its spectrum (Figure 24-5A) matched that of a blackbody at 2.7 K. (We will call it the *3 K background radiation*. Latest results give a mean weighted temperature of 2.73 $\pm$ 0.05 K; Figure 24-5B.) Its discovery supports a hot Big Bang model (sometimes called the *primeval fireball*), which is the standard relativistic model accepted by most astronomers today.

Now, the energy density u for blackbody radiation is

$$u = aT^4 \qquad \textbf{(24-12)}$$

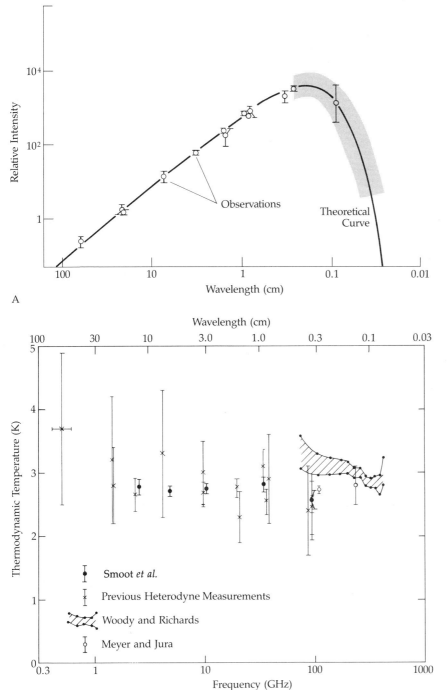

A

B

Figure 24–5

Spectrum of the cosmic background radiation. (A) Spectral observations; the shaded areas are infrared observations. The solid line is a 3 K blackbody curve. *(Adapted from a diagram by P. J. E. Peebles)* (B) Comparison of measurements of the temperature of the background radiation in the radio spectral region. *[G. F. Smoot, G. De Amici, S. D. Friedman, C. Witebsky, G. Sironi, G. Bonelli, N. Mandolesi, S. Cortiglioni, G. Morigi, R. B. Partridge, L. Danese, and G. De Zotti, Astrophysical Journal (Letters) 291:L23 (1985)]* and from $E = mc^2$, we can convert this energy density to an equivalent mass density ρ:

$$m = E/c^2$$

$$\rho_r = aT^4/c^2 \qquad (24-13)$$

where a, the radiation density constant, is

$$a = 4\sigma/c$$

with σ, the Stefan–Boltzmann constant, equal to 5.6697×10^{-8} W/m²·K⁴, so that

$$\begin{aligned} a &= (4)(5.6687 \times 10^{-8}\,\text{W/m}^2 \cdot \text{K}^4)/ \\ &\quad (2.998 \times 10^8\,\text{m/s}) \\ &= 7.564 \times 10^{-16}\,\text{W/m}^3 \cdot \text{K}^4 \end{aligned}$$

Then from Equation 24–13, the radiation density is

$$\begin{aligned} \rho_r &= (7.564 \times 10^{-16})(2.7^4)/(2.998 \times 10^8)^2 \\ &= (7.564 \times 10^{-16})(53)/(8.988 \times 10^{16}) \\ &= 4.5 \times 10^{-31}\,\text{kg/m}^3 \end{aligned}$$

Note that this density is much less than that of luminous matter, $\rho_m \approx 4 \times 10^{-28}$ kg/m³. Hence, we say that the Universe is now *matter-dominated*.

This was not always the case, however. Consider shrinking the Universe. Then the matter density, since lengths scale as $a(t)$, increases as

$$\rho_m \propto a^{-3} \qquad (24-14)$$

In contrast, the radiation density, from Equation 24–13, goes as T^4. Now, the wavelength of a photon is proportional to $a(t)$, so that

$$\lambda \propto a$$

and because a photon's energy is $E = h\nu = hc/\lambda$,

$$E = h\nu \propto a^{-1}$$

and for blackbody radiation,

$$T \propto a^{-1}$$

so that

$$\rho_r \propto a^{-4} \qquad (24-15)$$

At some time in the past, the energy density of radiation exceeded that of matter and the Universe was *radiation-dominated*.

Another way to compare the cosmic radiation with matter is to compute the ratio of number of photons to number of protons or neutrons. Blackbody radiation peaks at a wavelength

$$\lambda_{max} = 2.9 \times 10^{-3}/T$$

The energy of a photon of wavelength λ is

$$E_{ph} = hc/\lambda$$

and so for photons of wavelength $\lambda = \lambda_{max}$,

$$E_{ph} = hcT/(2.9 \times 10^{-3})$$

With $h = 6.6 \times 10^{-34}$, $c = 3 \times 10^8$, and $T = 2.7$ K, the energy of the photons at the peak of the blackbody curve is

$$\begin{aligned} E_{ph} &= (6.6 \times 10^{-34})(3 \times 10^8)(2.7)/(2.9 \times 10^{-3}) \\ &= 2.0 \times 10^{-22}\,\text{J/photon} \end{aligned}$$

The photons making up the blackbody radiation have a variety of energies, but if we take this peak energy as typical, the number of photons needed to produce the 4×10^{-14} J of radiation contained in every cubic meter of the cosmic radiation at 2.7 K is

$$\begin{aligned} n_{ph} &= (4 \times 10^{-14}\,\text{J/m}^3)/(2.0 \times 10^{-22}\,\text{J/photon}) \\ &\approx 2 \times 10^8\,\text{photons/m}^3 \end{aligned}$$

Now consider the matter. The mass of one nucleon (a proton or neutron) is 1.7×10^{-27} kg. The 4×10^{-28} kg of matter contained in 1 m³ is equivalent to

$$\begin{aligned} n_{nu} &= (4 \times 10^{-28}\,\text{kg/m}^3)/ \\ &\quad 1.7 \times 10^{-27}\,\text{kg/nucleon}) \\ &\approx 2 \times 10^{-1}\,\text{nucleons/m}^3 \end{aligned}$$

The ratio of these two numbers is

$$n_{ph}/n_{nu} = 2 \times 10^8/2 \times 10^{-1} = 10^9$$

So there are 1 billion times as many photons in the Universe as nucleons.

One of the characteristics of cosmic blackbody radiation is that the total number of photons remains the same if account is taken of the expanding volume. Similarly for matter: the total number of nucleons remains the same. So the above ratio of photons to nucleons will stay constant as the Universe expands.

Now to recast our equations of motion for the matter-dominated and radiation-dominated cases. To do so, we need to use the conservation of energy for a sample volume V. We do so in the form of the first law of thermodynamics:

$$dE + P\,dV = 0$$

where P is the pressure and E is the matter–energy density in V, so that $E = \rho c^2$. Now $V \propto a^3(t)$, so that

$$dE/dt + P(dV/dt) = 0$$

gives (with dots representing the derivative with respect to time)

$$d/dt\,(\rho c^2 a^3) + P(d/dt)(a^3) = 0$$

$$c^2 a^3 \dot{\rho} + 3\rho c^2 a^2 \dot{a} + 3P a^2 \dot{a} = 0$$

$$\dot{\rho} = -3(\rho + P/c^2)(\dot{a}/a)$$

$$\dot{\rho} = -3(\rho + P/c^2)H \qquad \textbf{(24-16)}$$

Let $P = 0$ (noninteracting matter) and assume that matter dominates ($E_{\text{mat}} = \rho_{\text{mat}} c^2 > E_{\text{rad}} = \rho_{\text{rad}} c^2$), so that we have

$$\dot{\rho} = -3\rho H$$

For a flat geometry, however (which is the transitional case),

$$H = [(8\pi G/3)\rho]^{1/2}$$

so that

$$\dot{\rho} = -3\rho[(8\pi G/3)\rho]^{1/2} = 3\rho^{3/2}(8\pi G/3)^{1/2}$$

$$\rho^{-3/2}\dot{\rho} = -(24\pi G)^{1/2}$$

Now integrate this equation with respect to time:

$$\int \rho^{-3/2}\,d\rho = (24\pi G)^{1/2} \int dt$$

$$2\rho^{-1/2} = (24\pi G)^{1/2}t$$

$$t = (1/6G\pi\rho)^{1/2} \quad \text{(matter-dominated)} \quad \textbf{(24-17)}$$

This equation gives the relationship of time and density for a matter-dominated, zero-pressure model, that is, it gives $\rho(t)$.

Now we take Equation 24-16 to see what happens in the radiation-dominated case. Then

$$P = (1/3)(\text{energy density}) = (1/3)u = (1/3)\rho c^2$$

where T, ρ, and P are the temperature, density, and pressure of the *radiation*. Equation 24-16 becomes

$$\dot{\rho} = -3[\rho + (1/3)\rho]H = -4H\rho$$
$$= -4(8\pi G\rho/3)^{1/2}\rho$$

so that we have

$$\rho^{-3/2}\dot{\rho} = -(128\pi G/3)^{1/2}$$

and, as before, we integrate with respect to t:

$$2\rho^{-1/2} = (128\pi G/3)^{1/2}t$$

$$t = (3/32\pi G\rho)^{1/2} \quad \text{(radiation-dominated)}$$
$$\textbf{(24-18)}$$

Now that we have time and density related, we can

substitute temperature for radiation density because

$$P = (1/3)\rho c^2 = E/3 = aT^4/3$$

so that

$$\rho = aT^4/c^2$$

and Equation 24-18 becomes

$$T = (3c^2/32\pi Gat^2)^{1/4} \qquad \textbf{(24-19)}$$

or, if we substitute for the constants,

$$T(K) \approx (1.5 \times 10^{10})t^{-1/2} \qquad \textbf{(24-20)}$$

You should see that this equation relates every time t in a radiation-dominated Universe to a temperature T. So we can trace out the thermal history of the Universe as it evolves.

24-4 The Standard Big Bang Model

This section outlines the physical evolution of the cosmos in the standard Big Bang model—that of the primeval fireball. You will find that this model has some surprising strengths and also some crucial weaknesses. Section 24-5 presents the latest revision of the standard model, which patches up some of its weak points.

In all reasonable cosmological models based upon Einstein's general theory of relativity, an initial expansion began about 2×10^{10} years ago. Cosmologists refer to this event as the *initial singularity* because at this time the volume of the Universe was zero and the density of mass–energy was infinite. From the equations that govern the *hyperbolic, flat,* and *closed* models, we find that all three models behave in precisely the same manner near this singularity. As time passes, the evolution of the three models changes because each is characterized by a different curvature; as we look out at the Universe today, these curvature effects become observable only at redshifts of $z \geq 1$.

For simplicity, we consider the flat cosmological model of our Universe and follow the astrophysical consequences of its evolution from creation to the present epoch (Figure 24-6). In this model, the creation event—the primeval fireball—produced such a rapid expansion that the temperature and density dropped quickly. The contents of the Universe—initially a flood of energetic light and heavy particles,

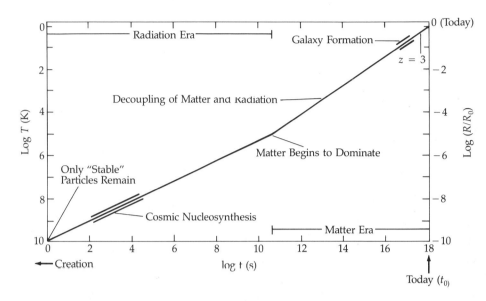

Figure 24–6

Thermal history of the Universe.

such as protons and their respective antiparticles — changed with temperature and time. In the early Universe, temperature played a crucial controlling role. Each period of time since creation can be matched with a corresponding temperature. Roughly, the Universe's thermal history divides into four eras: (1) a *heavy-particle era* when massive particles (protons and neutrons) and their antiparticles dominated, (2) a *light-particle era* when electrons and positrons were continually created and destroyed, (3) a *radiation era* when most particles vanished and radiation became the main form of energy, and (4) a *matter era,* in which we now live, when matter is the dominate form of energy.

Before we tell the story, we need to consider the creation of matter and antimatter from photons. One consequence of the fact that there are about 10^9 photons per nucleon in the Universe now is that, at some time in the past, the energy density of radiation exceeded that of matter. Though the *number* density of photons relative to nucleons has not changed, the *energy* density of radiation in the past was higher than the energy density of matter. When the temperature of the radiation was higher, each photon had a higher energy. At some point, the radiation temperature was so high that photon collisions produced matter. This process can happen when the energy contained in two colliding photons equals or exceeds the rest-mass energy of the particles produced. This production process is just the reverse of matter – antimatter annihilation, which completely converts

rest mass to photons. So when photons make matter, they must do so in the form of a matter–antimatter pair, such as an electron and positron.

Let's see what minimum temperature the radiation had to be in order for photons to produce an electron and a positron; this is called the *particle threshold temperature.* The mass of an electron or positron is 9.1×10^{-31} kg, and the equivalent energy of the two together is

$$
\begin{aligned}
E &= (m_{e-} + m_{e+})c^2 \\
&= (2 \times 9.1 \times 10^{-31})(3 \times 10^8)^2 \\
&= 1.6 \times 10^{-13} \text{ J}
\end{aligned}
$$

For blackbody radiation, the typical energy of a photon is roughly

$$E \approx kT$$

where k is Boltzmann's constant, 1.4×10^{-23} J/K. For a photon to have an energy of 1.6×10^{-13} J requires

$$
\begin{aligned}
T &\approx E/k \\
&\approx (1.6 \times 10^{-13} \text{ J})/(1.4 \times 10^{-23} \text{ J/K}) \\
&\approx 1.2 \times 10^{10} \text{ K}
\end{aligned}
$$

For protons and antiprotons, a greater temperature and energy are needed. Because a proton's mass is about 1800 times greater than that of an electron, the energy needed to form a proton–antiproton pair is $(1800)(1.6 \times 10^{-13})$ J $= 3.0 \times 10^{-10}$ J, and the threshold temperature is therefore 2.2×10^{13} K. To make a proton–antiproton pair requires two photons, each having energy 1.5×10^{-10} J, which corresponds to a photon wavelength of

$$\begin{aligned}
\lambda &= c/\nu \\
&= (3 \times 10^8)(2.3 \times 10^{23}) \\
&= 1.3 \times 10^{-15} \text{ m}
\end{aligned}$$

These photons are gamma rays. In the Big Bang model, high-energy light produced matter very early in the Universe's history.

(a) The Heavy-Particle Era

Our Universe emerged from a singularity at a very high temperature about 2×10^{10} years ago. At such temperatures, all matter behaves like photons (since particles move at essentially the speed of light); hence, the initial state was a chaotic, gaseous inferno of high-energy elementary particles and photons. An abundance of relativistic particles and antiparticles existed in equilibrium at these temperatures. As the Universe expanded, the temperature dropped and the heavier particles (hyperons and mesons) annihilated and decayed into the less-massive, stable particles (protons, neutrons, and neutrinos). Note that the neutron decays (by beta decay; $n^0 \rightarrow {}^1_1H + e^- + \bar{\nu}$ with a halflife of about 10^3 s. This halflife is long relative to the characteristic expansion time of the Universe at this stage, so that the neutron was essentially stable here.

(b) The Light-Particle Era

At a time of about 10^{-4} s, the temperature had dropped to below that needed to make heavy particles, such as protons. Only low-mass ("light") particles—electrons—could be made. At a time of 1 s, this era ended when temperatures fell below roughly 10^{10} K.

Early in this era, the interactions of electrons, protons, neutrons, and neutrinos produced a balance between the number of protons and neutrons. As the temperature decreased, the number of neutrons fell while the number of protons rose. Neutrons, totaling 16 percent of the particles, could no longer be produced by interaction of protons with electrons or of antiprotons with positrons. The neutrons were now free to decay into protons and electrons.

(c) The Radiation Era and Cosmic Nucleosynthesis

After particle formation, most of the energy in the Universe was in the form of light. We call this epoch the *radiation era*. Near the beginning of the radiation era, when the temperature was about 10^9 K, *cosmic nucleosynthesis* took place (Figure 24–7). In this epoch, however, composite nuclei could not exist, for they would immediately have dissociated into their constituent nucleons owing to the ferocity

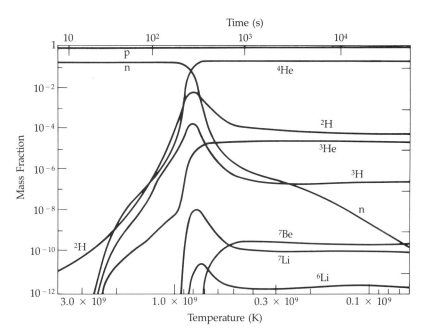

Figure 24–7

Nucleosynthesis in the Big Bang. The top axis gives the age of the Universe; the bottom gives temperature; the vertical axis is abundance in terms of the fraction of total mass. *(Adapted from a diagram by R. V. Wagoner)*

of the radiation. At about 10^9 K, nucleosynthesis began with the production of deuterium by the thermonuclear reaction

$$^1_1H + n \longrightarrow \, ^2_1H + \gamma$$

The normal proton–proton reaction that occurs in stellar interiors is far too slow to have operated at this time. Moreover, there was an abundance of neutrons. The presence of deuterium then led to other, even faster reactions in which deuterons combined to produce helium.

The decay of the original neutrons provided additional protons for this reaction, and at the end of about 5 min, the nucleosynthesis was complete. The end product of this phase of cosmic evolution was hydrogen (^1_1H), deuterium (^2_1H) and helium (^4_2He), in the relative *mass* fractions 0.75:0.01:0.25. Traces of lithium and beryllium were also made, but not heavier elements. The Universe expanded through the nucleosynthesis phase too fast to do any more fusion than this.

(d) The Matter Era

After 2000 years, when matter (hydrogen and helium) began to dominate the Universe and radiation became a secondary constituent, the *matter era* began. The evolution of this phase is governed by the standard flat-model relationship $R \propto t^{2/3}$. The matter existed in ionized form (a plasma) until the temperature dropped to the point where recombination to the neutral form could predominate, which finally occurred at $T \approx 3000$ K ($t \approx 10^6$ years). At this stage, the matter and radiation decoupled, and both evolved independently thereafter. The radiation cooled to become the 2.7 K cosmic blackbody radiation observed today, and the gaseous matter underwent a critically important transformation—it could now form the first galaxies. Prior to decoupling, radiation pressure kept the matter distribution smooth. Afterward, small inhomogeneities could grow and condense into the first gravitationally bound systems.

(e) Evidence for the Big Bang Model

The standard model has gained a following because it makes some crucial testable predictions. First, it predicts that the Universe expands. Second, it predicts that blackbody radiation pervades the Universe. Within the limits of observational errors, measurements confirm a blackbody spectrum with a tempera-

ture of roughly 3 K. Measured in many directions of space, the radiation comes to the Earth with a very uniform intensity. The background radiation has the attributes expected of a cosmic hot origin. Third, the model predicts that the primeval helium abundance should be 25 to 30 percent by mass. It sets this helium abundance as the basement level; any observation of a substantially lower amount would call the model into question. Because helium is formed in stars and some of it is ejected back into space, the present helium abundance must be larger than 25 to 30 percent.

The best objects to search for primeval helium are the oldest stars now surviving, the Population II stars. Unfortunately, most Population II stars are much too cool to excite helium lines in their spectra. However, indirect evidence from star models indicates that these stars have helium abundances of about 30 percent. Though the range of determinations is moderately large, it is remarkable that the helium abundance falls so close to the number predicted by the Big Bang model.

(f) The Future

Recall that Einstein's general theory of relativity allows the cosmos to have one of three general geometries: hyperbolic, closed, or flat. If hyperbolic or flat, the Universe will expand forever and time will never end. If closed, however, the Universe must eventually collapse, running backwards through the history outlined here.

The observational test indicates a hyperbolic Universe. The measured value of the Hubble constant ($H_0 = 50$ km/s · Mpc) and Einstein's theory give a critical density for a closed Universe of about 5×10^{-27} kg/m³. The observed density seems to be much less, and so on the surface, this evidence implies that the Universe is open (unless the missing mass exists).

The standard hot Big Bang model provides another test, one that rests on the observed present cosmic abundance of deuterium. To do Big Bang model calculations requires the *present* value of the Hubble constant, the *present* temperature of the cosmic radiation, and the *present* average density of the Universe. The first two items are reasonably well known, but the third is not. The amount of helium that comes out of the Big Bang calculations does not depend very much on the value used for the present density of the Universe, but the amount of deuterium depends very sensitively on this value (Figure 24–8). This is so because, in a low-density Universe, the deuterium nuclei do not collide as often, and so few get converted to helium before the nuclear reactions cease. So we can use the model to turn the argument around. If we can measure the present cosmic abundance of deuterium, then we can check this number against that predicted by the Big Bang model for various densities. The model then gives the present average density of the Universe, and we can compare that number with the critical density.

Unlike helium, deuterium easily burns up in fusion reactions in stars, even in cool main-sequence stars. So stars do not reveal the primordial deuterium abundance. Even if we find deuterium in the interstellar medium, we have to estimate how much of that has been processed by stars. Ultraviolet observations of the interstellar medium indicate a deuterium to hydrogen ratio that indicates that something like half of the primordial deuterium was burned in stars. Thus the original ratio may be seen to be about $4 \times$

10^{-5}, which implies a cosmic density of 4×10^{-28} kg/m^3. So the ratio of actual density to critical density may be less than unity. If so, the Universe has a flat geometry and will expand forever.

24–5 The Inflationary Universe

The successes of the hot Big Bang model pertain to events that took place more than 1 s after creation. Most of the problems arise in the chaos of the first second. A revision of the standard model called the *inflationary universe* aims to solve some of these problems by a natural incorporation of elementary particle physics with cosmology. The connection between the large and the small took place in the Big Bang (Figure 24–9). To explore this connection, we need to deal a bit with contemporary high-energy physics and Grand Unified Theories—fondly known as GUTs.

(a) Forces and Particles

Let's focus on how particles relate to each other —their interactions. We generally think of these relationships in terms of forces between particles. You are very familiar with two of these: gravitation and electromagnetism. These forces have one property in common: they both work over infinite distances. Electromagnetic forces are *much* stronger than gravity even though the range of interaction is the same (Table 24–1).

The two other forces that operate in nature work in the subatomic domain. One, called the *strong force*, holds the nuclei of atoms together; it overcomes the electromagnetic repulsive force of the protons in the nucleus. The strong force has a range of only 10^{-15} m; the diameter of a nucleus. The other, called the *weak force*, crops up in radioactive decay. Without it, fission would not take place. Like the strong force, the weak force operates over very short distances, 10^{-17} m and less.

These four forces are all that are known. As its name implies, the strong force is the strongest of them (Table 24–1), electromagnetism second, the weak force next, and gravity takes the bottom as the weakest of them all. Now, although these forces appear to operate very differently, might they have an underlying unity? The quest for such a unified theory has tempted physicists for most of this century. And

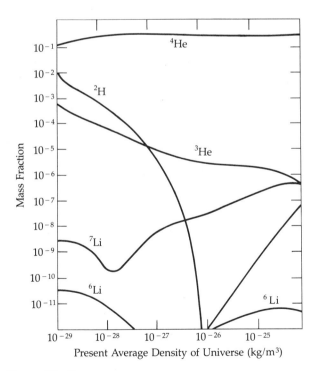

Figure 24–8

Helium formation in the Big Bang. The mass fraction has been calculated as a function of the average density of the Universe now. *(Adapted from a diagram by R. V. Wagoner)*

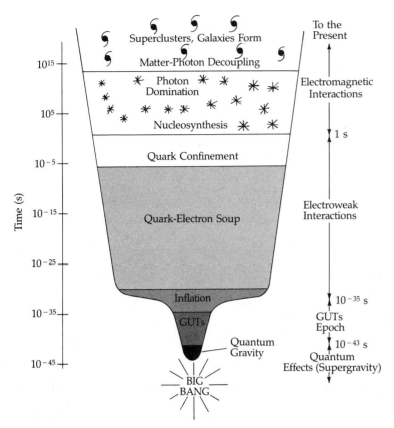

Figure 24–9

A schematic history of the early Universe. The first second of the Universe's history involved uncertain physical processes. General relativity and quantum mechanics have not yet been unified; that attempt is called supergravity. During the first 10^{-43} s, quantum fluctuations of spacetime may have generated the seeds for the growth of later structures. After this time, gravity froze out, while the other three forces remained unified. After 10^{-35} s, the electroweak forces froze out; this process somehow drove the inflation. After the inflation era, the Universe contained a hot soup of interacting quarks and electrons. At 10^{-6} s, the quarks became confined in protons that then partook in nucleosynthesis. *(J. Burns)*

recently they have had some success in struggling to find a Grand Unified Theory. The development of GUTs (there are more than one) has curiously enough connected to the Big Bang cosmology and helped to shore up some of the model's weaknesses.

How is that connection made? First, in the 1970s, theoreticians were able to unify the weak and electromagnetic forces. Buoyed by their success, they next took aim at unifying the electroweak and strong forces. As part of that work, they developed the con-

cept that a new elementary particle, called a *quark*, makes up other so-called elementary particles. (A proton consists of three quarks, for example.) GUTs also predict that the unification of the strong and electroweak forces won't be apparent until energies of greater than 10^5 J. Such energies cannot be made in particle accelerators on Earth, but they did occur in the Big Bang at a time of 10^{-35} s, when the temperature was 10^{26} K. The Big Bang serves as a way to test GUTs, and GUTs have important implications for understanding the Big Bang.

Table 24–1   Properties of the Fundamental Forces

Force	Strength (relative to strong)	Range (m)
Strong	1	10^{-15}
Electromagnetic	1/137	Infinite
Weak	10^{-5}	10^{-17}
Gravity	6×10^{-39}	Infinite

(b) Flaws in the Big Bang Model

We have so far glossed over the drawbacks of the standard model. Here we address three of them: (1) the imbalance of matter over antimatter—the *antimatter problem,* (2) the implication from observations that the geometry of spacetime is possibly flat —the *flatness problem,* and (3) the need to make galaxies rapidly in the young Universe—the *galaxy-formation problem.* Each relates to conditions during the first second of the Universe's existence—special conditions that are *assumed* to be true in the standard model rather than arising naturally from known physical ideas.

The antimatter problem arises from the fact that particle production from photons takes place in matter–antimatter pairs, exactly 50 percent of each. Yet little of the observable Universe appears to contain that percentage of antimatter. The standard model assumes that, during the radiation era, a slight imbalance of matter over antimatter was set up. This imbalance increases as the Universe ages, but no reason is given for the cause of this initial imbalance.

The flatness problem relates to the issue of whether or not the Universe's geometry is open or closed. We focus on this problem with the parameter Ω, where

$$\Omega = (\text{actual mass–energy density})/\text{critical density}$$

If $\Omega = 1$, the geometry is flat; if $\Omega < 1$, it is hyperbolic; and if $\Omega > 1$, it is spherical and closed. In relativistic models, if $\Omega = 1$, it will remain exactly so as the Universe expands. If $\Omega \neq 1$ at creation, even slightly, such a deviation would increase rapidly and be much greater than unity today. Observations indicate, however, that $0.1 \leq \Omega \leq 1$, which requires a perturbation of no more than 10^{-15} during the first second. The implication that Ω was very close to unity then is also assumed as an initial condition of the standard model.

The galaxy-formation problem relates to the observational evidence that galaxies were formed in the first few billion years of cosmic history. Yet, the radiation (before decoupling) kept matter from clumping and also smoothed out any clumpiness already present. Theoretical calculations indicate that, in the standard model, if clumpiness eventually initiated galaxy formation, gravitational collapse occurred so slowly that galaxies could have just barely formed by now! The standard model gets around this problem by assuming that a certain set of inhomogeneities occurred during the first second—again without a good physical reason.

(c) GUTs and the Cosmos

In the earliest times of the Universe, matter was in the form of elementary particles. So GUTs integrate with cosmology when the energies in the early Universe were just those appropriate to elementary-particle interactions and to the unification of the basic forces. The unification requires a symmetry among the forces, and a crucial concept is that of *spontaneously broken symmetry*. Consider, for example, freezing a liquid. Prior to freezing, a liquid can be viewed from any direction and look the same—it has rotational symmetry. After a liquid has crystallized, however, the structure looks different from different directions. The rotational symmetry is broken by the natural phase transition of freezing. The symmetry can be re-established by heating the solid so that it again changes phase. GUTs predict that a spontaneously broken symmetry of fundamental forces took place at a temperature of about 10^{27} K, corresponding to a time of 10^{-35} s in the Universe's history.

GUTs provide possible natural solutions to the three flaws of the standard-model we mentioned. First, the antimatter problem, which rests on the conservation of baryons—heavy particles such as protons and neutrons. The number of baryons minus the number of antibaryons is not supposed to change. That is true now, but before symmetry was broken, GUTs indicate that net baryon number was *not* conserved. Elementary particles interacted at a temperature just below 10^{27} K to provide the excess of matter over antimatter.

The flatness problem also relates to the symmetry breaking. The phase transition at the breaking released an enormous amount of energy, which drove a rapid and large expansion of the Universe—a factor of 10^{50} in some 10^{-32} s. (This means that the distance between two particles increased by a factor of 10^{50}.) This inflationary period neatly and naturally solves the flatness problem. Imagine that, before inflation, spacetime contains strongly curved regions. The inflation caused these regions to grow to flatness automatically. As an analogy, consider the curved surface of a partially inflated balloon. The surface is clearly curved when compared with the overall size of the balloon. Now rapidly blow up the balloon, keeping a close eye on the curvature of the surface. It becomes distinctly flatter (less curved). Similarly, when the Universe inflated, curved regions became flat. So the ratio of the actual to critical density naturally reached a value very close to unity without the need for special assumptions. This period of inflation gives this model its name: the *inflationary Universe*.

Finally, the galaxy-formation problem may also be solved by the phase transition at symmetry breaking. Consider as an analogy the freezing of a pond:

the ice sheet forms not all at once but in patches. That is, the freezing process is not perfect but has defects. These defects have mass and survive for long times — long enough so that at the decoupling time, when matter could clump, the defects served as the cores — the lumps — on which gravitational instability occurred.

Another factor involved in galaxy formation relates to neutrinos. Some GUTs predict that neutrinos should have a very small mass (and a few experiments, yet unconfirmed, support this claim) — perhaps 0.001 percent that of an electron. At the creation of the Universe, neutrinos froze out at a time of 1 s, and those relic neutrinos should be with us today. If massless, they have little effect. If they have even a slight mass, however, they can promote galaxy formation. Neutrinos with mass can start clumping by gravity well before other particles can. After recombination, atoms would gather around these neutrino clumps and so speed up the instability process. Also, massive neutrinos may congregate in clusters of galaxies and bind them gravitationally. Neutrinos can also gather in the halos of galaxies to make up the so-far invisible matter there.

Models of the clustering of hot dark matter, such as massive neutrinos, have gotten around the most serious objection to galaxy formation in the standard model. Observations rule out ordinary matter's having large disturbances in it at the time of decoupling, but neutrino matter could be lumpy and would not affect the smoothness of the background radiation. Computer simulations show that massive neutrinos would gather into pancake-like shapes hundreds of millions of light years in diameter. Where the pancakes push into each other, knots and strings of ordinary material can collect, cool, and condense. So this pancake model for galaxy formation predicts that galaxies should occur in long chains and filaments, with pancake-like voids in between. And that's what we just may see in the structure of superclusters.

Problems

1. Assume that dust grains, whose characteristic size is 1 μm, are uniformly distributed throughout intergalactic space at a mean mass density of 10^{-27} kg/m^3 (the *critical* mass density).
   (a) What is the number density (number per cubic meter) of this dust, and what is the average separation between grains?
   (b) Compare your answers to (a) with the number density and separation of the interstellar dust grains in our Galaxy.
   (c) Show that this hypothetical intergalactic dust will drastically *redden* the stars observed in the Andromeda galaxy (such reddening is *not* observed in practice).

2. If intergalactic space is filled with H II at a temperature of 10^6 K (a *plasma*),
   (a) what are the mean speed and mean kinetic energy (per particle) of these protons?
   (b) to what wavelength of electromagnetic radiation does this individual kinetic energy correspond? Could such radiation be detected from the surface of the Earth?

3. What is the approximate volume of our Galaxy (express your answer in cubic kiloparsecs)? By what scale factor must the dimensions of our Universe shrink if there is to be no empty space between the galaxies? Is this stage of cosmic expansion a reasonable time for galaxy formation?

4. Planck's law for the intensity of *blackbody radiation* (Chapter 8) is
$$I_\lambda = (2hc^2/\lambda^5)(e^{hc/\lambda kT} - 1)^{-1}$$
   As the Universe expands with a scale factor ("radius") $R(t)$, the intensity varies as $I_\lambda \propto R^{-5}$ while the wavelength goes as $\lambda \propto R$.
   (a) Show that $T \propto R^{-1}$ if the blackbody formula is to remain valid.
   (b) At what wavelength does the blackbody curve reach a maximum for the observed 2.7 K background radiation?

5. If the Hubble constant is observed to be $H_0 = 50 \pm 5$ km/s · Mpc, what are the permissible ranges for the Hubble time ($t_0 \approx H_0^{-1}$), the "size" of the Universe ($r_{hor} \approx cH_0^{-1}$), and the critical mass density ($\rho_0 \propto H_0^2$)?

6. The diagram below illustrates the famous expanding-balloon analogy for our Universe. All of space is represented by the surface of the spherical balloon, and clusters of galaxies are represented by spots painted on this surface; the radius of the balloon corresponds to $R(t)$ — the "radius" of the Universe.
   (a) As the balloon expands, the spots remain at constant angular separations (θ) from one another. Let

the balloon expand at a *constant* rate and verify that

$$\Delta s/\Delta t = (1/R)(\Delta R/\Delta t)s$$

where s is the separation between any two spots on the surface and $\Delta s/\Delta t$ is the speed of recession of one spot from another. (Note that this is Hubble's law.)

(b) The photons from distant galaxies may be represented by ants crawling along the balloon's surface at speed 1. Show that, for uniform cosmic expansion ($\Delta R/\Delta t = $ constant), there is a distance s from beyond which these ants cannot ever reach our Galaxy (this distance is called the *horizon*).

(c) Discuss the effects that take place if the balloon's expansion is decelerated [the increase of $R(t)$ is slowed down].

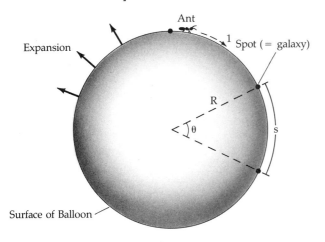

7. Consider an expanding gaseous sphere of uniform mass density ρ, total mass M, and radius $R(t)$. A gas particle at the surface of this sphere will move radially outward in accordance with the *vis-viva equation* (Chapter 1):

$$v^2/2 = GM/R + \text{constant}$$

where $v = \Delta R/\Delta t$ is the radial speed.

(a) Show that this equation may be written in the form

$$[(1/R)(\Delta R/\Delta t)]^2 = 8\pi G\rho/3 + 2(\text{constant})/R^2$$

Note that this is the equation that governs the expansion of our Universe and leads to the three cosmological models discussed in this chapter.

(b) From your knowledge of the vis-viva equation, show that the constant can be either positive, zero, or negative; illustrate the evolution of $R(t)$ in each case by drawing an approximate graph of R versus t. Comment upon your results.

8. Demonstrate that the Universe is now matter-dominated. Argue that in the past it must at some time have been radiation-dominated. (*Hint:* Background radiation at about 3 K.)

9. **(a)** We don't understand matter at $T \geq 10^{12}$ K. To what time does this correspond in the Big Bang model?

(b) Hydrogen recombines at 3000 to 4000 K. To what time does this correspond in the history of the Universe?

(c) Electrons are the lightest stable particles made in the Big Bang. What is the *latest* time they could have been formed?

10. For a flat ($k = 0$) Universe, show that

$$t_0 = (2/3)H_0^{-1}$$

where t_0 is the age of the Universe. Evaluate t for the uncertainty in H_0.

11. Consider the growth of gravitational instabilities at the time of decoupling ($T \approx 3000$ K, $\rho \approx 10^{-24}$ kg/m^3) for a pure hydrogen gas using the concept in problem 19-14 of a minimum size and mass (called the *Jeans length* and *Jeans mass*). Find the Jeans length at that time and the mass contained within it. Compare that mass with the masses of galaxies. Conclusions?

Halley's Comet Update

With an international network of telescopes (the International Halley Watch) and an armada of spacecraft, astronomers probed Halley's Comet like no other comet before. The wealth of data will take a few years to digest; here we will highlight some of the key results (as of June 1986).

The intercept missions, especially the Soviet Vega 1 and 2 and the European Space Agency (ESA) Giotto, aimed to penetrate the coma for measurements of the physical environment there and also to image the nucleus, which had never been seen — for any comet! One of the great surprises was how black the surface of the nucleus is; its albedo is only about 0.03, making it one of the darkest objects in the solar system. That the dust particles in the nuclear region were largely carbon (90 percent) and only a small part silicates (10 percent) was expected from the very low albedo. The surface of the nucleus, in fact, resembles that of a C-type asteroid (Section 7–2).

The sunlit side of the nucleus is the source of the material outflow, as expected from Whipple's dirty-iceberg model (Section 7–3). The surprise is that the outflow is concentrated in a few narrow jets (about 9 from the Giotto results), whose source regions are not more than a few kilometers in diameter. These jets spewed out dust (20 percent) and water compounds (80 percent) at speeds of about 1 km/s from about 10 percent of the total surface area. They showed up as the bright regions in CCD images (Figure U–1), and the scattered light from the dust in the jets provided the background for seeing the dark nucleus. The dust particles are very small, most of them with diameters of 10 nm and smaller. Roughly shaped like a potato,

the nucleus is about 10 by 15 km in size and rotates once every 52 hours. It has a scalloped surface, with a few crater-like features about 1 km in diameter and one hill some 1 by 2 km at its base and a few hundred meters high.

Figure U–1

The nucleus of Halley's Comet. Taken by the multicolor camera of the Giotto spacecraft, this image shows the bright regions of the jets and a dark area that is the surface of the irregularly shaped nucleus. The "sun pointer" in the upper corner indicates by the arrow the direction of the sun when the image was made. (*Max-Planck-Institute for Aeronomy and the European Space Agency*)

Figure U–2

The tail of Halley's Comet. This photo, taken by the United Kingdom Schmidt telescope in Australia, shows the coma and tail on 22 February 1986. Note the fan-shape of the dust tail, and the twisted, wavy structure in the ion tail. *(Royal Observatory, Edinburgh)*

Key instruments onboard the spacecraft were designed to detail the interaction of the comet's gases with the solar wind. A hydrogen corona was detected 10 million km out. At 1 million km, heavy ions were measured. At about this same distance, a thick bow shock was identified. It was much different from the Earth's bow shock (Section 4–6), which marks a sharp transition from the supersonic flow of the solar wind. Halley's shock was thick (roughly 10,000 km) and slowed down the solar wind gradually. Essentially, this deceleration occurred as the solar wind picked up heavy ions from the comet. Deep within the coma, the ions were mostly species of water (H_3O^+ and H_2O^+ were especially prominent) with a surprising amount of singly ionized sulfur (source unknown) and carbon (from the dust of the nucleus). The jets alone do not account for the ion production rate; some leave the surface simply because the escape velocity there is so low.

Ground-based observers also contributed to the intense, coordinated observations that gave a better chance than ever before of understanding the physics of comets. Especially interesting were the wide-angle photographs of the comet's ion tail (Figure U–2). They showed kinks and spirals as the comet's plasma interacted with the solar wind, which carried along a magnetic field. At least one series of observations indicated a magnetic disconnection (as the solar wind magnetic field flips direction) and a reconnection as the old tail drifts downstream with the wind. (Similar physical processes in the Earth's magnetosphere generate the aurorae—Section 4–6.)

Halley's Comet now heads out to aphelion before another plunge sunward. But it has left its legacy for science—the first comprehensive view of a comet, now less mysterious but also more puzzling than ever before.

Appendix 1
Bibliography

A1–1 General Astronomy Books

Abell, George. *Exploration of the Universe,* 4th ed. Philadelphia: Saunders College, 1982.

Aller, Lawrence H. *Atoms, Stars and Nebulae.* Cambridge, MA: Harvard University Press, 1971.

Audouze, Jean, and Israel, Guy. *The Cambridge Atlas of Astronomy.* Cambridge: Cambridge University Press, 1985.

Beatty, J. Kelly, O'Leary, Brian, and Chaikin, Andrew. *The New Solar System,* 2d ed. Cambridge, MA: Sky, 1982.

Bok, B. J., and Bok, P. F. *The Milky Way,* 5th ed. Cambridge, MA: Harvard University Press, 1981.

Dufay, Jean. *Introduction to Astrophysics: The Stars.* New York: Dover, 1964.

Field, G., and Chaisson, E. *The Invisible Universe.* Boston: Birkhauser, 1985.

Harrison, E. *Cosmology.* Cambridge: Cambridge University Press, 1981.

Hartmann, William K. *Astronomy: The Cosmic Journey.* Belmont, CA: Wadsworth, 1978.

Jastrow, Robert, and Thompson, M. H. *Astronomy: Fundamentals and Frontiers,* 4th ed. New York: Wiley, 1983.

Minnaert, M. G. J. *Practical Work in Elementary Astronomy.* New York: Springer-Verlag, 1969.

Noyes, R. *The Sun, Our Star.* Cambridge, MA: Harvard University Press, 1982.

Pasachoff, Jay M. *Contemporary Astronomy,* 3d ed. Philadelphia: Saunders College, 1985.

Pasachoff, J. M., and Kutner, M. L. *University Astronomy.* Philadelphia: Saunders College, 1978.

Rose, W. K. *Introduction to Astrophysics.* New York: Holt, Rinehart and Winston, 1973.

Russell, H. N., Dugan, R. S., and Stewart, J. Q. *Astronomy.* Boston: Ginn, 1926. (a classic)

Shu, Frank H. *The Physical Universe: An Introduction to Astronomy.* Mill Valley, CA: University Science Books, 1982.

Swihart, Thomas L. *Astrophysics and Stellar Astronomy.* New York: Wiley, 1968.

Unsold, Albrecht. *The New Cosmos,* 3d ed. New York: Springer-Verlag, 1983.

Zeilik, Michael. *Astronomy: The Evolving Universe,* 4th ed. New York: Harper and Row, 1985.

Zeilik, Michael, and Gaustad, John. *Astronomy: The Cosmic Perspective.* New York: Harper and Row, 1983.

A1–2 Journals and Periodicals

(a) General Astronomy (often suitable for the interested layperson)

Astronomy, monthly. AstroMedia Corp., Box 92788, Milwaukee, WI.

The Irish Astronomical Journal, quarterly. Observatories of Armagh and Dunsink, Armagh, Northern Ireland.

Journal of the British Astronomical Association, bimonthly. 303 Bath Road, Hounslow West, Middlesex, England.

Journal of the Royal Astronomical Society of Canada, bimonthly. 252 College St., Toronto, Ontario.

Mercury, bimonthly. Astronomical Society of the Pacific, 1240 24th Ave., San Francisco, CA.

The Observatory, monthly. Royal Greenwich Observatory, Herstmonceux Castle, Hailsham, Sussex, England.

Physics Today, monthly. American Institute of Physics, 335 E. 45th St., New York, NY.

Proceedings of the Astronomical Society of Australia. Sydney University Press, Sydney, Australia.

Publications of the Astronomical Society of the Pacific, bimonthly. Astronomical Society of the Pacific, 1240 24th St., San Francisco, CA.

Scientific American, monthly. 415 Madison Ave., New York, NY.

Sky and Telescope, monthly, Sky Publishing Corp., 49 Bay State Rd., Cambridge, MA.

(b) Publications of a More Technical Nature

The Astronomical Journal, monthly. American Institute of Physics, 335 E. 45th St., New York, NY.

Astronomy and Astrophysics, monthly. Springer-Verlag, 175 Fifth Ave., New York, NY.

The Astrophysical Journal, semimonthly. University of Chicago Press, 5750 Ellis Ave., Chicago, IL.

Icarus, International Journal of Solar System Studies, bimonthly. Academic Press, Orlando, FL.

Monthly Notices of the Royal Astronomical Society, monthly. Blackwell Scientific, 5 Alfred St., Oxford, England.

Space Science Reviews, monthly. D. Reidel, Singel 419–421, Dordrecht, Holland.

(c) Symposia and Continuing Series Publications

Advances in Astronomy and Astrophysics. Academic Press, Orlando, FL.

Annual Review of Astronomy and Astrophysics. Annual Review, 4139 El Camino Way, Palo Alto, CA.

Astrophysics and Space Science Library. D. Reidel, Singel 419–421, Dordrecht, Holland.

International Astronomical Union Symposia (and Colloquia). D. Reidel, Singel 419–421, Dordrecht, Holland.

Appendix 2
Constellations

Table A2–1 The Constellations

Name	Genitive Form of Name	Abbreviation	Approximate Position			
			Equatorial Coordinates		Galactic Coordinates	
			α	δ	l	b
			h	°	°	°
Andromeda	Andromedae	And	1	+40	135	−25
Antlia	Antilae	Ant	10	−35	270	+15
Apus	Apodis	Aps	16	−75	315	−15
Aquarius	Aquarii	Aqr	23	−15	50	−60
Aquila	Aquilae	Aql	20	+5	45	−15
Ara	Arae	Ara	17	−55	335	−10
Aries	Arietis	Ari	3	+20	160	−35
Auriga	Aurigae	Aur	6	+40	175	+10
Boötes	Boötis	Boo	15	+30	45	+65
Caelum	Caeli	Cae	5	−40	245	−35
Camelopardalis	Camelopardalis	Cam	6	+70	145	+20
Cancer	Cancri	Cnc	9	+20	210	+35
Canes Venatici	Canum Venaticorum	CVn	13	+40	110	+80
Canis Major	Canis Majoris	CMa	7	−20	230	−10
Canis Minor	Canis Minoris	CMi	8	+5	215	+20
Capricornus	Capricorni	Cap	21	−20	30	−40
Carina	Carinae	Car	9	−60	270	−10
Cassiopeia	Cassiopeiae	Cas	1	+60	125	−5
Centaurus	Centauri	Cen	13	−50	305	+10
Cepheus	Cephei	Cep	22	+70	110	+10
Cetus	Ceti	Cet	2	−10	170	−65
Chamaeleon	Chamaeleontis	Cha	11	−80	300	−20
Circinus	Circini	Cir	15	−60	320	0
Columba	Columbae	Col	6	−35	240	−25
Coma Berenices	Comae Berenices	Com	13	+20	320	+85

Table A2–1 *(Continued)*

Name	Genitive Form of Name	Abbreviation	Approximate Position			
			Equatorial Coordinates		Galactic Coordinates	
			α	δ	l	b
			h	°	°	°
Corona Australis	Coronae Australis	CrA	19	−40	355	−20
Corona Borealis	Coronae Borealis	CrB	16	+30	50	+50
Corvus	Corvi	Crv	12	−20	290	+40
Crater	Crateris	Crt	11	−15	270	+40
Crux	Crucis	Cru	12	−60	295	0
Cygnus	Cygni	Cyg	21	+40	85	−5
Delphinus	Delphini	Del	21	+10	60	−25
Dorado	Doradus	Dor	5	−65	275	−35
Draco	Draconis	Dra	17	+65	95	+35
Equuleus	Equulei	Equ	21	+5	55	−30
Eridanus	Eridani	Eri	3	−20	205	−60
Fornax	Fornacis	For	3	−30	225	−60
Gemini	Geminorum	Gem	7	+20	195	+10
Grus	Gruis	Gru	22	−45	355	−55
Hercules	Herculis	Her	17	+30	50	+35
Horologium	Horologii	Hor	3	−60	380	−50
Hydra	Hydrae	Hya	10	−20	260	+25
Hydrus	Hydri	Hyi	2	−75	300	−40
Indus	Indi	Ind	21	−55	340	−40
Lacerta	Lacertae	Lac	22	+45	100	0
Leo	Leonis	Leo	11	+15	230	+60
Leo Minor	Leonis Minoris	LMi	10	+35	190	+55
Lepus	Leporis	Lep	6	−20	225	−20
Libra	Librae	Lib	15	−15	345	+35
Lupus	Lupi	Lup	15	−45	325	+10
Lynx	Lyncis	Lyn	8	+45	175	+30
Lyra	Lyrae	Lyr	19	+40	70	+15
Mensa	Mensae	Men	5	−80	290	−30
Microscopium	Microscopii	Mic	21	−35	10	−40
Monoceros	Monocerotis	Mon	7	−5	210	0
Musca	Muscae	Mus	12	−70	300	−10
Norma	Normae	Nor	16	−50	330	0
Octans	Octantis	Oct	22	−85	305	−30
Ophiuchus	Ophiuchi	Oph	17	0	30	+15
Orion	Orionis	Ori	5	+5	195	−15
Pavo	Pavonis	Pav	20	−65	330	−30
Pegasus	Pegasi	Peg	22	+20	80	−25
Perseus	Persei	Per	3	+45	145	−10
Phoenix	Phoenicis	Phe	1	−50	300	−70
Pictor	Pictoris	Pic	6	−55	260	−30
Pisces	Piscium	Psc	1	+15	125	−45

Table A2 – 1   *(Continued)*

Name	Genitive Form of Name	Abbreviation	Approximate Position			
			Equatorial Coordinates		Galactic Coordinates	
			α	δ	l	b
			h	°	°	°
Piscis Austrinus	Piscis Austrini	PsA	22	−30	20	−50
Puppis	Puppis	Pup	8	−40	255	−5
Pyxis	Pyxidis	Pyx	9	−30	255	+10
Reticulum	Reticuli	Ret	4	−60	270	−45
Sagitta	Sagittae	Sge	20	+10	50	0
Sagittarius	Sagittarii	Sgr	19	−25	10	−15
Scorpius	Scorpii	Sco	17	−40	345	0
Sculptor	Sculptoris	Scl	0	−30	10	−80
Scutum	Scuti	Sct	19	−10	25	−5
Serpens	Serpentis	Ser	17	0	20	+5
Sextans	Sextantis	Sex	10	0	240	+40
Taurus	Tauri	Tau	4	+15	180	−30
Telescopium	Telescopii	Tel	19	−50	350	−20
Triangulum	Trianguli	Tri	2	+30	140	−30
Triangulum Australe	Trianguli Australis	TrA	16	−65	320	−10
Tucana	Tucanae	Tuc	0	−65	310	−50
Ursa Major	Ursae Majoris	UMa	11	+50	160	+60
Ursa Minor	Ursae Minoris	UMi	15	+70	110	+45
Vela	Velorum	Vel	9	−50	260	0
Virgo	Virginis	Vir	13	0	310	+65
Volans	Volantis	Vol	8	−70	280	−20
Vulpecula	Vulpeculae	Vul	20	+25	65	−5

Appendix 3
Solar System Data

Table A3 – 1   Planetary Orbits

| Planet | Symbol | Synodic Period (Days) | Sidereal Period | | Semimajor Axis | | Eccentricity | Inclination to Ecliptic |
			Tropical Years	Days	AU	10^6 km		
Mercury	☿	115.9	0.241	87.96	0.387	57.9	0.206	7.00°
Venus	♀	583.9	0.615	224.70	0.723	108.2	0.007	3.39
Earth	⊕	—	1.000	365.26	1.000	149.6	0.017	0.00
Mars	♂	779.9	1.88	686.98	1.524	228.0	0.093	1.85
Jupiter	♃	398.9	11.86	—	5.203	778.3	0.048	1.31
Saturn	♄	378.1	29.46	—	9.54	1427	0.056	2.49
Uranus	♅	369.7	84.01	—	19.18	2871	0.047	0.77
Neptune	♆	367.5	164.8	—	30.06	4497	0.009	1.77
Pluto	♇	366.7	248.6	—	39.44	5913	0.249	17.15

Table A3 – 2   Planetary Rotation

Planet	Sidereal Rotation Period	Oblateness	Obliquity*
Mercury	58.6 days	0	0.0°
Venus	243 days	0	177.4
Earth	23^h 56^m 4.1^s	0.0034	23.5
(Moon)	27.3 days	0.0006	6.7
Mars	24^h 37^m 22.6^s	0.0052	25.2
Jupiter	9^h 50^m–9^h 55^m	0.062	3.1
Saturn	10^h 14^m–10^h 38^m	0.096	26.7
Uranus	17^h	0.06	98
Neptune	17^h 50^m	0.02	29
Pluto	6.4 days	?	65

* Obliquity is defined as the inclination of the equator to the orbital plane. Obliquities greater than 90° imply retrograde rotation.

Table A3 – 3   Planetary Physical Data

Planet	Mass 10^{24} kg	Mass $\oplus = 1$	Equatorial Radius km	Equatorial Radius $\oplus = 1$	Average Density (kg/m^3)	Surface Gravity ($\oplus = 1$)	Albedo	Escape Speed (km/s)	Temperature (K) Equilibrium Blackbody	Temperature (K) Observed	Temperature (K) Subsolar Blackbody
Terrestrial											
Mercury	0.33	0.055	2,440	0.38	5.4 × 10^3	0.38	0.06	4.2	445	100–700	633
Venus	4.87	0.815	6,050	0.95	5.2	0.91	0.76	10.3	325	700	464
Earth	5.98	1.000	6,378	1.00	5.52	1.00	0.3–0.5	11.2	277	250–300	395
Moon	0.07	0.012	1,738	0.27	3.34	0.16	0.07	2.4	277	120–390	395
Mars	0.64	0.107	3,394	0.53	3.9	0.39	0.16	5.1	225	210–300	319
Pluto	0.01	0.002	1,700	0.27	0.5(?)	0.03	0.5	1.3	44	40	63
Jovian											
Jupiter	1900	318	68,700	11.19	1.40	2.74	0.51	61	122	110–150	173
Saturn	569	95	57,550	9.01	0.69	1.17	0.50	36	90	95	127
Uranus	87	14.5	25,050	3.93	1.19	0.94	0.66	22	63	58	90
Neptune	103	17.2	24,700	3.87	1.66	1.15	0.62	23	50	56	72

Table A3 – 4   Satellites of Terrestrial Planets

Planet	Moon	Distance from Planet (10^3 km)	Sidereal Period (days)	Orbital Eccentricity	Orbital Inclination (Degrees)	Radius (km)	Mass (Planet = 1)	Bulk Density (kg/m^3)
Earth	Moon	384	27.32	0.055	5.1	1738	0.012	3300
Mars	Phobos	9	0.32	0.021	1.1	14 × 11 × 9	1.5 × 10^{-8}	1900
	Deimos	23	1.26	0.003	1.6	8 × 6 × 5	3.1 × 10^{-9}	2100

Table A3–5   Satellites of Jupiter

Name	Number	Distance from Jupiter		Orbital Period (days)	Radius (km)	Mass (planet = 1)	Bulk Density (kg/m³)
		10³ km	Jupiter radii				
Metis	J16	128	1.79	0.29	20	—	—
Andrastea	J14	129	1.80	0.30	20	—	—
Almathea	J5	181	2.55	0.49	130×80	2×10^{-9}	3000
Thebe	J15	222	3.11	0.68	40	—	—
Io	J1	422	5.95	1.77	1820	4.7×10^{-5}	3530
Europa	J2	671	9.47	3.55	1500	2.6×10^{-5}	3030
Ganymede	J3	1070	15.10	7.15	2640	7.8×10^{-5}	1930
Callisto	J4	1880	26.60	16.70	2500	5.6×10^{-5}	1790
Leda	J13	11,110	156	240	≈ 4	5×10^{-13}	—
Himalia	J6	11,470	161	251	85	8.5×10^{-10}	1000
Lysithea	J10	11,710	164	260	≈ 10	1×10^{-12}	—
Elara	J7	11,740	165	260	≈ 30	4×10^{-11}	—
Ananke	J12	20,700	291	617	8(?)	7×10^{-13}	—
Carme	J11	22,350	314	692	12(?)	2×10^{-12}	—
Pasiphae	J8	23,300	327	735	14(?)	8×10^{-12}	—
Sinope	J9	23,700	333	758	10(?)	2×10^{-12}	—

Table A3–6   Satellites of Saturn

Name	Distance from Saturn		Orbital Period (days)	Radius (km)	Mass (planet = 1)	Bulk Density (kg/m³)
	10³ km	Saturn Radii				
Atlas	137.67	2.28	0.602	10×20	—	—
1980 S27	139.35	2.31	0.613	$70 \times 50 \times 40$	—	—
1980 S26	141.70	2.35	0.629	$55 \times 45 \times 35$	—	—
Epimetheus	151.42	2.51	0.694	$70 \times 60 \times 50$	—	—
Janus	151.47	2.51	0.695	$110 \times 100 \times 80$	—	—
Mimas	185.54	3.08	0.942	195	6.6×10^{-8}	1200
Enceladus	238.04	3.95	1.370	250	1×10^{-7}	1200
Tethys	294.67	4.88	1.888	530	1.3×10^{-6}	1200
Telesto	294.87	4.88	1.888	$17 \times 14 \times 13$	—	—
Calypso	294.87	4.88	1.888	$17 \times 11 \times 11$	—	—
Dione	377.42	6.26	2.737	560	1.85×10^{-6}	1400
1980 S6	377.42	6.26	2.737	$18 \times 16 \times 15$	—	—
Rhea	527.07	8.74	4.518	765	4.4×10^{-6}	1300
Titan	1221.86	20.25	15.945	2560	2.36×10^{-4}	1880
Hyperion	1481.00	24.55	21.277	$205 \times 130 \times 110$	—	—
Iapetus	3560.80	59.02	79.33	730	3.3×10^{-6}	1200
Phoebe	12,954	214.7	550.45	≈ 110	—	—

Table A3 – 7   Satellites of Uranus, Neptune, and Pluto

Planet	Satellite	Distance from Center of Planet (10^3 km)	Orbital Period (days)	Radius (km)	Mass (planet = 1)	Bulk Density (kg/m³)
Uranus	Ariel	191.8	2.52038	580	1.5×10^{-5}	
	Umbriel	267.3	4.14418	595	6×10^{-6}	
	Titania	438.7	8.70588	805	5×10^{-5}	
	Oberon	586.6	13.46326	775	3×10^{-5}	
	Miranda	130.1	1.414	160	10^{-6}	
Neptune	Triton	653.6	5.87683	1700	3×10^{-3}	
	Nereid	5570	365	470	10^{-6}(?)	
Pluto	Charon	17.5	6.39	750	0.1	500(?)

Appendix 4
Stellar Data

Table A4−1   The Nearest Stars (within 5 parsecs)

Star	α (1975)	δ (1975)	Distance* (parsecs)	Proper Motion ("/yr)	A			B			C		
					m_v	M_v	Spectral Type	m_v	M_v	Spectral Type	m_v	M_v	Spectral Type
α Centauri	14ʰ 38.0ᵐ	−60°44′	1.31	3.68	0.01	4.4	G2 V	1.4	5.8	K5 V	10.7	15	M5 eV
Barnard's Star	17 56.6	+4 37	1.83	10.34	9.54	13.2	M5 V	Unseen companion					
Wolf 359	10 55.3	+7 10	2.32	4.71	13.66	16.8	M6 eV						
BD +36° 2147	11 2.0	+36 8	2.49	4.78	7.47	10.5	M2 V	Unseen companion					
Sirius	6 44.0	−16 41	2.65	1.32	−1.47	1.4	A1 V	8.7	11.5	wd			
Luyten 726-8	1 37.7	−18 5	2.74	3.35	12.5	15.4	M5.5 eV	12.9	15.8	M6 eV			
Ross 154	18 48.3	−23 52	2.90	0.72	10.6	13.3	M4.5 eV						
Ross 248	23 40.7	+44 3	3.16	1.60	12.24	14.7	M5.5 eV						
ϵ Eridani	3 31.8	−9 33	3.28	0.97	3.73	6.1	K2 V						
Luyten 789-6	22 37.2	−15 27	3.31	3.25	12.58	14.9	M5.5 eV						
Ross 128	11 46.4	+0 57	3.32	1.40	11.13	13.5	M5 V						
61 Cygni	21 5.8	+38 37	3.43	5.22	5.19	7.5	K5 V	6.02	8.3	K7 V	Unseen companion		
ϵ Indi	22 1.4	−56 53	3.44	4.69	4.73	7.0	K5 V						
Procyon	7 38.0	+5 17	3.48	1.25	0.34	2.7	F5 IV–V	10.7	13.0	wd			
BD +59° 1915	18 42.5	+59 35	3.52	2.29	8.90	11.1	M4 V	9.69	11.9	M5 V			
BD +43° 44	0 16.9	+43 53	3.55	2.91	8.07	10.3	M2.5 eV	11.04	13.2	M4 eV			
CD −36° 15693	23 4.2	−36 0	3.59	6.90	7.39	9.6	M2 V						
τ Ceti	1 42.9	−16 4	3.67	1.92	3.50	5.7	G8 Vp	Unseen companion					

Table A4 – 1 The Nearest Stars (within 5 parsecs) (continued)

Star	α (1975)	δ (1975)	Distance* (parsecs)	Proper† Motion ("/yr)	A			B			C		
					m_v	M_v	Spectral Type	m_v	M_v	Spectral Type	m_v	M_v	Spectral Type
BD + 5° 1668	7 26.1	+5 18	3.76	3.73	9.82	11.9	M4 V						
CD − 39° 14192	21 15.8	−38 58	3.85	3.46	6.72	8.7	M0 V						
CD − 45° 1841	5 10.6	−45 0	3.91	8.72	8.81	10.8	M0						
Kruger 60	22 27.1	+57 34	3.94	0.87	9.77	11.8	M3 V	11.43	13.4	M4.5 eV			
Ross 614	6 28.1	−2 48	4.02	1.00	11.13	13.1	M4.5 eV	14.8	16.8	?			
BD − 12° 4523	16 28.9	−12 36	4.02	1.18	10.13	12.0	M4.5 eV						
v. Maanen's Star	0 47.9	+5 16	4.28	2.98	12.36	14.3	wd						
Wolf 424	12 32.1	+9 10	4.37	1.78	12.7	14.4	M5.5 eV	12.7	14.4	M6 eV			
CD − 37° 15492	0 3.9	−37 29	4.45	6.11	8.59	10.3	M3 V						
BD + 50° 1725	10 9.9	+49 35	4.61	1.45	6.59	8.3	M0 V						
CD − 46° 11540	17 26.8	−46 52	4.63	1.06	9.34	11.3	M4 V						
CD − 49° 13515	21 31.9	−49 7	4.67	0.81	9	11	Me V						
CD − 44° 11909	17 35.3	−44 18	4.69	1.14	11.2	12.8	M5 V						
Luyten 1159-16	1 58.7	+12 58	4.72	2.08	12.3	13.9	M7?						
BD + 15° 2620	13 44.5	+15 2	4.80	2.30	8.6	10.2	M2 V						
BD + 68° 946	17 36.6	+68 22	4.83	1.31	9.1	10.7	M3 V	Unseen companion					
Luyten 145-141	11 44.2	−64 42	4.85	2.69	11	12.5	wd						
Ross 780	22 51.9	−14 24	4.85	1.12	10.2	11.8	M5 V						
o² Eridani	4 14.2	−7 42	4.87	4.08	4.5	6.0	K0 V	9.2	10.7	wd	11.0	12.5	M5 eV
BD + 20° 2465	10 18.2	+20 0	4.95	0.49	9.4	10.9	M4.5 V	Unseen companion					

* Distances are from P. van de Kamp
† Proper Motions from W. Gliese
BD refers to Bonner Durchmusterung
CD refers to Cordoba Durchmusterung

Table A4-2 The 25 Brightest Stars

Star	α (1975)	δ (1975)	m_v	Distance (parsecs)	Proper Motion ("/yr)	Spectral Type	M_v
Sirius, α CMa	6ʰ44.0ᵐ	−16°41′	−1.5*	2.7	1.32	A1 V	+1.4
Canopus, α Car	6 23.6	−52 41	−0.7	55	0.02	F0 Ib	−3.1
α Centauri	14 38.0	−60 44	−0.3*	1.3	3.68	G2 V	+4.4
Arcturus, α Boo	14 14.5	+19 19	−0.1	11	2.28	K2 III	−0.3
Vega, α Lyr	18 36.0	+38 46	0.0	8.1	0.34	A0 V	+0.5
Capella, α Aur	5 14.8	+45 52	0.0*	14	0.44	G2 III	−0.7
Rigel, β Ori	5 13.3	−8 14	0.1*	250	0.00	B8 Ia	−6.8
Procyon. α CMi	7 38.0	+5 17	0.3*	3.5	1.25	F5 IV–V	+2.7
Achernar, α Eri	1 37.8	−57 22	0.5	20	0.10	B5 V	−1.0
β Centauri	14 02.1	−60 15	0.6*	90	0.04	B1 III	−4.1
Altair, α Aql	19 49.5	+8 48	0.8	5.1	0.66	A7 IV–V	+2.2
Betelgeuse, α Ori	5 53.8	+7 24	0.8†	150	0.03	M2 Iab	−5.5
Aldebaran, α Tau	4 34.0	+16 28	0.9*	16	0.20	K2 III	−0.2
α Crucis	12 25.2	−63 00	0.9*	120	0.04	B1 IV	−4.0
Spica, α Vir	13 23.9	−11 01	1.0†	80	0.05	B1 V	−3.6
Antares, α Sco	16 27.8	−26 22	1.0*†	120	0.03	M1 Ib	−4.5
Pollux, β Gem	7 43.8	+28 05	1.2	12	0.62	K0 III	+0.8
Formalhaut, α PsA	22 56.2	−29 45	1.2	7	0.37	A3 V	+2.0
Deneb, α Cyg	20 40.6	+45 11	1.3	430	0.00	A2 Ia	−6.9
β Crucis	12 46.2	−59 33	1.3	150	0.05	B0.5 IV	−4.6
Regulus, α Leo	10 7.0	+12 5	1.4*	26	0.25	B7 V	−0.6
ε Canis Majoris	6 57.7	−28 56	1.5	240	0.00	B2 II	−5.4
Castor, α Gem	7 33.0	+31 56	1.6	14	0.20	A1 V	+0.9
λ Scorpii	17 31.8	−37 5	1.6	96	0.03	B2 IV	−3.3
Bellatrix, γ Ori	5 23.8	+6 20	1.6	210	0.02	B2 III	−3.6

* Multiple-star apparent magnitude is integrated magnitude, other data are brightest component.
† The star is a variable.
Distances for more distant stars are from spectroscopic parallaxes.

Table A4-3 Stellar Characteristics by Spectral Type and Luminosity Class

Spectral Type	M_v V	M_v III	M_v Ib*	$B-V$ V	$B-V$ III	$B-V$ I	T_{eff}(K) V	T_{eff}(K) III	T_{eff}(K) I	BC V	$R/R_\odot$ V	$R/R_\odot$ III	$R/R_\odot$ I	$M/M_\odot$ V	$M/M_\odot$ III	$M/M_\odot$ I
O5	−6.0			−0.32	−0.32	−0.32	50,000			−4.30	18			40		100
B0	−4.1	−5.0	−6.2	−0.30	−0.30	−0.24	27,000			−3.17	7.6	16	20	17		50
B5	−1.1	−2.2	−5.7	−0.16	−0.16	−0.09	16,000			−1.39	4.0	10	32	7		25
A0	+0.6	−0.6	−4.9	0.00	0.00	+0.01	10,400			−0.40	2.6	6.3	40	3.6		16
A5	+2.1	+0.3	−4.5	+0.15	+0.15	+0.07	8200			−0.15	1.8		50	2.2		13
F0	+2.6	+0.6	−4.5	+0.30	+0.30	+0.24	7200			−0.08	1.3		63	1.8		13
F5	+3.4	+0.7	−4.5	+0.45	+0.45	+0.45	6700	6500	6200	−0.04	1.2	4.0	80	1.4		10
G0	+4.4	+0.6	−4.5	+0.60	+0.65	+0.76	6000	5500	5050	−0.06	1.04	6.3	100	1.1	2.5	10
G5	+5.2	+0.3	−4.5	+0.65	+0.86	+1.06	5500	4800	4500	−0.10	0.93	10	126	0.9	3	13
K0	+5.9	+0.2	−4.5	+0.81	+1.01	+1.42	5100	4400	4100	−0.19	0.85	16	200	0.8	4	13
K5	+8.0	−0.3	−4.5	+1.18	+1.52	+1.71	4300	3700	3500	−0.71	0.74	25	400	0.7	5	16
M0	+9.2	−0.4	−4.5	+1.39	+1.65	+1.94	3700	3500	3300	−1.20	0.63		500	0.5	6	16
M5	+12.3	−0.5	−4.5	+1.69	+1.85	+2.15	3000	2700		−2.10	0.32			0.2		

* All class Ia stars have an absolute visual magnitude of −7.0ᵐ.
BC is Bolometric Correction

Appendix 5
Atomic Elements

Table A5 – 1   The Periodic Table

Element	Symbol	Atomic Number	Atomic Weight*
Hydrogen	H	1	1.008
Helium	He	2	4.003
Lithium	Li	3	6.9
Beryllium	Be	4	9.0
Boron	B	5	10.8
Carbon	C	6	12.0
Nitrogen	N	7	14.0
Oxygen	O	8	16.0
Fluorine	F	9	19.0
Neon	Ne	10	20.2
Sodium	Na	11	23.0
Magnesium	Mg	12	24.3
Aluminum	Al	13	27.0
Silicon	Si	14	28.1
Phosphorus	P	15	31.0
Sulfur	S	16	32.1
Chlorine	Cl	17	35.5
Argon	A	18	39.9
Potassium	K	19	39.1
Calcium	Ca	20	40.1
Scandium	Sc	21	45.0
Titanium	Ti	22	47.9
Vanadium	V	23	51.0
Chromium	Cr	24	52.0
Manganese	Mn	25	54.9
Iron	Fe	26	55.9
Cobalt	Co	27	58.9
Nickel	Ni	28	58.7
Copper	Cu	29	63.5

Table A5 – 1   *(Continued)*

Element	Symbol	Atomic Number	Atomic Weight*
Zinc	Zn	30	65.4
Gallium	Ga	31	69.7
Germanium	Ge	32	72.6
Arsenic	As	33	74.9
Selenium	Se	34	79.0
Bromine	Br	35	79.9
Krypton	Kr	36	83.8
Rubidium	Rb	37	85.5
Strontium	Sr	38	87.6
Yttrium	Y	39	88.9
Zirconium	Zr	40	91.2
Niobium	Nb	41	92.9
Molybdenum	Mo	42	96.0
Technetium	Tc	43	(99)
Ruthenium	Ru	44	101.1
Rhodium	Rh	45	102.9
Palladium	Pd	46	106.4
Silver	Ag	47	107.9
Cadmium	Cd	48	112.4
Indium	In	49	114.8
Tin	Sn	50	118.7
Antimony	Sb	51	121.8
Tellurium	Te	52	127.6
Iodine	I	53	126.9
Xenon	Xe	54	131.3
Cesium	Cs	55	132.9
Barium	Ba	56	137.4
Lanthanum	La	57	138.9
Cerium	Ce	58	140.1

Table A5 – 1   *(Continued)*

Element	Symbol	Atomic Number	Atomic Weight*
Praseodymium	Pr	59	140.9
Neodymium	Nd	60	144.3
Promethium	Pm	61	(147)
Samarium	Sm	62	150.4
Europium	Eu	63	152.0
Gadolinium	Gd	64	157.3
Terbium	Tb	65	158.9
Dysprosium	Dy	66	162.5
Holmium	Ho	67	164.9
Erbium	Er	68	167.3
Thulium	Tm	69	168.9
Ytterbium	Yb	70	173.0
Lutetium	Lu	71	175.0
Hafnium	Hf	72	178.5
Tantalum	Ta	73	181.0
Tungsten	W	74	183.9
Rhenium	Re	75	186.2
Osmium	Os	76	190.2
Iridium	Ir	77	192.2
Platinum	Pt	78	195.1
Gold	Au	79	197.0
Mercury	Hg	80	200.6
Thallium	Tl	81	204.4
Lead	Pb	82	207.2
Bismuth	Bi	83	209.0

Table A5 – 1   *(Continued)*

Element	Symbol	Atomic Number	Atomic Weight*
Polonium	Po	84	(209)
Astatine	At	85	(210)
Radon	Rn	86	(222)
Francium	Fr	87	(223)
Radium	Ra	88	226.1
Actinium	Ac	89	(227)
Thorium	Th	90	232.1
Protoactinium	Pa	91	(231)
Uranium	U	92	238.1
Neptunium	Np	93	(237)
Plutonium	Pu	94	(244)
Americium	Am	95	(243)
Curium	Cm	96	(248)
Berkelium	Bk	97	(247)
Californium	Cf	98	(251)
Einsteinium	E	99	(254)
Fermium	Fm	100	(253)
Mendeleevium	Md	101	(256)
Nobelium	No	102	(253)
Lawrencium	Lw	103	(256)
Rutherfordium	Rf	104	(261)
Hahnium	Ha	105	(260)

* Where mean atomic weights have not been well determined, the atomic mass numbers of the most stable isotopes are given in parentheses.

Appendix 6
Conversion of Units

Astronomers have traditionally used a cgs system of units, while physicists have pretty much adopted SI units (but not without complaints!). Here we give the basic SI units and some useful conversions to cgs and English units. SI stands for Système International, the International System of units.

SI Basic Units

Length: meter (m)
Time: second (s)
Mass: kilogram (kg)
Current: ampere (A)
Temperature: kelvin (K)
Luminous intensity: candela (cd)

SI Derived Units

Force: newton (N)	$1 \text{ N} = 1 \text{ kg} \cdot \text{m/s}^2$
Work and energy: joule (J)	$1 \text{ J} = 1 \text{ N} \cdot \text{m}$
Power: watt (W)	$1 \text{ W} = 1 \text{ J/s}$
Frequency: Hertz (Hz)	$1 \text{ Hz} = \text{s}^{-1}$
Charge: coulomb (C)	$1 \text{ C} = 1 \text{ A} \cdot \text{s}$
Magnetic induction: tesla (T)	$1 \text{ T} = 1 \text{ N/A} \cdot \text{m}$
Pressure: pascal (Pa)	$1 \text{ Pa} = 1 \text{ N/m}^2$

Conversion

Length
1 km = 0.6215 mi
1 mi = 1.609 km
1 m = 1.0936 yd = 3.281 ft = 39.37 in.
1 in. = 2.54 cm
1 ft = 12 in. = 30.48 cm
1 yd = 3 ft = 91.44 cm
1 lightyear = 9.461×10^{15} m
1 Å = 0.1 nm

Area
$1 \text{ m}^2 = 10^4 \text{ cm}^2$
$1 \text{ km}^2 = 0.3861 \text{ mi}^2$
$1 \text{ in.}^2 = 6.4516 \text{ cm}^2$
$1 \text{ ft}^2 = 9.29 \times 10^{-2} \text{ m}^2$
$1 \text{ m}^2 = 10.76 \text{ ft}^2$

Volume
$1 \text{ m}^3 = 10^6 \text{ cm}^3$
$1 \text{ L} = 1000 \text{ cm}^2 = 10^{-3} \text{ m}^3$
$1 \text{ gal} = 3.786 \text{ L} = 231 \text{ in.}^3$

Time
1 h = 60 min = 3.6 ks
1 day = 24 h = 1440 min = 86.4 ks
1 year = 365.24 day = 31.56 Ms

Speed

   1 km/h = 0.2778 m/s = 0.6215 mi/h

   1 mi/h = 0.4470 m/s = 1.609 km/h

Angle and Angular Speed

   π rad = 180°

   1 rad = 57.30°

   1° = 1.745 × 10⁻² rad

   1 rev/min = 0.1047 rad/s

   1 rad/s = 9.549 rev/min

Mass

   1 g = 0.035 oz

   1 kg = 1000 g

   1 tonne = 1000 kg = 1 Mg

Density

   1 g/cm³ = 1000 kg/m³ = 1 kg/L

Force

   1 N = 0.2248 lb = 10⁵ dyn

   1 lb = 4.4482 N

Pressure

   1 Pa = 1 N/m²

   1 atm = 101.325 kPa = 1.01325 bars

   1 atm = 14.7 lb/in.² = 760 mmHg

   1 torr = 1 mmHg = 133.32 Pa

   1 bar = 100 kPa

Energy

   1 kW · h = 3.6 MJ

   1 Btu = 778 ft · lb = 252 cal = 1054.35 J

   1 eV = 1.602 × 10⁻¹⁹ J

   1 erg = 10⁻⁷ J

Power

   1 hp = 550 ft · lb/s = 745.7 W

   1 Btu/min = 17.58 W

   1 W = 1.341 × 10⁻³ hp

Magnetic Induction

   1 G = 10⁻⁴ T

   1 T = 10⁴ G

Appendix 7
Constants
and Units

Table A7–1 Astronomical Constants

Astronomical unit	$AU = 1.496 \times 10^{11}$ m
Parsec	$pc = 206{,}265$ AU
	$= 3.26$ ly
	$= 3.086 \times 10^{16}$ m
Lightyear	$ly = 6.324 \times 10^{4}$ AU
	$= 0.307$ pc
	$= 9.46 \times 10^{15}$ m
Sidereal year	$1 \text{ yr} = 365.26$ days
	$= 3.16 \times 10^{7}$ s
Mass of Earth	$M_{\oplus} = 5.98 \times 10^{24}$ kg
Radius of Earth at equator	$R_{\oplus} = 6378$ km
Orbital velocity of Earth	$V_{\oplus} = 30$ km/s
Mass of Sun	$M_{\odot} = 1.99 \times 10^{30}$ kg
Radius of Sun	$R_{\odot} = 6.96 \times 10^{5}$ km
Luminosity of Sun	$L_{\odot} = 3.90 \times 10^{26}$ W
Effective temperature of Sun	$T_{\text{eff}} = 5800$ K
Mass of Moon	$M_{\mathrm{D}} = 7.3 \times 10^{22}$ kg $= 0.0123 M_{\oplus}$
Radius of Moon	$R_{\mathrm{D}} = 1738$ km $= 0.273 R_{\oplus}$
Radius of Moon's orbit	$d_{\mathrm{D}} = 3.84 \times 10^{5}$ km
Sidereal month	$P_{\mathrm{D}} = 27.3$ days
Synodic month	$= 29.5$ days
Distance of Sun from center of Galaxy	$R_{\odot} = 8.5$ kpc
Velocity of Sun about galactic center	$V_{\odot} = 220$ km/s
Diameter of Galaxy	$= 120$ kpc
Mass of Galaxy	$M = (3.4 \times 10^{11}) M_{\odot}$

Table A7–2   Physical and Mathematical Constants

Velocity of light	$c = 3.00 \times 10^8$ m/s
Constant of gravitation	$G = 6.67 \times 10^{-11}$ N · m²/kg²
Planck constant	$h = 6.625 \times 10^{-34}$ joule · s
Boltzmann constant	$k = 1.38 \times 10^{-23}$ joule/K
Rydberg constant	$R = 1.097 \times 10^7$/m
Stefan–Boltzmann constant	$\sigma = 5.67 \times 10^{-8}$ W/m² K⁻⁴
Wien's law constant	$\lambda_{max}T = 2.898 \times 10^7$ Å K
Mass of hydrogen atom	$m_H = 1.67 \times 10^{-27}$ kg
Mass of electron	$m_e = 9.11 \times 10^{-31}$ kg
Charge of electron	$e = 1.60 \times 10^{-19}$ C
Electron volt	1 eV $= 1.602 \times 10^{-9}$ J
Wavelength equivalence of eV	1 eV $\rightarrow 1.24 \times 10^4$ Å

$\pi = 3.1416$

$e = 2.7183$; $\log_{10}e = 0.4343$

Table A7–3   Units and Conversions

giga $= 10^9$

mega $= 10^6$

kilo $= 10^3$

centi $= 10^{-2}$

milli $= 10^{-3}$

micro $= 10^{-6}$

nano $= 10^{-9}$

pico $= 10^{-12}$

Temperature

K $=$ C $+ 273$

°F $= (5/9)$C $+ 32$

Angular measure; degrees and time

$360° = 24^h = 2\pi$ rad; 1 rad $= 57°17'45'' = 206,264.8''$

$1° = 60' = 3600''$; $1° = 0.01745$ rad

$= 4^m$ $1'' = 4.848 \times 10^{-6}$ rad

$15° = 1^h$ Solid angle: sphere $= 4\pi$ sr

Appendix 8
The Greek Alphabet

Alpha	A	α	Nu	N	ν
Beta	B	β	Xi	Ξ	ξ
Gamma	Γ	γ	Omicron	O	o
Delta	Δ	δ	Pi	Π	π, ϖ
Epsilon	E	ϵ	Rho	P	ρ
Zeta	Z	ζ	Sigma	Σ	σ
Eta	H	η	Tau	T	τ
Theta	Θ	θ	Upsilon	Υ	υ
Iota	I	ι	Phi	Φ	ϕ, φ
Kappa	K	$\kappa, \varkappa$	Chi	X	χ
Lambda	Λ	λ	Psi	Ψ	ψ
Mu	M	μ	Omega	Ω	ω

Appendix 9
Mathematical
Operations

In the following seven sections, we briefly review the basic mathematical methods of astronomy and astrophysics: trigonometry, exponential notation, analytical geometry, vector analysis, series, the calculus, and mensuration formulas. The most useful results are placed in boxes and in tables for handy references.

A9–1 Trigonometry

(a) Angular Measure

Figure A–1 depicts a circle of unit radius. The angular measure θ may be specified in three ways. The most ancient and familiar procedure is to divide the circle's circumference into 360 equal parts and to term that θ corresponding to one of these parts an *arc-degree* (°). Each arc-degree is further subdivided into 60 *arc-minutes* (′) and each arc-minute into 60 *arc-seconds* (″). Hence, there are $360 \times 60 \times 60 = 1{,}296{,}000''$ in the full circle.

Astronomically, one rotation of the Earth requires 24 *hours* (^{h}) of time; we are accustomed to dividing the hour into 60 *minutes* (^{m}) and each minute into 60 *seconds* (^{s}). Hence, there are $24 \times 60 \times 60 = 86{,}400^s$ per rotation. A complete rotation (24^h) corresponds to the full circle (360°), however, so that we may say $1^h = 15°$, $1^m = 15′$, and $1^s = 15″$.

Finally, we may define a *radian* (rad) as that angle θ corresponding to a unit distance along the circumference of our unit circle. Since the entire circumfer-

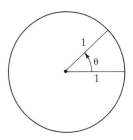

Figure A–1

ence is 2π units in length ($\pi \equiv 3.141593 \ldots$), there are 2π rad in the full 360°. Therefore: 1 rad $= 360°/2\pi = 57.2958° = 206{,}264.81.''$ [Radian measure is extended to *angular areas* by noting that the surface area of a sphere of unit radius is 4π square units, that is, 4π *steradians* (sr). Since a steradian is one square radian, there are 41,252.96 square arc-degrees on the sphere.]

(b) The Right Triangle

The triangle *OHA* in Figure A–2 is a *right triangle* since the angle at vertex *H* is 90°. With respect to the

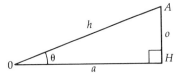

Figure A–2

angle θ, the three sides of this triangle are labeled a (adjacent), o (opposite), and h (hypotenuse). The fundamental trigonometric functions, *sine* (sin) and *cosine* (cos), are defined as

$$\sin \theta = o/h \qquad \cos \theta = a/h$$

A dependent function, the *tangent* (tan), then follows as

$$\tan \theta = o/a = (o/h)/(a/h) = \sin \theta/\cos \theta$$

Also occasionally encountered are the three reciprocal functions:

$$cosecant \longrightarrow \csc \theta = h/o = 1/\sin \theta$$
$$secant \longrightarrow \sec \theta = h/a = 1/\cos \theta$$
$$cotangent \longrightarrow \cot \theta = a/o = 1/\tan \theta$$

The following trigonometric identities are extremely useful:

Pythagorean	$\sin^2 \theta + \cos^2 \theta = 1 \qquad 1 + \tan^2 \theta = \sec^2 \theta$
Sum-and-difference	$\begin{cases} \sin(\theta \pm \phi) = \sin \theta \cos \phi \pm \cos \theta \sin \phi \\ \cos(\theta \pm \phi) = \cos \theta \cos \phi \mp \sin \theta \sin \phi \end{cases}$
Double-angle	$\begin{cases} \sin 2\theta = 2 \sin \theta \cos \theta \\ \sin^2 \theta = (1/2)(1 - \cos 2\theta) \qquad \cos^2 \theta = (1/2)(1 + \cos 2\theta) \end{cases}$

The trigonometric functions may be extended to the full circle ($0° \leq \theta \leq 360°$) by using the signs given in Table A9–1, the special values listed in Table A9–2, and the values for every arc-degree from $0°$ to $90°$. The following practical identities are needed in this extension:

(c) The Planar Triangle

Figure A–3 illustrates the general *planar triangle* ABC, with vertex angles A, B, and C and corresponding opposite sides a, b, and c. For any such triangle, the following formulas obtain:

$$\sin \theta = +\cos(\theta - 90°) = -\sin(\theta - 180°) = -\cos(\theta - 270°)$$
$$\cos \theta = -\sin(\theta - 90°) = -\cos(\theta - 180°) = +\sin(\theta - 270°)$$
$$\tan \theta = -\cot(\theta - 90°) = +\tan(\theta - 180°) = -\cot(\theta - 270°)$$

Table A9–1

Region	sin	cos	tan
$0°–90°$	+	+	+
$90°–180°$	+	−	−
$180°–270°$	−	−	+
$270°–360°$	−	+	−

Table A9–2

Angle					
Arc-Degrees	Rad	sin	cos	tan	cot
0	0	0	1	0	∞
30	$\pi/6$	1/2	$\sqrt{3}/2$	$\sqrt{3}/3$	$\sqrt{3}$
45	$\pi/4$	$\sqrt{2}/2$	$\sqrt{2}/2$	1	1
60	$\pi/3$	$\sqrt{3}/2$	1/2	$\sqrt{3}$	$\sqrt{3}/3$
90	$\pi/2$	1	0	∞	0

$$\text{Area} = \sqrt{s(s - a)(s - b)(s - c)}, \text{ where } s = (1/2)(a + b + c)$$

$$\text{Law of sines} \quad \begin{cases} \dfrac{a}{\sin A} = \dfrac{b}{\sin B} = \dfrac{c}{\sin C} \end{cases}$$

$$\text{Law of cosines} \quad \begin{cases} a^2 = b^2 + c^2 - 2bc \cos A \\ b^2 = c^2 + a^2 - 2ca \cos B \\ c^2 = a^2 + b^2 - 2ab \cos C \end{cases}$$

A9-2 Exponential Notation

(a) Powers and Roots

When a positive number a is multiplied against itself an integer m number of times, the result is the *mth power of a:*

$$a \times a \times a \times \cdots (m \text{ times}) \cdots \times a = a^m$$

When several powers of the same number are multiplied together, their *exponents* add: $a^m a^n = a^{m+n}$. If we define $a^0 \equiv 1$, then negative exponents are admitted and are called *reciprocals:* $a^{-m} = 1/a^m \Rightarrow a^m a^{-m} = a^{m-m} = a^0 = 1$.

Similarly, we term $a^{1/m}$ the *mth root of a* since we recover a when its root is raised to the mth power: $(a^{1/m})^m = a^{m/m} = a^1 = a$. Note that when a power or root is raised to a power, the two exponents involved multiply. These results are readily generalized to *any real exponent* (not necessarily an integer or a rational fraction) by the following formulas:

$$a^0 = 1 \qquad a^{-m} = 1/a^m$$
$$(ab)^m = a^m b^m$$
$$a^m a^n = a^{m+n} \qquad (a^m)^n = a^{mn}$$

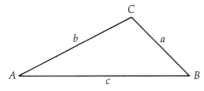

Figure A-3

We define the *factorial* of an integer n as the product of n with all smaller integers (down to 1): $n! \equiv n(n - 1)(n - 2) \cdots (3)(2)(1)$. It is conventional to also define $0! \equiv 1$.

The following simple examples illustrate these manipulations:

$$3^4 = 3 \times 3 \times 3 \times 3 = 81$$
$$2^{-3} = 1/2^3 = 1/(2 \times 2 \times 2) = 1/8$$
$$15^2 = (3 \times 5)^2 = 3^2 \times 5^2 = 9 \times 25 = 225$$
$$6^2 \times 6^3 = 6^{2+3} = 6^5$$
$$(\sqrt{2})^3 = (2^{1/2})^3 = 2^{3/2}$$
$$4! = 4 \times 3 \times 2 \times 1 = 24$$

(b) Exponentials and Logarithms

When the *base a* is given, the *exponential* formula

$$y = a^x = \text{ the base } a \text{ to the power } x$$

yields a value of y for every value of x *(exponent)* we choose. However, if we know both a and y and desire to learn x, we must invert this relationship to obtain the *logarithmic* formula

$$x = \log_a y = \text{ the exponent of } a \text{ that yields } y$$

For example, given $8 = 2^x$, we know that $x = 3$, since $2^3 = 2 \times 2 \times 2 = 8$; hence, $\log_2 8 = 3$.

The general properties of powers and roots lead to the following useful relationships for logarithms:

Product	$\log_a (xy) = \log_a x + \log_a y$
Quotient	$\log_a (x/y) = \log_a x - \log_a y$
Power	$\log_a (y^n) = n \log_a y$
Change of base	$\log_a y = (\log_a b)(\log_b y)$

In this text, we most frequently encounter the *decimal* base, $a = 10$; logarithms with respect to this base are termed *common* logarithms (written "log"). Every common logarithm consists of two parts: an integer (the *characteristic*) and an "endless" decimal (the *mantissa*). For example,

$$\log 33.7 = \log 10^{1.5276} = 1.5276$$

characteristic ⎵ ⎵ mantissa

When we use the powers-of-ten notation, $33.7 = 3.37 \times 10^1$, the characteristic 1 is immediately evident.

Important in the calculus (Section A9–6), although infrequently encountered in this text, are exponentials to the base $e \equiv 2.71828. \ldots$. The associated *natural*, or *Naperian*, logarithms are denoted "ln." In all practical computations, we will make a change of base to the decimal system (common logarithms) using the relationships:

$$e^x = 10^{0.4343x}$$
$$\ln x = (2.3026) \log x$$

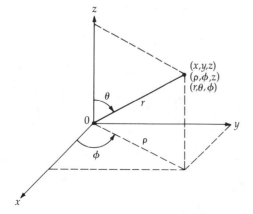

Figure A–4

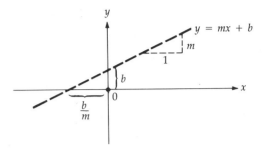

Figure A–5

A9–3 Analytical Geometry

(a) Coordinate Systems

There are three common coordinate systems used to locate a point in three-dimensional space. The most familiar system is *rectangular Cartesian* coordinates (x, y, z). Beginning at the *origin* O ($x = 0$, $y = 0$, $z = 0$), we move out the x axis x units, across parallel to the y axis y units, and up parallel to the z axis z units (Figure A–4).

In *cylindrical polar* coordinates (ρ, ϕ, z), the point is found by moving out from the origin in the xy plane the distance ρ at the angle ϕ to the x axis, then up z units parallel to the z axis (Figure A–4). These coordinates are clearly related to Cartesian coordinates by

$$x = \rho \cos \phi \qquad y = \rho \sin \phi \qquad z = z$$

Finally, in *spherical* coordinates (r, θ, ϕ), we move the distance r from the origin at the angle θ to the z axis; the projection of this motion on the xy plane is inclined the angle ϕ to the x axis and has the length

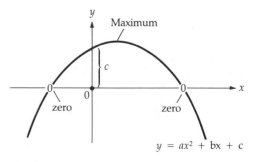

Figure A–6

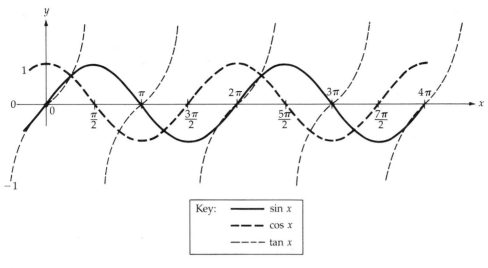

Key: ————— sin x

——— ——— cos x

———— tan x

Figure A–7

$\rho = r \sin \theta$ (Figure A–4). The connection to Cartesian coordinates is therefore given by

$$x = r \sin \theta \cos \phi \quad y = r \sin \theta \sin \phi \quad z = r \cos \theta$$

(b) Graphs

We define "y as a *function* of x" by the algebraic equation $y = y(x)$. Therefore, for each value of x, the function yields a value of y; we have an (x, y) pair. To better illustrate the properties of the function, let us *graph* every (x, y) pair as a point in a two-dimensional Cartesian coordinate system; the result is a *curve*.

Consider the *linear* equation $y = mx + b$, where m and b are constants. When $x = 0$, $y = b$. When $x = -b/m$, $y = 0$. And for every unit increase of x, y "increases" by m units; we say that the *slope* is m. The graph of this function is the *straight line* shown in Figure A–5.

Now consider the *quadratic* equation $y = ax^2 + bx + c$, where a, b, and c are constants. When $x = 0$, $y = c$; the two zeroes of the equation (where $y = 0$) are given by the *quadratic formula*:

$$x = \frac{-b \pm \sqrt{b^2 - 4ac}}{2a}$$

The graph of this function (Figure A–6) is a *parabolic curve*.

The usefulness of graphs is most evident when we consider more complicated functions. Figure A–7 shows the trigonometric functions $\sin x$, $\cos x$, and $\tan x$. Figure A–8 depicts the exponential function $y = a^x$; the logarithmic function $x = \log_a y$ may be seen by rotating the diagram 90° counterclockwise.

(c) The Conic Functions

In planar polar coordinates (ρ, ϕ), all gravitational orbits may be described by the single equation $\rho = d(1 + e)/(1 + e \cos \phi)$, where $\rho = d$ is the dis-

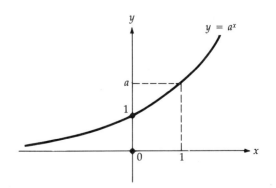

Figure A–8

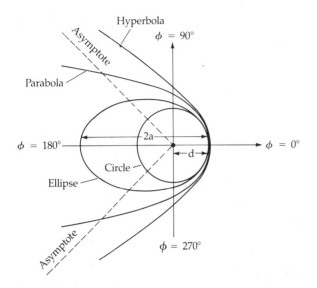

Figure A–9

tance of *closest approach* to the origin (at $\phi = 0°$). The graph of this function yields a variety of curves called the *conic sections* (Figure A–9).

When $e = 0$, we have a *circle* of radius d. When e lies in the range $0 < e < 1$, the curve is an *ellipse*; we usually write $d = a(1 - e)$, so that the *major axis* (longest dimension) of the ellipse is $2a$. When $e = 1$, the curve is a *parabola* that is "open" to the left at $\phi = 180°$. Finally, we speak of a *hyperbola* when $e > 1$; this curve exhibits $\rho \to \infty$ at the two angles where $\cos \phi = -1/e$ (along lines called *asymptotes*).

A9–4 Vector Analysis

(a) Vectors

A *vector* is like an *arrow*, for it has both a *magnitude* (length) and a *direction*. The magnitude is a *scalar*, a simple number without a direction (like temperature or mass). We denote a vector by a letter printed in boldface, **c**, and its magnitude by the same letter printed in italic type, c.

Two vectors are added, $\mathbf{c} = \mathbf{a} + \mathbf{b}$, using the *parallelogram rule of vector addition* illustrated in Figure A–10. Conversely, a vector may always be decomposed into two *component* vectors. For convenience, we decompose along the coordinate axes and write the vector as $\mathbf{c} = (c_x, c_y)$. Now the rule of vector addi-

tion may be stated in terms of components as

$$c_x = a_x + b_x \qquad c_y = a_y + b_y$$

From the Pythagorean theorem and Figure A–10, it is clear that the magnitude of **c** is just $c = (c_x^2 + c_y^2)^{1/2}$. In terms of the angle α between **c** and the x axis, the direction of **c** is given by $\tan \alpha = c_y/c_x$. Finally, as a consequence of vector addition, the magnitude of **c** may be written as

$$\begin{aligned}
c &= [c_x^2 + c_y^2]^{1/2} \\
&= [(a_x + b_x)^2 + (a_y + b_y)^2]^{1/2} \\
&= [(a_x^2 + a_y^2) + (b_x^2 + b_y^2) + 2(a_x b_x + a_y b_y)]^{1/2} \\
&= [a^2 + b^2 + 2(\mathbf{a} \cdot \mathbf{b})]^{1/2}
\end{aligned}$$

(see the vector dot product in the following section) or from the law of cosines

$$c^2 = a^2 + b^2 + 2ab \cos \beta$$

where β is the smallest angle between **a** and **b**.

As an example, consider the vectors $\mathbf{a} = (1, 1)$ and $\mathbf{b} = (3, -4)$. Their magnitudes are

$$\begin{aligned}
a &= (a_x^2 + a_y^2)^{1/2} = (1^2 + 1^2)^{1/2} \\
&= (1 + 1)^{1/2} = (2)^{1/2} = \sqrt{2} \\
b &= (3^2 + 4^2)^{1/2} = (9 + 16)^{1/2} = 5
\end{aligned}$$

Their vector sum is

$$\begin{aligned}
\mathbf{c} = \mathbf{a} + \mathbf{b} &= (a_x + b_x, a_y + b_y) \\
&= (1 + 3, 1 - 4) = (4, -3) = (c_x, c_y)
\end{aligned}$$

with the magnitude

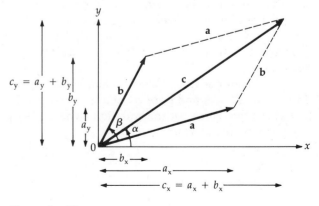

Figure A–10

$$c = (4^2 + 3^2)^{1/2} = 5$$

(Construct a diagram like Figure A–10 for these vectors and show that **a**, **b**, **c** form an *isoceles* triangle!)

(b) Dot Product

In three-dimensional Cartesian coordinates, the *vector dot product* of **a** and **b** is defined as the scalar

$$\mathbf{a} \cdot \mathbf{b} \equiv a_x b_x + a_y b_y + a_z b_z$$

If ψ is the smallest angle between **a** and **b**, then we can easily show that

$$\mathbf{a} \cdot \mathbf{b} \equiv ab \cos \psi$$

Therefore, the dot product is a measure of the component of **a** *in the direction of* **b** (or vice versa), and $\mathbf{a} \cdot \mathbf{b} = 0$ when the two vectors are perpendicular ($\psi = 90°$; Figure A–11).

Consider the example from the previous section, $\mathbf{a} = (1, 1)$ and $\mathbf{b} = (3, -4)$. Now, $\mathbf{a} \cdot \mathbf{b} = a_x b_x + a_y b_y = (1)(3) + (1)(-4) = 3 - 4 = -1$. The angle ψ satisfies

$$\begin{aligned}\cos \psi &= \mathbf{a} \cdot \mathbf{b}/ab = (-1)/(\sqrt{2})(5) \\ &= -\sqrt{2}/10 = -0.1414\end{aligned}$$

so that Table A9–1 implies that $\psi \approx 98°$.

(c) Cross Product

The *vector cross product* of **a** and **b**, denoted **a** x **b**, is *another vector* and is perpendicular to both **a** and **b**. The direction of the resultant vector is given by the *right-hand rule:* "Align the fingers of your right hand along **a**, and then rotate this hand through the smallest angle (ψ) between **a** and **b** toward **b**; your thumb will point in the direction of the cross-product vector." In terms of components, the cross product is defined by

$$\mathbf{a} \times \mathbf{b} \equiv (a_y b_z - a_z b_y,\ a_z b_x - a_x b_z,\ a_x b_y - a_y b_x)$$

The cross product is essentially a measure of the component of **a** *perpendicular to* **b** (or vice versa), and it is therefore also given by

$$|\mathbf{a} \times \mathbf{b}| \equiv ab \sin \psi$$

Note that when **a** and **b** are parallel (or antiparallel), $\mathbf{a} \times \mathbf{b} = 0$; also, it is true that $\mathbf{b} \times \mathbf{a} = -\mathbf{a} \times \mathbf{b}$ (check this by using the right-hand rule and the component definition of cross product).

Figure A–11 illustrates some of the properties of the dot product and the cross product of two vectors **a** and **b**.

We conclude with the calculation of the cross product of $\mathbf{a} = (1, 1)$ and $\mathbf{b} = (3, -4)$ (see the preceding sections):

$$\begin{aligned}\mathbf{a} \times \mathbf{b} &= (0, 0, -4 - 3) \\ &= (0, 0, -7) \quad \text{since } a_z = 0 = b_z\end{aligned}$$

Therefore, **a** x **b** is directed in the negative z direction (perpendicular to both **a** and **b**, which lie in the xy plane) and the area of the parallelogram in Figure A–11 is $|\mathbf{a} \times \mathbf{b}| = 7$. An alternate method for finding **a** x **b** is the following. First discover its direction by using the right-hand rule; then find its magnitude via $|\mathbf{a} \times \mathbf{b}| = ab \sin \psi$:

$$\begin{aligned}|\mathbf{a} \times \mathbf{b}| &= (\sqrt{2})(5) \sin 98° = (1.414)(5)(0.99) \\ &= 6.999 = 7\end{aligned}$$

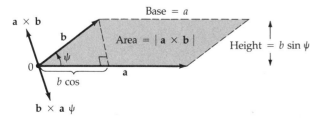

Figure A–11

A9–5 Series

In the functional relationship $y = y(x)$, we term x the *argument*. In many practical applications of astronomy and astrophysics (and particularly in the calculus; Section A9–6), we need to know the behavior of certain functions for very small values of the argument ($0 \leq x \ll 1$). Hence, we expand the function in a *series* of powers of x; useful series expansions are listed below (together with the precise range of applicable x values):

Binomial	$\begin{cases} (1 \pm x)^n = 1 \pm nx + (1/2)n(n-1)x^2 \pm (1/6)n(n-1)(n-2)x^3 + \cdots \\ \qquad\qquad (x^2 < 1;\ \text{all } n) \end{cases}$	
Trigonometric	$\begin{cases} \sin x = x - (1/6)x^3 + (1/120)x^5 - \cdots & (x^2 < 1) \\ \cos x = 1 - (1/2)x^2 + (1/24)x^4 - \cdots & (x^2 < 1) \\ \tan x = x + (1/3)x^3 + (2/15)x^5 + \cdots & (x^2 < \pi^2/4) \end{cases}$	
Exponential	$e^x = 1 + x + (1/2)x^2 + (1/6)x^3 + \cdots$	$(x^2 < 1)$
Logarithmic	$\ln(1+x) = x - (1/2)x^2 + (1/3)x^3 - (1/4)x^4 + \cdots$	$(x^2 < 1)$

Three simple examples illustrate the use of these series. First, let us evaluate $\sqrt{e}$. We have (approximately)

$$e^{1/2} = 1 + (1/2) + (1/2)(1/2)^2$$
$$+ (1/6)(1/2)^3 + \cdots$$
$$= 1 + 1/2 + 1/8 + 1/48 + \cdots$$
$$= 79/48 + \cdots \approx 1.65$$

Second, consider the very narrow triangle used in stellar parallax (Chapter 12), with short side $= 1$ AU, adjacent side $= d$ AU, and included angle at the star $= \pi(\text{rad}) \ll 1$. Then we compute

$$(1 \text{ AU})/(d \text{ AU}) = \tan \pi(\text{rad}) \approx \pi(\text{rad}) \Rightarrow$$
$$d(\text{pc}) \approx 1/\pi''$$

since there are $206{,}265''$ per radian and 1 pc $= 206{,}265$ AU.

Finally, when computing tidal accelerations (Chapter 3), we seek the very small difference between two large quantities: $GM/[r \pm (d/2)]^2$. Extract the r^2 in the denominator, and use the binomial series on the remaining denominator (since $d \ll r \Rightarrow x = d/2r \ll 1$):

$$\left(1 \pm \frac{d}{2r}\right)^{-2} = 1 \mp 2\left(\frac{d}{2r}\right) + 3\left(\frac{d}{2r}\right)^2 \cdots$$

$$\approx 1 \mp \frac{d}{r} + \cdots$$

Therefore, we quickly find

$$\frac{GM}{\left(r - \dfrac{d}{2}\right)^2} - \frac{GM}{\left(r + \dfrac{d}{2}\right)^2}$$

$$= \frac{GM}{r^2}\left[\left(1 - \frac{d}{2r}\right)^{-2} - \left(1 + \frac{d}{2r}\right)^{-2}\right]$$

$$= \frac{GM}{r^2}\left[\left(1 + \frac{d}{r} + \cdots\right) - \left(1 - \frac{d}{r} + \cdots\right)\right]$$

$$\approx 2GMd/r^3$$

A9–6 The Calculus

(a) Derivatives

We seek the *derivative* (or *instantaneous slope*) of the function $y(x)$ at the point x. As illustrated in Figure A–12, we select a nearby point $x + \Delta x$, evaluate $y(x + \Delta x)$, and in the limit as Δx becomes infinitesi-

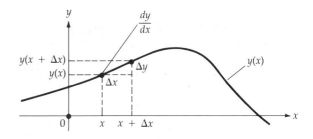

Figure A–12

mally small $\left(\lim\limits_{\Delta x \to 0}\right)$, we *define* the derivative as

$$\frac{dy}{dx} \equiv \lim_{\Delta x \to 0} \frac{y(x + \Delta x) - y(x)}{\Delta x}$$

Let us use this definition to derive two simple derivatives. Consider $y(x) = x^2$; then

$$y(x + \Delta x) = (x + \Delta x)^2 = x^2 + 2x(\Delta x) + (\Delta x)^2$$

Therefore

$$\frac{dy}{dx} = \lim_{\Delta x \to 0} \frac{x^2 + 2x(\Delta x) + (\Delta x)^2 - x^2}{\Delta x}$$

$$= \lim_{\Delta x \to 0} 2x + \Delta x = 2x$$

Hence the derivative of x^2 at x is $2x$.

Secondly, consider $y(x) = \sin x$. Then (using the sum-of-angles identity) $y(x + \Delta x) = \sin(x + \Delta x) = \sin x \cos \Delta x + \cos x \sin \Delta x$. Since Δx becomes very small, however, we may use the series expansions for $\sin \Delta x \approx \Delta x$ and $\cos \Delta x \approx 1$. Therefore,

$$\frac{dy}{dx} = \lim_{\Delta x \to 0} \frac{\sin x + (\Delta x) \cos x - \sin x}{\Delta x} = \cos x$$

the sought-for derivative of $\sin x$.

Proceeding in just this fashion, one may easily verify the following useful formulas for derivatives [where a and n are constants and $u = u(x)$ and $v = v(x)$]:

Definitions	$da/dx = 0 \qquad dx/dx = 1$
Linearity	$\left\{ \begin{array}{l} d(au)/dx = a(du/dx) \\ d(u + v)/dx = (du/dx) + (dv/dx) \end{array} \right.$
"Chain rule"	$d(uv)/dx = u(dv/dx)$ $+ v(du/dx)$
Powers	$d(u^n)/dx = nu^{n-1}(du/dx)$
Trigono-metric	$\left\{ \begin{array}{l} d(\sin u)/dx = \cos u(du/dx) \\ d(\cos u)/dx = -\sin u(du/dx) \\ d(\tan u)/dx = \sec^2 u(du/dx) \end{array} \right.$
Exponential	$\left\{ \begin{array}{l} d(a^u)/dx = a^u(\ln a)(du/dx) \\ d(e^u)/dx = e^u(du/dx) \end{array} \right.$
Logarithmic	$d(\ln u)/dx = (1/u)(du/dx)$

For example, here are the steps in finding $d(x \sin x)^2/dx$:

1. Note that this is in the form of $d(u^n)/dx$:

$$\frac{d}{dx}(x \sin x)^2 = 2(x \sin x) \frac{d}{dx}(x \sin x)$$

2. Apply the chain rule:

$$= 2(x \sin x)\left[x \frac{d(\sin x)}{dx} + \sin x \left(\frac{dx}{dx}\right)\right]$$

3. Note that $\dfrac{dx}{dx} = 1$ and $\dfrac{d(\sin x)}{dx} = \cos x$:

$$= 2(x \sin x)(x \cos x + \sin x)$$

4. Expanding:

$$= 2x^2 \sin x \cos x + 2x \sin^2 x$$

[Note that the derivative of a vector is defined in terms of the derivatives of its components: $d\mathbf{a}/dx = (da_x/dx, da_y/dx, da_z/dx)$.]

(b) Integrals

The *integral* of the function $y(x)$ may be either indefinite or definite. The *indefinite integral*, denoted by $\int y(x)\,dx$, is to be thought of as "that function of x whose derivative is $y(x)$." Hence, it is clear that $\int \cos x\,dx = \sin x$, since $d(\sin x)/dx = \cos x$. Therefore, the indefinite integral is the *inverse* of the derivative, in the sense that $\int [dy(x)/dx]\,dx = y(x)$.

The *definite integral*, denoted by $\int_a^b y(x)\,dx$, is the net area under the curve $y(x)$ between $x = a$ and $x = b$ (Figure A–13). If we have $y(x) = df(x)/dx$, then by definition it follows that

$$\int_a^b y(x)\,dx = \int_a^b (df/dx)\,dx = f(x)\Big]_a^b$$

$$= f(b) - f(a)$$

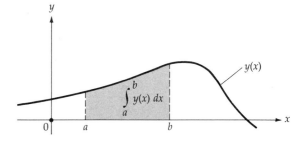

Figure A–13

In general, indefinite integrals are found by trial and error, but we can tabulate some useful known results (see the list of derivatives above):

Linearity	$$\int ay(x)\, dx = a \int y(x)\, dx$$ $$\int (u + v)\, dx = \int u\, dx + \int v\, dx$$
"By parts"	$$\int u\, dv = uv - \int v\, du$$
Powers	$$\int x^n\, dx = x^{n+1}/(n + 1)$$ (except for $n = -1$)
Trigono-metric	$$\int \sin x\, dx = -\cos x$$ $$\int \cos x\, dx = \sin x$$ $$\int \sec^2 x\, dx = \tan x$$
Exponential	$$\int e^{ax}\, dx = e^{ax}/a$$
Logarith-mic	$$\int [(dy/dx)/y(x)]\, dx = \ln y(x)$$ $$\int (1/x)\, dx = \ln x$$

For a vastly more extensive tabulation, look in any standard *table of integrals*.

We may illustrate the usefulness of this brief table of integrals by considering $\int (\sin^2 x) \cos x\, dx$. Let $u = \sin x$; then $du = \cos x\, dx$ and our integral is $\int u^2\, du = u^3/3$. Substituting back in for $u = \sin x$, we have the answer: $(1/3) \sin^3 x$. If this had been the definite integral, $\int_0^{\pi/2} (\sin^2 x) \cos x\, dx$, we would find

$$\int_0^{\pi/2} (\sin^2 x) \cos x\, dx = [(1/3) \sin^3 x]_0^{\pi/2}$$
$$= (1/3)[\sin^3 (\pi/2) - \sin^3 (0)]$$
$$= (1/3)(1^3 - 0^3) = 1/3$$

[Note that the integral of a vector is *another vector*, defined in terms of the components $\int \mathbf{a}\, dx = \int (a_x(x),\ a_y(x), a_z(x))\, dx = (\int a_x\, dx, \int a_y\, dx, \int a_z\, dx).$]

A9–7 Mensuration Formulas

Such things as lengths, areas, and volumes are given by *mensuration formulas;* a typical example is the area of a circle of radius R: $A = \pi R^2$. In Section A9–7(a), we show how to derive these formulas using the integral calculus; those interested only in the answers should proceed at once to Section A9–7(b).

(a) Multiple Integrals

At a given point in a coordinate system, we make infinitesimal changes in the three coordinates and define (a) infinitesimal lengths, (b) infinitesimal surface areas, and (c) infinitesimal volumes. By appropriately summing (that is, integrating) these, we obtain finite lengths, areas, and volumes. In general, we will be dealing with *multiple integrals*.

In rectangular Cartesian coordinates (x, y, z), the infinitesimal extensions are (dx, dy, dz). The distance along the x axis from $x = 0$ to $x = L$ is then just $\int_0^L dx = x]_0^L = L$. The infinitesimal surface areas are $dx\, dy$ [in the xy plane at (x, y, z)], $dy\, dz$, and $dz\, dx$. Therefore, the area in the xy plane bounded between $0 \leq x \leq L$ and $0 \leq y \leq W$ is $\int_0^L dx \int_0^W dy = x]_0^L y]_0^W = LW$. Finally, at (x, y, z), the infinitesimal volume is $dx\, dy\, dz$. The volume of a rectangular parallelopiped of dimensions $L \times W \times H$ is clearly $\int_0^L dx \int_0^W dy \int_0^H dz = LWH$.

In cylindrical polar coordinates, the elementary lengths are $(d\rho, \rho\, d\phi, dz)$, the elementary areas are $(\rho\, d\rho\, d\phi, \rho\, d\phi\, dz,$ and $d\rho\, dz)$, and the elementary volume is $\rho\, d\rho\, d\phi\, dz$. Therefore, the *circumference* of a circle of radius $\rho = R$ is $\int_0^{2\pi} R\, d\phi = R \int_0^{2\pi} d\phi = R\phi]_0^{2\pi} = 2\pi R$; the *area* of this circle is $\int_0^R \rho\, d\rho \int_0^{2\pi} d\phi = (1/2)\rho^2]_0^R \phi]_0^{2\pi} = (R^2/2)(2\pi) = \pi R^2$; and the *volume* of a right cylinder (of radius $\rho = R$ and height $z = H$) is $\int_0^R \rho\, d\rho \int_0^{2\pi} d\phi \int_0^H dz = \pi R^2 \int_0^H dz = \pi R^2 H$.

In spherical coordinates, the basic lengths are $(dr, r\, d\theta, r \sin \theta\, d\phi)$, the basic areas are $(r\, dr\, d\theta, r^2 \sin \theta\, d\theta\, d\phi,$ and $r\, dr \sin \theta\, d\phi)$, and the basic volume element is $r^2\, dr \sin \theta\, d\theta\, d\phi$. Therefore, the *surface area* of a sphere of radius $r = R$ is $\int_0^\pi R^2 \sin \theta\, d\theta \int_0^{2\pi} d\phi = 2\pi R^2 \int_0^\pi \sin \theta\, d\theta = 2\pi R^2[-\cos \theta]_0^\pi = 4\pi R^2$, and the volume is $\int_0^R r^2\, dr \int_0^\pi \sin \theta\, d\theta \int_0^{2\pi} d\phi = [(1/3)r^3]_0^R[-\cos \theta]_0^\pi[\phi]_0^{2\pi} = 4\pi R^3/3$.

A final example illustrates how these techniques are extended to more complex cases. Suppose we want to know the surface area of a sphere of radius $r = R$ in the range $0 \leq \theta \leq \theta_0$. The appropriate multiple integral is $R^2 \int_0^{\theta_0} \sin \theta \, d\theta \int_0^{2\pi} d\phi = 2\pi R^2 [-\cos \theta]_0^{\theta_0} = 2\pi R^2 (1 - \cos \theta_0)$. Note that the area is $2\pi R^2$ when $\theta_0 = \pi/2$ (half the sphere's surface) and $4\pi R^2$

when $\theta_0 = \pi$ (the entire surface), as should be the case.

(b) Useful Mensuration Formulas

Using methods similar to those shown in section A9–7(a), it is relatively straightforward to verify the following handy formulas:

Planar

   Arbitrary Triangle
      Area $= (1/2)$(base length) $\times$ (vertical height)
           $= \sqrt{s(s - a)(s - b)(s - c)}$ $\left\{ \begin{array}{l} \text{where } s = (12)(a + b + c) \text{ and} \\ \text{the sides have lengths } a, b, c \end{array} \right.$

   Parallelogram and Rhombus
      Area $=$ (base length) $\times$ (vertical height)

   Trapezoid
      Area $= (1/2)(a + b) \times$ (vertical height) $\left\{ \begin{array}{l} \text{where } a \text{ and } b \text{ are the} \\ \text{lengths of top and bottom} \end{array} \right.$

   Circle
      Circumference $= 2\pi$(radius) $= \pi$(diameter)
      Area $= \pi$(radius)$^2 = (\pi/4)$(diameter)^2
      Area of segment $= (1/2)$(radius)$^2(\theta - \sin \theta)$ $\left\{ \begin{array}{l} \text{where } \theta \text{ is the central} \\ \text{angle in radians} \end{array} \right.$
        Area of sector $= (1/2)$(radius)$^2\theta$
      Area of thin annulus $= 2\pi R(\Delta R)$ $\left\{ \begin{array}{l} \text{where } R \text{ is the radius and} \\ \Delta R \text{ the radial thickness} \end{array} \right.$

   Ellipse
      Area $= \pi ab$ ($a =$ semimajor axis, $b =$ semiminor axis)

Solid

   Rectangular Parallelopiped
      Volume $= abc$ (the sides have lengths a, b, c)
   Pyramid and Cone
      Volume $= (1/3)$(base area) $\times$ (vertical height)
   Right Cylinder
      Volume $= \pi R^2 H$ ($R =$ radius, $H =$ height)
   Sphere
      Surface area $= 4\pi$(radius)$^2 = \pi$(diameter)^2
      Volume $= (4\pi/3)$(radius)$^3 = (\pi/6)$(diameter)^3
      Surface area of segment $= 2\pi$(radius) $\times$ (height of segment)
      Volume of segment $= (\pi/3)$(height)$^2 \times$ (3 radius $-$ height)

   Ellipsoid
      Volume $= (4\pi/3)abc$ $\left\{ \begin{array}{l} \text{where } a, b, c \text{ are the lengths} \\ \text{of the three semiaxes} \end{array} \right.$

Appendix 10
The Celestial
Sphere

To map the sky, we assign locations to each of the celestial phenomena we study. We specify the three-dimensional spatial position of an event by its Cartesian (rectangular), polar, and spherical coordinates. Because angular positions are our primary interest in positional astronomy, we discuss almost exclusively *spherical coordinate systems*. On the surface of a sphere, the circumference and radius of a circle are not related by 2π and the sum of the interior angles of a triangle is always greater than 180°. Hence, the familiar planar geometry and trigonometry are not applicable; they must be replaced with spherical geometry and trigonometry, for which the most important formulas are given in the Appendix 9.

To begin, consider the surface of a sphere of arbitrary radius. Any plane passing through the center of the sphere intersects the surface in a *great circle*. We select one plane — usually the one perpendicular to an axis of rotation — and designate its great circle the *primary circle*. All great circles intersecting the primary circle perpendicularly are called *secondary circles;* all the secondary circles meet at only two points, the *poles*. We define one intersection point of the primary circle and a given secondary circle (the reference circle) as the *point of origin*. A coordinate system may now be set up on the spherical surface as follows: the position of a point A is specified by (1) the angular distance in a conventional direction along the primary circle from the point of origin to the point of intersection closest to A of the secondary circle passing through A and (2) the shortest angular distance along this secondary circle from the primary circle to point A. Before proceeding to celestial coordinates, let us illustrate these ideas using the surface of the Earth.

A10-1 Longitude and Latitude on the Earth

Figure A-14 shows the familiar longitude–latitude system of terrestrial coordinates. The *equator* is the primary circle, defined by the central plane perpendicular to the Earth's axis of rotation; the rotational axis intersects the surface of the Earth at the north and south poles. Secondary circles pass through the poles, and each semicircle terminating at both poles is termed a *meridian*. The reference semicircle, the *prime meridian*, passes through Greenwich, England, and meets the equator at the point of origin (0° longitude). *Longitude* is the shortest angular distance along the equator from the prime meridian to a given meridian; it is measured eastward or westward from 0° to 180°. The International Date Line is located essentially at longitude 180° E (or W). *Latitude* is the angular distance north or south from the equator, measured along a meridian in degrees from 0° (the equator) to 90° (the poles). Notice that planes

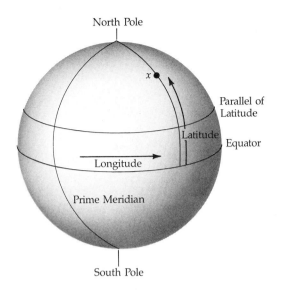

Figure A – 14

parallel to the equator slice the Earth's surface in small circles—the *parallels* of latitude. Some examples of approximate locations specified by this system are New York City (73°58′W, 40°40′N) and Sydney, Australia (151°17′E, 33°55′S).

A10–2 The Horizon System

Primary observations are clearly location-dependent, and so let us outline the observer-based *horizon,* or *azimuth – altitude,* system of coordinates. On the celestial sphere, we construct a spherical coordinate system with the observer at its center (Figure A – 15). The point vertically overhead is termed the *zenith,* and the opposite point (directly underfoot) is the *nadir.* Together these two points define an axis. The plane passing through the observer perpendicular to this axis meets the sky at the *celestial horizon,* which is everywhere 90° distance from both zenith and nadir. Because of natural and artificial obstructions, the actual horizon is seldom the celestial horizon; the closest approximation occurs for a sea-level observer in the middle of a calm ocean. Planes parallel to the axis and passing through the observer cut the celestial sphere in great circles called *vertical circles.* The reference circle is that vertical circle which includes the observer's zenith and the north and south points on her or his horizon—this circle we designate the ob-

server's *celestial meridian.* The north point on the horizon is the point of origin, and east and west lie on the horizon midway between the north and south points.

The position of a celestial phenomenon is specified in the horizon system by giving its instantaneous azimuth and altitude. *Azimuth* is the angular distance along the horizon measured eastward from the north point to the foot of the vertical circle containing the phenomenon; the foot closest to the phenomenon is the one intended, and the azimuth ranges from 0° to 360°. *Altitude* is the shortest angular distance upward along this vertical circle from the horizon to the phenomenon, and it ranges from 0° (the horizon) to 90° (the zenith). The complement of a body's altitude is its *zenith distance* (90° minus the altitude). Two events on the observer's celestial meridian are of particular interest: a celestial body is said to be in *upper transit* when it crosses the celestial meridian moving westward and in *lower transit* when it crosses moving eastward. In the cases of meteors and artificial satellites, which may rise in the west and set in the east, another criterion for upper transit is necessary: usually upper transit is that crossing of the celestial meridian visible to the observer.

A10–3 Celestial Equatorial Coordinates

We come now to the most important astronomical coordinate system—the *celestial equatorial system.* Recall that the *celestial sphere* is centered upon

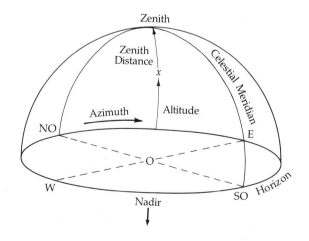

Figure A – 15

the Earth's center and that its radius is indefinitely large. This last stipulation follows because we intend to map the entire sky onto the surface of this sphere and because the lines of sight to any star (other than our Sun) are essentially perfectly parallel for any two earthbound observers. There are many ways to project a spherical coordinate system onto the interior surface of the celestial sphere; the Earth's rotation is the basis of the present method. Though much of the terminology is different, the celestial equatorial coordinate system is almost completely analogous to terrestrial longitude and latitude.

To a given observer, the apparent rotation of the celestial sphere causes all stars to circumnavigate the heavens once each day; hence, the azimuth and altitude of each star are constantly changing with time. By transforming to a spherical coordinate system that rotates with the celestial sphere — the celestial equatorial system — we can obtain positions that are *fixed* to an accuracy of one part in 10^4 per year. The chief cause of the remaining variations is the Earth's precession, which results in the westward precession of the equinoxes by about 50″ per year.

Visualize a stationary celestial sphere at the center of which the Earth rotates eastward upon its axis once each day. Imagine halting the Earth's rotation and projecting the terrestrial longitude–latitude coordinate mesh onto the surface of the celestial sphere, so that the Earth's equatorial plane cuts the celestial sphere in the great circle of the *celestial equator* and extending the Earth's rotational axis so that it intersects the sphere at the north and south *celestial poles* (Figure A–16). Meridians of longitude are mapped into *hour circles* on the celestial sphere, and parallels of latitude appear as small circles concentric to the poles. If we now permit the Earth to resume its rotation, this celestial equatorial coordinate mesh fixed to the celestial sphere appears, to any observer on the Earth's surface, to rotate westward as an entity once each day.

In the celestial equatorial coordinate system, the celestial equator is the primary circle and the hour circles are the secondary circles. The position of a celestial body is specified by its declination (DEC or δ) and its right ascension (R.A. or α). *Declination,* the analog of terrestrial latitude, is the smallest angular distance (measured in degrees, minutes, and seconds of arc) from the celestial equator to the body along the hour circle passing through the body. Positions between the celestial equator and the north celestial

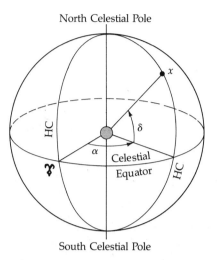

Figure A–16

pole have positive declination by convention, and those between the celestial equator and the south celestial pole have negative declination; hence, declination ranges from 0° (the celestial equator) to 90° (the north celestial pole) or −90° (the south celestial pole). *Right ascension,* the analog of terrestrial longitude, is the angular distance (measured in hours, minutes, and seconds of *time,* or ^{hms}) eastward along the celestial equator from the prime hour circle (see below) to the hour circle containing the body. Right ascension ranges from $0^h0^m0^s$ to $23^h59^m59^s$. To understand how the time units of right ascension come about, consider the prime hour circle to exactly coincide with an observer's local celestial meridian. Since the Earth rotates through 360° in 24 h, the observer's celestial meridian will lie 15° east of the prime hour circle after 1 h. Therefore, we call this right ascension $1^h0^m0^s$; a rotation of 1° corresponds to 4 min of time, of 1 arc-min to 4 s of time, and of 1 arc-sec to 1/15 s of time.

The point of origin of right ascension is the *vernal equinox* (♈), historically known as the first point of Aries. This origin, fixed on the celestial sphere, is defined by the celestial equator and the ecliptic. The *ecliptic,* the apparent annual path of the Sun in the sky, is the great circle where the orbital plane of the Earth intersects the celestial sphere. Figure A–17 is a map of the celestial sphere showing the celestial equator and the sinuous ecliptic. These two great circles are inclined at the *obliquity* angle of 23°26.5′ to one another, so that they intersect at only two points

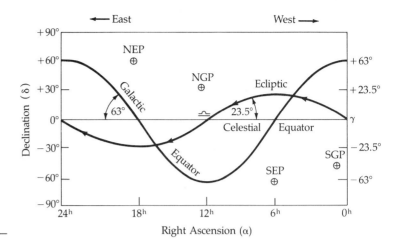

Figure A–17

—the *equinoxes*. As the Sun progresses eastward along the ecliptic, it crosses the celestial equator moving northward at the vernal equinox (spring) and again six months later moving southward at the *autumnal equinox* (fall). By definition, the right ascension–declination of the vernal equinox is $(0^h, 0°)$.

For observational purposes, we sometimes use the celestial equatorial system but take as our reference circle the local celestial meridian. Declination measures north–south angular positions, whereas *hour angle* tells us how far west of the celestial meridian a celestial body is in units of time. The hour angle of an astronomical object depends both on time and on the observer's location, being $0^h0^m0^s$ at upper transit and $2^h30^m0^s$, 2.5 h later.

A10–4 Ecliptic and Galactic Coordinates

For describing the motions of bodies within the Solar System, the *ecliptic coordinate system* is extremely useful. Here the ecliptic is the primary circle and the poles are called the *north ecliptic pole* (that pole closest to the north celestial pole) and the *south ecliptic pole*. *Celestial longitude* (λ) is the angular distance (from 0° to 360°) eastward along the ecliptic from the vernal equinox (the point of origin). *Celestial latitude* (β) is the angle from the ecliptic, measured positively toward the north ecliptic pole and negatively toward the south ecliptic pole; it ranges from 0° to $\pm90°$. Hence, when the Sun's center lies at the autumnal equinox, its coordinates are $(\lambda, \beta) = (180°, 0°)$.

When we discuss phenomena associated with our Galaxy, we find it convenient to employ modern *galactic coordinates*. Here the primary circle is defined by the central plane of the Milky Way and is called the *galactic equator*. The center of the Galaxy (in Sagittarius), which lies on the galactic equator, is the point of origin. *Galactic longitude* (l or l^{II}) is measured eastward along the galactic equator from the direction to the galactic center and ranges from 0° to 360°. As viewed from the *north galactic pole*, galactic longitude increases in the counterclockwise direction. *Galactic latitude* (b or b^{II}) is the angle from the galactic equator toward the north or south galactic pole and ranges from 0° to $\pm90°$. To avoid confusion, it should be noted that prior to August 1958, a different system of galactic coordinates (l^I, b^I) was in use; we shall have no occasion to refer to these obsolete coordinates.

In Figure A–17 we have also indicated the poles and equators of the ecliptic and galactic coordinate systems. There we see that the great circle of the galactic equator is inclined 63° to the celestial equator. Hence, we already know of three different coordinate systems, each of which completely covers the celestial sphere.

INDEX

Key:

- Definitions and principal references are in **boldface.**
  This is generally only done when there are multiple page listings.
- Page numbers for figures are *italicized.*
  This often results in listing the same page twice, once for text, once for figure.
- Page numbers for tables are followed by t, except for those in Appendix.
- References to the Appendix are followed by A.
- References to problems are followed by p.

Aaronson, Marc, 389
Aaronson-Huchra-Mould method, 389, 390
  infrared radiation, 390
Abell classification, 397
Abell, George, 395, 398
Aberration of starlight, 42, 43, *43*, 259
Absolute magnitude, **207**, 233
Absolute time, 10
Absorption, *see* Interstellar absorption
Absorption coefficient, 339, 342
  interstellar dust, 339
  interstellar medium, 342
Absorption lines, 153, 155, *155*; *see also*
      Electromagnetic radiation;
      intensity; Fraunhofer absorption
      spectrum
  equivalent width, 159, *159*, 228
  formation, 155, *156*
  profile, *158*, 159
  stellar atmospheres, 228
  temperature dependence, 160, 232, *233*
Abundance of elements, 185t
  asteroids, 128
  clusters, 236
  cosmic rays, 370, 371
  deuterium, 442
  Galaxy, 252, 364
  heavy elements, 364
  intergalactic matter, 406
  interstellar gas, 344
  magnetic variables, 314
  metals, 252
  stars, 238–240
  stellar populations, 252
  sun, 184, 185t

Acceleration, 11–14
  centripetal, *13*
Accretion
  impact cross section, *137*
  planetary, 139
  protostar, 279, 280
  solar nebula, 137, 138
  X-ray bursters, 331
Accretion disk
  Cygnus X-1, 329
  novae, 322
  SS 433, 330
Acoustic waves, 180
Active galaxies, 409–419, 425
  BL Lac objects, 417–419, *417*, *418*
  jets, *410*, *414*, 427,
  radio galaxies, 409–415, *409*, *410*, *411*, *412*, *413*, *414*
  Seyfert galaxies, 415–417, *416*
  starburst galaxies, 418–419
Active regions, *see* Solar activity
Adams, J.C.
  discovery of Neptune, 109
Adams, W.S., 237
Age of Universe, 435
Air mass, 210
Albedo, 27, 28
  asteroids, 127, 128, 139
  Earth, 59
  Phobos and Deimos, 114
Aldebaran (diameter), 224
Alpha particle, 276
Alpha Centauri, 43
Alphonsus (crater), 62
Altitude, 485A
Amalthea, 119, 120

Analytical geometry, 474–476A
Andromeda galaxy (M31), 247, *248*, 365, 385, 395
Angular momentum, 16
  galactic disk, 371
  gravitational collapse, 281
  novae, 322
  solar system, 33
  spin angular momentum, 51p
Anomalistic month, 54, 56
Anorthosites (lunar), 65, *66*
Antarctic Circle, 39
Aperture synthesis, 172
Aphelion, 24
Apogalacticon, 255p, 268
Apogee, 20
Apollo Program, 57, 64
Apparent magnitude, 207
Apparent solar time, 36, *37*
Appenine Mountains (lunar), 63
Arctic Circle, 39
Arcturus, *258*
Ariel, 124
Aristotle, 11
Arp, Halton, 426
Artificial satellites
  apogee, 20
  energy equation, 20
  escape speed, 20
  parking orbits, 19
  perigee, 20
Associations, 251, 252; *see also* OB
      Associations
Asteroid belt, 30
Asteroidal moons, 119
Asteroids, 30, 114, 119, 127, 139

albedo, 127
Apollo, 30
  compared with Halley's Comet, 448
  composition, 128
  origin, 139
  physical characteristics, 127
  Trojans, 30
Astrometry, 205
Astronomical constants, 468A
Astronomical seeing, *see* Seeing
Astronomical Unit, 6, **205**, 468A
  distance scale, 389
Astrophysical jets, 422, 425, 426, *427*, 428
Atmosphere, *see also* Earth, atmosphere
  effect on electromagnetic spectrum, 170
  ionosphere, 170
  optical depth, 177p
  transmission of electromagnetic
    radiation, 147, 148
  water vapor, 177p
Atmosphere (unit), 68
Atmosphere(s)
  planets, 28–30, *29; see also* individual
    planets
  stars, *see* Stellar atmospheres
Atmospheric extinction, 209, *209*
Atmospheric refraction, 69, *70*, 206
Atmospheric seeing, *see* Seeing
Atmospheric transmission, 147, 148t,
    170, *170*
Atomic elements, 464–465A
Atomic mass, 152
Atomic number, 152
Atomic transition, 153, 154
Atoms,
  metastable states, 156
  nuclei, 274
  spectra, 157–158; *see also* spectral lines
  structure, 152–157; *see also* Bohr atom
  subatomic particles, 152t
  subshells, 157
  valence electrons, 157
Aurorae, 71, 72, 73, *74*
  Jupiter, 104
Azimuth, 484A

B-emission stars, 317
Baade, Walter, 300
Background radiation, 436, 438
  Big Bang, 442
Bailly (lunar crater), 62
Balmer line(s), *see* Hydrogen, Balmer
    lines
Bandpass (UBV), 208
Barnard's Star, 237, 259, *259*
Barometric equation, 67
Barringer Meteor Crater, *32*
Barycenter,
  binary system, 214
  Earth-Moon system, 53, *54*
Baryons, 407
Batuski, David, 405
Becklin-Neugebauer objects, *342*, 352,
    353
Bessel, Friedrich Wilhelm, 43, 205
Beta Regio (Venus), 85, *86*

Bethe, Hans, 274
Biermann, Ludwig
  solar wind, 130
Big Bang, **392**, 402, 407, 432, 436, 438,
    444–445
  evidence of, 442
  expansion redshift, 436
  model, 436–442
Binary galactic systems, 417
Binary pulsars, 332, 333
Binary systems, 213–224, 221t
  apparent binary, 213
  apparent orbit, *214*
  astrometric binary, 213
  center of mass, 215, 217
  close binaries, 223
  contact binaries, 222, 223, *223*
    Roche lobes, 223
  eclipsing binaries, 213, 218–223; *see
    also* Eclipsing binaries
  light curve, 219, *219*, 220
  mass function, 217, **218**
  mass-luminosity relation, 215, 216
  novae, 322
  pulsars, 332
  Roche lobes, 223, 322
  RS CVn stars, 314, 315
  semidetached, 223, 322,
  spectroscopic binary, 213, 216–218, *216*
  spectrum binary, 213
  stellar data, 221t
  stellar masses, 213–219
  stellar radii, 219, 220
  stellar rotation, 222
  supernova, 325
  tidal effects, 222, 223
  velocity curve, 216–218, *217*, *218*
  visual binaries, 213–216, *214*
  white dwarfs, 294
  Wolf-Rayet stars, 318
  X-ray binaries, 328–331, 328t
Bipolar Magnetic Regions (BMR), **196**,
    197, 198
  flare model, 200
  prominences, 196
  solar cycle model, 201
  solar, 196
BL Lacertae objects, 417–419, *417, 418*
  comparison to quasars, 425
  redshift, 418
  spectrum, 418
  variability, 418
Black dwarfs, **285**, 296
  missing mass, 407
Black hole(s), **303–306**,
  Cygnus X-1, 329
  density, 303
  energy, 303
  elliptical galaxies, 385
  escape velocity, 303
  galactic center, 253, 365
  gravitational redshift, 303
  mass, 303
  missing mass, 407
  quasars, 422
  radius, 303
  rotation, 306

Schwarzschild radius, 303, 305, 422
  singularity, 303
  spacetime diagram, *304, 305*
  X-ray sources, 328
Blackbody radiation, **162–165**; *see also*
    sun, photosphere
  binary systems, 220
  color index, 233, *233*
  effective temperature, 210
  energy density, 164
  Jovian planets, 101, 101t
  Planck radiation law, 163
  planets, 27
  radio corona, 190
  Rayleigh-Jeans distribution, 164
  solar photosphere, 183
  spectral distribution, 436
  Stefan-Boltzmann law, 165
  temperature, 165
  Wien's law, 164
BMR, *see* Bipolar Magnetic Regions
Bode's law, 24, 30
Bode, Johann, 24
Bohr atom, 152–157
  continuum, 153
  energy level diagrams, 154, *154*
  ionic spectra, 158
  quantized radiation, 153
  spectral lines, 154
  wave number, 153
Bohr, Neils
  atomic theory, 152
Bolometer, 173, 174
Bolometric correction (BC), **210**, 436A
Bolometric magnitude, 210
  relation to temperature and radius,
    211p
Boltzmann constant, 159, 469A
Boltzmann equation, **159**, 165t, 228, 229
  chromospheric temperature, 185
  molecular clouds, 349
  stellar atmospheres, 220, 227, 230
Boltzmann, Ludwig, 159
Boltzmann-Saha temperature, 160, 162,
    232
Bombardment, 120; *see also* cratering
Bow shock, 449
Brackett alpha line, 352
Bradley, James, 42
  nutation, 48
Braginsky, V.B., 433
Brahe, Tycho, 205
Breccia, 65, *66*
  meteorites, 134
Broadening, *see* Spectral line, broadening
Brown dwarfs, 296
Burnell, Jocelyn Bell, 298
Burns, Jack, 405

3C 48, 419
3C 206, 426, *426*
3C 273, 419, *420*, 422, *423*
Ca I spectral lines, 161
Ca II spectral lines, 161
  H & K lines, 186
Calculus, 478–484A

Callisto, 118, *119*, 457A
    craters, 118
    ringed basin
Caloris Basin (Mercury), 81, *81*, 83
Cannon, Annie J., 230
Capilla Peak Observatory, 168, 177
Capture model, 140
Carbon (CNO) cycle, *see* CNO cycle
Carbon monoxide (CO)
    bipolar flow, 355, *355*
    distribution in Galaxy, 360
    galactic center, 365
    galactic rotation curve, 266, *266*, 267
    interstellar molecules, 349
    molecular clouds, 355
    radio astronomy, 265
    spiral arm structure, 363
Carbonaceous chondrites, 132
Cassini division, 121, 122
Cassini, Giovanni, 121
Cataclysmic variables, 309, 320–328, 321t
Cavendish experiment, *15*
Cavendish, Henry, 14
CCD, *see* Charge-coupled devices
cD galaxies, 378, 397, **399–400**; *see also*
        Galaxies
    clusters of galaxies, 407
    galactic cannibalism, 400
    gravitational lenses, 425
    radio sources, 409
Celestial coordinates, 36, 483–486A
Celestial mechanics, 4
Celestial sphere, *36*, 260, 483A
Cen X-3, *see* Centaurus X-3
Centaurus A, 411, 413, *413*, *414*
    jets, 422, 427
Centaurus X-3, 329, 330, *330*
    accretion disk, 329
Center of mass, *16*
    binary systems, 215, 217
    Earth-Moon system, 53
Centripetal acceleration, 13, *13*, 42
Centripetal force, 13, 19
Cepheid variables, 265, 309, 310, **311–312**
    classical Cepheids, 312
    distance scale, 385, 389
    instability strip, 310
    period-luminosity relationship, 311,
        312
    stellar evolution, 283, 285
    W Virginis stars, 312
Ceres, 127
    discovery, 30
Chandrasekhar limit, 294
Characteristic height, 75p
Charge–coupled devices (CCD), 176, 177
    images, *175*, *412*, 449
Charon, 110, *110*, 126, 458A; *see also*
        Pluto
    orbit, 111, *111*
Chemical composition; *see also*
        Abundances of elements; mean
        molecular weight
    role in stellar evolution, 281
    role in stellar structure, 277
    Russell-Vogt theorem, 277
Chondrites, 132

origin, 139
Chondrules, 132, 139
Christy, James
    discovery of Charon, 111
Chromosphere, 180, 185, 186, *see also*
        Sun, chromosphere
    fine structure, 186
    He II lines, 185
    Hydrogen Balmer lines, 185
    network, 186, 187, *187*
        Lyman alpha, 187
    plages, 186, *186*, 196
    spectrum, 185, 186
    temperature, 182, *182*, *183*
    X-ray and radio bursts, 200
Clark, Alvin, 294
Clavius (lunar crater), 62
Closed models, 439
Clumping, 445, 446
Clustering, *401*, *406*; *see also*
        Superclusters
    length, 404
Clusters, *see* Open clusters; Globular
        clusters; Star clusters
Clusters of galaxies, 386, 395–405, 431,
        432
    Abell classification, 395
    cannibalism, 399–401
    Coma, 397, *397*
    dark-matter, 399,407
    Hercules, 397, *398*
    Hubble diagram, *391*
    intergalactic matter, 405
        hot gas, 406, 407
    irregular, 395
    Local Group, 395, 395t, *396*, 397, 402
    luminosity function, 386, 395, 398, *399*
    mass distribution, 398, 407
    mass-luminosity ratio, 399
    masses, 399
    and quasars, 426
    radio galaxies, 413
    regular, 395
    superclusters, 401–405, *401*; *see also*
        Superclusters
    tidal effects, 400, *400*
    virial theorem, 399, 406
    wedge diagram, 403, *403*
    X-ray emission, 406, 407
CNO cycle, 275, *275*, 276, 282
    massive stars, 285
    WN stars, 318
CO, *see* Carbon Monoxide
Coal Sack, 336, *336*
Collisional excitation, 155, 156
    neutral-Hydrogen 21-cm line, 348
    planetary nebulae, 347
    solar transition region, 187
Color excess, 209, **338**
Color index, **208**, 209, *209*, 233, *233*, 235
    relation to temperature, 209
Color temperature, 227, 228
Color-magnitude diagram, **235**; *see also*
        Hertzsprung-Russell diagram
Coma (comet), 128, 448
Coma cluster of galaxies, 397, *397*
Coma supercluster, 403, *403*

Comet Arend-Roland, *31*
Comet Giacobini-Zinner, 130
Comet Halley, 31, *129*, 130, 448–449, *448*,
        449
Comet Ikeya-Seki, 131
Comet, Kohoutek, *129*
Comet West, 131, *131*
Comets, 30, 31, **128–131**, 132
    albedo, 448
    brightness, 130
    coma, 128, 448
    composition, 130t, 131
    dirty iceberg cometary model, 131
    escape from solar system, 50
    flourescence, 130
    formation, 139
    halo, 128, 449
    head, 128
    ion tail, 128, 449
    jets, 448
    magnetic field, 449
    meteoroid material, 132
    non-gravitational forces, 131
    nucleus, 128, 448, *448*
    Oort cloud, 31, 131, 139
    orbits, 31, 131
    spectrum, 130
    structure, 129–130, *129*
    tail, 31, 128, *449*
Comparative planetology, 78, 92
    Mars, 89
    Mercury, 82
    terrestrial planets, 94, 95
    Venus, 85, 86, 87, 88
Composition, *see* Abundances of
        elements
Computers
    image processing, 175, *410*
Conduction, 272
Conic sections, 10, *10*, 475–476A
Conjunction, 5
Conservation of energy, 17
Constellations, 452–454A
    distribution of Milky Way Galaxy, 247
Contact binaries, 222, *223*
Continental drift, 58
Continuum, *see* Electromagnetic
        radiation
Contour map, 170, *171*, 172
Contraction, *see* Gravitational
        contraction,
Convection, 180, 181, 272
    Earth's atmosphere, 41
    stars, 272, 277, 283
    sun, 180–181
    T Tauri stars, 313
    Wolf-Rayet stars, 319
Convergent point, 261, *261*
Conversion of units, 466A, 469A
Coordinate systems, *249*, 258, 483–486A
    galactic, 248
    galactocentric system, 260
Copernicus, Nicolaus, 4
    Heliocentric model, 4, 43, 205
Copernicus (lunar crater), 61, *62*, 63, *63*
Core (stellar), 277, 279, 280, 283, 328
    core burning, 286

helium burning, 285, 287
planetary nebula, 319
supernova, 325, 327
Wolf-Rayet stars, 319
Coriolis effect, 39, 40, *40*
Coriolis, Gaspard Gustave de, 40
Corona, 180, 189–191, *189, 190*
    coronal streamers, 189, *190,* 196
    electron scattering, 189
    emission lines, 190
    F corona, 189
    forbidden lines, 190
        Fe X, 190
        Fe XIV, 190
    free-free transitions, 190
    K corona, 189
    magnetic fields, 190
    opacity at radio wavelengths, 190
    radio emission, 190, 200
    temperature, *186,* 189, 191,
    transition region, *186,* 187
    X-ray, 191, *191,* 200-201, *200*
Coronal holes, 191
    solar wind, 192
Coronal interstellar gas, 351
Coronal loops, 191, *192,* 315, 316, *316*
    RS CVn stars, 315, 316
Coronal streamers, 189, *190,* 196
Coronium lines, 190
Cosmic background radiation, 436, *437,*
        438, 442
Cosmic expansion, *see* Expansion of
        Universe
Cosmic parameter (Omega), 445
Cosmic rays, **371,** 372, 373
    acceleration, 372
    composition, 371
    energies, 371
    galactic, 371–373
    galactic magnetic field, 372, 373
    interstellar gas, 343
    lifetime, 371
    secondary, 371
    solar, 198
    source, 372
Cosmological redshift, 387, 392; *see also*
        Hubble's law, Redshift
    quasars, 421
Cosmology, 433–446
    models, *435*
    Newtonian, 434–435
    relativistic, 435–436
Coulomb's law, 152
Coulombic barrier, 274
Coulomb (unit), 72
Crab Nebula, 300, 301, 325–327, *326*
    cosmic rays, 372
    distance, 325
    energy, 325
    expansion rate, 325
    jets, 427
    magnetic field, 325
    polarization of, 325
    pulsar, 299, 300–303, *301,* 325
    radio emission, 326
    spectrum, 325
    Taurus A, 325

X-ray emissions, 325, 326
Crab Nebula pulsar, 300–303, 326
    energy, 300–302
    optical pulses, 300, *301*
    slowdown, 300–302, 327
Cratering, *see also* individual planets and
        moons
    ejecta, *63*
    Mercury, 83
    meteoric impact, 62
    Moon, *95,* 140
    terrestrial planets, 95
    Venus, 88
Craters, *see* cratering; *see also* individual
        planets and moons by name
Curvature of Universe, 435, 439
Cyclone, 41
Cyclotron frequency radiation
    Jupiter, 103
Cygnus A, 411, *412*
Cygnus X-1, 328, 329; *see also* X-ray
        binaries
    accretion disk, 329
    mass, 329

DAM (Jupiter), 102–103
Dark Nebulae, 336, *336, 337,* 338, 354
    star formation, 354
Dating
    radioactive decay, 59
Daughter atoms, 59
Davis, Robert, 276
De-excitation, *see* excitation of atoms
Decameter radio bursts (DAM)
    Jupiter, 102–103
Decay constant (radioactivity), 59
Decimeter radiation (DIM)
    Jupiter, 102
Declination, 485A
Decoupling of matter and energy, 442,
        446
Degenerate electron gas, **283,** 294
    neutron star, 297
    white dwarf, 294
Degenerate neutron gas, 328
Deimos, 114, *114,* 456A
Density waves, 253, 369–371
Derivatives, 478–479A
Descartes, Rene, 11
Detectors, 174–177
Deuterium, 152, 275
    abundance of, 442, 443
    primordial, 442, 443
Differential forces, *see* Gravitation, Tidal
        forces, or Roche limit
Differential galactic rotation, 262–264,
        369; *see also* Galactic rotation
Differential solar rotation, 196
Diffraction, 149, *150,* 169
Diffraction limit, 149
DIM (Jupiter), 102
Dione, 120, *121*
Disk Population, 364
Dispersion, 149
Distance determinations; *see also*
        Distance scale

cluster diameters, 250, 251
geometric methods, 205, 206
luminosity methods, 206
main-sequence fitting, *240,* 241, 251
moving clusters, 261, *261, 262*
parallax, **205–206,** *206*
period-luminosity relation, 312, 386t
spectroscopic parallax, 240, *240*
Distance modulus, **208,** 240, 250
    effect of interstellar absorption, 250,
        338, 339
Distance scale, 262, 385–392, 386t, *387,*
        397, 402, 431
    Aaronson-Huchra-Mould method, 389
    Hubble law, 387–390
    Local Group, 397
    Sandage-Tamann method, 389
    Tully-Fisher method, 390
Doppler broadening, 162
    corona, 189
Doppler effect, **150,** 162, *258*
    binary systems, 216
    cosmological redshifts, 388
    Earth's revolution, 44
    galactic center, 365
    galactic rotation, *360*
    galaxies
        redshift, 388
        rotation, 383, 384
    and gravitational redshift, 296
    Mercury
        radar mapping, 78
        rotation, 79
    moons of Saturn, 124
    nonrelativistic, 150, 165p
    planetary rotation, 24, 108, *108*
    pulsars, 332, *332*
    pulsating stars, 310
    quasars, 419
    radar, 25, *79*
    radial speed, 258
    relativistic, **150–151,** 388, 419–420
    rings of Saturn, 124
    shell stars, 317
    solar oscillations, 180, 185
    SS 433, 331
    21-cm line, 360
    Venus, 83
    X-ray binaries, 328
Doppler shift, *see* Doppler effect
Doppler, C.J., 150
Draconic month, 54
Draper, Henry, 230
Dust; *see also* Interstellar dust
    comets, 449
    galactic center, 365
    intergalactic, 405
Dwarf novae, 309, 320

Earth
    accretion, 140
    age, 60
    albedo, 60
    atmosphere, 67–70, 140
        composition, 67t
        evolution, 75, 140

refraction, *70*
  scale height, 68
  scattering of light, 69
  secondary atmosphere, 67
  structure, 68, *68*
  transparency, 69, *170*
atmospheric seeing, *see* Seeing
aurorae, 72–74
  composition, 56
  core, 56, 57, 70
  crust, 59, *59*
  density, 56, 59
  eccentricity of orbit, 38
  equatorial bulge, 42
  evolution, 75, 95
  formation, 139, 140
  hydrosphere, 59
  interior, 56–57, *58*
  internal heat, 75
  lithosphere, 59
  magnetic field, 70
    dynamo, 70
    field reversal, 70
  magnetosphere, 70, *71*, *72*
  mantle, 56, 57
  nutation, 48
  oblateness, 41, 42, *42*, 45
  orbit, 205
  physical data, 456A
  plate tectonics, 59, *59*, 140
  precession, 47, 48, *48*, 54
  revolution, 42
  rotation, 39, 42
    tidal evolution, 46
  seasons, 38, *38*
  surface, 59–60, *59*
  tidal forces, 45, *45*, 46, *46*
  wind patterns, *41*
Earth-Moon system, 47, *54*, 456A
  angular momentum, 46, 53
  dimensions, 53, 54
  evolution, 75
Eccentricity of planetary orbits, 9, 455A
Eclipses, 55
  annular, 55
  contact, 219
  lunar, 55, *56*, *57*
  partial, 55
  Saros cycle, 56
  solar, 55, *56*
  total, 56
Eclipsing binaries, 218–223, *219*, 329
  close binaries, 223
  contact binaries, 222, 223
  ellipticity effect, 222
  geometry, *220*
  light curve, 219–222, *219*, *222*
  mass radius correlation, 220, *221*
  orbits, 220
  partial, 220, *220*
  reflection effect, 221
  spectroscopic binaries, 217, 218, 220
  stellar radii, 219, 220
  stellar rotation, 222
  tidal effects, 222, 223
Eclipsing-spectroscopic binaries, 220
Ecliptic, 4, 39, 485A

Ecliptic coordinate system, 486A
Eddington, Arthur S., 216
Eddy, John, 194
Edlen, B., 190
Effective temperature, 210, 233, 277
  binary systems, 220
  stellar atmospheres, 227, 228
Einstein absorption coefficient, 163
Einstein X-ray Observatory, 174, 329,
        367; *see also* X-ray astronomy;
        Space astronomy
  radio galaxies, 410, 413
  supernova remnants, 345
Einstein's theory of relativity, 150, 295–
        296, 304–306, **433–434**
  models of Universe, 435–439, 442
Einstein, Albert
  blackbody radiation, 163
  Doppler effect, 150
  mass-energy equivalency, 275
  special relativity, 150
Einstein-Rosen bridge, 306
Ejecta blanket, 62
Ejected matter, *see* expansion
Electric field, 149; *see also*
        Electromagnetic wave
  vector **E**, 147, *147*
Electromagnetic radiation, 146, 147, 148t
  absorption, 153
  angular resolution, 149
  blackbody radiation, 163, *164*
  continuum, 153
  cyclotron frequency
    Jupiter, 103
  DAM decameter radio bursts
    Jupiter, 103
  DIM synchrotron radiation
    Jupiter, 103
  dispersion, 149
  Doppler effect, *see* Doppler effect
  emission, 153
  energy flux, 165
  flux, 151, 152
  intensity, 149, 151
  Maxwell's theory, 163
  monochromatic intensity, 151
  particle theory of light, 151
  photons, 151
  Planck radiation law, 163–164; *see also*
        Planck radiation curve
  synchrotron radiation, *see* Synchrotron
        radiation
  wavelength, 147, *147*
Electromagnetic spectrum, 147, 148t; *see
        also* electromagnetic radiation; &
        specific spectral regions
  detectors, 170
  gamma rays, 148t
  infrared, 148t
  radio, 148t
  ultraviolet, 148t
  visible, 148t, 168
  X-rays, 148t
Electromagnetic wave, 147, *147*; *see also*
        Electromagnetic radiation and
        Wave nature of light
  amplitude, 146

  equation, 147
  frequency, 146
  phase, 147
  phase shift, 149
Electron, 152, 469A
Electron degeneracy, 283, 294
  neutron star, 297, 325
  white dwarf, 294
Electron density
  pulsar, 299
Electron pressure, 229
Electron scattering, 273
  corona, 189
  Sun, 273
Electron volt, 154, 469A
Elementary particles, 152t, 441
Elements, 152, 464–465A; *see also*
        Abundance of elements
  isotopes, 152
Ellipse, 8, 9, *9*
Elliptical galaxies, 378, *378*, 381–383,
        382t; *see also* Galaxies, cD galaxies
Elongation, 4, 5
Emission lines, 153, 155–156, *155*, *155*;
        see also Electromagnetic radiation,
        Spectral lines
  profile, *158*, 159
Emission nebulae, 340, 344; *see also* H II
        region
Emission-line stars, 240
Enceladus, 120
Energy, **17**
  total, 17
  flare stars, 313, 314
  pulsating stars, 311
  solar flares, 200
  units, 466–467A
Energy density
  blackbody radiation, 164
  radiative excitation, 163
Energy flux
  planets, 28
  Stefan-Boltzmann law, 165
Energy generation, 273–277, *275*
  CNO cycle, 276
  gravitational contraction, 274
  proton-proton (pp) chain, 275, 276
  thermonuclear reactions, 274–277
  triple alpha, 276
Energy level(s)
  diagram, 154, *155*
  forbidden transitions, *347*
  hydrogen, 154
  maser transitions, 350, *350*
  molecules, 158
  spectral lines, 228
  21-cm, 347, *348*
Energy transport, 272
Ephemeris time, 38
Equation of State, 67, 277
  neutron star, 297
  stellar atmospheres, 229
  white dwarf, 294
Equation of time, 37
Equatorial coordinates, **36**, 484–486A
Equinoxes, 39, 485A
Equivalent width, 159, *159*, 228, 229

Eratosthenes, 53
Escape velocity
  atmospheric gas, 29, *29*
  black holes, 303
  comets, 449
  neutron star, 297
  projectile, 19, 20, *20*
  Universe, 435
Europa, 116, 117, *117*, 457A
  craters, 116
  ice, 116, 117
Evening star, 4
Event (spacetime), 433
Event horizon, 305
Evolution, *see* Stellar evolution, Solar
    nebula
Evolutionary tracks, 278–290; *See also*
    Stellar evolution
  clusters, 289, 290
  low metal abundance, *287*
  massive star, *282*
  Population I, 282–286, 289, 290
  Population II, 287–290
  pre-main-sequence star, *280*
  solar mass star, *284*
Excitation of atoms, 155; *see also* atoms
  absorption lines, 155, 228
  Boltzmann's equation, 159
  collisional, 155
  collisional de-excitation, 156
  de-excitation, 155
    spontateous transition, 159
  excitation potential, 155
  forbidden lines, 156
  potentials, 157t
  radiative, 155, 163
  radiative de-excitation, 156
Excitation curves, *161*
Excitation potential, 155, 157t
Exosphere, 67, 68
Expanding arm, 362
Expanding atmosphere(s), 239, 240, *318*
  planetary nebulae, 319
  Wolf–Rayet stars, 318
Expansion broadening, 162
Expansion of Universe, 392, 431, 432, 435
Explosive stars, 309
Exponential notation, 473A
Extended atmosphere, 316, 316t, *317*, *318*
Extended radio galaxies, 409, 411, 413;
    see also Radio galaxies
  bending sequence, 413, *414*
  bursts, 413
  classification, 411
  doubles, 413
  tails, 411, 413
Extinction,
  Earth's atmosphere, 68
  interstellar, 338–340
Extinction coefficient, 210, 339
Extragalactic radio galaxies, 372
Extreme UV (corona), 191

F corona, 189
F ratio, 168
Faculae, 196; *see also* Plages

Faraday rotation, 372, 373
  pulsar, 299
Filaments (solar), 196
  chromospheric, 186
Fireball, 32
Fisher, J. Richard, 390, 402
Fizeau, Armand
  Doppler effect, 150
Flare stars, 309, 312, 313, *314*
  RS CVn stars, 315, *315*
Flares (solar), **198–201**, *199*
  energy, 200
  magnetic fields, 200
  model, 200, 201, *201*
  and prominences, 196
  radio bursts, 200
  solar cosmic rays, 198, 200
  solar wind, 192, 200
  X-ray bursts, 200–201, *200*
Flourescence, 319, 344
  comets, 130
  planetary nebulae, 319
Flux, **151–152**, *151*, 183; *see also*
    Electromagnetic radiation and
    apparent magnitude
  apparent magnitude, 207
  bolometric, 210
  solar, 183
  stellar, 210
  visual, 210
Focal length, 168
Focus, 168
Forbidden lines, **156**, 190
  novae, 320
  planetary nebulae, 346, *347*
  quasars, 419
  Seyfert galaxies, 416
  solar corona, 190
  T Tauri stars, 313
  21-cm line, 348
Force, 11, 12, *12*
  units, 466–467A
Fornax cluster, 397
Foucault, Bernard Leon, 41
Foucault's pendulum, 41, *42*
Fourier analysis
  speckle interferometry, 224
Fraunhofer absorption spectrum, 184
  corona, 189
Fraunhofer, Joseph von, 184
Free-fall collapse, *see* Gravitational
    collapse
Free-free transitions, 190, 344
Frequency, 146
Fusion reactions, *see* Thermonuclear
    reactions

Galactic, *see also* Galaxy
Galactic cannibalism, 399–401
Galactic center, 255, 263, 360, 361, 364,
    365–367, *367*
  black hole, 253, 365, 367
  distance, 251
  dust, 365
  gamma rays, 366
  H II regions, 365

  infrared observations, 365
  ionized gas, 366
  model, *369*
  radio sources, 365, 366, *366*
  rotation, 366
  Sagittarius A, 365, 366
  size, 366
  21-cm observations, 360, 361
  X-ray emission, 366, *368*
Galactic cluster(s), 235, 236, 252, 289, 364
    (Same as Open clusters)
  distances, 250, 389
Galactic coordinate system, 248, *249*,
    486A
Galactic equator, 248, 486A
Galactic nuclei, 425, 427
  BL Lac objects, 418
  Cen A, 413
  Cygnus A, 411
  M 82, 418–419
  quasars, 421, 422, 425
  Seyfert galaxies, 417
Galactic plane, 250, 253, 255, 364
Galactic rotation, 252, **262–268**, *263*, *264*,
    265 359, 360, 368, 370
  density wave, 253
  Doppler shifts, *360*
  Local Standard of Rest (LSR), 263
  orbits, 263
  rotation curve, 266–267, *266*
  solar orbit, 263
  21-cm line, 359–362, *359*, *361*; *see also*
    21-cm line
Galactic structure, 253–254, *254*, 359–373
Galactocentric system, 260
Galaxies, 247, 249, 378–392, 382t
  active, 410, 425; *see also* Active
    galaxies
  angular momentum, 381
  Barred spiral (SB), 378, 381
  binary galactic systems, 384, 392p, 417
  black holes, 385
  cannibalism, 399–401, *400*
  cD galaxies, 378, 381, 385, 397, 399,
    400, 404
  characteristics, 382t
  classification, 378, *379*, *380*
  clusters, *see* Clusters of galaxies
  density waves, 381
  distance scale, 385, 386t, 389, 390
  dust content, 382t
  dwarf, 381, 382
  elliptical (E), 378, 383, 387
    giant, *See* cD galaxies, *see also* Radio
      galaxies
  evolution, 425
  gas distribution, *362*
  halos, 400
  H I content, 384
  Hubble sequence, 378, 385
  interacting, 401, *400*
  irregular, 381, 383
  Local Group, 385, 395–397, 395t, *396*
  luminosities, 381, 382t, 385, 386, 389,
    390
  magnetic fields, 381
  mass, 381, 383, 384, 382t

mass-luminosity ratios, 382t, 385
nuclei, *see* Galactic nuclei
peculiar, 381, *382*, 401
radio galaxies, *see* Radio galaxies
redshifts, 389
rotation curve, 383
SO galaxies, 381
Seyfert, 415–417, *see also* Seyfert
    galaxies
sizes, 382t, 384
spectra, 382t, 383, 384
spirals (S), 378, *379*, *380*, 381, 383, 386,
    410
    nuclei, 381
    radio emission, 409
starburst, 418; *see also* Starburst
    galaxies
supergiant ellipticals, *see* cD galaxies
supernovae, 323, 372
tidal interactions, *400*, 419
21-cm observations, 384
Galaxy, **246–255**, 260, 338; *see also entries*
    under Galactic
abundances, 252, 364
age, 369, 370, 371
anticenter, 248
center, 248, *367*
central bulge, 365
CO observations, 360, *362*
coronal gas, 351, 352
cosmic rays, 370–371
density wave, 253
density-wave model, 369, 370
diameter, 468A
disk, 249, **252–253**, 255, 364, 370
distribution of stars, 249, 250
dynamics, 252–253
evolution, 249, 369, 370, 371
gas distribution, 362, *362*, 367–368
gravitation, 252
halo, 252, 253, *254*, 267, 364, 365, 368,
    370
Hertzsprung-Russell diagram, 235
hydrogen distribution, 360, 362, *362*
magnetic field, 340, 372, 373
mass, 252, 253, 255, 263, 267, 367–368,
    468A
molecular hydrogen, 361, 362
nucleus, 248, 253, 260, 365–367, 368,
    372
populations, *see* Population I,
    Population II
rotation, 252, **262–268**, 359, 368, 369,
    370
rotation curve, 266, *266*, 267
spiral arms, 252–255, *253*, *254*, 359,
    363–364, *364*, 370
structure, 247, 252, *254*, **359–373**
tidal interaction, 362
3-kpc arm, 361–362
21-cm observations, 359–362, *361*
warp, 360, *361*
Galilean moons, **114–118**, 115t, *116*, 138,
    457A
Galilei, Galileo, 5, 11, 193
discovery of Neptune, 109
gravity experiments, 433

lunar craters, 61
lunar librations, 54
moons of Jupiter, 114
star clouds, 247
Galle Johann G.
discovery of Neptune, 109
Gamma-ray astronomy, 174, 178p, 302
Gamma rays, 147
early Universe, 441
galactic center, 367
Ganymede, 117, *118*, 457A
craters, 117
ice, 118
Gas law, 28, 227, 272
General obscuration, 336–338
General relativity, 295, 296, 433–444
black holes, 304–306
cosmology, 435–436, 439, 442
gravitational lenses, 423
Geometry (analytical), 474A
Giant elliptical galaxies, *see* cD galaxies
Giant molecular clouds, *see* Molecular
    clouds
Giants, 235, 236, *237*, 238t, *239*, 240
pressure broadening, 238
Gibbous phase, 6
Giotto (ESA) spacecraft, 448
Globular clusters, 236, 251, 252, 267, 364,
    370, 371
age, 432
H-R diagram, 236, *237*, 289, 290
orbits, 370
system of, 251, *254*
X-ray bursters, 331
Globules, 336, *337*
Graboske, Harold, 138
Grains, *see* Interstellar grains
Grand Unified Theories (GUTs), 443–446
Gravitation, 13–15, 433, 443
constant, 469A
differential forces, 45, 48
Gravitational collapse, **279–281**
black holes, 303
infrared radiation, 279
massive stars, 353
molecular clouds, 353
neutron stars, 297
protostars, 279–281
rotation effect, 280, 281, *281*
supernovae, 325
Gravitational constant, 14, 469A
Gravitational contraction, 138, 274, 279,
    282
brown dwarfs, 296, 297
Jovian planets, 138, 139
stellar core, 282
white dwarf, 293
Gravitational cross section, 138
Gravitational lenses, 423–425, *424*
Gravitational mass, 433
Gravitational redshift, 295
black hole, 303, 304
neutron star, 297
Greek alphabet, 470A
Greenhouse effect, 83
Greenwich mean time, 37
Gregorian calendar, 38

Grindley, Jonathan, 331
Grotian, W., 190
Ground state of atom, 153
Gum Nebula, 300, 302, *302*, 345, *346*
Gursky, Herbert, 331
GUT's, 443–446

H, *see* Hubble constant
H alpha line, *see* Hydrogen, Balmer lines
H I; *see also* 21-cm line
distribution in Galaxy, 266, 359–364
intergalactic, 405
rotation of galaxies, 384
H I regions, 350, **351**, 359, 363
H II
intergalactic matter, 405
H II regions, 280, *337*, 351, 343–345, *344*,
    352, 363, 370
Balmer lines, 344
continuous radio emission, 344–345,
    344
distance scale, 389
fluorescence, 344
galactic center, 365, 366
Lyman continuum, 343–344
molecular clouds, 348, 349
protostar, 352
radio recombination lines, 363
spiral arms tracers, 363
spiral galaxies, 381
Stromgren sphere, 343
thermal radio emission, 344, *344*, 345
h and Chi Persei cluster, 262
H- (negative hydrogen ion), 184
photosphere, 183
H–R diagram, *see* Hertzsprung-Russell
    diagram
H_2O (water) emission, 349
molecular clouds, 351
Hadley cells, 41
Venus, 84
Hale telescope, 169; *see also* Mt. Palomar
Hale, George E., 196
Halflife, 60, *60*
Hall, Asaph, 114
Halley's comet, 31, *129*, 130, 448–449,
    448, *449*
Halo, 364, 368
Halo Population II, 364
Harmonic law, *see* Kepler's laws
Harvard spectral classification system,
    230–233, *231*, 232t, 329
Hawaiian volcanoes, 86
He + ionization, 311
HEAO 2, *see* Einstein Observatory
Heavy particle era, 440, 441
Heisenberg uncertainty principle, 154
broadening of spectral lines, 162
Heliocentric model, 4, 43
Heliocentric parallax, *see* Trigonometric
    parallax
Helioseismology, 180
Helium
abundance, 442
fusion of hydrogen, 275
primordial, 442, *443*

sun, 185
Helium burning, 283, 284
　long-period red variables, 312
　pulsating stars, 311
　WC stars, 318
　X-ray bursters, 332
Helium flash, 283, 285
Henderson, T., 43
Henry Draper Catalog, 230
Hercules supercluster, 403, *404*
Hercules cluster, 397, *398*
Herculina, 128
Herschel, William, 213, 260
　magnitude scale, 206
Hertz, 146
Hertzsprung, Ejnar, 233, 237
　H–R diagram
　narrow line stars, 237
Hertzsprung–Russell diagram, 229, **233–
　241**, *234*, *235*, *239*, 262
　bright stars, *234*, *235*
　clusters, *236*, *237*, *288*, *289*, 289, *290*
　distance determinations, 240
　evolutionary tracks, 278–288
　globular clusters, 237, 290, *290*
　heavy element stars, 287
　Hyades cluster, *237*
　instability zone, 283, 285–287, 310
　long-period red variables, 312
　nearby stars, *234*
　open clusters, *288*, *289*, 289, 290
　Pleiades, *236*
　Population II, 290, *290*
　pulsating stars, 311
　Russell–Vogt theorem, 278
　T Tauri stars, 313, *313*
　variable stars, 309, *309*
　white dwarf, 295, *295*
Hewish, Anthony, 298
Hey, John S., 170
High energy, *see* X-rays
High velocity clouds, 362
High-velocity objects, 267, *267*, 268, 364
　globular clusters, 267
　RR Lyrae stars, 268
Hipparchus
　magnitude scale, 206
　age of universe, 432
Homestake mine, 276
Horizon system, 484A
Horsehead nebula, 336, *337*
Houk, Nancy, 230
Hour angle, 36, 485–486A
Hubble Atlas of Galaxies, 378
Hubble constant, **388**, *389*
　age of universe, 392, 436
　cosmic expansion, 432, 435–436
　as distance indicator, 390–392, 402, 421
　obtaining value, 389–390, *390*, *391*,
　　402, 432, 435–436
Hubble diagram, *see* Hubble's law
Hubble sequence, 378, 385
Hubble Space Telescope, 174, 389
　limiting magnitude, 207
　parallax measurements, 205
　resolution
　　problem, 178

sunspot observations, 196
Hubble time, 392
Hubble's law, **387–388**, *389*, *391*, 431,
　432, 435
Hubble, Edwin, 378, 387, 402
Huchra, John, 389
Hulse, Russell, 333
Humanson, Milton, 387
Huygens, Christian
　rings of Saturn, 121
Hyades cluster, 236, 261, 262
　color-magnitude diagram, *237*
　distance scale, 389
　moving cluster method, 206, *262*
Hydrogen, *see also* H I and H II
　atom, 153
　Balmer lines, 154, *155*, 160, 230, 413
　　excitation/ionization, 160–161, *161*
　　H II regions, 344
　　quasars, 419
　　sun, 184–186, *186*, *187*, 188
　　white dwarf, 295
　Bohr model of atom, 153, *154*
　distribution in Galaxy, 360
　early Universe, 42
　energy level diagram, 154, *155*
　hyperfine structure, 347
　Lyman lines, 154, *155*
　　sun, 187
　magnetic moment, 347
　mass, 469A
　metallic
　　Jupiter, 100
　negative ion (H-), 183
　nuclear fusion, 274–277
　radio astronomy, 265
　radio recombination lines, 363
　spectral lines
　　Balmer series, 154, 230, 419
　　Bracket series, 352
　H II regions, 352
　　Lyman series, 154
　21-cm transition, 347–348, *348*
Hydrogen burning, 275, 277, 281, 286
Hydrogen corona (comets), 449
Hydrosphere, 58
Hydrostatic equilibrium, **67–68**, 227, 229,
　271, *271*, 277–280
　Jupiter, 100
　pulsating stars, 311
　white dwarf, 294
Hyperbola, 10
Hyperfine splitting, 347, *347*

Iapetus, 120, *121*
Image, 168, *168*
Image processing, 175
Impact parameter, 137
Impacts, *see* cratering
Index of refraction, 148
　Earth's atmosphere, 68
Inertia, 11
Inertial mass, 433
Inferior conjunction, 5
　moon, 55
Inferior planets, 4

Inflationary Universe, 443, 445
Infrared astronomy, 173, 336; *see also*
　　Space astronomy
　blackbody radiation, 174
　bolometer, 173
　interstellar dust, 173
　interstellar matter, 326
　Wein's law, 174
Infrared observations, 170
　galactic center, 365
　interstellar dust, 341
　molecular cloud, 354
Infrared radiation, 147, 170
　gravitational collapse, 279
　interstellar grains, 341–342, *342*
　long-period red variables, 312
　Orion Nebula, *353*
　T Tauri stars, 313
　Tully–Fisher relation, 390, *390*
　stellar spectra, 232
Insolation, 39, *39*
Instability limit, *see* Roche limit
Instability zone (H-R diagram), 283, 285–
　287
　variable stars, 310
Integrals, 479–481A
Intensity of radiation, 149, 163
　monochromatic, 151
　relation to flux, 151, *151*
Interference, 149, *150*
Interferometry, 171–173; *see also* VLA &
　　VLBI
　aperture synthesis, 171
　lunar occultation, 223, *223*
　maximum resolution, 177p
　Michelson interferometer, *224*
　radio interferometer, *172*
　speckle, 224
　　resolution, 224
　　brown dwarf, 297
　stellar, 224
　　stellar diameters, 223
Intergalactic medium, 405–407
　clouds, 420
　extended radio galaxies, 413
　jets, 428
International Astronomical Union (IAU),
　265
International Cometary Explorer (ICE)
　Comet Giacobini-Zinner, 130
International Ultraviolet Explorer (IUE),
　174
Interplanetary medium; *see also* Solar
　　wind
　dust and gas, 32, 134
　plasma, 298
Interstellar absorption, **250**, 336, 337,
　338, 339
　color excess, 338, 339
　effect on distance modulus, 250, 338,
　339
　galactic latitude effect
Interstellar dust, 250, 336–341, *338*; *see
　also* Interstellar grains
　color index, 209
　irregular galaxies, 381
　galactic center, 365

long-period variables, 312
polarization, 339
spiral arm structure, 363
T Tauri stars, 313
Interstellar extinction, **341**
    coefficient, 339
    optical depth, 339
    wavelength dependence, 338, *338*, 339,
        341, *341*
Interstellar gas, 342–352, 349t
    absorption lines, 342, 343, *343*
    distribution, 343
    electrons, 299, 344–345
    galactic center, 365
    galactic structure, 359
    infrared observations, 365, 366
    ionization, 343
    molecules, 348–351, 349t; *see also*
        Molecular clouds
    opacity, 342
    spiral arm structure, 363
Interstellar grains, 279, 280, 339–341
    absorption bands, 341, *342*
    color index, 209
    composition, 341, 342
    extinction, 341, *341*
    infrared observations, 341, 342, *342*
    interstellar molecule formation, 348,
        349
    irregular galaxies, 381
    models, 341
    polarization, 339
    spiral arm structure, 363
    T Tauri stars, 313
    UV observations, 341
    sites of molecular formation, 349
Interstellar medium, 250, 279, 299, 336–
        356, 370, 372; *see also* other entries
        under Interstellar
    cosmic rays, 372
    dark nebulae, 336, *337*, 338
    dust to gas ratio, 336, 342
    electron density, 299, 345
    extinction, 338, 339
    galactic disk, 371
    general obscuration, 336, 337, 341
Interstellar molecules, 348, 349t, 362
    as building blocks of life, 348
    grains, 348
    protostars, 348
    radio lines, 348
Interstellar obscuration, 336, 337, 341
Interstellar absorption lines, 342, 343, *343*
Interstellar polarization, 339–340, 373
    correlation with reddening, 339
    dust grains, 341
Interstellar radio lines, 347–350
Interstellar reddening, 209, **338**, *338*, 339;
        see also Color excess
    relation to polarization, 339
Inverse square law of light, 152
    stellar distances, 206, 208
Io, 116, 457A
    atmosphere, 115
    sodium cloud, 116, *116*
    volcanoes, *117*
Ionization, **156,**

continuum of states, 156
equilibrium, 160–161, *161*
recombination, 160
potential(s), 156, 157t, 228
stage of ionization, 160, 161, 228
Ionosphere, 28, 68, 170
    solar flares, effect of, 198
Ions, **156**
IRAS (Infrared Astronomical Satellite),
    341
Irregular galaxies, 381–383, *381*, 382t
Isochromes, 289
Isophotal contours, 385
Isotopes, 152
    nucleosynthesis, 327
    nuclides, 152
IUE, 174

Jansky, Karl, 170
Jets
    bipolar outflow, 356, *355*
    comet, 448
    galaxies, 426–428, *427*
    M87, *410*
    quasars, 422, 425
Johnson, Harold, 208
Jolly, Phillip von, 14
Joss, Paul, 331
Jovian planets, 98–110, 455–456A; *see
        also* individual planets
    atmospheric composition, 28
    formation, 138
    internal heat sources, 101t
    physical data, 456A
    thermal conductivity, 112p
Jupiter, 98–104, *98*, 455–456A; *see also*
        Jovian planets
    albedo, 98
    asteroidal moons, 119
    atmosphere, 98, 99, *101*
    aurorae, 104
    blackbody radiation, 101, 141p
    composition, 98
    core, 100, 101
    cyclotron frequency radiation, 103
    DAM radiation, 103
    density, 26, 98, 101
    differential rotation, 98
    DIM synchrotron emmision, 103
    excess energy, 101
    formation, 138
    Galilean moons, 114–118, 457A
    gravitational contraction, 102
    Great Red Spot, 98, 99, *100*
    interior, 100, *101*
    lightning, 103
    magnetic field, 102
    magnetosphere, 103, *102*
    mass, 98
    metallic hydrogen, 100, 101
    moons, 114; *see also* individual moons
    motions, 98
    perturbation of comets orbits, 131
    physical data, 456A
    radiation belts, 102
    rings, 119

radio map, *103*
satellites, 457A
structure, 100, *101*
synchrotron radiation, 102–104, *103*
temperature, 102
zones, 98

K corona, 189
Keenan, P.C., 237
Kepler's laws, **7–8**, *8*
    Earth–Moon system, 47
    first law, 7
    Roche limit, 48
    second law (law of areas), 8, 15, *15*, 36
    third law (harmonic law), 8, 214, 218,
        303; *see also* Keplerian motions
        Earth–Moon system, 53
        galactic rotation, 252, 263
        gravitational collapse, 279
        mass determination, 18, 19
        moons of Saturn, 124
        Newton's form, 16
        Pluto, 111
        pulsating stars, 311
        rotation of galaxies, 383
        white dwarf, 295
        X-ray binaries, 328
Kepler, Johannes, 6
    distance determination, 6, 7
    laws of planetary motion, 7–8
Keplerian motions, 263, 266, 368
Kinetic energy, 18
    projectile, 19
Kirkwood gaps, 30
Kleinmann-Low nebula, 352
Kohlschutter, A., 237

Latitude, 483A
Lava flows
    Io, 116, *117*
    lunar, 62
Law of Areas, *see* Kepler's laws, second
Least energy orbit, 21p
Leavitt, Henrietta, 311
Leverrier, U.J.
    discovery of Neptune, 109
Librations (lunar), 54, *55*
Light, 146; *see also* Electromagnetic
        radiation; Electromagnetic
        spectrum; Radiation
    inverse square law, 152
    speed of, 146
    velocity, 469A
    visible spectrum, 147
    wave nature, 146–151, *146*, *147*, *150*
Light curve(s)
    eclipsing binaries, 219, 220
    flare stars, 313
    pulsating stars, 310
    RS CVn stars, 315
    supernovae, 324
    X-ray bursters, 331
Light particle era, 440, 441
Light wave, 146–151
Light-gathering power, 169

Lighthouse model of pulsars, 299, *300*, 303
Lightyear, **205**, 468A
Limb, **78**
Limb brightening, 190
Limb darkening, 182, *183*
    eclipsing binaries, 221
Lin, C.C., 369
Line blanketing, 228, *228*, 232, 238
Line of nodes, 54
Line profiles, *see* Spectral lines
Linear momentum, 11, 12
Lithosphere, 58
Local Group of galaxies, 385, 395–397, 395t, *396*, 402
Local standard of rest (LSR), 258, **260–261**, 262–265, 267, 268
Local Supercluster, 402, *402*, 404
Logarithms, 473A
Long-period red variables, 312
Longitude, 483A
Longitudinal compression (P) waves, 56, 57
Loop nebula, 345, *345*
Lorentz factor, 423
Lorentz force, 72, *73*
Los Alamos National Laboratory, 428
Low mass stars, 313
Lowell Observatory, 90
Lowell, Percival, 90
    discovery of Pluto, 110
LSR, *see* Local standard of rest
Luminosity, 235, 273, 283
    stellar, 210
    stellar distances, 208
Luminosity classifications, *233*, 236–238, *239*, 463A
Luminosity function, 250, *250*, 251
Lunar; *see also* moon
Lunar eclipses, 56, *56*, *57*
Lunar librations, 54, *55*
Lunar occultation (stellar diameter), 223
Lunar phases, 55, *55*
Lunar rock samples, *see* Moon
Lyman lines, 154, *155*
    intergalactic, 405
    solar transition region, 187

M 101, 389
M 31 (Andromeda Galaxy), 365
M 67, *289*
M 81, *400*, 401, 419
M 82, *400*, 401, 418, 419
M 87, *409*, 410, 411
    black hole possibility, 425
    jet, 410, *410*, 427
    radio source, 410, *411*
M–K luminosity classification (Morgan–Keenan), 237, 238t
M–L law, *see* Mass luminosity relationship
M/L ratio, *see* Mass luminosity ratios
Mach number, 427
Mach, Ernst, 427
Magellanic clouds, 253, *254*, 362, 384
    period–luminosity relationship, 311

Magellanic stream, 362
Magnetic buoyancy, 201
Magnetic field vector **B**, 147, *147*
Magnetic field(s)
    aurorae, 72–74
    charged particle motion 72, *73*
    comets, 130
    coronal loops and holes, 191
    Crab nebula, 326
    Earth, 70–74
    flare stars, 314
    Galaxy, 299, 341
    interstellar dust, 340, 341
    Jupiter, 102
    Lorentz force, 72, *73*
    magnetosphere, 70–72, *71*, *72*
    Mars, 93
    Mercury, 81
    moon, 70
    Peculiar A stars, 240
    pulsars, 299
    radio galaxies, 411
    RS CVn stars, 315
    Saturn, 106
    solar active regions, 196
    solar cycle, 193, 201
    solar flare model, 201
    solar wind, 192
    SS 433, 331
    sunspot polarity, 196
    sunspots, 193
    T Tauri stars, 313
    Venus, 88
    white dwarfs, 296
    Zeeman effect, 162
Magnetic Induction units, 466–467A
Magnetic variables, 309, 312, 313, 314
    oblique rotator, 314
    star spots, 314
Magnetograph, 196
Magnetosphere, 70
    Earth, 70–74, *71*, *72*
    Jupiter, *102*, 103
    pulsars, 299
    Saturn, *107*
Magnifying power, 169
Magnitude, **206–211**
    absolute, 207
    apparent, 206, 207
    bolometric, 210
    infrared, 208
    photographic, 208
    photovisual, 208
    systems, 208
    UBV system, 208
    visual, 208, 210
Main sequence, 235, 236, 275, 306
    evolution, 281, 289
    globular clusters, 236, *237*
    open clusters, 235, *236*, 237
    turnoff, 289
Main sequence fitting, *240*, 241, 251
Mare Ibrium, 63
Mare Orientale, 62, 73
Margon, Bruce, 330
Maria, *see* Moon, maria
Mariner 9 (Mars), 91

Mariner 10 (Mercury/Venus), 80
Mars 88–94, *90*, 455–456A
    arroyos, 91, *93*
    atmosphere, 92
        composition, 89
    canals, 90
    craters, 91, 92, 94, *94*
    evolution, 94, 95
    interior, 88, *89*, 93
    lava flows, 91
    magnetic field, 93
    moons, 88, 114
    motions, 88
    Olympus Mons, 92, 93
    physical characteristics, 88, 456A
    polar caps, 90, *90*
    sandstorms, 89
    soil, 91
    surface features, 89, *92*
    temperature, 89
    Tharsis ridge, 92, 94
    Valles Marineris, 94
    Viking landers, 89
    volcanoes, 91
    water, 90, 91, 94
Mascons, 64
Maser, **349–350**
    energy level diagram, *350*
    molecular clouds, 349, 350
Mass
    gravitational, 433
    inertial, 12, 433
    units, 466–467A
Mass defect, 275
Mass exchange, *see* Mass loss
Mass fraction, 227, 272, 442
Mass function, 217, 218
    Cygnus X-1, 329
Mass loss, 240, 285, 288, 293, 306, 316–319
    extended atmospheres, 316–319
    giants, 317
    novae, 322
    planetary nebulae, 319
    Population II, 287
    protostars, 356
    shell stars, 317
    source of interstellar dust, 342
    sun, 203p
    supergiants, 317, 342
    T Tauri stars, 313
    variable stars, 309
    Wolf–Rayet stars, 319
Mass–luminosity (M/L) ratio
    clusters of galaxies, 399
    galaxies, 385
    missing mass, 406
Mass–luminosity relationship, 215, *215*, **216**, 229, 278, 281
    X-ray binaries, 328
Mass–radius relationship, 294, *297*
Mathematical operations, 471–481A
Matter era, 438, 440, 442
Matthews, Thomas, 419
Maunder minimum, 194
Maury, Antonia, 237
Maxwell Montes (Venus), 85, *85*

Maxwell's theory of radiation, 147, 163
Maxwellian distribution, 29, *29*
Mean free path, **373p**
Mean molecular weight, **227**, 272, 277
  Stellar evolution, 281
Mean parallax, 206
Mean solar time, 36, 37
Mensuration formulas, 480–482A
Mercury, 78–83, 455–456A
  albedo, 80
  atmosphere, 80
  basin formation, 83
  Caloris Basin, 81, *81*
  comparison to Moon, 82
  cratering, 82
  craters, 80, *81*, 81
  elongation, 4
  evolution, 81, 82
  interior, 80, *80*, 82
  lava flows, *81*, 82
  magnetic field, 81, 82
  physical characteristics, 80, 456A
  radar mapping, 78, 79
  revolution, 78
  rotation, 78, 79
  scarp, 80, *82*
  solar day, 79, *79*
  solar wind, 80
  surface features, 80
  synodic period, 79
Meridian, 483A
Mesosphere, 67, *68*
Messier's catalog, 236
Messier, Charles, 236, 247
Metal abundances, 252, 364; *see also*
    Abundances
  stellar populations, 252
Metallic Hydrogen, 100, 106
Metastable states, **156**, 190
  corona, 190
  spectral line broadening, 162
Meteors, 32, 132, *132*
Meteor crater, 32, *32*
Meteor shower, 32
  radiant, 32
Meteorites, 32–34, 132
  carbonaceous chondrites, 114
  chondrites, 132
  chondrules, 132
  classification, 132
  compostiton, 132
  iron, 132, *133*
  nebular model, 139
  orbits, 132
  origin, 132, 134
  planetesimals, 139
  radiometric dating, 134
  stony, 132, *133*
  stony irons, 132
  structure, 132
Meteorite impact, 62
Meteoroids, 32, 132
Metonic cycles, 56
Michelson, Albert A., 45
  stellar interferometer, 224
Micrometeorites, 32, 134
Microwave radiation, 147; *see also* Radio
    waves

Milky Way Galaxy, 246–255, *246*, *247*,
    336; *see also* Galaxy
  star clouds, 247, *247*
Minkowski, R., 300
Minor planet, *see* asteroid
Mira stars, 309, 312, 319
Miranda, 124
Missing mass, 406, 407
  intergalactic space, 436
Model stellar atmosphere, 227
Molecular cloud complex, 366
Molecular clouds, 342, 348, 349t, 350,
    351, 352, 365, 366, 370; *see also*
    Carbon monoxide, OH
  dark cloud, 354
  density, 349
  giant clouds, 351, 352
  H II regions, 348, 349
  maser excitation, 349
  molecular hydrogen, 348
  origin of interstellar dust, 342
  protostars, 356
  radio emissions, 352
  star formation, 352–355
  stimulated emission, 350
  temperature, 349
  water, 349
Molecules
  angular momentum, 158
  comets, 130
  diatomic, 158
  disassociation, 158
  electronic transitions, 158
  energy levels, 158
  interstellar gases, 349t
  long period red variables, 312
  rotational energy states, 158
  spectral bands, 158
  vibrational energy states, 158
Momentum, 11–12, 16; *see also* Angular
    momentum
  angular, **16**
  linear, **11–12**
Monochromatic intensity, 151
Moon, 57–76, *64*, 139, 456A
  age, 66, 74
  albedo, 61
  anorthosites, 65
  Apollo samples, 65, *66*
  Appenine Mountains, 64
  atmosphere, 66
  binary accretion model, 140
  breccias, 65
  capture model, 140
  composition, 58, 61, 65–66, 140
    surface, 65–66
  Copernicus crater, *62. 63*
  cratering, 95, *95*, 140
  craters, 62
  eclipses, 56
  ejecta blanket, 63
  escape velocity, 66
  evolution, 74–75, 75t, 95
  fission model, 140
  formation, 139
  giant impact model, 140

gravitational perturbations, 65
highlands, 62, 63, *65*, 75
igneous rocks, 65
interior, 57–58, *59*
internal heat, 75
iron abundance, 66
lava, 63, 66, 74
librations, 54
line of nodes, 54, 55
magnetic field(s), 70
mare basins, 64
maria, 62, 63, *65*, 74
mascons, 65
meteoric impacts, 74
meteorite impact craters, 62
mountains, 62
obsidian, 65
origin of meteorites, 134
phases, 55, *55*
physical data, 456A
rays, 62, 63
regolith, 65
rock samples, 65–66, *66*
secondary craters, 62
seismology, 57–58
soil, 65
structure, 57–58, 62–66
surface, 60–66
  composition, 65, 66
synchronous rotation, 46, 54
synodic orbit period, 38
temperature, 61
tidal evolution, 46
tidal forces, 45, *45*, 46, *46*
water, 66
Moons, 30, 114–127, 456–457A
  of asteroids, 128
  Jovian planets, 30
  Jupiter, 114–119, 115t, *116*, *117*, *118*,
    119
  Mars, 114, *114*, *115*
  Neptune, 126–127, *126*, *127*
  Pluto, 110–111, *110*, *111*
  relative sizes, *53*
  Saturn, 120, *121*, *122*
  synchronous rotation, 114
  terrestrial planets, 30
  Uranus, 124–126, *125*
Morgan–Keenan Luminosity
    classification (MK), 237, 238t, *239*
Morgan, W.W., 237
Mould, Jeremy, 389
Moving cluster(s), 261, *261*, 262
  convergent point, 261, *262*
  distance scale, 389
  measure of distances, 206, 240
Multiplicity, 159

Nadir, 484A
Nebulae, **247**
  dark, 336–338, *336*, *337*
  emission, 343–345, *337*, *345*
  nova shell, 322, *323*
  planetary nebulae, 345–347, *347*
  reflection, 340, *340*
  supernova remnants, 325–326, *326*,

345, *345, 346*
Neptune, 109, *109*, 110, 455–456A; *see also* Jovian Planets
  blackbody radiation, 110
  composition, 139
  density, 110
  discovery, 109
  interior, *108*
  moons, 126–127, *126, 127*, 458A
  motions, 109
  orbit, *25*, 109
  physical data, 456A
  rings, 126, 127
Nereid, 126
Neutral hydrogen, *see* H I and 21-cm line
Neutrinos, **275**, 276, 446
  detection, 276
  sun, 180, 276
Neutron, 152
Neutron star(s), **297–303**, 306; *see also* Pulsars
  escape velocity, 297
  gravitational redshift, 297, 298
  mass, 297
  missing mass, 407
  pulsars, 298–300
  SS, 433, 330
  surface gravity, 297
  X-ray bursters, 331
Newton (unit of force), 14, 466A
Newton's laws, 11–13
  first law (inertia), 11
  second law (force), 12, 433
  third law (action–reaction), 12
  universal gravitation, 13–15, *15*, 213
Newton, Isaac, 10–13
Newtonian Cosmology, 434
NGC 188, *290*
NGC 1265, *415*
NGC 1295, *416*
NGC 2682 (M67), *288, 289*, 290
NGC 4038, 401, *404*
NGC 4039, 401,*400*
NGC 6624, 332
NGC 7023, 355, *355*
Nodical month, 54, 56
Nonthermal radiation; *see also* Synchrotron radiation
  quasars, 420
  radio galaxies, 410–413
  solar flare, 200
  Sgr A, 365
  Taurus A (Crab Nebula), 325–327
Norman, Michael, 428
Nova(e), 309, 320–323, *320*, 321t, *322, 323*
  accretion disk, 322
  as binary stars, 320, 322
  distance determination, 322
  distribution in Galaxy, 320
  dwarf novae, 320, 321t
  expansion, 322, *323*
  light curve, 320, *322*
  luminosity, 320, 322
  mass loss, 322
  recurrent novae, 320, 321t
  shell, 322, *323*

spectra, 320, *321*
  supernovae, *see* Supernovae
  velocity of ejection, 320, 322
  X-ray emission, 322
Nova Aquilae, 320
Nova Cygni, 323, *320, 324*
Nova Persei, 322, *323*
Nuclear force, 274
Nuclear reactions, 274–277; *see also* Nucleosynthesis; Proton-proton cycle; CNO cycle
  novae, 322
  r process, 327
  s process, 327
  supernovae, 327–328
  time scales, 331
Nucleosynthesis, **274**, 327–328
  early Universe, 441, *441*, 442
  formation of heavy elements, 327
  red giants, 328
  supernovae, 327, 328
Nucleus, atomic, 152
Nutation, 48

O and B stars, *231*, **232t**, 233, 238t, 240, 351, 352, 363, 365
  distance measurements, 385
  H II regions, 343, 351, 352
  spiral arm tracers, 363, 370
  ultraviolet spectra, 233
OB Associations, **251**, 252
  molecular clouds, 352, 353, 354
  spiral arm tracers, 363
Oberon, 124
Objective, 168
Oblateness, **25**, 41, 456A
  Earth, *42*
Oblique rotator, 314
Obliquity, 456A, **485A**
Occultations, 26
  by asteroids, 128
  lunar, 223, *223*
  Uranus, 126
Ocean of Storms (moon), 62
OH (interstellar), 348–350
  maser, 349–350, *350*
  molecular clouds, 350
Olympus Mons (Mars), 85, 91, 92, 93, *93*
Omega, 445
Oort comet cloud, 31, **131**, 139
Oort constants, 264–267, *265*
Oort, Jan,
  comet cloud, 31, 131
  differential galactic rotation, 263
Opacity, **182**, 229, 273; *see also* Optical depth
  chromosphere, 184, 186
  corona, 190
  electron scattering, 273
  mass-luminosity relationship, 278
  photoionization, 273
  photosphere, 182–183
  Pre-Main Sequence star, 280
  pulsating stars, 311
  radio corona, 190
  red giant, 283

stellar atmospheres, 229
  sun, 182–184, 186, 190, 273
Open clusters, 235, 236, 252, 289, 364, (*same as* Galactic clusters)
  distances, 250
  distance scale, 389
  H-R diagram, *288*, 289, 290
  moving cluster distances, 261
Opposition, 4, 5
  moon, 55
Optical depth, 177p, **183**, 279, 339; *see also* Opacity
  interstellar medium, 339
  stellar atmospheres, 227
Optics, 148, 149, 168–169
Orbits, *see also* Kepler's laws and Binary systems
  aphelion, 24
  artificial satellites, 19, 20
  eccentricity, 24
  escape speed, 19
  globular clusters, 370
  orbital velocity, 17
  perihelion, 24
  period, 18, 19
  planets, 7–10, 455A
  spacecraft, 19–20, *20*
Orion Nebula, 336, *337*, 345, 348, 352, *353*
  OB association, 353
  infrared emission, 352, *353*
  maser process, 350
Ozone (Earth's atmosphere), 69

P Cygni stars
  novae shells, 322
  profiles, 314
  supernovae, 324
P waves, 56, 57
P-L relationship, **311–312**, *312*
Palomar Sky Survey, 211
Panov, V.I., 433
Parabola, 9
Parallactic orbit, 43, 44, *44*
Parallax, 43–44, **44**, **206–207**, *206*, 259, 386t
  mean, 206
  statistical, 206, 386t
  trigonometric, 205
Parent atoms, 59
Parsec, 205, 468A
Particle physics, 443
Particle threshold temperature, 440
Pauli exclusion principle, 157, 294
Peculiar A stars, 240, 314
Peculiar galaxies, 381, *382*, 401
Peculiar motions, 260, 261
Penzias, Arno, 436
Perfect gas law, 28, 227, 272
Perigalacticon, 255p, 268
Perigee, 20
Perihelion, 9
Period–density relationship
  pulsating stars, 311
Period–luminosity relationship, **311–312**, *312*

as distance indicator, 386t
zero point, 312
Periodic table, 464–465A
Perseid meteor shower, 32
Perseus supercluster, 403, *405*
Perturbations, 49, 50
Phase shift, 149
Phases
  moon, 55, *55*
  planets, 6
Phobos, 114, *115*, 456A
  craters, 114
  orbit, *114*
Phoebe, 120
Photoelectric effect, 175
Photoelectric photometry, 176, 208
  UVB system, 208–209
Photography, 174, 175
Photoionization, 273
Photons, 151
Photosphere, **180–185**, 193, 227
  granulation, 181, *182*, 185
  limb darkening, 182, *182, 183*
  opacity, 182–184
  spectrum, 184–185, *230*
  sunspots, 193–196
  temperature, 182–183, *183*
Phototubes, 175–176
Piazzi, Guiseppe, 30
Pic-du-Midi Observatory, 78
Pioneer 11, 124
Pioneer Venus spacecraft, 85
Pixel, 177
Plages, 186, *186*, 196
Planck constant, **151**, 153, 469A
Planck radiation curve, **163–164**, *164*,
    165t
  color index, 209, *209*
  color temperature, 27, 227–228, *228*,
    232–233
  long-period red variables, 312
  planets, 27, *27*
Planck, Max K.E.L., 163
Planetary bombardment, 139; *see also*
    Cratering
Planetary configurations, 4–5, *5*
Planetary formation, *see* Solar system
Planetary nebula(e), 293, *293*, **319**, *319*,
    345–347, *347*; *see also* H II regions
  central star, 319
  cooling mechanisms, 346, 347
  forbidden lines, 346
  mass loss, 319
  spectra, 346
  temperature, 346
  white dwarf, 293
Planetary orbits, 7–10, 23–26, 455A; *see
    also* Kepler's laws
Planetary rotation, 24–25, 78–79, *79*,
    456A
  synchronous, 25
Planetary systems formation, 356
Planetesimals, 136, 138, 139
Planetology, 78
Planets, 455–456A; *see also* individual
    planets by name
  accretion, 137–138

atmospheres, 28–30, 95
  composition, **28**
  escape speed, 29
  ionosphere, 28
  retention, 29–30, *29*
  speed of particles, 29
blackbody radiation, 27, *27*
color, 27
density, 26
distances, 24, *24*
evolution, 94–97, 138–140
heat loss, 75, 82
interiors, 26–27
internal heat, 75, 82
masses, 26
moons, *see* Moons
motions, 23–26
observations, 78
orbital characteristics, 24, 25
orbits, 23–26, *24*, 455A
physical data, 456A
Planck curve, 27–28, *27*
resonance, 25
rotation, 25, 456A
satellites, 456–458A; *see* Moons
sizes, 26, *23*
spin-orbit coupling, 25
surfaces, 27–28
symbols, 455A
temperature, 27, 28, 455A
Plasma, 180, 272
  solar wind, 192
Plate scale, 168
Plate tectonics, 59, *59*, 140
Pleiades, 235, *289*, 290, *340*
Pluto, 110–111, *110, 111*, 126, 456A
  albedo, 110
  Charon, 111
  diameter, 110
  mass, 111
  motions, 110
  orbit 24, *25*, 111, *111*, 126
Pluto-Charon system, 111, 458A
PMS, *see* Pre-Main Sequence
Pogson, N.R., 206
Polar icecaps, 39
Polaris, 48
Polarity reversals (sunspots), 196, 202
Polarization, **147**
  Faraday rotation, 299, 372
  interstellar, 339–341
  galactic magnetic field, 340, 372, 373
  phase, 147
Polarized radiation
  corona, 189
  Crab nebula, 326, *326*
  pulsar, 299
  reflection nebulae, 340
Pollack, James, 139
Population I, **235–236**, 237, 238t, 252,
    255, 310, 365, 370; *see also* Open
    clusters; H II region; O B stars;
    Stellar Populations
  characteristics, **363t**
  clusters, 290
  evolution, 282–286, 289, 290
  Extreme Population I objects, 363

irregular galaxies, 381
Old Population I, 364
spiral galaxies, 381, 383
Population II, **235–236**, 238t, 252, 253,
    287, 310, 364, 370; *see also* Stellar
    Populations
  abundance of elements, 252
  characteristics, **363**
  dwarf galaxies, 382
  evidence for Big Bang, 442
  evolution, 287–290
  globular clusters, 290
  Intermediate Population II, 364
  novae, 322
  spiral galaxies, 383
Positron, 275
Potential energy, 18
Power (units), 466–467A
Poynting-Robertson effect, 134, 135
PP chain, *see* Proton-proton chain
Praesepe Cluster, 262
Pre-Main Sequence (PMS), 278, 279, 280,
    280
  T Tauri stars, 313
Precession, 38, **47–48**, *48*, 54
Pressure; *see also* Hydrostatic equilibrium
  stellar structure, 271–272
  units, 466–467A
Pressure broadening, 162, 238
Primeval fireball, 436, 439
Primordial deuterium, 443
Principle of equivalence, 433
Procyon B, 295
Prominences (solar), 196–197, *197, 198*
  active, 197
  loop, 196, *199*
  quiescent, 197
Proper motion, **258–259**, *259*, 260, 261
  nova shell, 322
Proton, 152
Proton-proton (PP) chain, 275, 276, *275*
Protoplanets, 136, 138
Protostars, **278–281**, *280*, 289, 352
  Becklin-Neugebarer objects, 352
  formation, 352
  gas flows, 356
  interstellar molecules, 348
Proxima Centauri, 43
Ptolemy (magnitude scale), 206
Pulsar(s), **298–303**, 326; *see also* Crab
    Nebula pulsar
  age, 299
  binary, 332, *332*
  Centaurus X-3, 329, 330
  change of period, 298
  cosmic rays, 372
  Crab Nebula pulsar, 300–303, *301*, 325
  density, 300
  distance, 299
  Doppler shifts, 332, *332*
  energy, 301, 303
  galactic magnetic fields, 372
  lighthouse model, 299, *300*, 303
  magnetic fields, 299
  magnetosphere, 299
  model, 299–300, *299*
  optical, 300, *302*

period, 298
polarized light, 299
pulses, 298, *298, 299, 301, 302*
velocity dispersion, 299
rotation, 299
slowdown, 298, 300–302
Vela X, 345
Pulsating stars, 283, 285, 309, **310–312**;
see also Cepheid variables, RR
Lyrae stars
Doppler shifts, 310
characteristics, **310t**, *310*
light curves, 310, *310*
mechanism, 311
period–density relationship, 311
period–luminosity relationship, 311–
312, *312*
radial velocity curves, 310, *310*
Quadrature, 5, 55
Quanta, 151
Quantum efficiency, 175
Quantum number, 153
Quantum theory, 151–154
Bohr's atomic theory, 152–154
broadening of spectral lines, 162
molecular spectra, 158
Quark, 444
Quasars, 372, **419–426**, *419*,
absorption lines, 419, 426
and nearby galaxies, 426
black hole model, 422
3C 273, 419, *420, 422, 423*
comparison to Seyfert galaxies, 425
double, 423–425, *424*
emission lines, 419, 425
energy, 421–423
forbidden lines, 419
gravitational lens, 423–425
Hydrogen lines, 419
infrared radiation, 421
jets, 422, 425, 427
noncosmological redshifts
interpretation, 426
polarization, 421
radio emissions, 421
redshift, 419, 425, 426
relativistic Doppler shift, 419
Schwarzschild radius, 422
size, 421
spectral index, 420
spectrum, 419, *420*, 425
speed, 422
superluminal motions, 422, 423, *423*
sychrotron radiation, 421, 422
variability, 421
Quasi-stellar object, *see* Quasars

R process, 327
Radar astronomy, 170
Earth-Moon distance measurement, 54
planetary mapping, 79, *79*
planetary motion studies, 78
Radial speed, 258, 259, 261, 263, 264
space motion, 259
galactic rotation, 264, *264*
Radiant (meteor), 32

Radiation, *see* Electromagnetic radiation
Radiation era, 438, 439, 440, **441–442**
Radiation pressure, 134, 135, 272
Radiative excitation, 155
Radiative transport, 273, 277
Radio astronomy, 170–173, 336; *see also*
Radio telescope(s)
aperture synthesys, 172
contour map, 170, *171*
Radio bursts
Cygnus X-1, 328
flare stars, 313
Jupiter, 102–103
solar flares, 200
Radio continuum; *see also* Synchrotron
radiation
HII regions, 344–345
Radio galaxies, 409–415; *see also* Active
galaxies
bending sequence, 413, *414*
BL Lac objects, 417–419
Centaurus A, 411, *412*, 413
compact, 410
Cygnus A, 411, *412*
doubles, 409, 411, 413, 415
energy, 409–411, 413
extended radio galaxies, 409, 411, 413
extended, 411
flux, 410–411
jets, 410, 411, 422
lobes, 411, 413
M 87, 410, 411
magnetic fields, 411
nonthermal emission, 410–411
polarization, 410
spectral index, 411
spectrum, 413
starburst galaxies, 418
synchrotron emission, 411
tails, 411, 413
variability, 413
X-ray emission, 410
Radio hydrogen recombination, 365
Radio interstellar polarization, 340
Radio recombination lines, 344
H II regions, 344, 363
Radio source(s)
galactic center, 365
Milky Way, 170
Sagittarius A, 365
scintillation, 298
SS 433, 330
Radio telescope(s), 170–173, *171, 173*
interferometers, 171–172
resolution (VLBI), 172
resolving power, 171
Very Large Array (VLA), 172, *173*
Very-Long-Baseline-Interferometer
(VLBI), 172
Radio waves, 147, 148t
Radioactive dating, **60–61**
age of universe, 432
Earth, 60–61
meteorites, 134
Moon, 74
Radioactive decay, 60–61, *60*
planetesimals, 139

Rayet, G., 239
Rayleigh scattering law, 68
Rayleigh–Jeans distribution, 164
Rays (lunar), 61
Reber, Grote, 170
Recombination, 157, 160; *see also*
Ionization
early universe, 442
H II regions, 343
Red giants, 237–238, **238t**
evolutionary phase, 283–287
mass loss, 285, 306
nucleosynthesis, 328
Red variables, 312
Reddening, 209, **338**, *338*, 339
Redshift, **387**; *see also* Doppler effect
galactic distances, 387, 389, 390
Hubble's law, 387, 432
quasars, 419, 426
superclusters, 402
Reflection, 148, *149*
Reflection nebulae, 340, *340*
Refraction, 148, *149*, 168
Earth's atmosphere, 68
index of, 148
Regolith, 65
Regression of nodes, 56
Reifenstein, Edward C., 300
Relativistic beaming, 423
Relativistic cosmology, 435–436, 442
Relativistic Doppler shift, **150**
quasars, 419–420
Relativity
special relativity, 150
general, 295–296, 304–306, **433–434**,
439, 442
Resolution (angular), 149, 169
Resolving power, 169
radio telescopes, 171–172
Resonance line, 317
Retention of atmospheres, 29–30, *29*
Retrograde motion, 4, *4, 6*
rotation, 25, *26*
Riccioli, 62
Richter scale, 57
Right ascension, 485A
Rigid body rotation, 255p, 263
Rings, 114–127
Jupiter, 119–120
Neptune, 127
Saturn, 121–124, *122, 123*
Uranus, 126
Roche limit, **48–50**
Jupiter, 119
Roche lobes, 223
novae, 322
Roche, Edouard, 48
Romer, Ole, 34p
Rotation curve (Galaxy), 266–267, *266*
Rotational broadening, 162
RR Lyrae stars, 309, 310, **312**, 364
instability strip, 287, 310
RS CVn stars, 313, **314–316**, *315*
binary characteristics, 314
coronal loops, 316
flares, 314, *315*
magnetic activity, 314

spectrum, 315
Russell, Henry Norris, 233, 277
Russell-Vogt theorem, 277, 278
Rutherford, Ernest, 153
RV Tauri stars, 309
Rydberg constant, 153–154, 469A

S process, 327, 328
S stars, 240
S waves, 56, 57
SO galaxies, 381
Sagittarius A, 365–366, *367*
Saha equation, **160**, 161, 227–229, 238
Saha, Meghnad N., 160
Sandage, Allan, 386, 389, 419
Sandage-Tamann method, 389
Sanduleak, Nicholas, 330
Saros cycle, 56
Satellites, 456–458A, *see* Moons
Saturn, 104–106, *104*, 455–456A; *see also*
    Jovian planets
    albedo, 105
    atmosphere, 105, *105*
    blackbody radiation, 105
    composition, 106
    density, 26, 105
    excess energy, 105
    formation, 138
    interior, 106, *106*
    magnetic field, 106, 122
    magnetosphere, 106, *107*
    mass, 105
    moons, 120, *121, 122*, 457A
    motions, 104
    orbits, 104
    physical data, 456A
    radius, 105
    rings, 120–124, *122, 123*
        Cassini division, 122
        infrared observations, 124
        Roche limit, 49
        shepherd satellites, 124
    satellites, 457A
    structure, 106
SB galaxies, 381
Scale (photographic plate), 168
Scale height, 67
    planets, 96
Scattering
    interstellar dust, 338
    reflection nebulae, 340
Schechter, Paul, 398
Schiaparelli, 90
Schmidt, Maarten 419
Schwarzschild, Karl, 303
Schwarzschild radius, *303*, 305, 422
Scintillation (radio sources), 298
Seasons, 38, *38*, 39
Secchi, Angelo, 230
Seeing, **69**, 78, 169, 206, 223, 224
    interferometry, 223, 224
Seismology, 56–57, *58,*
Semimajor axis, 9
Seyfert, Carl, 415
Seyfert galaxies, **415–417**, *416*
    Balmer lines, 417

comparison to quasars, 425
Doppler broadening, 415–417
emission lines, 415, 416
energy source, 416, 417
forbidden lines, 416
galactic binaries, 417
nuclei, 415, 417
radio sources, 417
spectrum, 415–417
starburst process, 419
Sgr A, 365–366, *367*
Shane, Donald, 401
Shapley, Harlow, 251, 311
Shell burning, **283**, 284t, 285, 286, 287
    long-period red variables, 312
Shell stars, 316, 317, *318*
Shield volcanoes, 91, 92, *93*
Shklovski, I. S., 325
Shock wave, 370
    cosmic rays, 372
    density wave, 370
    jets, 427–428, *427*
    molecular cloud, 352, 353
    supernova, 325, 345, 372
Shu, Frank, 369
SI units, 466A
Sidereal month, 54, 468A
Sidereal period, 5, 7, 38
Sidereal rotation period, 24, 456A
Sidereal time, **36**, 37,*37*, 38
Sidereal year, 38, 468A
Siderius Nuncius, 5
Singularity, 303, 305, 436
    Big Bang, 439, 441
Sirius B, 293, 294
    gravitational redshift, 296
Slipher, Vesto M., 387
Smarr, Larry, 428
Snell's law, 148, *149*
Solar, *see also* Sun
Solar activity, 193t, **193–202**, 371; *see also*
    Flares; Sun; Sunspots; Solar cycle
    active regions, 193t, 196–198, *200*
    bipolar magnetic regions, 196, *199*
    coronal holes, 191
    coronal streamers, *190*, 196
    cosmic rays, 200
    flares, 198–201, *199, 200, 201*
    flux loop, 201
    location, 193t
    magnetic fields, 193, 201
    plages, 193t, 196, *186, 197*
    polarity reversal, 202
    prominences, 196, 197, 198, *198, 199*
    radio bursts, 200
    solar cycle, 193–196
        model, 201–202, *202*
    solar maximum, 193
    solar rotation, 193
    solar wind, 192, *192*
    sunspot numbers, 193, *195*
    X-ray bursts, 200, *200*
Solar antapex, 260
Solar apex, 260
Solar constant, 210
Solar corona, 189–191; *see also* Corona
Solar cosmic rays, 198, 200

Solar cycle, 193–196, 201–202; *see also*
    Solar activtiy
Solar flares, 198–201; *see also* Solar
    activity; Flares
    compared with flare stars, 313, 314
Solar insolation, 39, *39*
Solar luminosity, 210, 273
Solar motion, 260, 261, *260*
    mean parallax, 206
    space velocity 259, 260, *260*
Solar nebula, 135–139
    accretion, 137
    angular momentum, 136
    chemistry, 136
    condensation 136, 137, *137*
    meteorites, 139
    magnetic fields, 136, *136*
    temperature, 136, *137*
Solar neighborhood, 233, 264
    luminosity function, 250
Solar prominences, 196–198; *see also*
    Prominences
Solar rotation, 191, **196**
Solar system, 23–33, *23, 24*, 135; *see also*
    Solar nebula
    age, 134, 432
    angular momentum, 33, 135–136
    center of mass, 202p
    coplanar orbits, 136
    data, 455A
    dynamics, 135–136
    formation, 135–140
Solar wind, 134, 180, **192**, *192*
    comets, 130, 449
    coronal holes, 191
    cosmic ray modulation, 371
    Earth's magnetosphere, 69
    flares, 192, 200
    magnetic field, 192
    mass loss, 285
    Mercury, 80
    plasma, 192
    relation to Earth's aurorae, 72
    solar activity, 192
    temperature, 192
Solid angle, 151
Solid body rotation, 263
Solstices, 39
Space astronomy, 170, 174; *see also*
    Infrared astronomy; ultraviolet
    astronomy; X-ray astronomy;
    gamma ray astronomy
    bolometric magnitudes, 210
    Hubble telescope, 174
    Very-Long-Baseline Interferometry, 172
Space probes, *see* individual named
    programs, e.g., Apollo
Space velocity, 259, 260, *260*
Spacetime, **303–306**, *304*, 433, 434, 445
Spallation, 371, 372
Speckle interferometry, 224
    brown dwarf, 297
Spectral classification, **229–233**, *231*, 236–
    240, 463A; *see also* Hertzsprung-
    Russell diagram
    abundance effects, 238–240
    color index, 233

Harvard, 230, *231*, 232t
luminosity effects, 236–238
M–K classification, 237
temperature effects, 232–233
white dwarfs, 295
Spectral energy curves, *see* Planck
    radiation curves
Spectral index, 411
    quasars, 420
Spectral lines, 155, 157–162
    absorption line spectrum, 155
    Balmer lines,
        strength, 159–161
    broadening, 161, 162
    chromosphere, 185–186
    collisional broadening, 162
    corona, 190–191
    emission line spectrum, 155
    equivalent width, 159, *159*
    forbidden lines, 156
    Fraunhofer lines, 184–185, *230*
    intensities, 158–161
    line blanketing, 228, *228*
    Lyman series, 157
    metastable states, 162
    molecules
        electronic transitions, 158
        rotational transitions, 158
        vibrational transitions, 158
    natural broadening, 162
    photosphere, 184–185, *184*
    pressure broadening, 162
    profiles, 158, *158*, 159, *184*
    solar oscillations, 180
    solar transition region, 187
    stellar atmospheres, 228–229, *231*
    thermal Doppler broadening, 162
    Zeeman effect, 162
Spectral type, *see* Spectral classification
Spectrograph, 229
Spectroheliograms, 196, *197*
Spectrometer, 229
Spectroscopic binaries, 213, 216–218, *216*
    eclipsing, 217, 218, 220
    mass function, 218
    velocity curve, 216–218, *217*, *218*
Spectroscopic parallaxes, 240, *240*
Spectrum, 147, 148t; *see also*
        Electromagnetic radiation; spectral
        classification; blackbody radiation
Spectrum variables, 309
Spectrum binary, 216
Speed, 11; *see* Velocities
Speed of light, 34p, 146, 469A
Spicules, **186**, 188
Spin–orbit coupling, 25, *26*
Spiral galaxies, 378, 381–383, *379*, *380*,
        382t; *see also* Galaxies
Spiral structure, 252–255, *253*, *254*, 359,
        360, **363**, *364*; *see also* Galaxy
    density-wave model, 369, 370
    hydrogen distribution, 360
SS 433, 330–331, *331*, 427
Staelen, David H., 300
Standard time, 37
Star clouds, 247
Star clusters, 250, **251–252**; *see also*

Globular clusters, open clusters
    ages, 289
    evolution, 288–290
    H–R diagram, 288, *288*, 289, 290
    moving-cluster distances, 261
Star counts, 249, 250, 338
Star distribution, 249–251
Star formation, 252, 278–281, **352–356**,
        369; *see also* Stellar evolution
    infrared observations, 352, *353*
    massive stars, 280, 352–354
    molecular clouds, 352, 353, 354, *354*
    solar-mass stars, 279, 354–356
Star spots (magnetic variables), 314
Starburst galaxies, 418
    tidal interactions, 419
Stars, 460–463A; *see also* entries under
        Stellar
    atmosphere, 227–229; *see also* Stellar
        atmosphere
    binary systems, 213–224; *see also*
        Binary systems
    blackbody radiation, 162–165
    brightest, *234*, 233, *235*, 462A
    chemical composition, 238–240, 272,
        277
    color, 227,228
    convection zone, 272, 313, 319
    core, *see* Core (stellar)
    degenerate, 294, 297
    density, 271, 277
    distances, 205–206
    early-type, 230
    energy generation, 273–277; *see also*
        Energy generation
    evolution, *see* Stellar evolution
    flare, *see* Flare stars
    flux, 210
    giants, 235
    gravitational contraction, *see*
        Gravitational contraction
    high-velocity, 267
    instability zone, 283, 285, 287
    late-type, 230
    lifetime, 281
    luminosity, 235, 273, 277
    luminosity class, 236–238, 463A
    magnitude systems, 206–211; *see also*
        Magnitude
    mass, 271, 277, 279
        binary systems, 213
    mass fractions, 272
    mass loss, 285, 309, 313, 316
    mean molecular weight, 277
    motions, *see* Stellar motions
    nearest, 233, *234*, 460–461A
    populations, *see* Stellar Populations
    pressure, 271, 277
    protostars, *see* Protostars
    pulsating, *see* Pulsating stars
    shell burning, 283, 285
    spectral type, *see* Spectral classification
    subdwarfs, 238
    supergiants, 235
    temperature, 161, 235, 271, 277
        blackbody radiation, 165
    variable, *see* Variable stars

white dwarfs, 235, 293–296
Statistical parallaxes, 206, **386t**
Stefan's law
    long-period red variables, 312
    planets, 28
Stefan, Josef, 165
Stefan-Boltzmann constant, 469A
Stefan-Boltzmann relation, **165**, 165t
    bolometric magnitude, 210
    eclipsing binaries, 220
    photospheric temperatures, 182
    stellar atmospheres, 228
Stellar associations, 251, 253; *see also* OB
        associations
Stellar atmospheres, 227–230
    absorption lines, 228
    abundances, 227, 238–240
    color temperatures, 228
    composition, 227, 228
    density, 227
    line blanketing, 228, *228*
    model, 227
    molecules, 228
    opacity, 227, 229
    optical depth, 227, 229
    pressure, 227, 228, 229, 237
    spectral lines, 228
    surface brightness, 228
    surface gravity, 238
    temperature, 227, 228, 233, 238
    X-ray emission, 241
Stellar classification, 236–240, 463A; *see
        also* Spectral classification
Stellar clusters, 235, 251, 261; *see also*
        Open clusters; globular clusters
Stellar composition, 227, 272, 277; *see
        also* Abundances
    evolution, 281
    Russell–Vogt theorem, 277
Stellar core, 277, 279, 280, 283, 285–287,
        328
    planetary nebulae, 319
    supernovae, 325, 327, 328
    Wolf–Rayet stars, 319
Stellar diameters
    interferometry, 223
    lunar occultation, 223
Stellar evolution, 236, 252, **278–290**, 282t,
        284t, 309, 310, 313
    evolutionary track, **278**; *see also*
        Evolutionary tracks
    instability zone, 287, 310
    interstellar gas, 351
    massive stars, 280–281, *282*, 282–283,
        285, *286*, 352–354
    novae, 322
    planetary nebulae, 319
    pre-main sequence, 278–279, *280*
    protostars, 279–281
    solar mass star, 279–280, 283–285, *284*,
        284t
    stages, 282t
    supernovae, 325, 328
    T Tauri stars, 313
    turnoff, 289
    Wolf–Rayet stars, 318
    zero-age main sequence (ZAMS), **281**

Stellar interferometer (Michelson), 224
Stellar interiors, *see* Stellar structure
Stellar magnitude, 206–211; *see also*
    Magnitude
Stellar masses, 213–219, 238, 277, 279
    binary systems, 213–219
    eclipsing binaries, 220
    effect on evolution, 286, 287
    visual binaries, 214, 215
    white dwarfs, 295
Stellar motions, 259–262
    peculiar motions, 258, 259
    proper motions, 260
    radial speeds, 259
Stellar parallax, 43
Stellar populations, 235–239, 252, 364;
    see also Population I and
        Population II
    abundances, 252
    characteristics, 363t
    galaxies, 385
    metal abundances, 370
Stellar radii, 238
    interferometry, 223
    light curve, 219
    lunar occultation, 223
Stellar rotation, 317
    binary systems, 222
    effect on gravitational collapse, 280
Stellar spectra, 227, **229–240**, *231, 233*
Stellar structure, 271–276; *see also* Energy
        generation; stellar evolution
    boundary conditions, 277
    models, 277
Stellar velocities, 258–262
    high-velocity stars, 267–267
    Z components, 267
Stellar winds, 313
    mass loss, 285
    protostars, 356
    Wolf–Rayet stars, 318
Stephenson, Bruce, 330
Steradian, 151
Stimulated emission, 349–350
    molecular clouds, 350
Stony irons (meteorites), 132
Stratosphere, 67, *68*
Stromgren sphere, 343
Strong force, 443, 444t
Struve, F., 205
Struve, F.G.W., 43
Subdwarfs, 235, 236, 238
Subsolar temperature, 28
Sun
    abundances, 185, 185t
    active regions, *see* Solar activity
    age, 255, 273
    Balmer lines, 184–187, *186, 187, 188*
    bipolar magnetic fields, 196
    blackbody continuum, *27*, 183
    center, 271, 272
    chromosphere, 180, **185–188**; *see also*
        Chromosphere
    chromospheric network, 186, 187, *187*
    composition, 185, 185t
    convection zone, 180, 181, 272, 277
    core, 180

corona, **189–191**; *see also* Corona
cosmic rays, 200
density, 180, 271
differential rotation, 196
    solar cycle model, 201, *202*
eclipses, 55, 185
emission lines, 184, 187
faculae, 196
filaments, 186
flares, 198–200; *see also* Flares
Fraunhofer absorption lines, 184, *230*
    corona, 189
galactic orbit, 255
granulation, 181, *182*, 193
H-continuum, 183, 184
helioseismology, 180
H-alpha, 186, *186, 187, 197*
limb darkening, *182*, 183
luminosity, 180, 273
luminosity class, 237
Lyman alpha radiation, 187
magnetic fields , 197
    flux tubes, 201
    solar cycle model, 201, *202*
    solar flare, 201
    sunspots, 193
    supergranules, 185, 186
mass, 180
mass loss, 203p
neutrino flux, 180, 276
photosphere, **180–185**, 193; *see also*
    Photosphere
physical characteristics, 271–273
plages, 186, *186*, **196**, *197*
Planck curve, 27
Population I star, 255
prominences, 180, 196–197, *198, 199*
radiative zone, 180
radio burst, 200, 203p
radio emission, 190, 200
rotation, 136, 180, 191, 196
scale heights, 203p
Schwarzschild radius, 303
seismic waves, 180
solar activity, 193–202; *see also* Solar
    activity
solar cycle, 194, 201; *see also* Solar
    activity
solar oscillations, 180, 185
solar wind, *192*, 371; *see also* Solar
    wind
sound waves, 180
spectral type, 230
spicules, 180, **186**, 187, *188*
stellar models, 277
structure 180, *181*, 271–273
sunspots, 180, **193–196**; *see also*
    Sunspots
supergranules, 185, 186, *187*, 201
temperature, 180, *183*, *186*
    chromosphere, *186*, 187
    corona, *186*, 189
    photosphere, 182
    transition region, *186*, 187
transition region, *186*, 187
    spectrum, 187
    temperature, *186*, 187

ultraviolet radiation, 184, 187, 191
X-ray emission, 191, *191*, *197*, 200–201,
    200, 241
Sunspot maximum, *see* Solar activity
Sunspot numbers, 194, *195*
Sunspot polarity, 196
Sunspot(s), 180, **193–196**, *194*
    bipolar groups, 194
    latitude variation, 195, *195*
    magnetic fields, 193
    Maunder minimum, 194, *195*
    numbers, 194
    penumbra, 193
    polarity, 193, 196
    pores, 193, *195*
    solar cycle, 194–202, *195*; *see also* Solar
        activity
    structure, 193
    umbra, 193
Superclusters, 395, **401–405**, *401, 406*,
    431, 446
    Coma, 403, *403*
    Hercules, 403, 404, *404*
    Local Supercluster, 402, *402*, *403*, 404
    missing mass, 407
    model, *406*
    Perseus, 403, *403*
    space density, 404
    statistical analysis, 404
    X-ray emission, *406*
Supergiant galaxies, *see* cD galaxies
Supergiants, 235, **237**, 238, 238t, *239*
    surface gravity, 238
Supergranules, 185, 186, *187*, 201
Superior conjunction, 5
Superior planets, 4
Superluminal motions, 422–423
    quasars, 422
    relativistic beaming, 423
Supernova, 300, 306
Supernova explosion
    quasars, 426
Supernova of 1054, *see* Crab nebula
Supernova remnants, 300, **323–328**, 345,
    345, 354; *see also* Crab nebula
    cosmic rays, 372
    shock waves, 345
    synchrotron radiation, 345
    X-ray emission, 345
Supernovae, 309, 323–328
    Chinese chronicles, 323
    distribution in Galaxy, 325
    ejection of material, 325
    interstellar gas 351
    light curve, 323, *324*
    luminosity, 323
    mass, 325
    mass loss, 324, 325
    neutron stars, 325, 330
    spectrum, 324
    supernova remnants, 345
    temperature, 325
    Type I, 323, 324, 325
    Type II, 323, 324, 325
    white dwarf, 325
Surface gravity, 237
    neutron star, 297

supergiant, 238
Synchronous rotation, 54, 114
    binary systems, 222
    Galilean moons, 114
    Martian moons, 114
    Mercury, 78
    Moon, 55
    moons of Saturn, 120
Synchrotron radiation, 344, *327*
    BL Lacertae objects, 418
    cosmic rays, 372
    Crab nebula, 325–326
    jets, 427, 428
    Jupiter, 102–104
    Pulsars, 299
    quasars, 421, 422
    radio galaxies, 411
    RS CVn stars, 315
    solar flares, 200
    SS 433, 331
    supernova remnants, 345
Synodic month, 56
Synodic period, 5, 7

T Tauri stars, 309, 312, **313**, *313*, 363
Tammann, Gustav, 386, 389
Tangential speed, 259, **260**, 263, 264
Taurus A, 325
Taylor, Joseph, 33
Telescopes, 168–174
    focal length, 168
    infrared, 173–174
    light-gathering power, 169
    objective, 168
    optical, 168–169
    radio, 170–173
    reflecting, 168
    refracting, 168
    resolving power, 169
Television camera
    image detecting, 176
Temperature, **165**, 165t
    blackbody radiation, 165
    bolometric correction, 210
    color, 227, 232
    effective, 210, 227
    planets, 27–28
    relation to color index, 209
    solar corona, 189
    spectral lines, 161
    spectral sequence, 232, 233
    stellar interiors, 272–273
    stellar atmospheres, 227–228, 232–233
    sun, 182, *186*
    table of associated laws & phenomena,
        165t
    types, 165t
Temperature gradient, 272
Terrestrial planets, 78–96, 455–456A
    comparative evolution of, 94–95
    cratering, 95
Tethys, 120
Tharsis ridge (Mars), 91, 92, 94
Thermal bremsstrahlung, 344
    H II regions, 344–345
    spectrum turnover, 345
    Thermal conductivity, 112p

Thermal Doppler broadening, 162, 165t
Thermal equilibrium, 159, 274
Thermal radio emission, 344–345
Thermodynamic equilibrium, 227
Thermonuclear reactions, 274, 280; *see
    also* Nucleosynthesis
Thermosphere, 67, *68*
Third Cambridge Catalog, 419
3C 48, 419
3C 206, 426, *426*
3C 273, 419, *420*, 422, *423*
3 K background radiation, 436, *437*, 438,
    442
3-kpc arm, 362
Tidal bulge, 47, 54
Tidal evolution, 46, 47, *47*
Tidal friction, 46, 47
    effect on Earth's rotation, 47–48, *48*
Tidal interactions, 45–50
    asteroids, 139
    asteroids with moons, 128
    black hole, 303, 329
    clusters of galaxies, 399, 400
    Earth–Moon system, 45–50, *45*, *46*, *47*,
        48
    eclipsing binaries, 222, 223
    Galaxy and Magellanic Clouds, 362
    Jupiter-Io, 116
    planetary motion, 25
    Roche limit, 48, 49
    Seyfert galaxies, 417
    spin-orbit coupling, **25**
    starburst galaxies, 419
Tides, 45–46, *45*, *46*
Time, 36–38, 485A
    apparent solar, 36
    daylight savings, 38
    ephemeris, 38
    equation of, 37
    Greenwich mean, 37
    mean solar, 36
    standard, 37
    units, 466A
    universal, 37
Time zones, 37
Titan, 120, *121*, 457A
    atmosphere, 30, 120
Titania, 124, 458A
Titus of Wittenberg, 24
Titus–Bode rule, 24
Tombaugh, Clyde W., 110, 402
    discovery of Pluto, 110
Transit, 484A
Transition (atomic), 153, 154
Transition probability, 184
Transition region, *186*, 187
Transverse distortion (S) waves, 56, 57
Trapezium cluster, 352
Trigonometric parallax, **205–206**, *205*
Trigonometry, 471–473A
Triple alpha process, 276, 283, 286
Triton, 126, *126*, 458A
Tropic of Cancer, *38*, 39
Tropic of Capricorn, *38*, 39
Troposphere, 67, *68*
True anomaly, 9
Trumpler, Robert, J., 250
Tully, R. Brent, 390, 402

Tully–Fisher relationship, 390, *390*, *391*
Turbulence broadening, 162
Turnoff, 289
21-cm line, **347–348**, *348*, 352, 384
    galactic center, 361–362
    galactic distances, 390
    galactic structure, 359–362, *359*, *361*
    galaxies, 384
    high velocity clouds, 362
    profiles, 359, *359*, 360, *360*
    rotation of galaxies, 384
    warp of galactic plane, 360, *361*
Tycho Brahe, 205
Tycho's Supernova, 345, *346*
Tycho (crater), 61, 62
UBV magnitude system, 208, *208*, *209*,
    213
    bolometric correction, 210

Uhuru X-ray instrument, 328
    Centaurus X 3, 339
    X-ray bursters, 331
Ultraviolet astronomy, 174
    International Ultraviolet Explorer
        (IUE), 174
    primordial deuterium, 443
Ultraviolet radiation, 147, 148t; *see also*
        Electromagnetic radiation
    Fraunhofer absorption spectrum, 184
    intergalactic matter, 405
    interstellar dust, 341
    O and B stars, 317, 343
    observations, 170, 174
    planetary nebulae, 319
    solar transition ergion, 187
    T Tauri stars, 313
Ultraviolet spectra
    RS CVn stars, 315
    stellar atmospheres, 232
Umbriel, 126
Uncertainty principle, 154, 162
Units, 466–467A, 469A
Universal gravitation, 13–15, 433, 443
Universal time (UT), 37
Universe, 431–446
    age, 392, 432, *432*, 435
    big bank, 439–443; *see also* Big bang
    clumpiness, 445, 446
    critical density, 435
    curvature, 435, 439
    early, 404, 442, *444*
    expansion, 392, 432, 435
    geometries, 442
    GUT's, 443–446
    horizon, 432
    inflationary, 443–446
    mass density, 435, 436
    matter, 438
    missing mass, 406
    models, 434–436, *435*, 439–441
    nucleosynthesis, 441–442, *441*, *443*
    particle era, 440, 441
    radiation era, 438, 439
    size, 431
    strings, 446
    thermal history, 440, *440*
Uranus, 106–108, *107*, 455–456A; *see also*

Jovian planets
atmosphere, 108
aurorae, 108
composition, 139
density, 107
Doppler shift measurement, 108
interior, *108*
magnetosphere, 107, *108*
mass, 107
moons, 108, 124, *125*
motions, 26, 106
physical data, 456A
retrograde rotations, 25, 26
rings, 126
rotation, 26, 107, 108
satellites, 458A
shepherd satellites, 126
structure, 108, *108*
Voyager 2 spacecraft, 108, 125

Valence electrons, 157
Valles Marineris (Mars), 91, 94
Van Allen radiation belts, 71–72, *71*, *72*
mirror points, 71
Van Allen, James A., 70
Van Biesbroeck, 8, 297
Variable stars, 283, 309–319
evolution, 309
Hertzsprung–Russell diagram, *309*
nomenclature, 309
Vaucouleurs, Gerard de, 402
Vectors, 476–477A
cross product, 16
dot product, 18
Vega parallax, 43
Vega spacecraft, 448
Vela pulsar, 302, *302*
Vela X, 345
Velocities, 11, 29
escape, 19–20, 29
from Galaxy, 267
Maxwellian distribution, 29
most probable speed, 29
orbital, 15–17, *15*, *17*, 20, *20*
root mean square speed, 29
stellar, 258–261
Velocity curve (binary systems), 216, 217
Velocity dispersion, 299
Velocity of light, 34, 146, 469A
Venera spacecraft, 83, 87
Venus, 83–88, 455–456A
albedo, 83
atmosphere, 84, *84*, 85
composition, 83
cloud structure, 84, *84*
craters, 85
elongation, 4
evolution, 88, 95
greenhouse effect, 83
Hadley cells, 84
interior, 83, *83*
lava flows, 85
magnetic field, 88
mass determination, 83
motions, 83
mountains, 85
phases, Galileo's observations, 6

physical characteristics, 83, 456
radar mapping, 83, *85*, *86*
retrograde rotation, 25, 26
revolution, 83
rotation, 83
surface, 85–87, *87*
temperature, 83
volcanoes, 85, 86, 87
Vernal equinox, 36, 39, 485A
Very Large Array, 172; *see also* VLA
Very large Baseline Interferometer, 172;
see also VLBI
Viking landers (Mars), 89, 91
Virgo cluster, 395, 398, 402
Virial theorem, 274, **384**
clusters of galaxies, 399
masses of galaxies, 384
missing mass, 406
Vis viva equation, 18
Visible light, 147, 148t
Visible spectrum, *see also* Electromagnetic
radiation; spectrum
Visual binaries, 213–216
VLA observations, 172
gas flows, 356
radio contour map, *171*
radio galaxies, 410, *415*
RS CVn stars, 315
SS 433, 331
VLBI observations, 172
quasars, 421
superluminal motion, 422
Vogt, H., 277
Volcanoes (Vulcanism)
Io, 115, 116, *117*
Mars, 91
Mercury, 82
Moon, 64, 65
Venus, 85, 86, 87
von Jolly, Phillip, 14
Voyager(s), 114, 122
Jupiter, 99, 102
Jupiter's ring system, 119
rings of Saturn, 124
rings of Uranus, 126
Saturn, 105
Titan, 120
Uranus, 108, 124

W Virginis stars, 312; *see also* Cepheid
variables
W–R stars, *see* Wolf–Rayet stars
Water (molecular clouds), 439, 351
Wave nature of light, 146, *146*
Wave number, 153
Wavelength, 146, *146*, 147
WC stars, 240, 318; *see* Wolf–Rayet stars
Weak force, 443, 444t
Weight, 14
Weightlessness, 433
Whipple, Fred L.
cometary model, 131
Whipple's dirty iceberg model, 448
White dwarf(s), 235, 236, 285, **293–296**,
294, 306, 317
binary pulsars, 333
Chandrasekhar limit, 294

magnetic fields, 296
masses, 294, 295, 296
mass–radius relationship, 294
novae, 322
spectra, 295
supernova, 325
White hole, 305
Widmanstatten figures, 133
Wein's law, 27, **164–165**, 227, 469A
Wein, Wilhelm, 164
Wildt, Rupert, 89
Wilson, Robert, 436
Winkler, Karl-Heinz, 428
Wirtanen, Carl, 401
WN stars, 240, 318; *see* Wolf–Rayet stars
Wolf, C., 239
Wolf–Rayet stars, **239–240**, 285, 318, 319
binaries, 318
composition, 318
mass loss, 319
OB stars, 318
planetary nebulae, 319
WC, 240, 318
WN, 240, 318
Work, 17, 18
Wormhole, 306
X-ray astronomy, 170, 174
Einstein Observatory (HEAO 2), 174
X-ray binaries, 328–331
X-ray bursters, 331, 332, *332*
X-ray sources, 241, 329–333
binary, 422
Centaurus X-3, 329, 330
clusters of galaxies, 407
Crab nebula, 326
Cygnus X-1, 328, 329
flare stars, 314
galactic center, 367, *368*
intergalactic matter, 405–406, *406*
novae, 322
pulsars, 329, 330
radio galaxies, 413
RS CVn stars, 315
solar flares, 200
SS 433, 330, 331
supercluster, 405–406, *406*
X-rays, 147, 148t; *see also* X-ray
astronomy and X-ray sources

Year
anomalistic, 38
sidereal, 38
tropical, 38

Z components, 267, 364
ZAMS, 281, 283, 289
Zeeman effect, 162, 372
magnetograph, 196
sunspots, 193
21-cm line, 372

Zeeman, Pieter, 162
Zenith, 484A
Zenith angle, 209
Zero-age main sequence (ZAMS), 281,
283, 289
Zodiac, 4
Zodiacal light, 32, *33*, 134, 189

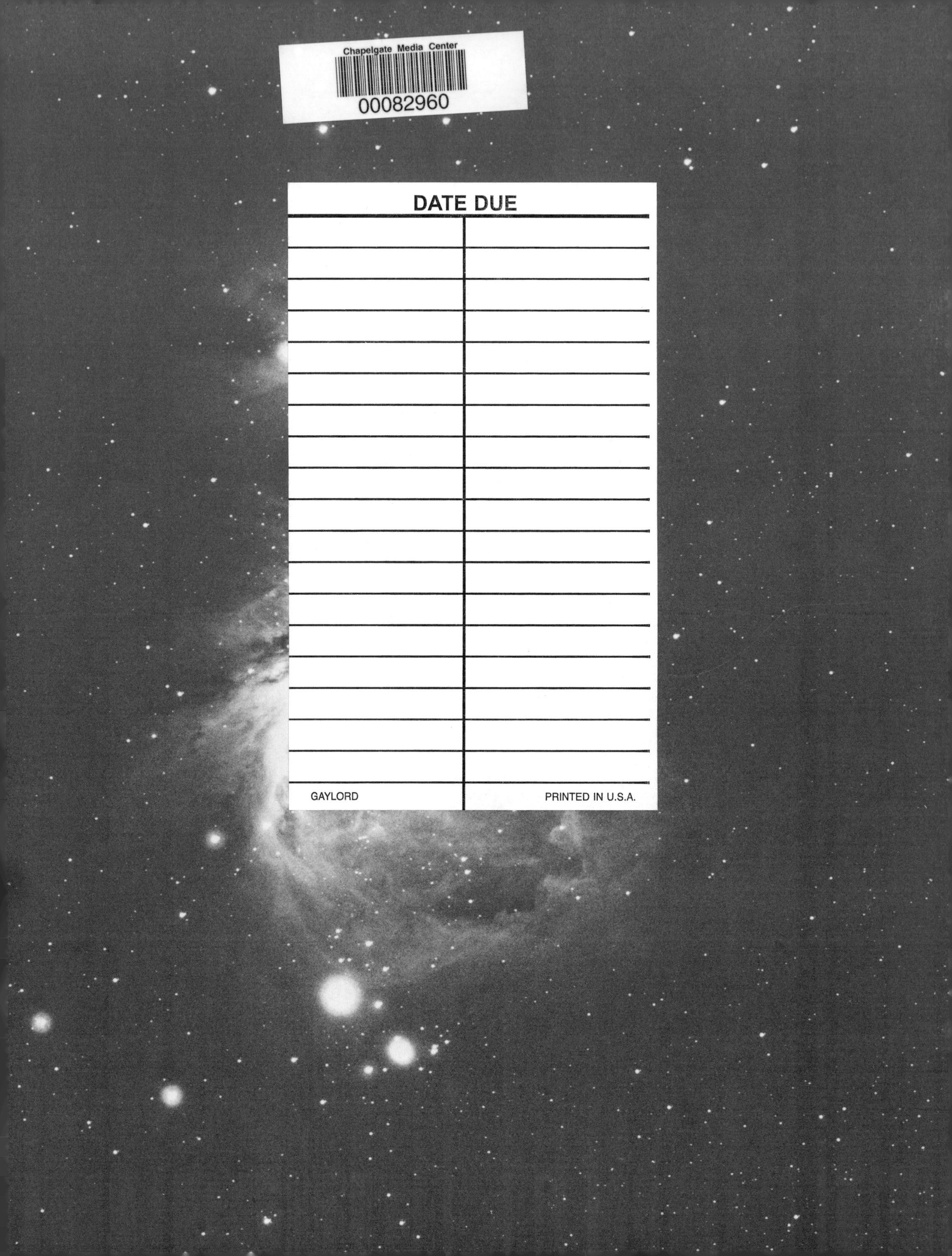